STRONG INTERACTIONS OF HADRONS AT HIGH ENERGIES

V. N. Gribov was one of the creators of high energy elementary particle physics and the founder of the Leningrad school of theoretical physics. This book is based on his lecture course for graduate students. The lectures present a concise, step-by-step construction of the relativistic theory of strong interactions, aiming at a self-consistent description of the world in which total hadron interaction cross sections are nearly constant at very high collision energies. Originally delivered in the mid-1970s, when quarks were fighting for recognition and quantum chromodynamics had barely been invented, the content of the course has not been 'modernized'. Instead, it fully explores the general analyticity and cross-channel unitarity properties of relativistic theory, setting severe restrictions on the possible solution that quantum chromodynamics, as a microscopic theory of hadrons and their interactions, has yet to find. The book is unique in its coverage: it discusses in detail the basic properties of scattering amplitudes (analyticity, unitarity, crossing symmetry), resonances and electromagnetic interactions of hadrons, and it introduces and studies reggeons and, in particular, the key player – the 'vacuum regge pole' (pomeron). It builds up the field theory of interacting pomerons, and addresses the open problems and ways of attacking them. This title, first published in 2009, has been reissued as an Open Access publication on Cambridge Core.

VLADIMIR NAUMOVICH GRIBOV received his Ph.D. in theoretical physics in 1957 from the Physico-Technical Institute in Leningrad, and became the head of the Theory Division of the Particle Physics Department in 1962. From 1971, when the Petersburg (Leningrad) Institute for Nuclear Physics was organized, Gribov led the Theory Division of the Institute. In 1980 he became Head of the particle physics section of the Landau Institute for Theoretical Physics, Moscow. From 1981 he regularly visited the Research Institute for Particle and Nuclear Physics in Budapest where he was a scientific adviser until his death in 1997. Vladimir Gribov was one of the leading theoretical physicists of his time, who made seminal contributions in quantum electrodynamics, neutrino physics, non-Abelian field theory, and, in particular, the physics of hadron interactions at high energies.

CAMBRIDGE MONOGRAPHS
ON PARTICLE PHYSICS,
NUCLEAR PHYSICS AND COSMOLOGY

General editors: T. Ericson, P. V. Landshoff

STRONG INTERACTIONS OF HADRONS AT HIGH ENERGIES

Gribov Lectures on Theoretical Physics

V. N. GRIBOV

Prepared by

Y. L. DOKSHITZER AND J. NYIRI

CAMBRIDGE
UNIVERSITY PRESS

Shaftesbury Road, Cambridge CB2 8EA, United Kingdom

One Liberty Plaza, 20th Floor, New York, NY 10006, USA

477 Williamstown Road, Port Melbourne, VIC 3207, Australia

314–321, 3rd Floor, Plot 3, Splendor Forum, Jasola District Centre, New Delhi – 110025, India

103 Penang Road, #05–06/07, Visioncrest Commercial, Singapore 238467

Cambridge University Press is part of Cambridge University Press & Assessment, a department of the University of Cambridge.

We share the University's mission to contribute to society through the pursuit of education, learning and research at the highest international levels of excellence.

www.cambridge.org
Information on this title: www.cambridge.org/9781009290272

DOI: 10.1017/9781009290227

First published 2009
Reissued as OA 2022

A catalogue record for this publication is available from the British Library.

ISBN 978-1-009-29027-2 Hardback
ISBN 978-1-009-29024-1 Paperback

Contents

Foreword

Quantum Chromodynamics (QCD) was in its infancy when Gribov delivered his lectures on strong interactions. Since then QCD had been established as the true microscopic theory of hadrons.

The main (though not the only) focus of these lectures is to present the 'old theory' of hadron interactions (known as *reggistics*). This theory has realized the 'Pomeranchuk–Gribov programme' of describing strong interactions without appealing to the internal structure of hadrons. The old theory was launched in 1958 by the Pomeranchuk theorem and reached a climax in Gribov's prediction of an asymptotic equality of hadron cross sections 15 years later. With the advent of QCD, it was abandoned by the great majority of theorists in the mid-1970s and has been neither taught nor learnt since.

QCD – the 'new theory' – is now in its fourth decade. The QCD Lagrangian approach did marvels in describing *rare* processes. This is the realm of *hard interactions* that occur at *small distances* where quarks and gluons interact weakly due to the asymptotic freedom. The domain of expertise of the old theory is complementary: it is about normal size hadron–hadron cross sections, *soft interactions* that at high energies are dominated by peripheral collisions developing at *large distances*. QCD only starts to timidly approach this domain, with new generations of researchers borrowing (sometimes improperly) the notions and approaches developed by the 'old theory'.

A few non-scientific comments are due before you start reading (better still, working through) the book.

The lectures you are about to encounter were given in early 1970s, and so they are presented here: no attempt has been made to 'modernize' the text. (Editor's comments are few and relegated to the footnotes marked (ed.).)

Let me mention two problems that emerged when preparing this text: one surmountable, another not. The first derived from the fact that the lectures were delivered twice (in 1972–1973 and then in 1974–1975). The only invariant in these two series was the format of lectures (four hours at the blackboard each Thursday; never a piece of paper with pre-prepared notes to guide the lecturer). The rest was subject to variability. So, a compromise often had to be found between two different presentations of the same topic.

The second problem is as follows. The equations of this book contain 3180 *equality signs*, while they seldom appeared on the blackboard. With Gribov-the-lecturer, the symbol = was clearly out of favour.

I think it was being done on purpose. Gribov was a generous teacher and always implied that his students were capable of deriving mathematically correct formulae, given the rules. He was trying to teach students, in the first place, how to think, how to approach a new problem, how to develop a 'picture' of a phenomenon in order to guess the answer prior to deriving it. And ignoring equality signs served as additional means for stressing 'what was important and what was not' in the discussion.

Unfortunately, this flavour of a live lecture is impossible to preserve in a printed text which has its specific, and opposite, magic of certainty. I am afraid that having debugged equations, the lectures may have lost in pedagogical impact.

I always looked upon these lectures as a treasure chest. I sincerely believe that when you open it, you will find it filled not with obsolete banknotes but with precious gold coins.

Yuri Dokshitzer

1

Introduction

The theory of strong interactions,
now that is quite something.

Elementary particles we know are *leptons* e and μ and their corresponding *neutrinos*, ν_e, ν_μ, a photon (γ) and a graviton, and then, hundreds of
strongly interacting particles – *hadrons*: proton p and neutron n, pions
$\pi^\pm$ and π^0, kaons $K^\pm$, K^0, $\bar{K}^0$, etc., etc.

1.1 Interaction radius and interaction strength

Electromagnetic interaction has two characteristic features. Firstly,
it is characterized by a small coupling,

$$\frac{e^2}{4\pi\hbar c} \simeq \frac{1}{137}.$$

Secondly, it is a long range force,

$$V = \frac{e_1 e_2}{r},$$

so that there is no typical distance, no characteristic interaction radius.

Gravitation behaves (at large distances!) similarly to the electromagnetic interaction,

$$V = G_{\mathrm{gr}}\frac{m_1 m_2}{r};$$

thus it has no radius either. To characterize the magnitude of the interaction one needs to construct a dimensionless parameter. Contrary to the
case of the electromagnetic charge, mass is not quantized, so that there is

1

no 'unit mass' to choose. Hence one usually takes the mass of the proton, m_p, to quantify the typical interaction strength:

$$G_{\text{gr}} \frac{m_p^2}{\hbar c} \simeq 7 \cdot 10^{-39}.$$

An important difference with electromagnetism is that here all the 'charges' have the same sign (mass is positive). Therefore the gravitation prevails over the electromagnetic interactions in the macro-world. Moreover, the gravitational interaction grows with energy, making the gravity essential at extremely small distances. This happens solely owing to the existence of the Planck constant $\hbar$, since by confining a system to small distances, Δr, we supply it with a large energy $\Delta E \sim \hbar c (\Delta r)^{-1}$.

Leptons, photons, graviton do not participate in strong interactions.

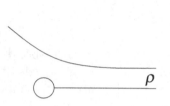

Rutherford was the first to observe (electromagnetic) scattering of strongly interacting particles. By comparing the scattering pattern of α-particles at large angles with classical formulae he concluded that the size of the gold nucleus was about 10^{-13} cm. By the way, to be able to describe the process as classical particle scattering, all the way down to $\rho \sim 10^{-13}$ cm, one has to have

$$k r_0 \gg 1.$$

However, the energies of α-particles in the Rutherford experiment, $E/m_\alpha = \mathcal{O}(\text{keV}/\text{GeV}) = 10^{-6}$, correspond to momenta k such that

$$k \cdot r_0 = \sqrt{2 m_\alpha \cdot E} \cdot \frac{1}{\mu} \ll 1, \quad \text{with} \quad (r_0)^{-1} \sim \mu = 140 \text{ MeV},$$

so that the scattering becomes *quantum*, rather than classical, already for the impact parameters much larger than r_0. It was fortunate for Rutherford that the scattering cross section in the Coulomb field happened to be identical to that in the classical theory!

The proton–proton cross section is very small, $\sigma_{pp} \sim 4 \cdot 10^{-26}$ cm^2. Why then do we refer to the 'strong interaction' as *strong*? To really evaluate the strength of interaction, one has to take into consideration the existence of the finite *interaction radius* since the interaction cross section is composed of the actual interaction strength and of the probability to hit the target, measured by the transverse area of the hadron $\sim \pi r_0^2$. This being said, if the interaction cross section turns out to be of the order of the geometric cross section,

$$\sigma \sim r_0^2,$$

we call the interaction *strong*; otherwise, if

$$\sigma \ll r_0^2,$$

such an interaction we consider as *weak*.

How can one determine experimentally the interaction radius r_0?

In the *classical theory* it is straightforward: from the angular dependence of the scattering cross section $d\sigma(\mathbf{q})$ one can reconstruct the potential, and, subsequently, extract the characteristic radius r_0.

In *quantum mechanics* we operate with the partial wave expansion of the scattering amplitude,

$$f(k,\theta) = \sum_{\ell=0}^{\infty} (2\ell + 1) f_\ell(k) P_\ell(\cos\theta).$$

Guided by quasi-classical considerations, we can define the interaction radius by comparing the magnitudes of the partial wave amplitudes f_ℓ with different orbital momenta ℓ:

$$f_\ell \sim \begin{cases} 1 & \text{for } \ell \lesssim kr_0, \\ 0 & \ell \gg kr_0. \end{cases}$$

Assume that the interaction radius is small, $kr_0 \ll 1$. Then, due to the fact that the partial waves with large orbital momenta are suppressed, $f_\ell \propto (kr_0)^{2\ell+1}$ (centrifugal barrier), the S-wave dominates,

$$f(k,\theta) \simeq f_0(k),$$

and the scattering pattern is spherically symmetric. Increasing the incident momentum we reach $kr_0 \sim 1$ where a few partial waves will start contributing and the corresponding Legendre polynomials with $\ell \neq 0$ will introduce angular dependence into the scattering distribution. Thus we can determine r_0 by studying at what energies the scattering ceases to be spherically symmetric. Alternatively, at large k, we can extract the interaction radius by measuring the characteristic scattering angle, $\theta_{\text{char}} \sim (kr_0)^{-1} \ll 1$.

Now that we know how to measure r_0 and may compare σ with r_0^2, let us ask ourselves another question: whether the situation when $\sigma \ll r_0^2$ really means that we are dealing with a *weak* interaction.

The answer is, yes and no!

Consider the scattering of a point-like neutrino off a proton, for example, the process $\nu_\mu + p \to \mu + X$. By examining the momentum dependence of the cross section we will extract that very same proton radius

$r_0 \sim 10^{-13}$ cm. At the same time, the interaction cross section is of the order of $\sigma_\nu \sim$
10^{-40} cm^2. Then, according to our logic we must
proclaim the neutrino a weakly interacting particle.

However, imagine that the proton has a tiny core, of the size 10^{-20},
which is smeared over the area of the radius $r_0 = 10^{-13}$. If so, the interaction of the neutrino with the proton actually turns out to be strong:
$(10^{-20})^2 \sim \sigma_\nu$. We can only state that ν interacts weakly at the distances
larger than 10^{-20} cm.

The most important property of the *weak interaction* is its *univer-
sality* with respect to hadrons and leptons. They get engaged in the
weak interaction in a similar manner and with the same universal *Fermi
constant*

$$G_F \simeq \frac{10^{-5}}{m_p^2}, \qquad m_p^{-1} \sim 10^{-14}\,\text{cm}.$$

Weak interaction increases with energy. At distances $10^{-3}/m_p \sim$
10^{-17} cm, corresponding to collision energies of the order of $10^3 m_p \simeq$
$1000\,\text{GeV}$, the weak interactions may become *strong*.

The main features of *strong interactions* of hadrons are the following:

(1) probability to interact is $\mathcal{O}(1)$ at the distances $r \lesssim r_0 = 10^{-13}$ cm;

(2) hadrons are intrinsically relativistic objects.

Indeed, to investigate the distances $r_0 = 1/\mu$, momenta $k \sim \mu$ are necessary, which correspond to the proton velocity $v \simeq \mu/m_p \sim 1/6$. (By the
way, it is this 1/6, treated as a small parameter, to which the nuclear
physics owes its existence.) At the same time, if we substitute for the
proton a π-meson (whose mass is $m_\pi = \mu$) we get $v \simeq 1$ and the very possibility of a non-relativistic approach disappears.

1.2 Symmetries of strong interactions

Imagine that we have an unstable particle whose decay time τ is much
larger than $r_0/c \sim 10^{-23}$ s. Does it decay due to the strong or weak
interaction? The answer lies in the symmetry of the decay process:
the *degeneracy* is much larger in the strong interaction; degeneracy
means symmetry, and symmetries, as you know, give rise to conservation
laws.

Electric charge Q. The hadrons have to know themselves about the electromagnetic interaction. Each hadron has a definite electric charge, and the strong interactions must respect its conservation, otherwise quantum electrodynamics would be broken.

Baryon charge B. This is another quantum number whose conservation is verified with a fantastic accuracy (stability of the Universe). The baryon charge equals $+1$ for *baryons* like p, n, Λ, Σ, Ξ, ... (and -1 for their antiparticles), and 0 for *mesons* (π, K, ρ, ω, φ, ...).

Isotopic spin I. Phenomenologically, hadrons split in groups of particles with close masses, and can be classified as belonging to *isotopic SU(2)* multiplets. For example, the doublet of the proton and the neutron, p, n ($I = \frac{1}{2}$); the triplet of pions, $\pi^{\pm}$ and π^0 ($I = 1$), etc. The relative mass difference of hadrons in one multiplet is 10^{-2}–10^{-3}, that is, of the order of the electromagnetic 'fine-structure constant':

$$\frac{m_n - m_p}{m_p} \sim \frac{m_{\pi^0} - m_{\pi^+}}{m_{\pi^+}} \sim \alpha \simeq \frac{1}{137}.$$

It looks that if we switched off the electromagnetic interaction, we would arrive at a complete degeneracy in the mass spectrum of strongly interacting particles. Independently of the hypothesis about the nature of this tiny mass splitting, these states can be treated as degenerate in the first approximation and therefore, there must be a symmetry and the corresponding conservation law.

Are the pn and pp scattering cross sections the same, if electromagnetic interactions are switched off? No – in the second case the particles are identical. In order to distinguish p from n, a new quantum number is introduced: the proton is treated as a *nucleon* with the isospin projection $I_3 = +\frac{1}{2}$, and the neutron with $I_3 = -\frac{1}{2}$. Thus, the nucleon wave function depends on coordinates, spin and *isospin* variables, $\psi(\mathbf{r}, \sigma, \tau)$. In strong interactions isospin is conserved.

For example, the lightest stable nuclei – the deuteron and the helium – consist of equal number of protons and neutrons, $D = (pn)$, $\text{He}^4 = (2p2n)$, and both have $I = 0$ (isotopic singlets). Therefore, the fusion reaction

$$D + D \not\rightarrow \text{He}^4 + \pi^0$$

is forbidden, since the pion has isospin $I = 1$.

Strangeness S. Any reaction takes place that is allowed by conservation laws. At the same time, it was observed that long-living hadrons like K-mesons, and Λ- and Σ-baryons, cannot be produced *alone* in the interactions of nucleons and pions. They always go *in pairs*, e.g.

$$\pi^- + p \to \Lambda + \bar{K}^0 \,,$$

while the reactions

$$\pi^- + p \to n + K^0, \quad \text{or} \quad \pi^- + p \to \Lambda + \pi^0$$

are forbidden.

By prescribing to these hadrons a new quantum number – *strangeness S*,

$$S(K^-, K^0) = S(\Lambda) = -1, \qquad S(\bar{K}^0, K^+) = S(\bar{\Lambda}) = +1,$$

we get the relation between the conserved quantities:

$$Q = I_3 + \frac{B}{2} + \frac{S}{2} \,.$$

There is one more approximate symmetry which combines strange and non-strange hadrons into of SU(3) multiplets, like *octets* of pseudoscalar mesons,

$$S = 1 \qquad (\bar{K}^0, K^+)_{I=\frac{1}{2}} \,,$$
$$S = 0 \qquad (\pi^-, \pi^0, \pi^+)_{I=1}, \quad \eta_{I=0},$$
$$S = -1 \qquad (K^-, K^0)_{I=\frac{1}{2}} \,,$$

and baryons,

$$S = 0 \qquad (n, p)_{I=\frac{1}{2}} \,,$$
$$S = -1 \quad (\Sigma^-, \Sigma^0, \Sigma^+)_{I=1}, \quad \Lambda_{I=0},$$
$$S = -2 \qquad (\Xi^-, \Xi^0)_{I=\frac{1}{2}} \,,$$

baryon *decuplet*, etc.

The isospin symmetry is broken by electromagnetic interactions. The weak interaction breaks *everything* except B, Q and, maybe, the lepton charge L. (Apparently, the electron and the muon lepton charges conserve separately, since $\mu^- \to e^- + \bar{\nu}_e + \nu_\mu$, but $\mu^- \nrightarrow e^- + \gamma$.)

The lightest strange particles are stable under strong interactions. However, K-mesons decay into pions, and the Λ-baryon into $\pi^- p$, due to the weak interaction, disrespecting the strangeness conservation. The weak forces violate spatial parity P, charge parity C, and even the time reflection symmetry T (the later equivalent to the 'combined parity' CP).

1.3 Basic properties of the strong interaction

1.3.1 Interaction radius

The question arises, what is r_0: is this an *interaction radius* specific for the strong interaction, or rather a real size of an object? This question can be answered using, for example, weak interactions as a short-range probe. It turns out that r_0 is the actual size of the proton that can be extracted, in particular, from the measurement of the spatial distribution of the electric charge inside the proton.

The hadron radius r_0 appears to be equal to the pion Compton wavelength,

$$r_0 \simeq m_\pi^{-1} \equiv \mu^{-1} \simeq 10^{-13}\,\text{cm}.$$

Is this coincidence an accident? In the past it was thought to be of fundamental importance; it is not so clear any more that it really is.

What is the problem with the description of the strong interactions?

As we have discussed above, a non-relativistic description does not make sense here. We have just one example which may help us to construct a relativistic theory: electrodynamics. In the quantum electrodynamics, the electron e and the photon γ are point-like, and so is the interaction between them.

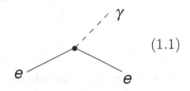

$$(1.1)$$

Now we want to describe hadrons: p, n, π. Are these particles point-like? The existence of the finite radius r_0 confirms, apparently, the opposite. There is no way, however, to give a *relativistic* description of a particle of finite radius. So we have to assume that the particles we consider are, in a sense, point-like.

Yukawa suggested that the point-likeness of a hadron does not contradict the existence of a finite interaction radius. Let us draw a pion–nucleon interaction $\quad N \overset{\pi}{} N \quad$ taking (1.1) for a model. The existence of this vertex means that there are processes of virtual emission and absorption of pions by the nucleon,

$$N \overset{\pi}{} N. \qquad (1.2)$$

Let us imagine now that this happens quite frequently. What will we see as a result of a scattering of an external particle off such a fluctuating nucleon? Estimating the energy uncertainty as

$$m \quad \overset{m+\mu}{\frown} \quad m \qquad \Delta E \simeq (m+\mu) - m = \mu, \qquad (1.3)$$

we conclude that the lifetime of the fluctuation is $\Delta t \sim (\Delta E)^{-1} \sim \mu^{-1}$. During this time interval, a pion (with a velocity $v \sim 1$) will cover the distance $\Delta r \sim \mu^{-1}$. Thus, our object, which was point-like in the beginning, is now spread over a distance μ^{-1}, and, in the process of scattering, it will interact with the projectile at impact parameters $\rho \sim \mu^{-1}$. In other

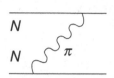

words, the scattering of an incident particle with our nucleon can be depicted as a pion exchange between the two nucleons – the process that has a characteristic radius $r_0 \sim \mu^{-1}$!

Without any theory, let us first calculate this amplitude in a naive way, by analogy. What would be the difference between the above process and the scattering of electrons that we have studied in the quantum electrodynamics,

$$\overset{e}{} \overset{q}{\diagup} \overset{}{\gamma} \quad = \frac{e^2}{q^2} \left(\bar{u} \gamma^\mu u \right) \left(\bar{u} \gamma_\mu u \right). \qquad (1.4)$$

We must replace the photon propagator $1/q^2$ in (1.4) by the Green function of the massive π meson:

$$D_\pi(q) = \frac{1}{\mu^2 - q^2}.$$

The corresponding scattering amplitude will have the form

$$A = \frac{g^2}{\mu^2 - q^2}, \qquad (1.5)$$

with g the pion–nucleon interaction constant, replacing the electric charge e in the QED amplitude (1.4).

What does this amplitude correspond to in the case of the non-relativistic scattering? The non-relativistic scattering amplitude reads

$$f = -\frac{2m}{4\pi} \int e^{i\mathbf{k}'\cdot\mathbf{r}} V(r) \, \psi(\mathbf{r}) \, d^3r. \qquad (1.6)$$

In the Born approximation, replacing the wave function $\psi(\mathbf{r})$ by a plane wave with momentum $\mathbf{k}$, we obtain

$$f_B = -\frac{2m}{4\pi} \int e^{i\mathbf{q}\cdot\mathbf{r}} V(r) \, d^3r, \qquad (1.7)$$

where $\mathbf{q}$ is the momentum transfer, $\mathbf{q} = \mathbf{k}' - \mathbf{k}$. For non-relativistic particles, the kinetic energy, $E = \mathbf{k}^2/2m$, is small, and the *energy* transfer component can be neglected:

$$|q_0| \sim \mathbf{q}^2/m \ll |\mathbf{q}|, \qquad \text{so that} \quad q^2 = q_0^2 - \mathbf{q}^2 \simeq -\mathbf{q}^2.$$

The scattering amplitude (1.5) becomes

$$A \simeq \frac{g^2}{\mu^2 + \mathbf{q}^2}.$$

What is the potential corresponding to this amplitude? Evaluating the inverse Fourier transform of the Born amplitude f_B in (1.7) we obtain the Yukawa potential,

$$V(r) = -\frac{4\pi}{2m} \int e^{-i\mathbf{q}\mathbf{r}} \frac{g^2}{\mu^2 + \mathbf{q}^2} \frac{d^3 q}{(2\pi)^3} = \frac{g^2}{2m} \cdot \frac{e^{-\mu r}}{r}. \tag{1.8}$$

So, indeed, the effective interaction is characterized by a finite radius $r_0 = 1/\mu$.

We conclude that the assumption of the point-like nature of the interaction does not exclude the finiteness of the interaction radius. Moreover, having adopted the point of view that the hadron has no intrinsic size (having no other option), we see that the interaction radius is not an independent quantity but is determined by the *masses* of the particles.

From the point of view of a relativistic theory the π-meson has to exist in nature, otherwise there would be no explanation for such a 'large' value of the proton radius.

1.3.2 Interaction strength

The other side of the strong interactions is their *strength*: once the particles approach each other to the distance r_0, the interaction is inevitable. Since a nucleon is always surrounded by a pion cloud, see (1.2), this means that the coupling constant g^2 (if it exists at all) is obliged to be *large*, $g^2 \sim 1$, contrary to the electromagnetic interaction, characterized by the small coupling $\alpha = 1/137 \ll 1$. Now that is bad indeed, because under these circumstances anything will go. For example, a virtual state with two pions will be there, having a typical lifetime $\Delta t \sim \frac{1}{2}\mu$ and, correspondingly, a spatial spread of the order of $\sim \frac{1}{2}r_0$.

We may even have a three-nucleon
state, $N \to N\pi$, $\pi \to N\bar{N}$. Since the
nucleon is much heavier than the
pion, this fluctuation is short-lived:
$\Delta t \sim 1/(2m_N) \ll r_0$. However, we
cannot state a priori that such a
process does not contribute to the

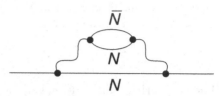

radius of the nucleon since these 'second-order' amplitudes may be actually *larger* than the one-pion emission amplitude (1.2), because the
coupling constant is not small.

It is clear that it will be certainly impossible to build a theory like
quantum electrodynamics to describe strongly interacting hadrons.

We can, however, introduce initial point-like objects, and then, in fact,
observe 'clouds', the radii of which are determined by the masses of the
hadrons.

This is the basic idea of the theory of the strong interaction.

We need to construct a framework which would allow us to draw *pictures* representing a formal series for the hadron interaction amplitudes.
From these pictures we will extract information without actually calculating the amplitudes, which would be, a priori, impossible. The Feynman
diagrams can be considered as a 'laboratory of theoretical physics'.

1.4 Free particles

We start by considering free particle states and their propagation. There
is a fantastic variety of hadrons with spins reaching up to $s = \frac{19}{2}$.

1.4.1 Particle states

$s = 0$. A free spinless (scalar) particle with a four-momentum p_μ is described by the wave function

$$s = 0 : \qquad \psi(x) = \frac{1}{\sqrt{2p_0}}\, e^{-ipx}. \qquad (1.9)$$

$s = \frac{1}{2}$. A spin-one-half particle has two states, $\lambda = 1, 2$,

$$s = \frac{1}{2} : \qquad \psi_\alpha^\lambda(x) = \frac{u_\alpha^{(\lambda)}}{\sqrt{2p_0}}\, e^{-ipx}; \qquad (1.10)$$

two states with definite parity are selected out of possible four spinors u_α
by the Dirac equation, $(\hat{p} - m)u^\lambda$.

$s = 1$. A spin one particle is described by a wave function ψ_μ bearing the Lorentz vector index:

$$s = 1: \qquad \psi_\mu^\lambda(x) = \frac{e_\mu^{(\lambda)}}{\sqrt{2p_0}} \, \mathrm{e}^{-ipx}. \qquad (1.11)$$

Here one has to single out three states, $\lambda = 1, 2, 3$, out of the four unit vectors. In the rest frame, the vector particle has three polarizations,

$$e_\mu = \begin{pmatrix} t \\ x \\ y \\ z \end{pmatrix}; \qquad e_\mu^{(1)} = \begin{pmatrix} 0 \\ 1 \\ 0 \\ 0 \end{pmatrix}, \qquad e_\mu^{(2)} = \begin{pmatrix} 0 \\ 0 \\ 1 \\ 0 \end{pmatrix}, \qquad e_\mu^{(3)} = \begin{pmatrix} 0 \\ 0 \\ 0 \\ 1 \end{pmatrix},$$

all having zero time component, $e_0^{(\lambda)} = 0$. In Lorentz invariant terms, the superfluous state is eliminated by the condition $e_\mu^{(\lambda)} p^\mu = 0$.

$s = 2$. A tensor particle has to have $2s + 1 = 5$ physical states. Its wave function is constructed with the help of a Lorentz tensor $T_{\mu\nu}$,

$$s = 2: \qquad \psi_{\mu\nu} = \frac{T_{\mu\nu}}{\sqrt{2p_0}} \, \mathrm{e}^{-ipx}.$$

This tensor can be simply constructed as a product of two *vector* states, $e_\mu^{\lambda_1}$ and $e_\nu^{\lambda_2}$. Since $p^\mu e_\mu^{\lambda_i} = 0$, in the rest frame we have

$$T_{00} = T_{i0} = T_{0k} = 0,$$
$$t_{ik} = e_i^{\lambda_1} e_k^{\lambda_2} + e_i^{\lambda_2} e_k^{\lambda_1}, \quad i, k = 1, 2, 3. \qquad (1.12)$$

The symmetric 3×3 tensor (1.12) has $3 \cdot 4/2 = 6$ independent components. Combining two spin 1 particles, we obtain not only a spin 2 state but also one with spin zero; to exclude the latter we must make the tensor T traceless,

$$T_{ik} = t_{ik} - \tfrac{1}{3}\delta_{ik} \sum_{\ell=1}^3 t_{\ell\ell}.$$

Finally, we have a symmetric Lorentz tensor, $T_{\mu\nu} = T_{\nu\mu}$,

$$T_{\mu\nu} = e_\mu^{\lambda_1} e_\nu^{\lambda_2} + e_\mu^{\lambda_2} e_\nu^{\lambda_1} - \frac{2}{3} \left(g_{\mu\nu} - \frac{p_\mu p_\nu}{m^2} \right) \left(e^{\lambda_1} e^{\lambda_2} \right),$$

which on the mass shell, $p_\mu^2 = m^2$, is traceless, $T_{\mu\nu} \cdot g^{\mu\nu} = 0$, and 'orthogonal' to the four-momentum of the particle, $p^\mu T_{\mu\nu} = 0$.

$s = \frac{3}{2}$. The wave function of a spin $\frac{3}{2}$ state bears simultaneously a spinor index α, and the vector index μ.

$$s = \tfrac{3}{2} : \qquad \psi_{\alpha,\mu}(x) = \frac{u_{\alpha,\mu}}{\sqrt{2p_0}}\, e^{-ipx}.$$

Initially, $u_{\alpha,\mu}$ has 16 degrees of freedom (4 spinors $\times$ 4 vector components). The first condition,

$$(\hat{p} - m)u_{\alpha,\mu} \equiv \sum_{\beta} (\hat{p} - m)_{\alpha,\beta} u_{\beta,\mu} = 0, \tag{1.13a}$$

selects two spinors ($16 \to 8$), and the second one,

$$\gamma^{\mu} u_{\alpha,\mu} = 0, \tag{1.13b}$$

(four equations) leaves us with $8 - 4 = 4 = 2 \cdot \frac{3}{2} + 1$ states. From the pair of conditions (1.13) it conveniently follows that

$$p^{\mu} u_{\alpha,\mu} = 0.$$

Indeed, from (1.13b) we get

$$\gamma^{\mu} u_{\alpha,\mu} = 0 \implies 0 = (\hat{p} + m)\gamma^{\mu} u_{\alpha,\mu} = \gamma^{\mu} \underbrace{(-\hat{p} + m)u_{\alpha,\mu}}_{=0} + 2p^{\mu} u_{\alpha,\mu}.$$

There exists a special technology how to move further to higher spins.

1.4.2 Particle propagators

Relativistic propagation of free particles is described by Green functions

$$G(x) = \int \frac{d^4p}{(2\pi)^4 i}\, e^{-ipx} G(p).$$

In the momentum space, the Green functions of scalar, fermion and vector particles are

$$s = 0 : \qquad G(p) = \frac{1}{m^2 - p^2}, \tag{1.14a}$$

$$s = \tfrac{1}{2} : \qquad G(p) = \frac{1}{m - \hat{p}} = \frac{1}{m^2 - p^2} \cdot (m + \hat{p}), \tag{1.14b}$$

$$s = 1 : \qquad G_{\mu\nu}(p) = \frac{1}{m^2 - p^2} \cdot \left(\frac{p_\mu p_\nu}{m^2} - g_{\mu\nu} \right). \tag{1.14c}$$

The factors in the numerator that accompany the pole $1/(m^2 - p^2)$ in the propagators of particles with spin originate from a summation over physical polarization states:

$$\sum_{\lambda=1}^{2} u^{\lambda}(p)\, \bar{u}^{\lambda}(p) \;=\; m + \hat{p},$$

$$\sum_{\lambda=1}^{3} e_{\mu}^{\lambda}(p)\, e_{\nu}^{*\lambda}(p) \;=\; \frac{p_{\mu}p_{\nu}}{m^2} - g_{\mu\nu}.$$

Analogously, propagators of particles with higher spins will contain the structures

$$s = \tfrac{3}{2}: \qquad G_{\mu\nu}(p) \propto (p_{\mu} + m\gamma_{\mu})(m - \hat{p})(p_{\nu} + m\gamma_{\nu}), \qquad (1.14d)$$

$$s = 2: \qquad G_{\mu\nu,\mu'\nu'} \propto \left(g_{\mu\mu'} - \frac{p_{\mu}p_{\mu'}}{m^2}\right)\left(g_{\nu\nu'} - \frac{p_{\nu}p_{\nu'}}{m^2}\right) + \text{perm.} \quad (1.14e)$$

The numerators (1.14) are polynomials in p, and for large p values the Green functions are growing as $p^{2(s-1)}$. This growth is unavoidable and leads to a large number of unpleasant problems.

One might imagine a scenario with mass degeneracy, such that a scalar and a vector state together would be described by a propagator

$$G_{1+0}(p) = \frac{-g_{\mu\nu}}{m^2 - p^2},$$

free of the increasing term $p_{\mu}p_{\nu}$ present in (1.14c). This, however, is not a solution, since such a scalar particle would be a *ghost* – a state with a negative transition probability, $-g_{00} = -1$. This would violate the rule according to which the amplitude near the pole has a definite sign, following from the unitarity.

Do we really need to ascribe a bare field and its own interaction to each of few hundred existing hadrons? Possibly, one can treat all these hadrons by expressing them by means of a few fundamental objects.

1.5 Hadrons as composite objects

In the language of field theory, we introduce some fundamental fields and describe them in terms of the wave function ψ. We construct the interaction Hamiltonian which may have *bound states*. The known example of appearance of such a composite particle in relativistic field theory is *positronium* – a bound state of an electron and a positron.

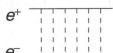

The QED coupling being small, the binding energy of positronium is much smaller than the electron mass, $\epsilon \ll m$. Consequently, in the average e^+ and e^- are at a relatively large distance $\Delta r \gg m^{-1}$ apart. However, even here there is a possibility to produce additional pairs etc., so that the field-theoretical nature of the state is rather rich and complex.

1.5.1 Scattering of composite states

How to describe the scattering of a bound system – positronium – in an external field?

In quantum mechanics the scattering amplitude has a structure

$$f \sim \int e^{-i\mathbf{p}'\cdot\mathbf{r}_c}\psi_f(\mathbf{r}_{12})[V(\mathbf{r}_1) + V(\mathbf{r}_2)]\, e^{i\mathbf{p}\cdot\mathbf{r}_c}\psi_i(\mathbf{r}_{12}), \qquad (1.15)$$

where $\mathbf{r}_c$ is the centre-of-mass coordinate, and $\psi(\mathbf{r}_{12})$ the relative motion wave function.

In terms of diagrams we have

$$(1.16)$$

The non-relativistic Green function of the e^+e^- system,

$$\mathbf{r}_c = \tfrac{1}{2}(\mathbf{x}_1 + \mathbf{x}_2), \qquad \mathbf{r}'_c = \tfrac{1}{2}(\mathbf{x}'_1 + \mathbf{x}'_2), \qquad (1.17a)$$

$$\mathbf{r}_{12} = \mathbf{x}_1 - \mathbf{x}_2, \qquad \mathbf{r}'_{12} = \mathbf{x}'_1 - \mathbf{x}'_2, \qquad (1.17b)$$

can be expressed as a sum over eigenstates of the product of the final and (conjugate) initial wave functions:

$$G(\mathbf{r}'_c, \mathbf{r}'_{12}, t'; \mathbf{r}_c, \mathbf{r}_{12}, t) = \sum_n \psi_n(\mathbf{r}'_c, \mathbf{r}'_{12}, t')\, \psi_n^*(\mathbf{r}_c, \mathbf{r}_{12}, t),$$

where $t \equiv x_{10} = x_{20}$, $t' \equiv x'_{10} = x'_{20}$. Among these terms there is one which corresponds to the bound state D:

$$G = \sum_{\mathbf{p}_c} \psi_D(\mathbf{r}'_c, \mathbf{r}'_{12}, t')\, \psi_D^*(\mathbf{r}_c, \mathbf{r}_{12}, t) \;+\; [\text{continuous } e^+e^- \text{ spectrum}].$$

In the mixed space–energy representation, the stationary state Green function,

$$G(\mathbf{r}, \mathbf{r}'; E) = \sum_n \frac{\psi_n(\mathbf{r}')\,\psi_n^*(\mathbf{r})}{E_n - E},$$

contains the *pole* in energy, corresponding to the positronium state. Let us single out this pole from the sum:

$$G(\mathbf{r}, \mathbf{r}'; E) \;\;\Longrightarrow\;\; G_D(\mathbf{r}, \mathbf{r}'; E) = \;\;\rangle\!\!-\!\!\overset{D}{}\!\!-\!\!\langle\;.$$

Then, the interaction diagram (1.16) will reduce to

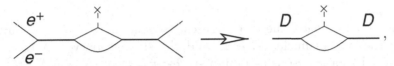

where we have cut off the electron ends since we are interested in a bound state at infinity, not a fermion pair. To calculate the interaction amplitude, we have to replace the positronium lines by the initial and final state wave functions. This way we arrive at the expression similar to (1.15) for the non-relativistic scattering amplitude.

In other words, the notion of the Green function of the positronium (or of the deuteron, for that matter) is unnecessary. It can, however, be introduced by separating from the product of the wave functions the dependence on the total four-momentum of the bound system,

$$\psi_D(x')\psi_D^*(x) = e^{-iE(t'-t)+i\mathbf{p}_c(\mathbf{r}'_c - \mathbf{r}_c)}\psi(\mathbf{r}')\psi^*(\mathbf{r}). \tag{1.18}$$

Attributing the exponent to the Green function of the free positronium, the wave function of the relative motion will have the meaning of an *exact vertex* γ describing the transition between the positronium and the free e^+e^- pair, which will determine the interaction of the bound state as a whole with the external field:

An interesting feature of γ is that this vertex does not contain any input bare value: it derives from the structure of the bound state.

In the non-relativistic theory, describing a bound state in terms of its proper Green function G_D and its proper interaction vertex Γ is merely a formal unification. However, the existence of two possibilities to construct a theory of particle interactions (with different results!) is important for us in the context of strong interactions, where a pion, for example, can

be looked upon as consisting of a pair of nucleons, π^+ ... $\bar{n}$ p .

1.5.2 Quarks

Thus, we start with the hypothesis (which may turn out to be wrong) that there exists a small number of fundamental objects. From the point of view of *hadrons*, these objects can be chosen almost arbitrarily. *Almost*, since one cannot build up a spin $s = \frac{1}{2}$ particle from spinless π-mesons, or strange hadrons like K-mesons and the Λ baryon out of non-strange nucleons.

In the *Sakata model*, the three lightest baryons were chosen as building blocks: the pair of nucleons, p, n, and the strange baryon Λ ($S = -1$): This model treats mesons as bound states: $\pi = N\bar{N}$, $K = N\bar{\Lambda}$, etc.,

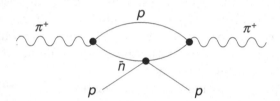

though it remains unclear what to do with a rich variety of baryons.

Gell-Mann put forward a deeper idea based on the existence of eight very similar baryons (n, p, Λ, $\Sigma^{\pm,0}$, $\Xi^{0,-}$) with close mass values. The underlying $SU(3)$ symmetry became the origin of the idea of *quarks*.

Tempted to consider all the hadrons as *composite* objects, one introduces three quarks, $q_{1,2,3} = u, d, s$, which, for some mysterious reason, are not observable as free particles but are confined* inside the mesons ($q\bar{q}$) and the baryons (qqq). Fractional electric charges of the quarks may not look too attractive. Still, it would be nice if they existed in nature, and this possibility should not be disregarded a priori.

Thus, we assume that there are some fundamental particles that we introduce as bare fields into the theory. These particles may or may not show up in the physical spectrum of the theory (it would be beautiful if not). If so, all the observable hadrons are composite objects, and all reactions between them can be represented as the strong interaction between the underlying quarks.

---

* This could be possible, for example, if the quark binding energy were so strong as to provide masses of the bound states *much smaller* than the large masses of the quarks.

1.6 Interacting particles

How to construct the interaction between our fundamental objects? Let us see what options quantum field theory (QFT) can offer us.

1.6.1 $\lambda\varphi^3$

Let us look at the simplest QFT that describes a scalar field φ with three-particle interaction. Given the interaction constant λ, and depicting by a straight line the free particle propagator,

$$G(p) = \frac{1}{m^2 - p^2} = \text{———},$$

we can draw Feynman diagrams for the exact particle Green function and for interaction amplitudes,

Evaluating the self-energy diagram, in the region of large virtual momenta we will have a logarithmically divergent integral,

$$\Sigma(p) = \text{———} = \lambda^2 \int \frac{d^4k}{(2\pi)^4 i} \frac{1}{m^2 - k^2} \frac{1}{m^2 - (p-k)^2} \sim \int_m \frac{d^4k}{k^4}.$$

This turns out to be the *only* divergence in the theory! The integrals with more than two propagators in the loop, converge:

$$\sim \lambda^3 \int \frac{d^4k}{k^6} \sim \frac{\lambda^3}{p^2},$$

with p^2 the characteristic virtuality of the external lines. This shows that (apart from the mass renormalization) the self-interaction effects vanish in the large momentum region.

The absence of divergences is clear from dimensional considerations. Let us look, e.g. at the particle number density operator, which has the dimension

$$\left[\varphi \frac{\partial\varphi}{\partial t}\right] = [\delta^{(3)}(\mathbf{r})] = [m^3].$$

Therefore, the field $\varphi(x)$ itself has the dimension of mass, $[\varphi] = [m]$. Since the *action* is dimensionless ($\hbar = 1$),

$$\mathcal{L} = \int \lambda \cdot \varphi^3(x)\, d^4x, \quad [\mathcal{L}] = [m^0],$$

we get $[\lambda] = [m]$. The coupling constant having the dimension of mass, at large momenta p (where the finite mass is unimportant) this gives us the real dimensionless expansion parameter $\sim\lambda^2/p^2$, vanishing in the ultraviolet region. Such theories are referred to as 'superconvergent'.

From the Born diagrams for the two particle scattering amplitude,

$$\ldots = \frac{\lambda^2}{m^2 - s} + \frac{\lambda^2}{m^2 - t} + \frac{\lambda^2}{m^2 - u},$$

we see that the interaction disappears when the energy and momentum transfer invariants (s, t, u) become large.

The widespread opinion according to which the $\lambda\varphi^3$ QFT is a bad one is owing to the non-positive definiteness of the energy density, $dE/dV \propto \lambda\varphi^3$, because of which this theory has no vacuum state.

Unfortunately, all this has nothing to do with Nature. It is, however, a useful QFT model for those cases when the spin is of no importance.

1.6.2 $\lambda\varphi^4$

Let us consider the next, more complicated example: the quartic interaction between spinless fields.

Pions are *pseudoscalar* particles, and therefore the transition $\pi \to \pi\pi$ is forbidden by parity conservation. This makes the $\lambda\varphi^4$ QTF closer to reality; it can be used to model interaction between pions. Now the coupling constant λ is *dimensionless*, so that we should expect logarithmic ultraviolet divergences, as in the case of QED.

The simplest correction to the Green function G now diverges *quadratically*:

$$\ldots \sim (d^4k)^2 \cdot \frac{1}{[k^2]^3}.$$

What concerns the effective charge, the first correction to the two-particle scattering amplitude λ consists of three graphs,

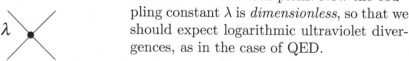

$$\ldots \sim \lambda^2 \int \frac{d^4k}{k^4} \sim \lambda^2 \ln \Lambda_{\mathrm{UV}}^2.$$

The situation is similar to that in electrodynamics, and the renormalization procedure is carried out in the same way. And, similarly, the 'zero charge problem' appears: in both theories the renormalized coupling tends to zero when the ultraviolet cutoff Λ is taken to infinity:

$$e_c^2 = \frac{e_0^2}{1 + \frac{e_0^2}{3\pi} \ln \frac{\Lambda^2}{m^2}}, \tag{1.19a}$$

$$\lambda_c^2 = \frac{\lambda_0^2}{1 + \frac{\lambda_0^2}{4\pi} \ln \frac{\Lambda^2}{m^2}}. \tag{1.19b}$$

In the QED context this was an 'academic' problem since the coupling was small, $e_c^2 \ll 1$, and the real contradiction appeared at fantastically large momenta and could be ignored. Not so when $\lambda = \mathcal{O}(1)$; the theory becomes unreliable at low momentum scales $p \gtrsim \Lambda \sim m$. Obviously, we are unable to get any information from such a theory.

1.6.3 Four-fermion interaction

It would be nice to start constructing the theory from *fermions*, since from fermions one can build bosons, but not vice versa. One can imagine a quartic interaction between fermions, in analogy with the $\lambda \varphi^4$ model for scalars. The vertex may look as follows (Fermi interaction):

$$= G_{\mathrm{F}} \, (\bar{u}_2 \mathbf{O} u_1)(\bar{u}_4 \mathbf{O} u_3),$$

where the operator $\mathbf{O}$ in each of the fermion brackets may contain Dirac matrices, $\mathbf{O} \propto 1$, γ_μ, $\gamma_5 \gamma_\mu$, $\sigma_{\mu\nu}$, etc. Here the coupling constant has a *negative* mass dimension, $[G_{\mathrm{F}}] = [m^{-2}]$, so that the interaction grows with energy, and the theory becomes non-renormalizable:

$$\sim G_{\mathrm{F}}^2 \int \frac{d^4 k}{(\hat{k})^2} \sim \Lambda_{\mathrm{UV}}^2.$$

1.6.4 A nucleon and a pion

Interaction between a fermion (nucleons) and a spinless field can be modelled as $\bar{\psi}\psi\,\varphi$:

$$= g\big(\bar{u}(p_2) i \gamma_5 u(p_1)\big).$$

(Here we have introduced in the vertex the factor $i\gamma_5$ to match the fact that the pion is a pseudoscalar.) In this theory nucleons interact via pion exchange,

$$A = g^2(\bar{u}i\gamma_5 u)\frac{1}{m^2 - q^2}(\bar{u}i\gamma_5 u).$$

The coupling g is dimensionless, and the essential divergences are just logarithmic. This is quite a nice theory, it differs from electrodynamics only in that the π-meson field that carries the interaction between fermions, unlike the photon, has a finite mass, $m_\pi \neq 0$.

1.6.5 A nucleon and a vector meson

We can also make the nucleon interact with a massive vector meson field V_μ, in a QED-like manner: $\bar{\psi}\gamma^\mu\psi V_\mu$.

$$= g\left(\bar{u}(p_2)\gamma^\mu u(p_1)\right) \cdot e^\lambda_\mu(k),$$

with $e^\lambda_\mu(k)$ the polarization vectors of the field V ($\lambda = 1, 2, 3$). However, the situation here is potentially dangerous. Recall that the Green function of a massive vector field contains the term with momenta in the numerator:

$$G(q) = \frac{1}{m^2 - q^2}\left(-g_{\mu\nu} + \frac{q_\mu q_\nu}{m^2}\right).$$

We must ensure that the term $q_\mu q_\nu/m^2$ drops out in the physical amplitudes, otherwise renormalizability of the theory would be lost. This is the case in QED, owing to the conservation of the electromagnetic current.

Conservation of current makes electrodynamics with a *massive photon* a perfectly legitimate renormalizable QFT, which construction can be borrowed to model strong interaction of point-like protons and neutrons with an *electrically neutral* vector meson. Such a theory would not be too bad; it would cause no objection apart from the 'zero-charge problem' that plagues it (together with the previously considered $NN\pi$ model).

$$\text{(1.20)}$$

However, in reality *neutral* vector mesons occupy no special place in the hadron world; *charged* mesons are plentiful and one sees no reason to discriminate between them.

1.6.6 Charged vector mesons

Assume that we want to describe a charged meson. Let us consider three vector mesons, V^0 and $V^\pm$ (like a triplet of ρ-mesons), and discuss how they might interact with p and n.

With the neutral meson V^0 everything is simple: it may be emitted (absorbed) either by a proton or a neutron as shown in (1.20), with λ_1 and λ_2 the corresponding coupling constants.

How will a charged meson interact with nucleons? A negative particle can be absorbed by a proton, which will turn into a neutron when absorbing a negative-charge meson V^-, or emitting V^+:

$$\text{(1.21a)}$$

Another possible process is absorption of V^+ (emission of V^-) by a neutron:

$$\text{(1.21b)}$$

Its amplitude is identical to that of (1.21a) if the theory is T-invariant. (Thus we have in principle three different coupling constants: two for the neutral meson in (1.20) and one for the charged, $\lambda = \lambda'$ in (1.21).)

Now we have a look at the diagram for scattering of nucleons via V exchange:

$$G(q) = \frac{1}{m^2 - q^2}\left(-g_{\mu\nu} + \frac{q_\mu q_\nu}{m^2}\right).$$

Will the 'bad term' now disappear? Convolution of the nucleon vertex with the meson momentum $q = k_n - k_p$ produces

$$q^\mu \cdot \left(\bar{u}_n \gamma_\mu u_p\right) = \bar{u}_p(\hat{k}_n - \hat{k}_p)u_n = (M_n - M_p)\cdot\left(\bar{u}_n u_p\right) \neq 0,$$

where we have used the Dirac equation for the on-mass-shell nucleons, $(\hat{k} - M)u = 0$. The result is not zero. And even if we set $M_p = M_n$ in zeroth approximation, this will not help in higher orders.

In the case of a neutral meson, electric charge of the fermion was preserved, and, as a result, the vector vertex of V^0 emission turned out to be identical to the conserved electromagnetic current, ensuring $q^\mu A_\mu = 0$. In graphs with $V^\pm$ emission this is no longer true.

Consider, for example, the elastic $V^- p$ scattering amplitude,

$$M_{\mu\nu} = \qquad\qquad\qquad\qquad\qquad\qquad\qquad\qquad (1.22)$$

We would like to have $q^\mu M_{\mu\nu} = 0$, with q_μ the vector meson momentum. In QED Compton scattering there were *two* graphs whose sum satisfied this property:

In our new context, the second contribution is absent: in a crossed diagram, an emission of V^- by a proton implies virtual exchange of a non-existent doubly charged nucleon.

We come to the conclusion that such a theory is always non-renormalizable. There is, however, a beautiful way to correct the situation provided by the *Yang–Mills theories*.

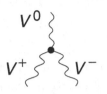

An incorporation of a specially chosen three-linear interaction between mesons allows one to construct a renormalizable theory of *massless* vector fields, $m_V = 0$. By adding another scalar field φ, vector mesons can be made *massive*, $m_V \neq 0$, without losing the renormalizability. This way the Glashow–Weinberg–Salam theory of weak interactions is constructed, with scalar 'Higgs' providing masses to the intermediate vector bosons Z^0 and $W^\pm$.

We postpone the discussion of the dynamics of Yang–Mills fields to the last lecture. Now let us turn to general features of relativistic particle scattering.

1.7 General properties of S-matrix: unitarity and crossing

Can we learn anything about the strong interactions given that there is no hope of employing perturbation theory?

Suppose we have some relativistic quantum field describing the objects that interact strongly, with a large *coupling constant* $g \sim 1$. In spite of the inapplicability of the perturbative methods, there is nevertheless a number of general statements that can be made.

1.7.1 S-matrix

First of all, in order to describe interaction processes we introduce the S-matrix whose elements S_{ab} quantify the transition from the initial state a to some final state b,

$$S = I + iT; \quad S_{ab} = \delta_{ab} + iT_{ab}. \tag{1.23}$$

Here I is a symbolic representation of the absence of interaction; δ_{ab} means that in the final state we find the incoming particles with unperturbed momenta. T is called the reaction matrix and takes care of the interaction. It contains the δ-function to ensure the energy–momentum conservation and the product of the factors $1/\sqrt{2p_0}$ that originate from the relativistic normalization of the wave functions of incoming (i) and outgoing particles (j):

$$T_{ab} = (2\pi)^4 \, \delta^4 \left(\sum_{i \in a} p_i - \sum_{j \in b} k_j \right) \prod_{i \in a} \frac{1}{\sqrt{2p_{0i}}} \prod_{j \in b} \frac{1}{\sqrt{2k_{0j}}} \cdot \mathcal{M}_{ab}. \tag{1.24}$$

So defined, the scattering amplitude $\mathcal{M}_{ab}$ is *Lorentz invariant*. Typically, the initial state consists of two particles with four-momenta p_1 and p_2.

To obtain the probability of the reaction one squares (1.24). Dropping the factor

$$(2\pi)^4\delta^4(0) = V \cdot T$$

which formally represents the full volume of the space–time, we get the measurable *probability density* of the reaction. The wave function normalization factors of the outgoing particles participate in forming the Lorentz invariant phase space volume element of the final state,

$$d\Gamma_j = \left(\frac{1}{\sqrt{2k_{0j}}}\right)^2 \frac{d^3\mathbf{k}_j}{(2\pi)^3} = \frac{d^3\mathbf{k}_j}{2(2\pi)^3 k_{0j}} = \frac{d^4 k_j}{(2\pi)^4} \cdot 2\pi\delta_+\left(k_j^2 - m_j^2\right),$$

$$(1.25)$$

where δ_+ selects among the two solutions of the on-mass-shell condition the *positive energy* (physical) one: $k_{0j} = \sqrt{m_j^2 + \mathbf{k}_j^2}$. The initial state normalization factors combine with the relative velocity of the incoming particles,

$$j \equiv |\mathbf{v}_1 - \mathbf{v}_2| = \left|\frac{\mathbf{p}_1}{p_{01}} - \frac{\mathbf{p}_2}{p_{02}}\right|,$$

to form the flux factor,

$$J = (\sqrt{2p_{10}})^2 (\sqrt{2p_{20}})^2 \cdot j = 4|p_{20}p_{1z} - p_{10}p_{2z}|, \qquad (1.26)$$

where we have chosen the direction $\mathbf{z}$ as the collision axis. The combination (1.26) is invariant under boosts along the z axis. Choosing the centre of mass system of reference (cms) in which the incoming momenta are equal and opposite, $\mathbf{p}_1 = -\mathbf{p}_2 = (0, 0, p_c)$,

$$p_c = p_c(s) = \frac{\sqrt{s^2 - 2s(m_1^2 + m_2^2) + (m_1^2 - m_2^2)^2}}{2\sqrt{s}}$$

$$= \frac{\sqrt{(s - (m_1 + m_2)^2)(s - (m_1 - m_2)^2)}}{2\sqrt{s}}, \qquad (1.27)$$

we get the Lorentz invariant flux

$$J = 4p_c(s)\sqrt{s}. \qquad (1.28)$$

Finally, the differential cross section of the process $p_1, p_2 \rightarrow \{k_j\}$ reads

$$d\sigma(a \rightarrow b) \equiv \frac{1}{J}|\mathcal{M}_{ab}|^2 (2\pi)^4 \delta^4\left(p_1 + p_2 - \sum_{j\in b} k_j\right) \cdot \frac{1}{[n!]} \prod_{j\in b} d\Gamma(k_j).$$

$$(1.29)$$

Among n produced particles there may be identical ones. The symmetry factor eliminates multiple counting of physically indistinguishable configurations produced by a permutation of identical particles. When in the final state b there are n_s particles of the type s,

$$\frac{1}{[n!]} \equiv \prod_s \frac{1}{n_s!}, \quad \sum_s n_s = n.$$

1.7.2 Unitarity

The S-matrix (1.23) is *unitary*,

$$S\,S^\dagger = 1 \quad \Longrightarrow \quad T_{ab} - T_{ab}^\dagger = i\left(T\,T^\dagger\right)_{ab},$$

or, deciphering a symbolic matrix multiplication,

$$\frac{1}{i}\left(T_{ab} - T_{ba}^*\right) = \sum_c T_{ac}\,T_{cb}^*. \tag{1.30a}$$

If the interaction is invariant with respect to the *time reversal*, T, which is the case for the strong interaction of hadrons, then the matrix $\mathcal{S}$ is *symmetric*, $T_{ab} = T_{ba}$, and the unitarity relation (1.30a) takes the form

$$\frac{1}{i}(T_{ab} - T_{ab}^*) = 2\,\mathrm{Im}\,T_{ab} = \sum_c T_{ac}\,T_{bc}^*. \tag{1.30b}$$

This expression implies an integration over momenta of all the particles in the intermediate state c. Therefore, the symmetry factor is present on the r.h.s. of (1.30), analogously to the differential cross section case (1.29).

In terms of the invariant amplitude $\mathcal{M}$ defined in (1.24),

$$\frac{\mathcal{M}_{ab} - \mathcal{M}_{ab}^*}{i} = \sum_n \frac{1}{[n!]} \int \mathcal{M}_{an}(\{p\}_a; \{k\}_n)\, \mathcal{M}_{bn}^*(\{p\}_b; \{k\}_n)$$

$$\cdot (2\pi)^4 \delta\left(\sum p_i^a - \sum_{\ell=1}^n k_\ell\right) \prod_{\ell=1}^n \left\{\delta_+(k_\ell^2 - m_\ell^2)\frac{d^4 k_\ell}{(2\pi)^3}\right\}, \tag{1.31}$$

where $\{p\}$ marks the set of momenta in the initial (a) and final states (b), and $\{k\}_n$ – momenta of n intermediate state particles.

If we take $a \equiv b$, the 'optical theorem' emerges which relates the imaginary part of the forward scattering amplitude to the total cross section:

$$2\,\mathrm{Im}\,A_{aa} = J \cdot \sigma_{\mathrm{tot}}^a. \tag{1.32}$$

1.7.3 Mandelstam plane for $2 \to 2$ scattering

Consider a two-particle interaction amplitude $1 + 2 \to 3 + 4$. How many
Lorentz invariant variables characterize
the process? We have three momentum
four-vectors, that is $3 \times 4 = 12$ independent components. Four on-mass-shell
conditions, $p_i^2 = m_i^2$, one per each participating particle, leave us with $12 - 4 = 8$.
Finally, we must subtract six parameters
(three rotations and three Lorentz boosts) which characterize the reference frame and do not affect the invariant amplitude, $8 - 6 = 2$.

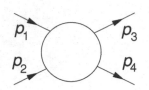

In a general case of the reaction $n_1 \to n_2$, the counting goes as follows,

$$4(n_1 + n_2 - 1) - (n_1 + n_2) - 6 = 3(n_1 + n_2) - 10. \tag{1.33}$$

For example, a $2 \to 3$ process depends on five independent Lorentz invariant combinations of momenta.

A convenient way to characterize $2 \to 2$ processes is provided by the
Mandelstam variables

$$s = (p_1 + p_2)^2 = (p_3 + p_4)^2, \tag{1.34a}$$

$$t = (p_1 - p_3)^2 = (p_2 - p_4)^2, \tag{1.34b}$$

$$u = (p_1 - p_4)^2 = (p_2 - p_3)^2. \tag{1.34c}$$

The variables (1.34) are not independent but satisfy an easy-to-verify
kinematic relation:

$$s + t + u = \sum_{i=1}^{4} m_i^2. \tag{1.35}$$

This relation makes it convenient to represent the kinematics of the process on the *Mandelstam plane*, exploiting the property of an equilateral
triangle as shown in Fig. 1.1. Where is the physical region of the reaction

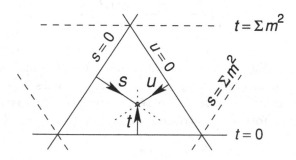

Fig. 1.1   Mandelstam plane.

on the Mandelstam plane? The meaning of the Mandelstam invariants
(1.34) is the most transparent in the centre-of-mass reference frame (cms)
of the reaction. Here $\mathbf{p}_1 + \mathbf{p}_2 = 0$, so that the variable s in (1.34a),

$$s = (p_{1\mu} + p_{2\mu})^2 \equiv (p_{10} + p_{20})^2 - (\mathbf{p}_1 + \mathbf{p}_2)^2 = (E_{1c} + E_{2c})^2 = E_c^2,$$

becomes the square of the total energy of the colliding particles.

t and u are invariant momentum transfers. In particular, t defined in
(1.34b) can be represented as

$$t = (p_{3\mu} - p_{1\mu})^2 \equiv (E_3 - E_1)^2 - (\mathbf{p}_3 - \mathbf{p}_1)^2$$
$$= (E_3 - E_1)^2 - (\mathrm{p}_3 - \mathrm{p}_1)^2 - 2\mathrm{p}_1\mathrm{p}_3(1 - \cos\Theta),$$

where $\mathrm{p}_i = |\mathbf{p}_i|$ and Θ is the scattering angle:

$$\cos\Theta = \frac{\mathbf{p}_1 \cdot \mathbf{p}_3}{\mathrm{p}_1 \mathrm{p}_3}.$$

In the centre-of-mass frame, the moduli of three-momenta of the incoming
particles, $\mathrm{p}_1 = \mathrm{p}_2 = p_c$, and of the produced ones, $\mathrm{p}_3 = \mathrm{p}_4 = p_c'$, are given
by (1.27) as a function of the energy and of the masses of the particles in
the initial and final state, correspondingly.

In the case of *elastic* scattering when $m_3 = m_1$ and $m_4 = m_2$ (as, e.g.
in the reaction $\pi N \to \pi N$), one has $\mathrm{p}_3 - \mathrm{p}_1 = E_{3c} - E_{1c} = 0$, and

$$t = -2p_c^2(1 - \cos\Theta_c). \tag{1.36a}$$

If *all* the masses are equal, $m_1 = m_2 = m_3 = m_4$, then the cms expression
for the variable u (1.34c) becomes simple also:

$$u = -2p_c^2(1 + \cos\Theta_c). \tag{1.36b}$$

In this case the physical region of the reaction $1 + 2 \to 3 + 4$,

$$s \geq 4m^2, \quad t \leq 0, \quad u \leq 0, \tag{1.37}$$

is shown by the shaded area in Fig. 1.2.

1.7.4 Crossing symmetry

One and the same diagram can be viewed differently. Let us 'rotate' our
scattering diagram by 90°:

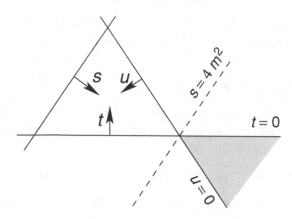

Fig. 1.2   Physical region of scattering of equal-mass particles.

The 'new' picture can be interpreted as an interaction between two particles with momenta p_1 and $-p_3$, producing particles p_4 and $-p_2$. To make this interpretation valid, we must take the energy components of the momenta p_3 and p_2 to be *negative*: $p_{30} \leq -m_3$ and $p_{20} \leq -m_2$. But as you know, in the relativistic theory the propagation of a negative energy particle 3 with a momentum p_3 corresponds to the propagation of its *antiparticle* ($\bar{3}$) with the four-momentum $\bar{p}_3 = -p_3$. Therefore, in this region of momenta the very same diagram describes another physical process, namely a collision between the particle 1 and the antiparticle $\bar{3}$ (having a four-momentum $\bar{p}_3 = -p_3$), which results in the production of particles 4 and $\bar{2}$ in the final state. This is called a *t-channel* reaction, since here the invariant

$$t = (p_1 - p_3)^2 = (p_1 + \bar{p}_3)^2 \geq (m_1 + m_3)^2 \qquad (1.38a)$$

has the meaning of the cms energy of colliding particles 1 and $\bar{3}$.

Analogously, in the region of momenta $p_{40} \leq -m_4$, $p_{20} \leq -m_2$ we obtain the amplitude of a *u-channel* process, $1 + \bar{4} \to 3 + \bar{2}$,

$$u = (p_1 - p_4)^2 = (p_1 + \bar{p}_4)^2 \geq (m_1 + m_4)^2. \qquad (1.38b)$$

Imagine that we have calculated the necessary diagrams and know the scattering amplitude as a function of the invariants in the physical region (1.37) of the reaction $1 + 2 \to 3 + 4$. If the amplitude were an analytic function of its variables s and t, we would be able to *analytically continue* the result into the physical region of either of the two crossing reactions (1.38a) or (1.38b). As we will shortly see, this is indeed the case: the analyticity is a direct consequence of *causality*.

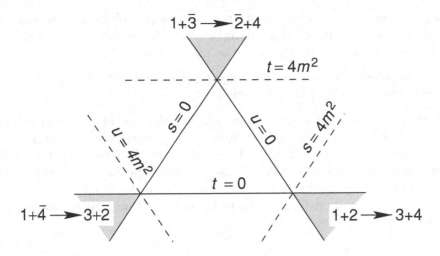

Fig. 1.3   Physical regions of crossing reactions on the Mandelstam plane.

Therefore, one function describes three different scattering processes that are related by crossing:

s-channel : $1 + 2 \rightarrow 3 + 4,$ $s = (p_1 + p_2)^2 \geq (m_1 + m_2)^2$;

t-channel : $1 + \bar{3} \rightarrow \bar{2} + 4,$ $t = (p_1 + \bar{p}_3)^2 \geq (m_1 + m_3)^2$;

u-channel : $1 + \bar{4} \rightarrow 3 + \bar{2},$ $u = (p_1 + \bar{p}_4)^2 \geq (m_1 + m_4)^2$.

Physical regions of the crossing reactions are displayed in Fig. 1.3 for the simplest case of equal particle masses.

It is important to remember that the *unitarity* seriously restricts the scattering amplitude. Moreover, these restrictions are different in each of the three crossing channels. Thus, one function has to satisfy three specific unitarity relations in complementary physical regions on the Mandelstam plane.

In non-relativistic quantum mechanics an interaction is described by means of a potential which can be chosen practically arbitrarily. Not so in the relativistic theory. If we were to introduce here a notion of 'relativistic potential', the latter would be severely restricted by the unitarity conditions in the cross-channels. This is a specifically relativistic feature since the crossing itself is of relativistic nature.

In the next lecture we will demonstrate that the causality ensures that the scattering amplitudes are analytic functions of momenta. An analytic function is identified by its singularities. The structure of these singularities may be studied, as it turns out, with the help of a (formally senseless) series of Feynman diagrams *as if* in the perturbation-theory framework.

This statement holds even for strongly interacting objects, in which case the very applicability of diagrammatic expansion is highly questionable.

Let us formulate straight away our main hypothesis:

Analytic properties of the exact amplitude coincide with those of the corresponding perturbation-theory diagrams.

To check that, we shall show that all singularities of Feynman graphs (their position, nature and strength) have a clear physical origin and are closely related to unitarity. This statement does not depend on the particular particle content of the theory or on specific properties of the interaction. The only important thing is to have the input objects – bare particles – to be point-like, that is to be included into some quantum field theory (QFT) scheme.

2

Analyticity and unitarity

Firstly, we are going to show that the property of *causality* results in *analyticity* of the scattering amplitude.

2.1 Causality and analyticity

Consider a four-point Green function $A(x_1, x_2; x_3, x_4)$, where the space–time points x_1, x_2 and x_3, x_4 lie in the remote past and future, correspondingly. Let y_1 (y_2) mark the point where the incident particle 1 (2) interacts for the first time, and y_3 (y_4) the point of the last interaction of the particle 3 (4). Then

$$A(x_1, x_2; x_3, x_4) =$$

$$= \int f(y_1, y_2, y_3, y_4) \left\{ \prod_{i=1}^{4} D(y_i - x_i)\, d^4 y_i \right\}.$$

(2.1)

Here $D(y - x)$ describes the propagation of a free particle which we take to be a scalar one since the spin play no rôle in the analysis that follows:

$$D(y_\mu - x_\mu) = \int \frac{d^3 \mathbf{p}}{(2\pi)^3} \int \frac{dp_0}{2\pi i} \frac{\exp\{-ip^\mu(y - x)_\mu\}}{m^2 - p^2 - i\epsilon}.$$

For the time ordering $y_0 > x_0$, one closes the integration contour in energy around the pole at $p_0 = \sqrt{m^2 + \mathbf{p}^2}$ in the lower half-plane, to obtain

$$D(y_\mu - x_\mu) = \int \frac{d^3\mathbf{p}}{(2\pi)^3} \frac{\exp\{-ip^\mu(y-x)_\mu\}}{2p_0}$$

$$= \int \frac{d^3\mathbf{p}}{(2\pi)^3} \, \psi_{\mathbf{p}}(y) \cdot \psi_{\mathbf{p}}^*(x), \qquad y_0 > x_0.$$

For the final state particles we have $x_{03} > y_{03}$, $x_{04} > y_{04}$, and their propagators take the form

$$D(y_\mu - x_\mu) = \int \frac{d^3\mathbf{p}}{(2\pi)^3} \, \psi_{\mathbf{p}}(x) \cdot \psi_{\mathbf{p}}^*(y), \qquad x_0 > y_0.$$

Thus, the 'truncated' interaction amplitude f in (2.1) gets multiplied by the product of the wave functions ψ of the incoming particles and of the conjugate wave functions ψ^* of the outgoing ones, evaluated at the 'entry' points y_1, y_2 and the 'exit' points y_3, y_4, correspondingly. The two-particle interaction amplitude in the momentum space, $\mathcal{M}(p_i)$, becomes the Fourier transform:

$$\mathcal{M}(p_i) = \int f(y_1, y_2, y_3, y_4) \, e^{-i(p_1y_1 + p_2y_2) + i(p_3y_3 + p_4y_4)} \prod d^4 y_i. \qquad (2.2)$$

An integral over the 'centre of gravity' of the four coordinates produces the energy–momentum conservation condition, and we are left with three integrations over the relative positions, $y_i - y_k$. For the sake of simplicity, let us restrict ourselves to to the *forward scattering* case, $p_1 \approx p_3$, $p_2 \approx p_4$. Then (2.2) reduces to

$$\mathcal{M} \implies (2\pi)^4 \delta(p_1 + p_2 - p_3 - p_4) \int e^{ip_1(y_3 - y_1)} f(y_{13}; p_2) d^4 y_{13}, \qquad (2.3)$$

where we have singled out the dependence on one of the momenta, namely p_1. This is sufficient since, because of the Lorentz invariance, the amplitude actually depends on the invariant energy, $\mathcal{M}(p_i) = \mathcal{M}(s)$,

$$s \equiv (p_1 + p_2)^2 = m_1^2 + m_2^2 + 2m_2 E_1,$$

which is proportional to the energy E_1 of one of the incident particles in the rest frame of the second one, $E_2 = m_2$.

Causality means that the function f in the integrand of (2.3) must have the form

$$f(y) = \vartheta(y_0)\vartheta(y_\mu^2) \cdot f_1(y) + f_0(y), \qquad (2.4a)$$

where f_0 does not contribute to the scattering:

$$\int d^4y \, f_0(y) \, \exp\{ip_1 y\} \; = \; 0. \tag{2.4b}$$

Mark that the condition (2.4b) does not imply $f_0 \equiv 0$, since it is only the Fourier components with a *physical* momentum p_1,

$$p_1 = \left(\sqrt{m^2 + \mathbf{p}^2}, \; \mathbf{p} \right),$$

that are required to vanish. How does such a decomposition emerge in the field theory? The easiest way to arrive at (2.4) is to invoke the general operator language.

The interaction amplitude in the coordinate space is related to the time-ordered product of operators describing an absorption of a particle in y_1 and a creation of another one in the space–time point y_3:

$$
\begin{aligned}
f(y_3, y_1) &\propto \left\langle T \, \psi(y_3) \bar{\psi}(y_1) \right\rangle \\
&\equiv \vartheta(\Delta y_0) \cdot \psi(y_3) \bar{\psi}(y_1) \; \pm \; \vartheta(-\Delta y_0) \cdot \bar{\psi}(y_1) \psi(y_3) \\
&= \vartheta(\Delta y_0) \left[\psi(y_3) \bar{\psi}(y_1) \mp \bar{\psi}(y_1) \psi(y_3) \right] \pm \bar{\psi}(y_1) \psi(y_3),
\end{aligned} \tag{2.5}
$$

where Δy stands for the relative coordinate,

$$\Delta y^\mu = y_3^\mu - y_1^\mu \,.$$

Alternative signs $\pm$ in (2.5) and below correspond to bosonic and fermionic operators, i.e. particles with integer and half-integer spin. For the space-like intervals, that is when $(\Delta y)^2 < 0$, by the virtue of *causality* our operators (anti)commute, so that

$$f(y_3, y_1) \propto \vartheta(\Delta y_0) \vartheta((\Delta y)^2) \cdot f_1 \pm \bar{\psi}(y_1) \psi(y_3) \,.$$

The latter piece (f_0) is given by a simple (not T-ordered) product of the two operators, which may be represented as a sum over all possible intermediate states:

$$\langle 0| \, \bar{\psi}(y_1) \psi(y_3) \, |0\rangle = \sum_n \langle 0| \, \bar{\psi}(y_1) |n\rangle \cdot \langle n| \psi(y_3) |0\rangle = \sum_n |C_n|^2 \, \mathrm{e}^{-i P_n (y_1 - y_3)}.$$

Here we have explicitly extracted the coordinate dependence in terms of the total intermediate state momenta, P_n. Substituting into the integral for the scattering amplitude, (2.4a), we immediately get

$$\sum_n |C_n^2| \int d^4 y_{31} \; \mathrm{e}^{i p_1 y_{31}} \cdot \mathrm{e}^{i P_n y_{31}} \propto \delta(p_{0,1} + P_{0,n}) \; = \; 0.$$

Vanishing of this contribution is due to the fact that both the incoming particle and any physical intermediate state n have a *positive energy*. This conclusion derives from the stability of the vacuum: any excitation must lie above the vacuum state, $P_{0,n} > 0$.

Finally, we arrive at the integral representation for the amplitude,

$$\mathcal{M}(E_1) = \int d^4y \, f_1(y) \cdot \vartheta(y_0)\vartheta(y_\mu^2) \, e^{ip_1 y} = \int d^3\mathbf{y} \int_{\sqrt{\mathbf{y}^2}}^{\infty} dt \, e^{iE_1(t - v_1 z)} f_1(y);$$

(2.6)

$$p_1 y \equiv E_1 t - \mathbf{p}_1 \cdot \mathbf{y} = E_1 \cdot (t - v_1 z),$$

where z is the coordinate projection on the direction of the momentum. The theta-functions in (2.6) ensure that the *phase* of the exponent is *positively definite*:

$$t > 0, \quad t > \sqrt{z^2 + \boldsymbol{\rho}_\perp^2} \geq |z| > |v_1 z| \implies (t - v_1 z) > 0.$$

As a consequence, $\mathcal{M}(E_1) \equiv \mathcal{M}(s)$ is a regular analytic function in the *upper* half-plane of complex energies E_1. Indeed, if the integral (2.6) exists (converges) for real values of the energy, it can be analytically continued onto the upper plane, $\operatorname{Im} E_1 > 0$, where it converges even better due to the additional exponentially falling factor, $\exp\{-\operatorname{Im} E_1(t - v_1 z)\}$.

2.1.1 *Causality and the polynomial boundary for* $\mathcal{M}(s)$.

Let us reverse the logic now. The inverse Fourier transform reads

$$f(t) = \int_{-\infty}^{+\infty} dE \, e^{-iEt} \mathcal{M}(E).$$

If $t < 0$, by moving the contour onto the upper half-plane where (as we have just established) $\mathcal{M}(E)$ is regular, we should get zero, to be in accord with causality. This is the case *provided* the amplitude does not increase exponentially along the imaginary axis:

$$|\mathcal{M}(\operatorname{Im} E \to +\infty)| < \exp(\gamma \operatorname{Im} E) \qquad \text{for arbitrary } \gamma > 0.$$

Otherwise, the exponentially decreasing factor $\exp(t \cdot \operatorname{Im} E)$ would not be sufficient to guarantee the vanishing of the response at small but finite negative times

$$-\gamma \leq t < 0.$$

That is why, to be on the safe side, we will impose an additional restriction on the scattering amplitude by bounding its possible growth with energy

$$|\mathcal{M}(s)| < |s|^N,$$

with N some finite (though maybe large) power.

2.2 Cross-channel singularities of Born diagrams from *s*-channel point of view

We will employ for simplicity the QFT model scalar particles with a $\lambda\varphi^3$ interaction. This is the simplest example of a renormalizable theory and, though not without a defect (it has no ground state – a stable vacuum), it is well suited for the qualitative analysis of singularities of scattering amplitudes.

Let m and λ be the renormalized mass and interaction constant. We consider a four-particle amplitude characterized by the Mandelstam variables s, t, u:

$$s = (p_1 + p_2)^2;$$
$$t = (p_1 - p_3)^2; \quad s + t + u = 4m^2,$$
$$u = (p_1 - p_4)^2.$$

What sort of singularities will we encounter at each order of perturbation theory? In the Born approximation we have three poles, in each of the Mandelstam invariants,

$$= \frac{\lambda^2}{m^2 - s}, \qquad = \frac{\lambda^2}{m^2 - t}, \qquad = \frac{\lambda^2}{m^2 - u}. \qquad (2.7)$$

Before moving further we will first discuss the meaning of these poles.

2.2.1 Pole in energy

The first one is the pole in the invariant collision energy s. Recall quantum mechanics. Here the amplitude of elastic scattering of a particle with initial momentum $\mathbf{p}$, $|\mathbf{p}| = |\mathbf{p}'| = \sqrt{2mE}$, has the following representation in terms of the potential V and the incoming state wave function:

$$f(E, \mathbf{q}) = -\frac{2m}{4\pi} \int d^3 r' \, e^{-i\mathbf{p}' \cdot \mathbf{r}'} V(\mathbf{r}') \psi_{\mathbf{p}}(\mathbf{r}'), \qquad (2.8)$$

with $\mathbf{q} = \mathbf{p}' - \mathbf{p}$ the momentum transfer.

In order to extract the energy dependence we are after, we invoke the Green function of the stationary Schrödinger equation:

$$(\widehat{\mathcal{H}} - E) G_E(\mathbf{r}', \mathbf{r}) = \delta(\mathbf{r}' - \mathbf{r});$$

$$G_E(\mathbf{r}', \mathbf{r}) = \sum_n \frac{\psi_n(\mathbf{r}') \psi_n^*(\mathbf{r})}{E_n - E}; \quad \widehat{\mathcal{H}} \psi_n = E_n \psi(n),$$

where E_n are the exact energy levels of the system. Now we express the exact wave function $\psi_{\mathbf{p}}(\mathbf{r}')$ as

$$\psi_{\mathbf{p}}(\mathbf{r}') = e^{i\mathbf{p}\cdot\mathbf{r}'} - \int d^3r\, G_E(\mathbf{r}',\mathbf{r})V(\mathbf{r})\,e^{i\mathbf{p}\cdot\mathbf{r}},$$

and substitute into (2.8) to derive

$$f(E,\mathbf{q}) = -\frac{2m}{4\pi}\left[V(\mathbf{q}) - \sum_n \frac{1}{E_n - E}\right.$$
$$\times \left. \left(\int d\mathbf{r}\, e^{-i\mathbf{p}'\,\mathbf{r}}V(\mathbf{r})\psi_n(\mathbf{r})\right)\left(\int d\mathbf{r}'\, e^{i\mathbf{p}\,\mathbf{r}'}V(\mathbf{r}')\psi_n^*(\mathbf{r}')\right)\right] \quad (2.9a)$$
$$= f_B + \sum_n \frac{C_n(\mathbf{p})C_n^*(\mathbf{p}')}{E_n - E}.$$

Here the Born scattering amplitude f_B,

$$f_B(\mathbf{q}) = -\frac{2m}{4\pi}\int d^3r\, e^{-i\mathbf{q}\cdot\mathbf{r}}V(\mathbf{r}), \qquad (2.9b)$$

depends only on the momentum transfer $\mathbf{q}$, while the dependence on the energy is contained in the sum over intermediate states n. If the system possesses a bound state – a discrete energy level E_n – its contribution to the amplitude can be depicted as a diagram with $1/(E_n - E)$ as the 'propagator' of the state n, and C as the 'coupling constant'. Thus, in non-relativistic

quantum mechanics a pole in energy corresponds to scattering via an intermediate state related to a discrete energy level. In the relativistic theory two particles can transfer into one, and it is this particle which plays the rôle of such an intermediate state.

2.2.2 Pole in momentum transfer

Having understood the physical meaning of the pole in s, we could repeat the same consideration in the crossing channels where t and u play, correspondingly, the rôle of energy, and thus would make sense of the two other poles in (2.7).

Still, what is the meaning of these poles from the s-channel point of view? The Mandelstam variable t measures the momentum transfer (for elastic scattering with $p_0 = p_0'$, $t = -\mathbf{q}^2$). Are there singularities in the *momentum transfer* in quantum mechanics? Let us examine the Born scattering amplitude (2.9b),

$$f_B(q) \propto \int d^3r\, e^{-i\mathbf{q}\cdot\mathbf{r}} V(\mathbf{r}).$$

It develops a singularity at a point where the integral diverges when we continue q onto the complex plane. If the potential has a power tail,

$$V(r) \propto r^{-n}, \qquad r \to \infty,$$

this happens for an arbitrary small value of $\operatorname{Im} q \neq 0$. This means that the singularity emerges at $t = 0$. By dimensional consideration,

$$\int \frac{d^3r}{r^n} \sim r^{-n+3} \implies q^{n-3},$$

for $n = 1, 2$ it is a pole; for an integer $n \geq 3$ it is a logarithmic singularity. It becomes clear that in order to have a singularity at some finite $t = m^2$, as this is the case of the amplitude (II) in (2.7), the potential has to fall exponentially at large distances. More precisely, it is the Yukawa potential,

$$V(r) = \frac{A}{r}\, e^{-mr}, \tag{2.10a}$$

whose Fourier image as we saw in the previous lecture gives the pole amplitude, see (1.8),

$$f_B \propto \frac{A}{m^2 + \mathbf{q}^2} \equiv \frac{\lambda^2}{m^2 - t}. \tag{2.10b}$$

We conclude that the relativistic Born amplitude (II) in (2.7) corresponds to a definite potential (Yukawa) with a definite strength ($A = \lambda^2$) and a definite sign (attraction).

Thus, singularities in momentum transfer are related to the *interaction radius*. Let us remark that our 'strong interaction potential' V has a finite radius $r_0 = 1/m$ owing to our basic supposition that all hadrons have non-zero masses.

2.2.3 Exchange potential

Finally, what is the pole in u?

The diagrams (II) and (III) in (2.7) differ by the exchange of final particles (momenta p_3 and p_4). In the non-relativistic quantum mechanics the *exchange potential* is a well known object:

$$V^{(\text{ex})}(\mathbf{r}_{ik}) = V(\mathbf{r}_{ik}) \cdot P_{ik},$$

with P_{ik} the particle permutation operator. Thus, the diagram (III) determines the exchange potential which in our case *coincides* with the direct one: the sum of the diagrams (2.7) automatically takes care of the identity of our scalar particles (Bose statistics).

2.3 Higher orders

In higher orders Feynman diagrams become more and more complex. In the second order in λ^2, we will have graphs like

$$+ \cdots + \qquad + \cdots + \qquad + \cdots \qquad (2.11)$$

(where the dots stand for the diagrams of the same type with the external lines transmuted). The first diagram seems to have a *double pole* in t. However, this 'self-energy insertion' into the propagator line modifies the particle mass, while we wanted m^2 to represent the true physical mass of the particle in the Green function $(m^2 - t)^{-1}$. Therefore a subtraction has to be made,

$$\frac{1}{m^2 - t} \Sigma(t) \frac{1}{m^2 - t} \implies \frac{1}{m^2 - t} \left[\Sigma(t) - \Sigma(m^2) \right] \frac{1}{m^2 - t},$$

after which the double pole disappears.

Moreover, the *simple pole* in t is effectively absent too. Indeed, by expanding the subtracted self-energy blob one step further,

$$\Sigma(t) - \Sigma(m^2) = (t - m^2) \cdot \Sigma'(m^2) + \Sigma_c(t),$$

we observe that the term proportional to $\Sigma'(m^2)$ must be dropped too, as modifying the value of the on-mass-shell coupling constant λ^2 in the corresponding t-channel Born diagram of (2.7). Since $\Sigma_c(t) \propto (t - m^2)^2$, the remaining contribution is finite. In the same way, the t-channel propagator pole cancels in the second (vertex correction) diagram of (2.11).

The remaining last ('box') graph in (2.11) has no poles either.

The fact that the second-order diagrams do not possess pole singularities does not mean that higher orders do not modify analytic properties of the amplitude. Far from that. This becomes immediately clear if we look at the unitarity condition for the elastic scattering amplitude:

$$2 \, \mathrm{Im} \, M_{aa} = \sum_c M_{ac} M_{ac}^* = \quad \text{} \quad . \qquad (2.12)$$

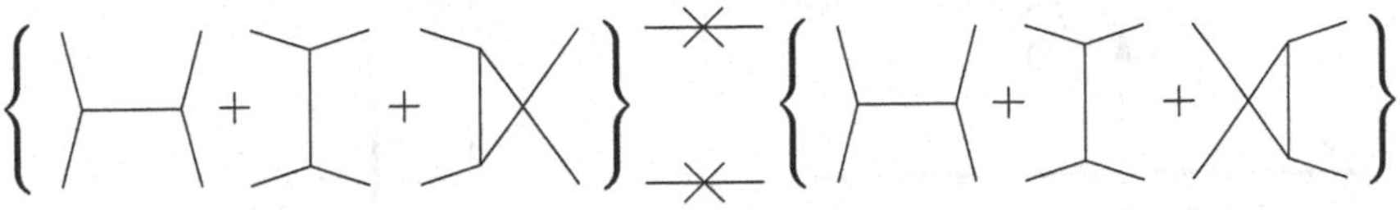

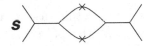

Fig. 2.1   Product of the Born amplitudes in the unitarity condition.

If in the r.h.s. we substitute the Born amplitude $M = \mathcal{O}(\lambda^2)$, see Fig. 2.1, then on the l.h.s. of the equation we will have Im M of the order of λ^4. This shows that starting from the second order in λ^2, the scattering amplitude must be complex above the two-particle threshold, $s > 4m^2$.

Among the products of the diagrams on the r.h.s. of the unitarity condition (2.12), there is a square of the s-channel Born graph. Let us examine the corresponding second-order diagram to see where the complexity comes from.

2.3.1 Two-particle thresholds

Consider the second-order diagram

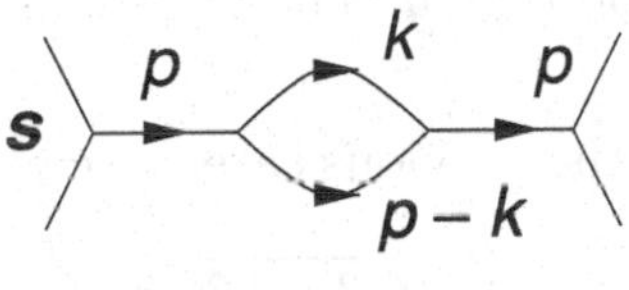

$$ \tag{2.13} $$

Looking for the origin of the complexity, we can drop the real pole factors and concentrate on the loop integral:

$$\Sigma(s) = \frac{1}{2!} \int \frac{d^4 k}{(2\pi)^4 i} \frac{\lambda^2}{(m^2 - k^2 - i\epsilon)(m^2 - (p-k)^2 - i\epsilon)},$$

(with $1/2!$ the symmetry factor characteristic for the loop of two identical particles). In the complex plane of the k_0 variable, singularities of the integrand are positioned at

$$ k_0 = \pm\sqrt{m^2 + \mathbf{k}^2} \mp i\epsilon, \tag{2.14a} $$

$$ k_0 = p_0 \pm \sqrt{m^2 + (\mathbf{p} - \mathbf{k})^2} \mp i\epsilon. \tag{2.14b} $$

The Feynman $i\epsilon$ prescription displaces the poles from the real axis to tell us on which side of the singularity the contour passes.

Fig. 2.2   Transformation of the k_0 integration contour; $s < 0$.

Consider the two cases.

$s < 0$. In this case we can always find a reference frame such that $p_0 = 0$, so that the pairs of the poles (2.14) are placed symmetrically around the imaginary axis, as displayed in Fig. 2.2(a). In this situation we can turn the integration path as on Fig. 2.2(b). Introducing a new integration variable, $k_0 = i\kappa$, the self-energy becomes

$$\Sigma(-\mathbf{p}^2) = \frac{1}{2!} \int \frac{d^3\mathbf{k}}{(2\pi)^3} \int \frac{d\kappa}{2\pi} \frac{\lambda^2}{(m^2 + \kappa^2 + \mathbf{k}^2)(m^2 + \kappa^2 + (\mathbf{p} - \mathbf{k})^2)},$$

where we have dropped the $i\epsilon$ terms since the denominators vanish nowhere in the integration region. The answer is obviously real valued.

$s > 0$. Now, contrary to the previous case, we can choose the frame $\mathbf{p} = 0$:

$$k_0 = \pm\sqrt{m^2 + \mathbf{k}^2} \mp i\epsilon,$$
$$k_0 = \pm\sqrt{m^2 + \mathbf{k}^2} + p_0 \mp i\epsilon.$$

With $p_0 > 0$ and increasing, the second pair of poles will move to the right, and at some point the trailing pole of the second pair will collide with the leading pole of the first one:

$$-\sqrt{m^2 + \mathbf{k}^2} + p_0 = \sqrt{m^2 + \mathbf{k}^2}.$$

Two poles *pinch* the contour, and the integral becomes complex.

$$\tag{2.15}$$

Since we are integrating over $\mathbf{k}$, this happens for the first time at

$$s = 4m^2 \leq \left(2\sqrt{m^2 + \mathbf{k}^2}\right)^2. \tag{2.16}$$

This is the value of s corresponding to the two-particle *energy threshold*.

It is straightforward to calculate the *imaginary part* of the self-energy graph in (2.13). To this end we close the contour around the two poles on the upper half-plane in (2.15) and look at the contribution of the pinching one at $k_0 = p_0 - \sqrt{m^2 + \mathbf{k}^2}$:

$$\Sigma(p_0^2) = \frac{1}{2!} \int \frac{d^3\mathbf{k}}{(2\pi)^3} \cdot \frac{1}{2(p_0 - \sqrt{m^2 + \mathbf{k}^2})}$$
$$\cdot \frac{\lambda^2}{m^2 + \mathbf{k}^2 - (p_0 - \sqrt{m^2 + \mathbf{k}^2})^2 - i\epsilon} + \text{real},$$

where the first factor is the residue, $2k_0$, and the second – the remaining Green function. Using

$$\text{Im} \frac{1}{a - i\epsilon} = \pi\delta(a),$$

we obtain

$$2\,\text{Im}\,\Sigma(p_0^2) = \frac{\lambda^2}{2!p_0} \int \frac{d^3\mathbf{k}}{2(2\pi)^2\sqrt{m^2 + \mathbf{k}^2}} \cdot \delta\left(2\sqrt{m^2 + \mathbf{k}^2} - p_0\right). \tag{2.17}$$

As expected, the integral has a non-vanishing support only when $p_0 > 2m$, cf. (2.16). The fact that the amplitude becomes complex means that in the complex plane of the energy variable $s = p_0^2$ it has a *branch cut* which starts at $s = 4m^2$ and runs to infinity. This is the *threshold singularity* corresponding to the production of two particles with equal masses m.

The integral (2.17) is easy to calculate; the calculation produces nothing but the phase-space volume for the production of two real particles with an aggregate four-momentum p, see (1.25). This is straightforward to see if we recast the answer in the Lorentz covariant form as

$$2\,\text{Im}\,\Sigma(p^2) = \frac{\lambda^2}{2!} \int \frac{d^4k}{(2\pi)^4} 2\pi\delta_+((p-k)^2 - m^2) \cdot 2\pi\delta_+(k^2 - m^2). \tag{2.18}$$

We conclude that in order to calculate $2\,\text{Im}\,\Sigma(p^2)$ one simply has to 'cut through' the diagram (2.13) and replace in the Feynman expression for the amplitude *each cut propagator* by (double) its imaginary part,

$$\frac{1}{m^2 - k^2 - i\epsilon} \implies (2\pi i)\delta_+\left(k^2 - m^2\right).$$

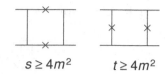

Fig. 2.3　The second-order box graph having branch cuts both in s and t.

This rule is in a perfect accord with the unitarity condition formally applied to the $1 \to 1$ transition via a two-particle intermediate state (take $n = 2$, $\mathcal{M}_{ac} = \mathcal{M}_{bc}^* = \lambda$ in (1.31)), and can be used to calculate imaginary parts of arbitrary diagrams.

When cutting a diagram in order to find its imaginary part, one must make sure that the two graphs that emerge correspond to a real physical process. For example, the diagram with the self-insertion into the t-channel particle exchange can be cut into two as

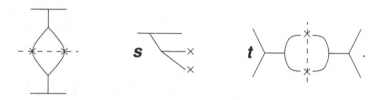

From the point of view of the s-channel, this looks like squaring an amplitude of the decay of the incoming particle into three, which process is kinematically forbidden. On the other hand, in the t-channel these very subgraphs correspond to a legitimate $2 \to 2$ scattering process, so that this cut describes the t-channel two-particle threshold and makes the amplitude complex for $t > 4m^2$.

Another example. Contrary to the self-energy insertion graph, the 'box' in Fig. 2.3 can be legitimately cut both in the s- and t-channels and therefore possesses two branch cuts. Another second-order graph – the 'crossed box' – shown in Fig. 2.4 has the same t threshold but cannot be cut in the physical region of the s-channel, since such a division would correspond to two-body decays of incoming particles. At the same time, it can be cut in the u-channel describing the $1\bar{4} \to 3\bar{2}$ transition. If we keep t fixed and study analytic properties of the amplitude as a function of s, the u-channel threshold at $u = 4m^2$ will manifest itself as another branch cut in $s = 4m^2 - u - t$ which starts at $s = -t$ and runs *to the left*.

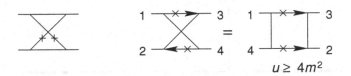

Fig. 2.4 The crossed box graph having branch cuts in t and u. Arrows mark the direction of the positive energy flow.

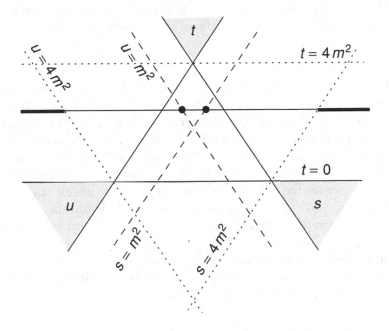

Fig. 2.5 Singularities of the second order amplitude on the Mandelstam plane.

2.3.2 Scattering amplitude as a function of s

Nine cut diagrams contained in Fig. 2.1 describe the imaginary part of the full two particle scattering amplitude in the order λ^4, due to the *s-channel* threshold. The same diagrams with permuted external particles give t- and u-channel complexities.

Thus, the amplitude on the Mandelstam plane is *real* inside the triangle marked by the dotted lines on Fig. 2.5. By fixing the variable t at some value in the interval $0 \leq t \leq 4m^2$, the regions where the amplitude is *complex* are displayed by two bold lines on Fig. 2.6. The physical s-channel scattering amplitude $A^{(s)}$ is defined for $s \geq 4m^2$ and is given by the value of the invariant amplitude on the *upper side* of the right cut, $\mathrm{Im}\, s = \to +i0$. Since for a fixed t we have $s + u = \mathrm{const.}$, the physical u-channel

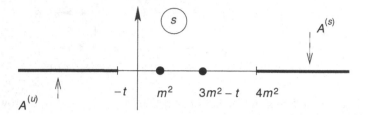

Fig. 2.6   Analytic structure of the amplitude in the complex s-plane and the definition of physical amplitudes of the s- and u-channel reactions.

amplitude $A^{(u)}$ that describes the process $1\bar{4} \to 3\bar{2}$ one obtains by approaching the left cut *from below* $\operatorname{Im} s \to -i0$, $\operatorname{Re} u > 4m^2$ (see Fig. 2.6).

As we have learnt, due to causality, $A^{(s)}$ is a regular function in the *upper* half-plane of the invariant energy s. Analogously, the *lower* half of the s-plane is free of any singularities as well, thanks to the causality property of the u-channel reaction (which makes the amplitude $A^{(u)}$ regular for $\operatorname{Im} u > 0$). Moreover, since the upper and the lower half-planes are analytically connected through a finite interval on which the amplitude is real, we conclude that the invariant amplitude is an analytic function on the entire complex s-plane apart from two poles and two *branch cuts* originating from two-particle thresholds.

In higher orders there appear higher thresholds, and related additional branching points, due to multi-particle thresholds in the intermediate states of the s- and u-channel reactions:

$$s_0 = \left(\sum_{i=1}^{n} m_i \right)^2 \to n^2 m^2. \qquad (2.19)$$

2.3.3 Dispersion relation

Once the imaginary part of the analytic function is known, we can restore its real part using the dispersion relation. We write a Cauchy integral around a point s in the complex plane where the amplitude is regular,

$$\int_C \frac{dz}{2\pi i} \frac{A(z)}{z - s} = A(s) \,,$$

and then inflate the contour to embrace the singularities of the amplitude $A(z)$ in the z plane, as shown in Fig. 2.7. If the function falls on the large circle, $|z| \to \infty$, we obtain s and u poles inherited from the Born amplitude, and a sum of integrals of the *discontinuity* across the right

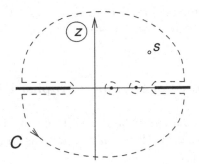

Fig. 2.7 Integration contour in the dispersion relation for the amplitude $A(s)$.

and left branch cuts:

$$\text{Im}_s A \;\equiv\; \tfrac{1}{2i}\big[A(s+i0,t)-A(s-i0,t)\big], \quad s>4m^2, \quad (2.20\text{a})$$

$$\text{Im}_u A \;\equiv\; \tfrac{1}{2i}\big[A(u+i0,t)-A(u-i0,t)\big], \quad u>4m^2. \quad (2.20\text{b})$$

The piece of the Born amplitude responsible for t-channel particle exchange, $\lambda^2/(m^2-t)$, does not fall with s. Therefore in order to eliminate the contribution of the large circle, one has to use the dispersion relation with one subtraction, that is to apply the Cauchy theorem to the function

$$A(s)-A(0) = \int_C \frac{dz}{2\pi i}\left[\frac{A(z)}{z-s}-\frac{A(z)}{z}\right] = \frac{s}{\pi}\int_C \frac{dz}{2i}\,\frac{A(z)}{z(z-s)}.$$

The s-independent piece then hides in the subtraction term $A(0,t)$. By combining this result with a complementary information coming from the t channel, we can restore the full second-order amplitude $\mathcal{O}(\lambda^4)$ from the Born one, $\mathcal{O}(\lambda^2)$. In principle, one can move further, order by order, and recursively build up full interaction amplitudes by exploiting unitarity in the cross-channels and analyticity.

It is worthwhile to mention that if the Born amplitude happens to *increase too fast* with energy, the subtraction trick fails to work (as it happens, e.g. in electrodynamics with an anomalous magnetic moment interaction vertex term $\Gamma_\mu \propto \sigma_{\mu\nu}q^\nu$). The number of arbitrary subtraction constants grows with the order of the perturbative expansion, which manifests, in the dispersive theory language, the *non-renormalizability* of the underlying interaction.

The dispersive programme was found indeed to be rather effective in quantum electrodynamics. There were times when it was being enthusiastically explored as a possible way of constructing the theory of hadrons without knowing the internal structure of the interaction. An attractive feature of such a scheme is that it operates only with experimentally accessible physical quantities – on-mass-shell amplitudes.

2.4 Singularities of Feynman graphs: Landau rules

Now we have to address an essentially technical problem, namely how to find singularities of arbitrary Feynman diagrams, and how to determine their position and character.

2.4.1 Position of singularities

Consider a diagram containing n internal lines with four-momenta k_1, $k_2, \ldots, k_n$ which may be expressed as linear combinations of the external momenta p_j and the integration momenta q_m:

$$k_i = \sum_{m=1}^{\ell} b_{im} q_m + \sum_j c_{ij} p_j, \tag{2.21}$$

where b_{im} are either 1 or 0. The number of the independent integration contours (*loops*) ℓ is a function of the topology of the given diagram.

The Feynman integral corresponding to our diagram has the structure

$$A_{n\ell} = \int \frac{d^4 q_1 \, d^4 q_2 \cdots d^4 q_\ell}{[(2\pi)^4 i]^\ell} \frac{1}{(m_1^2 - k_1^2)(m_2^2 - k_2^2) \cdots (m_n^2 - k_n^2)}. \tag{2.22}$$

We need to find the conditions under which this integral becomes singular in external variables $s_{ik} = (p_i + p_k)^2$ – Lorentz invariants formed by four-momenta of the external particles.

It is worthwhile to stress that for the case of particles with higher spins, matrices and different powers of momenta would appear in the *numerator* of the integrand. This, however, would not affect the singularities of the amplitude which depend exclusively on the structure of the scalar *denominator* in (2.22).

To study the appearance of singularities it is convenient to use the trick invented by R. Feynman in order to get rid of multiple four-vector integrations. Applying to the denominators $a_i = m_i^2 - k_i^2$ in (2.22) the Feynman identity

$$\frac{1}{a_1 a_2 \cdots a_n} = \frac{1}{n!} \int_{\alpha_i \geq 0} \frac{d\alpha_1 \, d\alpha_2 \cdots d\alpha_n}{(\alpha_1 a_1 + \alpha_2 a_2 + \cdots + \alpha_n a_n)^n} \delta\!\left(1 - \sum_{i=1}^n \alpha_i\right),$$

which is not difficult to prove by induction, we get

$$A_{n\ell} = \int \frac{d^4 q_1 \cdots d^4 q_\ell}{[(2\pi)^4 i]^\ell} \int_0^1 \frac{d\alpha_1 \, d\alpha_2 \cdots d\alpha_n}{\square^n} \, \delta\left(1 - \sum_{i=1}^n \alpha_i\right), \qquad (2.23)$$

where

$$\square = \square(\{\alpha\}, \{p\}; \{q\}) \equiv \sum_{i=1}^n \alpha_i (m_i^2 - k_i^2). \qquad (2.24)$$

The integrand in (2.23) depends analytically on the integration variables; the vanishing of the denominator inside the integration region, $\square = 0$, is the only potential source of singularities.

Substituting the decomposition (2.21), the characteristic function $\square$ of (2.24) becomes an inhomogeneous quadratic form. This form can be diagonalized by an orthogonal transformation:

$$\square(\alpha, p; q) = \Delta(\alpha, p) - \sum_{m=1}^\ell \delta_m \cdot \tilde{q}_m^2, \quad \delta_m = \delta_m(\alpha) > 0, \qquad (2.25)$$

with $\tilde{q} = \{\tilde{q}_m\}$ the set of new integration momenta.

In (2.25) $\Delta = \Delta(\alpha, s_{ik})$ is a function of invariant energies s_{ik}. Let us start from the kinematical domain where all these invariants are *negative*, $s_{ik} < 0$. (This is easy to achieve by taking all the energy components $p_j = 0$ in some reference frame.) Then the energy integrations can be transformed as we did above when we studied the self-energy graph,

$$\tilde{q}_{0m} = i\kappa_m, \quad \frac{d^4 \tilde{q}_m}{(2\pi)^4 i} = \frac{d\kappa_m \, d^3 \tilde{\mathbf{q}}_m}{(2\pi)^4} \equiv \frac{d^4 Q_m}{(2\pi)^4}.$$

The expression for the denominator becomes positively definite,

$$\square = \sum_i \alpha_i \left(m_i^2 + \left[\sum \kappa_s\right]^2 + \mathbf{k}_i^2\right) = \Delta(\alpha, s_{ik}) + \sum_{m=1}^\ell \delta_m \left[\kappa_m^2 + \tilde{\mathbf{q}}_m^2\right],$$

and consequently the integral (2.23) yields a regular, real amplitude $A_{n\ell}$. Rescaling the Q variables, the integral can be evaluated as

$$\begin{aligned}
A_{n\ell} &= \int_0^1 \frac{\delta(1 - \Sigma\alpha) \prod_i d\alpha_i}{(2\pi)^{4\ell}} \int \frac{d^4 Q_1 \, d^4 Q_2 \cdots d^4 Q_\ell}{[\Delta + \sum_{m=1}^\ell \delta_m Q_m^2]^n} \\
&= \int_0^1 \frac{\delta(1 - \Sigma\alpha) \prod_i d\alpha_i}{\prod_{m=1}^\ell \delta_m^2(\alpha)} \cdot \frac{1}{[\Delta(\alpha, s_{ik})]^{n-2\ell}} \times [\text{number}].
\end{aligned} \qquad (2.26)$$

As a function of the integration momenta $\tilde{q}$, $\square$ has a *minimum* at $\tilde{q}_m = 0$. At this point $\square = \Delta$, which permits us to determine Δ from the equation

$$\Delta(\alpha, s_{ik}) = \square\,(\alpha, p, q^{(0)}), \tag{2.27a}$$

where $\{q^{(0)}\}$ is the position of the *extremum* of $\square$, simultaneously in 4ℓ integration variables:

$$\left.\frac{\partial \square\,(\{q\})}{\partial q_k}\right|_{\{q\}=\{q^{(0)}\}} = 0, \quad k = 1, 2, \ldots, \ell. \tag{2.27b}$$

Now we start changing s_{ik} to see when (2.26) becomes singular. We are left with $n - 1$ integrations over Feynman parameters $0 \le \alpha_i \le 1$ restricted by the condition $1 - \sum_{i=1}^{n-1} \alpha_i = \alpha_n \ge 0$. An equation $\Delta(\alpha, s_{ik}) = 0$ determines a surface in the n-dimensional space of αs. A singularity appears when this surface *touches* for the first time the integration domain.

For each variable α_i this can happen in two ways: either a zero of Δ collides with the *endpoint* of the integration interval, $\alpha_i = 0$, or *two zeroes* simultaneously arrive from the complex plane and assume a common real value *inside* the interval, $\alpha_i > 0$:

$$\alpha_i = 0, \quad \text{or} \quad \frac{d\Delta}{d\alpha_i} = 0, \quad i = 1, 2, \ldots, n - 1. \tag{2.28}$$

By virtue of (2.27b),

$$\frac{d\Delta(\alpha; q^{(0)})}{d\alpha_i} = \frac{\partial \square}{\partial \alpha_i} + \sum_k \frac{\partial \square}{\partial q_k}\frac{dq_k^{(0)}}{d\alpha_i} = \frac{\partial \square}{\partial \alpha_i}. \tag{2.29}$$

Moreover, since $\square$, by its definition (2.24), is a homogeneous linear function of αs,

$$\square = \sum_{i=1}^{n} \alpha_i \frac{\partial \square}{\partial \alpha_i}. \tag{2.30}$$

By successively applying equations (2.27a), (2.30), (2.29) and (2.28) to the point $\Delta = 0$, we derive

$$0 = \Delta = \square = \sum_{i=1}^{n} \alpha_i \frac{\partial \square}{\partial \alpha_i} = \sum_{i=1}^{n-1} \alpha_i \frac{d\Delta}{d\alpha_i} + \alpha_n \frac{\partial \square}{\partial \alpha_n} = \alpha_n \frac{\partial \square}{\partial \alpha_n}.$$

Finally, the condition for the appearance of singularity reads

$$\frac{\partial \square}{\partial q_k} = 0, \qquad k = 1, 2, \ldots, \ell; \tag{2.31a}$$

$$\frac{\partial \square}{\partial \alpha_i} = 0 \quad (\text{or} \quad \alpha_i = 0), \quad i = 1, 2, \ldots, n. \tag{2.31b}$$

Equations (2.31b), combined with (2.30), guarantee that in this point $\square = 0$, so that we need not watch the relation $\square = \Delta = 0$ anymore.

The relations (2.31) together with the restriction $\sum \alpha_i = 1$ impose $4\ell + n + 1$ conditions on $4\ell + n$ variables q_k and α_i. This means that a solution may exist only for specific values of external momenta. The corresponding equation $f(s_{ik}) = 0$ determines the 'Landau surface' for the position of a singularity of the amplitude in the space of invariants s_{ik}. This equation may be resolved, e.g. to determine the position of a singularity in the invariant energy s for fixed momentum transfer variables: $s = s_0(t, u, \ldots)$, or vice versa.

Consider some closed contour inside the diagram:

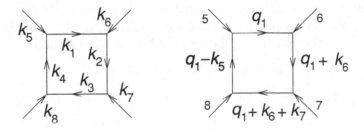

Introducing the loop momentum $q_1 = k_1$ and applying the first extremum condition (2.31a),

$$\frac{\partial}{\partial q_1}\left(\alpha_1 q_1^2 + \alpha_2(q_1 + k_6)^2 + \alpha_3(q_1 + k_6 + k_7)^2 + \alpha_4(q_1 - k_5)^2\right) = 0,$$

we obtain a system of linear equations stating that

$$\sum_i \alpha_i k_i^\mu = 0 \qquad \text{along each loop.} \tag{2.32a}$$

(It resembles Kirchhoff current law equations for electric circuits, with momentum k_i playing the rôle of the current, and α_i that of resistance.) In addition, the second condition (2.31b) tells us that each line either has to have an on-mass-shell momentum or should be dropped from consideration (short-circuited):

$$k_i^2 = m_i^2 \quad \text{or} \quad \alpha_i = 0.$$

(2.32b)

Let us consider some examples.

2.4.2 Threshold singularities

We first take the diagram which has a two-particle threshold singularity, in order to see how the *Landau equations* (2.32) reproduce the result that we have already learnt.

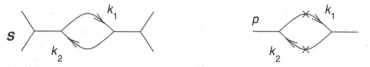

The Landau condition

$$\alpha_1 k_1 + \alpha_2 k_2 = 0$$

(2.33)

shows that the energy components of the momenta k_1 and k_2 have *opposite signs* ($\alpha_i > 0$), so that the singularity we are looking for corresponds physically to the decay of the external particle p into two.

Our elementary loop is too simple to sustain *reduction* (putting one of αs to zero), so the pinch singularity remains the only option:

$$k_1^2 = m_1^2, \quad k_2^2 = m_2^2; \quad \alpha_i > 0.$$

Projecting (2.33) onto k_i we obtain a system of linear equations for α_1 and α_2,

$$\alpha_1 k_1^2 + \alpha_2 (k_1 k_2) = 0$$
$$\alpha_1 (k_1 k_2) + \alpha_2 k_2^2 = 0,$$

(2.34)

whose determinant must be zero for a solution to exist. Substituting

$$2(k_1 k_2) = k_1^2 + k_2^2 - (k_1 - k_2)^2 = m_1^2 + m_2^2 - p^2,$$

we obtain the *solvability* condition for the system (2.34) in the form

$$\text{Det} = m_1^2 m_2^2 - \frac{1}{4}\left(m_1^2 + m_2^2 - p^2\right)^2 = 0,$$

that is,

$$\left(p^2 - (m_1 + m_2)^2\right) \cdot \left(p^2 - (m_1 - m_2)^2\right) = 0.$$

Only one of the two possible solutions,

$$p^2 = (m_1 + m_2)^2,$$

satisfies the condition $\alpha_i > 0$ (one has to have $(k_1 k_2) < 0$ in (2.34)). We conclude that the singularity emerges when the mass of the initial object (invariant energy s of the system) exceeds the threshold value $m_1 + m_2$ sufficient for a real decay into two particles m_1 and m_2 to take place.

2.4.3 Anomalous singularity and deuteron form factor

Now take a diagram with three lines in the loop:

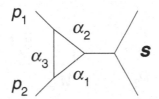

This graph can be reduced to a simpler one by setting one of αs to zero. For example, such a reduction

$$\alpha_3 = 0$$

gives us back the threshold singularity at $s = (m_1 + m_2)^2$. If we choose to nullify another Feynman parameter, e.g.

$$\alpha_1 = 0$$

the emerging singularity at $p_1^2 = (m_2 + m_3)^2$ has nothing to do with the energy s but corresponds to the situation when the external particle p_1 is *unstable* (acquires a 'complex mass').

Let us look for a genuine singularity of the triangle diagram that corresponds to three on-mass-shell lines:

$$\alpha_1 k_1^2 + \alpha_2 (k_1 k_2) + \alpha_3 (k_1 k_3) = 0 \,,$$

$$\alpha_1 (k_1 k_2) + \alpha_2 k_2^2 + \alpha_3 (k_2 k_3) = 0 \,, \qquad (2.35)$$

$$\alpha_1 (k_1 k_3) + \alpha_2 (k_2 k_3) + \alpha_3 k_3^2 = 0 \,.$$

We take masses of the internal particles to be the same,

$$k_1^2 = k_2^2 = k_3^2 = m^2,$$

and, at the same time, allow the virtual momenta of external particles to be different. For the sake of simplicity we take two of them equal,

$$p_1^2 = p_2^2 = M^2, \quad p_3^2 = Q^2,$$

and look for a singularity in the virtuality Q^2:

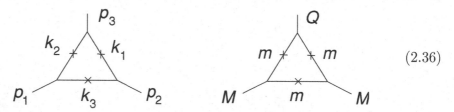

$$(2.36)$$

Kinematical relations between internal and external momenta result in

$$(k_1 k_2) = \tfrac{1}{2}\big[\, k_1^2 + k_2^2 - (k_1 - k_2)^2 \,\big] = m^2 - \tfrac{1}{2}Q^2,$$
$$(k_1 k_3) = (k_2 k_3) = m^2 - \tfrac{1}{2}M^2. \tag{2.37}$$

Bearing in mind that $\alpha_i > 0$, from the last line of the system (2.35),

$$(\alpha_1 + \alpha_2) \cdot \big(m^2 - \tfrac{1}{2}M^2\big) + \alpha_3 m^2 = 0,$$

we derive the necessary condition for the existence of singularity:

$$M^2 > 2m^2. \tag{2.38}$$

Substituting (2.37) into (2.35), the characteristic equation follows:

$$0 = \mathrm{Det}\begin{bmatrix} m^2 & m^2 - \tfrac{1}{2}Q^2 & m^2 - \tfrac{1}{2}M^2 \\ m^2 - \tfrac{1}{2}Q^2 & m^2 & m^2 - \tfrac{1}{2}M^2 \\ m^2 - \tfrac{1}{2}M^2 & m^2 - \tfrac{1}{2}M^2 & m^2 \end{bmatrix}$$
$$= \tfrac{1}{2}Q^2 \cdot \big[\, m^2\big(2m^2 - \tfrac{1}{2}Q^2\big) - 2(m^2 - \tfrac{1}{2}M^2)^2 \,\big]. \tag{2.39}$$

We obtain the so-called 'anomalous singularity' positioned at

$$Q_0^2 = 4M^2 \left[\, 1 - \left(\frac{M}{2m}\right)^2 \,\right]. \tag{2.40}$$

The graph (2.36) describes the scattering of an object of mass M with momentum transfer $p_3^2 = Q^2$. If this object is *stable* ($M < 2m$), the singularity (2.40) lies at $Q_0^2 > 0$, that is *outside* the physical region of the scattering reaction. However, if the 'mass defect' is small, $2m - M \ll M$, the singularity may occur very close to the physical region, $t = Q^2 \leq 0$, in which case it would strongly affect the behaviour of the amplitude. A

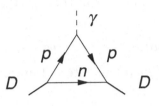

physical example is provided by a *deuteron D* – the lightest nucleus, a loosely bound state of a proton and a neutron, with the graph (2.36) describing, e.g. the electromagnetic electron–deuteron scattering:

$$M_D - (m_p + m_n) = \epsilon \ll m \quad (m = m_p \simeq m_n).$$

In non-relativistic quantum mechanics, the electron scattering amplitude

$$f(q) = \frac{e^2}{q^2} F(q)$$

is proportional to the *form factor* $F(q)$, given by the Fourier component of the distribution of electric charge inside the nucleus:

$$F(q) = \int d^3\mathbf{r}\, \psi^2(\mathbf{r})\, e^{i\mathbf{q}\cdot\mathbf{r}}, \quad F(0) = 1. \tag{2.41}$$

Here ψ is the proton wave function inside the deuteron:

$$\psi(\mathbf{r}) \propto e^{-r\sqrt{\epsilon m}},$$

with ϵ the binding energy. Expanding (2.41) at small momentum transfer,

$$F(q) = 1 - \mathbf{q}^2 \int d^3\mathbf{r}\, \frac{\mathbf{r}^2}{2} \psi^2(\mathbf{r}) + \mathcal{O}(\mathbf{q}^4) \simeq 1 - \frac{\mathbf{q}^2 \langle \mathbf{r}^2 \rangle}{2}, \quad \langle \mathbf{r}^2 \rangle \sim (\epsilon\, m)^{-1},$$

we see that the amplitude starts falling at characteristic momentum transfers of the order of the inverse deuteron radius.

On the other hand, in the relativistic-theory framework we have discussed that the interaction radius is determined by the t-channel singularities. If the triangle amplitude as a function of $t = Q^2$ had the *normal threshold* at $Q^2 = 4m^2 \gg \langle \mathbf{r}^2 \rangle^{-1}$ as the closest singularity to the physical region, this would contradict the non-relativistic expectation for a loosely bound large-size system. It is the *anomalous singularity* (2.40),

$$Q_0^2 \simeq 16\, m \cdot \epsilon \ll m^2,$$

that is responsible for a fast decrease of the elastic form factor with momentum transfer and reconciles the two approaches.

If, having started from the vicinity of $2m$, we decrease the mass M of the external particle down to $M = \sqrt{2}m$, the position of the anomalous singularity $Q_0^2(M)$ reaches its maximum, where it hits the two-particle threshold branch point, $Q_0^2 = 4m^2$, and disappears from the physical sheet of the amplitude, diving under the branch cut.

Thus, the anomalous singularity is present only inside a specific interval of masses, $2m^2 < M^2 < 4m^2$. In particular, it is absent when masses of all (external and internal) particles are the same (as in the $\lambda\varphi^3$ theory).

2.4.4 When imaginary part acquires imaginary part

For the last example consider the diagram with four internal lines:

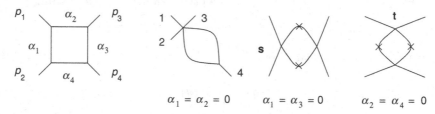

$$\alpha_1 = \alpha_2 = 0 \qquad \alpha_1 = \alpha_3 = 0 \qquad \alpha_2 = \alpha_4 = 0$$

By setting to zero two neighbouring α parameters, we get a singularity in the virtual mass of one of the external particles; setting $\alpha_1 = \alpha_3 = 0$ reproduces the s-channel two-particle threshold, and $\alpha_2 = \alpha_4 = 0$ – the corresponding t-channel singularity.

The reduction of one line leads us back to the triangle graphs which, as we already know, may possess anomalous singularities.

Let us examine a new genuine singularity of the box graph corresponding to $\alpha_i \neq 0$. Taking for simplicity all the masses to be equal, $p_i^2 = k_i^2 = m^2$, for the products of internal momenta we have

$$2k_1k_2 = 2k_2k_3 = 2k_3k_4 = 2k_4k_1 = m^2;$$
$$2k_1k_3 = 2m^2 - t, \qquad 2k_2k_4 = 2m^2 - s. \tag{2.42}$$

The system of Landau equations can be simplified using the symmetry of the solution following from the structure of the graph itself: $\alpha_3 = \alpha_1$, $\alpha_4 = \alpha_2$. Then the 4×4 system reduces to

$$\alpha_1 \cdot (m^2 + k_1k_3) + \alpha_2 \cdot 2k_1k_2 = 0,$$
$$\alpha_1 \cdot 2k_1k_2 + \alpha_2 \cdot (m^2 + k_2k_4) = 0.$$

Evaluating the determinant,

$$\mathrm{Det} = (m^2 + k_1k_3)(m^2 + k_2k_4) - 4(k_1k_2)^2 = 0,$$

and using the kinematical relations (2.42) we obtain the equation for the *Landau surface* in the form

$$(s - 4m^2)(t - 4m^2) = 4m^4. \tag{2.43}$$

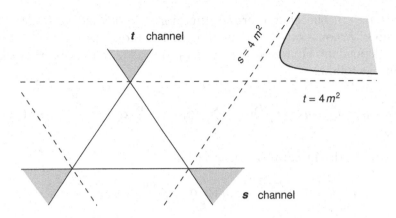

Fig. 2.8   Landau curve of the 's–t' box graph on the Mandelstam plane.

On the Mandelstam plane, it is a *hyperbola* (known as a *Karplus curve*), limited by the asymptotes $s = 4m^2$ ($t = \infty$) and $t = 4m^2$ ($s = \infty$). This curve is lying in the unphysical domain, 'between' the physical regions of s- and t-channel reactions, see Fig. 2.8. But for $s > 4m^2$, as well as for $t > 4m^2$, the amplitude is already complex due to the threshold singularities. Where then does the additional singularity come from and what is its meaning?

In the discussion of the analytic features of the second-order scattering amplitude and their relation to the unitarity in the $\lambda\varphi^3$ theory, we saw that for $s > 4\mu^2$ the two 'horizontal' lines may both turn on-shell, and the amplitude develops an imaginary part:

$$s > 4\mu^2 \; : \quad \text{[box diagram]} \; = 2i \, \text{Im}_s \, A(s,t) \equiv A(s + i0, t) - A(s - i0, t).$$

(2.44)

In the physical region of s-channel scattering, $t < 0$, the 'vertical' propagators stay always off-shell. However, the imaginary part (discontinuity) (2.44) is itself an analytic function of t. If we start to increase t, somewhere above $t = 4\mu^2$ the vertical lines will be able to go on-shell too. This happens precisely on the Landau curve (2.43) where the 'imaginary part' $\text{Im}_s \, A$ becomes complex, that is develops its own 'imaginary part': $\text{Im}_t \, \text{Im}_s \, A(s,t) \neq 0$.

2.4.5 Character of singularities

Let us find out the way the amplitude behaves near the singularity, that is the *character* of the latter. To do that we return to the original integral (2.23). As we know, when s approaches the Landau surface,

$s \to s_0(t, u, \ldots)$, the characteristic function $\square$ defined in (2.24) develops a simultaneous extremum in integration momenta, $q = q^{(0)}$, and in $n - 1$ independent Feynman parameters, $\alpha = \alpha^{(0)}$. It may be represented therefore as follows:

$$\square = \sum_{ij}^{4\ell} (q_i - q_i^{(0)})(q_j - q_j^{(0)}) a_{ij} + \sum_{ij}^{n-1} (\alpha_i - \alpha_i^{(0)})(\alpha_j - \alpha_j^{(0)}) b_{ij} + \gamma(s_0 - s).$$

Let us rescale the integration variables as

$$q_i - q_i^{(0)} = \sqrt{s_0 - s} \cdot y_i, \quad \alpha_i - \alpha_i^{(0)} = \sqrt{s_0 - s} \cdot \beta_i.$$

Then the s-dependence of the integral factors out,

$$A_{n\ell} \sim \frac{(s_0 - s)^{\frac{4\ell}{2}} (s_0 - s)^{\frac{n-1}{2}}}{(s_0 - s)^n} \int \frac{d^4 y_1 \cdots d^4 y_\ell \, d\beta_1 \cdots d\beta_{n-1}}{\square^n(y, \beta)},$$

$$\square(y, \beta) = \sum_{ij}^{4\ell} a_{ij} y_i y_j + \sum_{ij}^{n-1} b_{ij} \beta_i \beta_j + \gamma,$$

giving

$$A_{n\ell} = \left(\sqrt{s_0 - s}\right)^{4\ell - n - 1} \cdot \mathcal{N}. \tag{2.45}$$

Due to finite limits of the integrals over αs, the factor $\mathcal{N}$ here may have some residual s-dependence. With $s \to s_0$, however, the integration range in βs expands,

$$|\Delta \beta_i| \propto 1/\sqrt{s_0 - s} \to \infty.$$

Hence the integrals over β_i may be replaced by the s-independent integration over $-\infty < \beta_i < +\infty$, *provided* the multiple y–β integral so obtained does not diverge. A simple power counting shows that the integral for $\mathcal{N}$ converges when

$$(4\ell + n - 1) - 2n = 4\ell - n - 1 \; < 0.$$

Looking at (2.45) we conclude that the latter condition is equivalent to the amplitude $A_{n\ell}$ *increasing* towards the singularity. If this is the case, one has two possibilities for the character of the singularity, depending on the value of the characteristic exponent

$$\mathcal{E} \equiv 4\ell - n - 1, \tag{2.46}$$

namely:

(1) the branching point when $\mathcal{E}$ is *odd*, $\mathcal{E} = -(2k + 1)$,

$$A_{n\ell} \propto (s_0 - s)^{-k-\frac{1}{2}} \quad (k \geq 0); \qquad (2.47a)$$

(2) the pole in $(s_0 - s)$ of degree k for the case of *even* $\mathcal{E} = -2k$:

$$A_{n\ell} \propto (s^0 - s)^{-k} \quad (k \geq 1). \qquad (2.47b)$$

What kind of singularity appears when $\mathcal{E} \geq 0$ and the integral diverges? To answer the question we should treat not the amplitude $A_{n\ell}$ itself, but its s-derivative:

$$\frac{d^m A_{n\ell}}{ds^m} \sim \int \frac{d^4 q_1 \cdots d^4 q_\ell \, d\alpha_1 \, d \cdots d\alpha_{n-1}}{\square^{n+m}} (-\gamma)^m \frac{(n+m)!}{n!}.$$

By choosing a proper value of m, we make the integral convergent and can repeat the above analysis to obtain

$$\frac{d^m A_{n\ell}}{ds^m} = \text{const.} \cdot (s_0 - s)^{\frac{4\ell - n - 1}{2} - m}.$$

Now we integrate m times over s to restore the amplitude and face two cases as before:

(1) the branching point singularity when $\mathcal{E} = 2k + 1 \quad [m = k]$,

$$A_{n\ell} \propto (s_0 - s)^{k+\frac{1}{2}} \quad (k \geq 0); \qquad (2.48a)$$

(2) the logarithmic singularity for $\mathcal{E} = 2k \quad [m = k + 1]$,

$$A_{n\ell} \propto (s_0 - s)^k \ln(s_0 - s) \quad (k \geq 0). \qquad (2.48b)$$

Example 1. Return to multi-particle threshold singularities (2.19). Here we have n internal lines and $\ell = n - 1$ independent momentum integrations. The exponent (2.46) equals $\mathcal{E} = 3n - 5$, and when the number of particles in the intermediate state is *even*, the amplitude near the threshold singularity behaves as

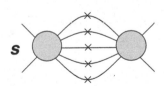

$$A \sim (s_0 - s)^{\frac{3}{2}(n-2)+\frac{1}{2}} \quad (n \text{ even}). \qquad (2.49a)$$

For $n = 2$ we recover the known square-root singularity characterizing the two-particle threshold. If the number of particle is *odd*, for $n = 1$ our formula gives a simple particle pole, $A \sim (m^2 - s)^{-1}$; otherwise the exponent is a positive integer, and the threshold singularity becomes logarithmic:

$$A \sim (s_0 - s)^{\frac{3}{2}(n-1)-1} \cdot \ln(s_0 - s) \qquad (n \text{ odd}). \qquad (2.49b)$$

This result is easy to get directly from the unitarity relation:

$$2 \operatorname{Im} A \propto \int \frac{d^4 k_1 \cdots d^4 k_n}{(2\pi)^{3n}} (2\pi)^4 \delta^4 \left(p - \Sigma k_i\right) \prod_{i=1}^{n} \delta(k_i^2 - m_i^2).$$

Near the threshold, three-momenta of all particles are small, and the integration produces the non-relativistic phase space volume:

$$\sim \prod_{i=1}^{n-1} d^3 \mathbf{k}_i \cdot \delta \left(\sqrt{s} - \sqrt{s_0} - \sum_{i=1}^{n} \frac{\mathbf{k}_i^2}{2m_i} \right) \sim |\mathbf{k}|^{3 \cdot (n-1)-2} ; \qquad \sqrt{s_0} = \sum_{i=1}^{n} m_i .$$

Each additional particle brings in the suppression factor $|\mathbf{k}|^3 \sim (s_0 - s)^{3/2}$, and the singularity weakens.

Example 2. Anomalous 'vertex' singularity:

$n = 3, \quad \ell = 1, \quad \mathcal{E} = 4 \cdot 1 - 3 - 1 = 0; \quad A \sim \ln(s_0 - s).$

Example 3. Box diagram:

$n = 4, \quad \ell = 1, \quad \mathcal{E} = 4 \cdot 1 - 4 - 1 = -1; \quad A \sim \dfrac{1}{\sqrt{s_0 - s}}.$

Example 4. A five-leg amplitude possesses a genuine *pole* singularity:

$n = 5, \quad \ell = 1, \quad \mathcal{E} = 4 \cdot 1 - 5 - 1 = -1; \quad A \sim \dfrac{1}{s_0 - s}.$

We observe that the strength of the singularity *grows* when we increase the number of lines in the loop. According to the Landau rules, multi-leg one loop amplitudes would seem to develop stronger and stronger singularities: $A_{61} \sim (s - s_0)^{-3/2}$, $A_{71} \sim (s - s_0)^{-2}$, etc.

Would they? Rather not. It is most likely that the strongest singularity a Feynman diagram may have is a *pole* and the reason is the following.

While studying the equation for the Landau surface that determines the position of singularities,

$$f(\{s_{ik}\}) = 0, \quad 1 \le i < k \le N - 1 \qquad (2.50)$$

(with N the number of external momenta), we were treating the Lorentz invariants s_{ik} that characterize the amplitude as *independent* variables. The number of *linearly independent* pair products of the external momenta $p_i p_k$ (not counting the masses, $i \ne k$) in (2.50) is

$$\frac{(N-1)(N-2)}{2} - 1, \qquad (2.51a)$$

where the subtracted unity stands for the additional on-mass-shell relation, $m_N^2 = s_{NN}$ for the last particle with momentum $p_N = -\sum_{i=1}^{N-1} p_i$. At the same time, we have calculated in (1.33) the total number of Lorentz invariant variables, the N-point amplitude depends on:

$$3N - 10. \qquad (2.51b)$$

The two numbers (2.51) coincide for $N = 5$, which is just the case of the pole singularity in the Example 4 above.

If we take $N > 5$, the number of the pair products (2.51a) takes over, which means that certain *kinematical relations* between s_{ik} appear, undermining our analysis of the character of singularities.

What remains to be done about the character of Landau singularities is to verify the case when some of the αs are zero.

Consider a singularity that emerges when $\alpha_1 = 0$. Then there is no extremum with respect to α_1, and the function $\square$ has the expansion

$$\square \simeq c\alpha_1 + \sum_{i,k}^{n-1}(\alpha_i - \alpha_i^{(0)})(\alpha_k - \alpha_k^{(0)})b_{ik} + \cdots + \gamma(s_0 - s). \qquad (2.52)$$

Near the singularity, the characteristic magnitude of α_1 under the integral, $\alpha_1 \sim (s_0 - s)$, is *much smaller* than all other deviations, $|\alpha_i - \alpha_i^{(0)}| \sim \sqrt{s_0 - s}$. Therefore α_1 can be neglected in the quadratic form in (2.52), as well as in the sum,

$$\delta\left(1 - \sum_{i=1}^{n}\alpha_i\right) \simeq \delta\left(1 - \sum_{i=2}^{n}\alpha_i\right).$$

Hence, we obtain an expression identical to that for the amplitude without the line k_1, α_1:

$$\int_0^1 d\alpha_2 \cdots d\alpha_n \int \frac{d\alpha_1\, \delta\big(1 - \sum_{i=1}^n \alpha_i\big)}{\big(c\alpha_1 + \square_{\,\alpha_1=0}\big)^n}$$

$$\implies \int_0^1 \frac{d\alpha_2 \cdots d\alpha_n}{\square^{\,n-1}}\, \delta\left(1 - \sum_{i=2}^n \alpha_i\right).$$

Thus, not only the position of the singularity but also its character can be derived from a *reduced graph* with the line α_1 contracted:

2.4.6 Amplitude near singularity

Now that we learnt how to determine the position and the character of the singularities, let us address the question of the *magnitude* of the amplitude near a singularity.

Let us take some complicated amplitude and set to zero all αs but four, to form a square graph. At the corresponding Landau singularity, all internal particles are on the mass shell, $k_i^2 = m_i^2$, therefore the full subamplitudes that determine their interaction vertices may each be equated with the renormalized on-mass-shell interaction constant g:

If we consider instead a simpler threshold branch-cut singularity, its magnitude will be determined by the square of the physical scattering amplitude near the threshold:

$$s \quad \text{} \quad \propto A^2\,(s = s_0). \tag{2.53}$$

In general, if we want to calculate the magnitude of an arbitrary Feynman diagram near the singularity, it will be always determined by the exact on-mass-shell interaction amplitudes.

2.5 Beyond perturbation theory: relation to unitarity

By using the language of Feynman diagrams we have arrived at the pattern of singularities of the interaction amplitudes.

Namely, for each singularity:

(1) its position is determined by masses of real hadrons;

(2) its character derives from the topology of the interaction process;

(3) the coefficient in front of a singularity is expressed in terms of the physical on-mass-shell amplitudes.

This conclusion goes beyond the perturbation theory which we have employed to derive it. The reason for that lies in the *unitarity* property: the series of Feynman diagrams (though having little sense in a theory of strongly interacting objects) formally solve the unitarity conditions.

The blocks in (2.53) are supposed to correspond to *exact* two-particle scattering amplitudes which become complex themselves when we move above the threshold. The particles in the intermediate state are allowed to interact many times: as a result, the threshold singularities overlay.

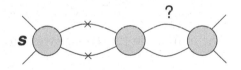

How to treat such an eventuality?

To take into account successive two-particle scatterings in the full amplitude, let us single out the block that has no two-particle intermediate state and, therefore, no threshold at $s = 4m^2$:

$$\text{(diagram equation)} \qquad (2.54)$$

According to Landau rules, to find the s-channel threshold singularity, we must pick one of the two-particle states and put the two lines on the mass shell. Then the chains of two-particle irreducible blocks sum up into the full amplitudes, on the left and on the right from the 'cut', resulting in (2.53). A branch cut singularity of the function is characterized by the *discontinuity* across the cut:

$$\Delta A(z) = \frac{1}{2i}\big[A(z + i0) - A(z - i0)\big].$$

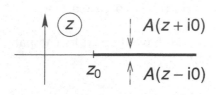

This is what we use to call the '*imaginary part*', cf. (2.20). The name is justified when the function below $z = z_0$ is *real*; hence it assumes complex conjugate values on the sides of the cut:
$A(z - i0) = [A(z + i0)]^*$.

Integrating over energy components of $n - 1$ loop momenta, we may close the contours around the positive energy poles of all but one intermediate state particles:

$$p \,\bigcirc\!\!\!\!\cdots\!\!\!\!\bigcirc\,k_n \propto \frac{1}{m_n^2 - k_{0n}^2 + \mathbf{k}_n^2}. \qquad (2.55)$$

The last line we must also put on the mass shell by replacing the remaining propagator (2.55) by $2\pi\delta(m_n^2 - k_n^2)$. Using the energy conservation, $k_{0n} = p_0 - \sum_{i=1}^{n-1} k_{0i} > 0$, this procedure is equivalent to evaluating *discontinuity* of the amplitude with respect to the incoming energy p_0:

$$\frac{1}{2i}\left[\frac{1}{m_n^2 - (p_0 + i\epsilon - \sum k_{0i})^2 + \mathbf{k}_n^2} - \frac{1}{m_n^2 - (p_0 - i\epsilon - \sum k_{0i})^2 + \mathbf{k}_n^2}\right].$$

The amplitude (2.54) has symbolically the structure of the product:

$$A(s) = \sum_{n=1}^{\infty} F_1 \cdot F_2 \cdot \dots \cdot F_n.$$

To find the discontinuity of the iterated amplitude,

$$\Delta\,\bigcirc = \boxed{\cdots} + \boxed{\cdots} + \boxed{\cdots} + \cdots$$

we must calculate

$$A_n(s + i0) - A_n(s - i0) = (F^+)^n - (F^-)^n, \quad \text{with } F^\pm = F(s \pm i0).$$

An evaluation of the discontinuity of the product of functions is algebraically similar to taking a derivative:

$$\Delta(AB) = A \cdot (\Delta B) + (\Delta A) \cdot B^*;$$

iterating this rule we obtain

$$\Delta A = \sum_{i,k=1} F_1^+ \cdots F_i^+ \left[\begin{array}{c} \times \\ \times \end{array}\right] F_1^- \cdots F_k^- = A(s + i0) \left[\begin{array}{c} \times \\ \times \end{array}\right] A(s - i0).$$

The r.h.s. of this expression is real, correcting (2.53).

Summing together the discontinuities of the amplitude across n-particle threshold branchings, we finally derive

$$\Delta A_{2 \to 2}(s) = \sum_n \tau_n(s) A_{2 \to n}(s) A^*_{2 \to n}(s), \qquad (2.56)$$

which is nothing but the unitarity relation, with τ_n the n-particle phase space volume.

2.6 Checking analytic properties of physical amplitudes

We will conclude the discussion of analyticity in this lecture by considering two practically important examples.

2.6.1 Dispersion relation for forward πN scattering

Consider pion–nucleon scattering. Due to the isotopic symmetry of strong interactions, πN interaction amplitude depends on the total isospin of the system, rather than on the individual isospin state (electric charge) of each of the participating particles.

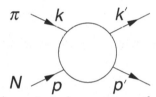

Pions π^+, π^0, π^- form an isotopic triplet ($I = 1$), and nucleons p, n – the doublet ($I = \frac{1}{2}$). Therefore, the full amplitude contains two independent functions describing the interaction: $1 \otimes \frac{1}{2} = \frac{1}{2} \oplus \frac{3}{2}$, or, from the t-channel point of view, $\pi\pi = 1 \otimes 1 = 0 \oplus 1$ (the $N + N$ pair cannot have isospin 2).

We will study the *forward* pion–nucleon scattering, $t = 0$:

$$s = (p + k)^2 = M^2 + \mu^2 + 2M\nu,$$

$$u = (p - k')^2 = M^2 + \mu^2 - 2M\nu = 2(M^2 + \mu^2) - s,$$

with ν the energy of the pion, $p_0 = p'_0$, in the rest frame of the target nucleon.

Let $\mathbf{U}$ denote the dublet of nucleons, and ϕ_α the isovector pion field, $\alpha = 1, 2, 3$. The general form of the scattering amplitude is

$$A = \overline{\mathbf{U}}(p')\phi_\alpha(k')\big(f_+(\nu)\delta_{\alpha\beta} \cdot \mathbf{I} + f_-(\nu)\varepsilon_{\alpha\beta\gamma} \cdot \boldsymbol{\tau}_\gamma\big)\phi_\beta(k)\mathbf{U}(p), \qquad (2.57)$$

where $\varepsilon_{\alpha\beta\gamma}$ is the anti-symmetric unit tensor, and $\boldsymbol{\tau}_\alpha$ is the triplet of Pauli matrices in the 2×2 space of nucleon isospinors. The diagonal term proportional to the unit matrix $\mathbf{I}$ takes care of elastic scattering, while

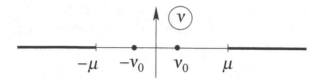

Fig. 2.9   Complex pion energy plane for forward πN scattering amplitude.

the τ term anti-symmetric in isospin indices is responsible for reactions with electric charge transfer, like $\pi^+ n \to \pi^0 n$.

The variable

$$\nu = \frac{s-u}{2M} = \frac{s-(M^2+\mu^2)}{M}$$

changes sign under the permutation of the initial and final pions, $s \leftrightarrow u$. Therefore, since pions are Bose particles, the amplitudes $f_\pm$ in (2.57) are, correspondingly, even and odd with respect to the crossing:

$$f_+(-\nu) = f_+(\nu), \quad f_-(-\nu) = -f_-(\nu).$$

We will study the symmetric part, f_+. The amplitude has a nucleon pole,

$$s = M^2 \implies \nu_0 = -\frac{\mu^2}{M},$$

and the branch cut that starts from the πN threshold:

$$s = (M+\mu)^2 \implies \nu_{\text{thresh}} = \mu.$$

The u-channel singularities mirror the s-channel ones on the complex plane of the variable ν, see Fig. 2.9. Let us try to write down the dispersion relation for $f = f_+$ as a function of the complex variable ν:

$$
\begin{aligned}
f(\nu) &= \frac{r}{\nu_0 - \nu} + \frac{1}{\pi} \int_\mu^\infty \frac{d\nu'\, \text{Im}\, f(\nu')}{\nu' - \nu} \\
&\quad + \frac{r}{\nu_0 + \nu} + \frac{1}{\pi} \int_{-\mu}^{-\infty} \frac{d\nu'\, \text{Im}\, f(\nu')}{\nu' - \nu}.
\end{aligned}
\tag{2.58a}
$$

Combining the contributions of two poles and s- and u-channel cuts, with account of $f(-\nu) = f(\nu)$, we get a more compact expression

$$f(\nu) = \frac{2r\,\nu_0}{\nu_0^2 - \nu^2} + \frac{1}{\pi} \int_\mu^\infty \frac{d\nu'^2\, \text{Im}\, f(\nu')}{\nu'^2 - \nu^2}. \tag{2.58b}$$

This relation makes sense only if the integral converges at $\nu' \to \infty$. The optical theorem (1.32) tells us that

$$\text{Im}\, A(\nu) = \tfrac{1}{2} J \sigma_{\text{tot}}(\nu) = 2M|\mathbf{k}| \cdot \sigma_{\text{tot}}(\nu), \quad |\mathbf{k}| = \sqrt{\nu^2 - \mu^2},$$

where we have calculated the invariant flux (1.26) in the nucleon rest frame, $J = 4M|\mathbf{k}_\pi|$. We conclude that the amplitude actually *grows* as $A \propto \nu$, since the total cross section is approximately constant at large collision energies. Therefore we must modify the dispersion relation by performing the subtraction, $f \to f(\nu) - f(0)$:

$$f(\nu) = f(0) + \frac{2r}{\nu_0}\frac{\nu^2}{\nu_0^2 - \nu^2} + \frac{\nu^2}{\pi}\int_\mu^\infty \frac{d\nu'^2}{\nu'^2}\frac{\text{Im}\, f(\nu')}{(\nu'^2 - \nu^2)}. \qquad (2.59)$$

Since the imaginary part of the amplitude is directly related to the total cross section, $\text{Im}\, f = 2Mk\,\sigma_{\text{tot}}$, once we have measured the total cross section, we know the integrand and may restore the amplitude which is a measurable quantity itself.

Importantly, information about the amplitude and cross section comes from essentially different sources. The total cross section $\sigma_{\text{tot}}(\nu)$ can be accessed by observing the *loss* of particles by an incident beam. On the other hand, the amplitude is extracted from a completely different experiment. One measures differential angular distribution of the elastic scattering and reconstructs the amplitude from the Legendre expansion:

$$f(\nu, \Theta) = \sum_{\ell=1}^{\infty} f_\ell(\nu)\, P_\ell(\cos\Theta).$$

The integral of the cross section converges, so that the main contribution comes from not too large energies.

Confronting the experimental information on the forward amplitude, $f(\nu, 0)$, and $\sigma_{\text{tot}}(\nu)$ the relation (2.59) was found to hold. This is a verification of the analyticity.

The dispersion relation allows one to determine experimentally the value of the residue r. But the residue in the pole of the amplitude is the *renormalized coupling constant*, in the field-theory framework.

Let us take a neutral meson π^0 and try to model the pion–nucleon interaction vertex in the QFT language:

$$= g \cdot \overline{\mathbf{U}}(p_2)\, i\gamma_5 \mathbf{U}(p_1) \cdot \boldsymbol{\phi}(k).$$

(We have included into the vertex the factor $i\gamma_5$ since π is pseudoscalar.) The pole diagram constructed on the base of this vertex reads

$$A_{\text{pole}} = \ \raisebox{-1ex}{\begin{array}{c}k\\ p\end{array}}\!\!\!\diagrammacr\!\!\!\raisebox{-1ex}{p'} \ = -g^2\,\overline{U}(p')\gamma_5\frac{1}{M - (\hat{p} + \hat{k})}\gamma_5 U(p)$$

$$= -g^2\overline{U}\frac{\gamma_5(M + \hat{p} + \hat{k})\gamma_5}{M^2 - s}U = -g^2\overline{U}\frac{M - \hat{p} - \hat{k}}{M^2 - s}U. \tag{2.60}$$

Using the Dirac equation, $(M - \hat{p})U(p) = 0$,

$$A_{\text{pole}} = \overline{U}(p')\frac{\hat{k}}{M^2 - s}U(p). \tag{2.61}$$

The pole amplitude has an interesting feature: why there is only a pion momentum in the numerator of the amplitude? The pion has a negative internal parity; to be allowed to fuse into a nucleon, the incident pion has to have an *odd* orbital momentum, $L = 1$, in order to match the spatial parity of the πN pair, $P_\pi P_N (-1)^L$, to that of the nucleon, P_N. The p-wave amplitude must be proportional to $\mathbf{k}$, explaining the structure of the pole amplitude (2.61).

In the forward scattering limit, $p \simeq p'$, we have

$$\overline{U}(p)\gamma_\mu U(p) = 2p_\mu,$$
$$\overline{U}(p)\hat{k}U(p) \equiv k^\mu\,\overline{U}(p)\gamma_\mu U(p) = 2kp = s - M^2 - \mu^2, \tag{2.62}$$

and in the numerator of the Born amplitude there appears the combination $s - M^2$ which cancels the pole. The remaining *true pole* contribution becomes

$$A_{\text{pole}} \implies -\frac{g^2\,\mu^2}{M^2 - s} = -\frac{g^2\,\mu^2}{M}\,\frac{1}{\nu_0 - \nu}, \quad \nu_0 = -\frac{\mu^2}{M}.$$

Comparing with the pole term in the dispersion relation (2.58a), we relate the residue r with the coupling constant g as

$$r = -g^2\mu^2/M. \tag{2.63}$$

We may roughly estimate the magnitude of the residue. Since the interaction is strong, its cross section is determined by the interaction radius, the latter being inverse proportional to the mass of the lightest hadron – the π meson: $\sigma_{\text{tot}} \sim \mu^{-2}$. Taking moderate pion energies of the order of its mass, $\nu \sim |\mathbf{k}| \sim \mu$, the estimate follows:

$$f \sim \mathrm{Im}\, f = 2M|\mathbf{k}|\sigma_{\text{tot}} \sim 2M\mu \cdot \frac{1}{\mu^2} = \frac{2M}{\mu}.$$

Suppose that the contributions of the pole and of the cut to the dispersion relation are of the same order:

$$\frac{r}{\nu_0 - \nu} \sim f; \quad -\frac{r}{\mu} \sim \frac{2M}{\mu} \quad \Longrightarrow \quad r \simeq -2M.$$

From (2.63) we then have

$$g^2 \simeq 2M^2/\mu^2 \simeq 100.$$

Real experimental measurement of the residue r yields

$$g^2/4\pi \simeq 14.$$

The closeness of the two numbers shows that indeed the pole term and the dispersion integral over the cut contribute equally.

Thus, although there is certainly no perturbation theory, we can obtain the pole term in the dispersion relation from the first graph (2.60).

It should be stressed that the pole term differs essentially from the Feynman diagram. Indeed, if we *did not drop* in the numerator of the Born amplitude (2.62) the piece $(s - M^2)$ which cancels the pole,

$$\frac{s - M^2 - \mu^2}{M^2 - s} = -\frac{\mu^2}{M(\nu_0 - \nu)} - 1,$$

the new estimate based on the *full amplitude* would have been $M/\mu \simeq 7$ times larger!

Is there a reason why the pure pole term gives a reasonable size contribution while the estimate based on the Feynman diagram fails?

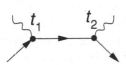

Recall the meaning of a Feynman graph, in the space–time language. The pion–nucleon interaction graph actually incorporates *two* space–time configurations, including the one with the *inverse time ordering*, $t_2 < t_1$, which corresponds to three coexisting nucleons. It is clear, however, that for moderate energies, such a state has nothing to do with the pro- 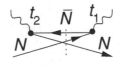 cess we are considering. The proton mass is very large (compared to that of the pion) and long before the $NN\bar{N}$ state there will be many additional pions present in the intermediate state.

When the coupling of the field theory is small, the expansion of the amplitude is organized in powers of this coupling. In particular, in QED it is easy to verify that the Compton scattering amplitude $\gamma e \to \gamma e$ is dominated just by the 'time inverted' Born diagram.

However, if the interaction is strong, so that the coupling is large and the perturbative expansion makes no sense, it is not the first-order graph but the *nearest singularities* that determine the answer.

This observation constitutes the core idea of the dispersive approach.

2.6.2 Chew–Low method

What can we measure directly? We actually have only one stable target – the proton. Plus a relatively stable neutron. All other hadrons are unstable. Nevertheless, we are able to prepare a beam of some unstable particles and scatter them off a proton, for example.

But how to measure an amplitude when the projectile and the target are both unstable, for example that of $\pi\pi$ scattering? One has to undertake a flanking manoeuvre in order to extract it from available data (Chew and Low, 1959).

Suppose a pion scatters off a nucleon, producing some hadron state. Could we single out some part of the process into which $\pi\pi$ interaction amplitude, of even cross section, would enter?

Such a diagram is easy to invent:

$$\pi\;N\;\text{(diagram)} = \pi\;N\;\pi\;\text{(diagram)} + \pi\;N\;\text{(diagram)} . \qquad (2.64)$$

The first diagram on the r.h.s. of (2.64) contains a pion exchanged in the t-channel, and the upper block of this diagram can be looked upon as pion–pion scattering amplitude.

The problem is, how to see and extract this particular term from the background of all other possible contributions. How to do that when the perturbative approach is not applicable? We must look for specific features of the pion exchange amplitude that might help us in our task.

Let us square the amplitude and look at the contributions to the cross section:

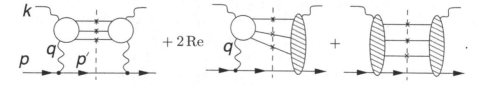

What is there remarkable about the first graph? If we do not integrate over the recoil nucleon momentum p' but measure the differential cross section in p', then this graph will contain the square of the pion propagator at given momentum transfer

$$\frac{d\sigma}{dt} \propto \left(\frac{1}{\mu^2 - t}\right)^2, \qquad t = (p' - p)^2.$$

(The interference term contains the pion propagator in the first power, $1/(\mu^2 - t)$, and the third one has none.) We may adopt the specific sharp t-dependence of the cross section as the means for extracting the pion exchange.

Let us focus on the t dependence in the sense of analytic properties.

Estimate of momentum transfer. For the momentum transfer we have

$$-q^2 = -(p - p')^2 = 2(p_0 p'_0 - \mathbf{p} \cdot \mathbf{p}' - m^2), \qquad (2.65)$$

where we made use of the on-mass-shell conditions $p^2 = p'^2 = m^2$. Here m is the proton mass, and the capital M we reserve for the invariant mass of the produced hadron system (minus the recoil nucleon): $M^2 = (k + q)^2$.

It is clear that the physical region corresponds to *negative* $t = q^2$. Indeed, since q^2 is Lorentz invariant, we can calculate it in an arbitrary reference frame. In the 'laboratory frame' where the nucleon target is at rest, $\mathbf{p} = 0$, (2.65) reduces to

$$-q^2 = 2m(p'_0 - m) > 0,$$

which expression is obviously positive (the energy p'_0 of the recoiling nucleon cannot be smaller than its mass m).

What is a typical momentum transfer? In particular, how small the virtuality of the exchanged pion may be? One can derive $\left|q^2\right|_{\min}$ from the kinematical relations at our disposal:

$$
\begin{aligned}
s &= (k + p)^2 = \mu^2 + m^2 + 2(kp); \\
M^2 &= (k + q)^2 = \mu^2 + 2(kq) + q^2, \\
m^2 &= (p - q)^2 = m^2 - 2(pq) + q^2.
\end{aligned}
$$

The resulting expression is rather cumbersome. It is important, however, to remark that $\left|q^2\right|_{\min}$ becomes *extremely small* at high collision energies.

To see this we introduce the *rapidity* variable,

$$p_0 = m \cosh \eta, \quad |\mathbf{p}| = m \sinh \eta; \qquad \eta = \ln \frac{p_0 + |\mathbf{p}|}{m},$$

and represent (2.65) as

$$|q^2| = 2m^2(\cosh\eta\cosh\eta' - \sinh\eta\sinh\eta' \cdot \cos\Theta - 1)$$
$$= 4m^2\big[\cosh^2\tfrac{1}{2}(\eta-\eta') + \sinh\eta\sinh\eta' \cdot \sin^2\tfrac{1}{2}\Theta\big], \tag{2.66}$$

where Θ is the nucleon scattering angle in the chosen frame of reference. In the πN centre of mass system the variation of Θ does not affect the modulus of the nucleon's three-momentum $|\mathbf{p}'| = p'_c$ and, therefore, the rapidity η'. This makes it obvious that the minimum of $|q^2|$ in (2.66) corresponds to $\Theta_c = 0$. Now we recall the expression (1.27) for the cms momentum as a function of masses:

$$p_c \cdot 2\sqrt{s} = \sqrt{s^2 - 2(m^2+\mu^2)s + (m^2-\mu^2)^2},$$
$$p'_c \cdot 2\sqrt{s} = \sqrt{s^2 - 2(m^2+M^2)s + (m^2-M^2)^2}.$$

Calculating the cms energies of the initial- and final-state nucleons,

$$E_N \cdot 2\sqrt{s} = s + m^2 - \mu^2, \quad E'_N \cdot 2\sqrt{s} = s + m^2 - M^2,$$

we then construct the difference of rapidities,

$$\eta_c - \eta'_c = \ln\frac{E_N + p_c}{E'_N + p'_c} = \ln\frac{(s+m^2-\mu^2+\cdots) + (s-m^2-\mu^2+\cdots)}{(s+m^2-M^2+\cdots) + (s-m^2-M^2+\cdots)}.$$

In the large s limit $(s \gg M^2, m^2)$ this gives

$$\frac{\eta_c - \eta'_c}{2} = \frac{M^2 - \mu^2}{s} + \mathcal{O}(s^{-2}),$$

so that

$$|q^2| \geq |q^2|_{\min} \simeq 4m^2\left(\frac{M^2-\mu^2}{s}\right)^2. \tag{2.67}$$

We see that the virtuality q^2 can actually be very small. Still, it is *negative* while in order to extract the $\pi\pi$ interaction amplitude we need to find the residue of the second-order pole at the *positive* virtuality $q^2 = \mu^2$. Could this be done? It is clear that, mathematically speaking, this is not a well defined problem. We have to find specific conditions under which the pole diagram gives a *significant* contribution to the physical cross section, because if it is small in the physical region, no extrapolation would help us to extract it!

Pion exchange contribution. We know how to calculate the cross section. We average over the initial (and sum over final) nucleon spin, sum over all final states and write

$$
\sigma_{\pi N} = \frac{1}{J} \sum_n \int \frac{d^4 p'}{(2\pi)^3} \delta_+(p'^2 - m^2) \prod_{i \leq n} \frac{d^4 k_i}{(2\pi)^3} \delta_+(k_i^2 - m_i^2)
$$

$$
\times (2\pi)^4 \delta\left(p + k - p' - \sum_i k_i\right) \cdot \left(\frac{1}{\mu^2 - q^2}\right)^2 \cdot g_{\pi N}^2 \quad (2.68)
$$

$$
\times \frac{1}{2} \text{Tr}\left[(m + \hat{p}) \, i\gamma_5 (m + \hat{p}') \, i\gamma_5\right] \cdot A_{kq}(\{k_i\}) A_{kq}^*(\{k_i\}).
$$

Here A_{kq} marks the amplitude of the $\pi\pi$ interaction with the production of n particles with momenta $\{k_i\}$.

The nucleon trace gives

$$
\tfrac{1}{2} \text{Tr}\left[(m + \hat{p}) \, i\gamma_5 (m + \hat{p}') \, i\gamma_5\right] = -q^2.
$$

Collecting all the ingredients that depend on the momentum q, we observe that by virtue of the unitarity relation they combine into the imaginary part of the forward scattering amplitude of a real pion k and the *virtual* pion q:

$$
2 \, \text{Im} \, A_{\pi\pi} = \quad \text{} \quad = \sum_n \int \prod_{i \leq n} \frac{d^4 k_i}{(2\pi)^3} \delta_+(k_i^2 - m_i^2)
$$

$$
\times (2\pi)^4 \delta^4\left(q + k - \sum_i k_i\right) \cdot A_{kq}(\{k_i\}) A_{kq}^*(\{k_i\}). \quad (2.69)
$$

We may write

$$
\frac{d\sigma_{\pi N}}{d^4 q} = \frac{g_{\pi N}^2}{(2\pi)^3 J} \delta(q^2 - 2pq) \frac{-q^2}{(\mu^2 - q^2)^2} \times 2 \, \text{Im} \, A_{\pi\pi}((k+q)^2; q^2), \quad (2.70)
$$

where $(k + q)^2$ is the invariant energy of the $\pi\pi$ collision. Strictly speaking, we must keep q^2 as the argument of the $\pi\pi$ amplitude which may depend on the pion virtuality. However, in the vicinity of the pole we may substitute $q^2 = \mu^2$ everywhere in the numerator of (2.70). Then enters the true physical pion scattering amplitude, $A_{\pi\pi}((k+q)^2; q^2) \rightarrow A_{\pi\pi}((k+q)^2; \mu^2)$, and we have

$$
\frac{d\sigma_{\pi N}^{\text{pole}}}{d^4 q} = \frac{g_{\pi N}^2}{(2\pi)^3 J} \delta(q^2 - 2pq) \frac{-\mu^2}{(\mu^2 - q^2)^2} \times 2 \, \text{Im} \, A_{\pi\pi}((k+q)^2). \quad (2.71)
$$

The first sad fact: in the physical region we had $-q^2 > 0$, and the cross section (2.68) was positive, while now we have $\sigma \rightarrow -\infty$ in the pole! So the residue in the form of (2.71) makes little sense.

Actually, the contribution of the pion exchange *vanishes* at $q^2 = 0$, as the original expression (2.70) for the cross section shows. This property is essential. If in the experiment it were observed that with the decrease of $|q^2|$ the cross section remained large, this would have meant that our pole graph was insignificant!

Another important property follows from the fact that the exchanged pion is spinless. Therefore it can transfer no information about the direction of $\mathbf{q}$ into the upper block. This means that the distribution of secondary particles must be *isotropic* in the cms of the $\pi\pi$ pair (the distribution in the so-called Treiman–Yang angle).

I have described two checks of whether the one-pion exchange term σ^{pole} contributes significantly in the physical region. If both these criteria are met, one can write

$$\frac{d\sigma}{dq^2\,d\Omega} \cdot (\mu^2 - q^2)^2 \simeq F(s_{12}, q^2), \qquad q^2 \leq - \left|q^2\right|_{\min},$$

makes fit to the data for the differential distribution and then extrapolate into the unphysical point $q^2 = \mu^2$ where

$$F(s_{12}, \mu^2) = \text{const.} \cdot \operatorname{Im} A_{\pi\pi}(s_{12}).$$

Strangely enough, this way one obtains reasonably consistent results from different experiments.

In spite of an error margin of the order of 100%, the knowledge of the $\pi\pi$ interaction amplitude, so obtained, is nevertheless extremely important. From an abstract position, the $\pi\pi$ interaction amplitude could differ significantly from directly measurable nucleon interaction amplitudes because, in principle, it could be determined by physical quantities that are totally different from those that govern the nucleon–nucleon interaction.

The Chew–Low method of studying the $\pi\pi$ interaction constitutes another example of how the knowledge of the analytic properties (pole at $q^2 = \mu^2$; distant other singularities) allows us to extract valuable information and to verify this way the basic concepts put in the foundation of the theory.

3

Resonances

3.1 How to examine unphysical sheets of the amplitude

In the previous lecture we have discussed in detail that owing to unitarity, analyticity and crossing symmetry – the general properties of the theory – all the physics of hadron interactions is determined, in principle, by the spectrum of real particles.

We saw that the interaction constant – the measure of the interaction strength – entered only as a residue in the pole of the scattering amplitude. Can it be true that the plethora of phenomena in the hadron world is described by a single quantitative characteristic – the residue?

This situation looks strange and profoundly unsatisfactory from a theoretical point of view. It makes one wonder whether there is not something *hidden* beyond the mass spectrum that we have introduced.

Philosophy aside, it is important to know how the amplitude behaves in the vicinity of the cut. We cannot say a priori that it changes smoothly there, since the question of smoothness of a *multi-valued* function is a delicate one.

If we position ourselves near the cut on the physical sheet then, rather close in the energy variable, we have an *unphysical* sheet about which nothing had been said so far. If there is a singularity (for example, a pole) on the unphysical sheet close to the physical one, the amplitude on the physical sheet would be changing fast.

Thus, knowing the analytic structure of the physical sheet alone turns out to be insufficient. We need information about what is happening on the other sheets of the scattering amplitude, what sort of singularities could be there.

We have to find means to extend – to *continue* – our knowledge of the amplitude '*under the cut*'. It is clear that perturbation theory would be of no help here. Nevertheless, there is a way to get under the cut.

Let us recall the unitarity condition:

$$A(s + i\epsilon, t) - A(s - i\epsilon, t) = i \int \frac{d^4 k}{(2\pi)^2} \delta(m_3^2 - k^2) \delta(m_4^2 - (p_1 + p_2 - k)^2)$$

$$\cdot A(p_1, p_2, k) A^*(p_5, p_6, k). \tag{3.1}$$

Since each of the block amplitudes A, A^* depends in fact on two invariants, it is convenient to rewrite the integral in terms of the Lorentz invariant momentum transfers,

$$A(s + i\epsilon, t) - A(s - i\epsilon, t)$$

$$= \int\int dt_1 \, dt_2 K(s, t_1, t_2) \cdot A(s + i\epsilon, t_1) A(s - i\epsilon, t_2), \tag{3.2}$$

where we have introduced K as the corresponding Jacobian transformation factor.

Now take $A(s - i\epsilon)$ to the r.h.s. and try to look upon (3.2) as an integral equation for $A(s + i\epsilon, t)$ with the kernel $\int dt_2 K(s, t_1, t_2) A(s - i\epsilon, t_2)$ and an inhomogeneity $A(s - i\epsilon, t)$. Imagine that we learned how to calculate the integrals and managed to solve the equation. What would have been the gain? We would have expressed the analytic function on the *upper* side of the cut, $A(+)$, in terms of that on the *lower* side of the cut:

$$A(s + i\epsilon) = F(A(s - i\epsilon)) \tag{3.3}$$

Till now we kept s real and used $i\epsilon$ to separate the points at the two sides of the cut. Let us now give an imaginary part to s itself, a negative one to be definite. Then the argument of $A(s - i\epsilon)$ would simply move onto the lower half-plane of the physical sheet, while the 'upper' function $A(s + i\epsilon)$ whose argument is tightly linked with that of $A(s - i\epsilon)$, will cross the cut and occur on the lower half-plane too, but on another – *unphysical* – sheet!

Under such continuation the relation (3.3) has acquired a new meaning: the value of the amplitude at a given point on the unphysical sheet is now (functionally) determined by the value of the physical amplitude. Along this way we would have solved a remarkable problem, namely, with the help of the unitarity condition we would have examined the content of the unphysical sheet and, in particular, found the singularities of the amplitude there (which is what concerns us in the first place).

3.2 Partial waves and two-particle unitarity

The programme that we have described is easy to carry out for the *first* unphysical sheet related to the two-particle unitarity condition (which holds for $4\mu^2 < s < 9\mu^2$).

Recall the partial wave expansion

$$A(s,t) = \sum_{\ell=0}^{\infty} (2\ell + 1) f_\ell(s) P_\ell(\cos\Theta). \tag{3.4}$$

It is clear that in these terms the unitarity condition (3.1) will simplify greatly. Indeed, ℓ – the total angular momentum in the cms – is a conserved quantity. Therefore, if we choose an initial state with a given ℓ, the intermediate state will be uniquely determined and the integration on the r.h.s. of (3.1) will have to disappear. Let us see how it actually happens.

First we attend to the momentum integration in (3.1):

$$d^4k = |\mathbf{k}|^2\, d|\mathbf{k}|\, dk_0\, d\Omega = \frac{1}{2}|\mathbf{k}|\, dk^2\, dk_0\, d\Omega, \quad \delta(m_3^2 - k^2)dk^2 = 1.$$

In the cms we have $\mathbf{p}_1 + \mathbf{p}_2 = \mathbf{0}$, $p_{10} + p_{20} = \sqrt{s}$, so that

$$\delta(m_4^2 - (p_1 + p_2 - k)^2) = \delta(m_4^2 - m_3^2 - s + 2k_0\sqrt{s}).$$

Taking off the integration over k_0 we arrive at

$$\begin{aligned}
\mathrm{Im}\, A(s,t) &= \frac{1}{2}\int \frac{d\Omega}{(2\pi)^2}\cdot\frac{1}{2}|\mathbf{k}|\cdot\frac{1}{2\sqrt{s}}A(s,t_1)A^*(s,t_2)\\
&= \frac{|\mathbf{k}|}{8\pi\sqrt{s}}\int \frac{d\Omega}{4\pi} A(s,\cos\Theta_1)A^*(s,\cos\Theta_2). \tag{3.5}
\end{aligned}$$

The modulus of the intermediate state momentum **k** is fixed by the on-mass-shell conditions. The partial wave expansion will help us to perform the remaining integration over its direction angles. Using (3.4) for the amplitudes on the l.h.s. and r.h.s. of (3.5) we have

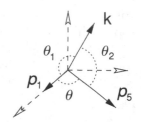

$$\text{Im } A(s,t) = \sum_{\ell=0}^{\infty}(2\ell+1)\,\text{Im } f_\ell(s)P_\ell(\cos\Theta) = \frac{|\mathbf{k}|}{8\pi\sqrt{s}}\sum_{\ell_1,\ell_2}f_{\ell_1}(s)f_{\ell_2}^*(s)$$

$$\times(2\ell_1+1)(2\ell_2+1)\int\frac{d\Omega}{4\pi}P_{\ell_1}(\cos\Theta_1)P_{\ell_2}(\cos\Theta_2) \qquad (3.6)$$

$$= \frac{|\mathbf{k}|}{8\pi\sqrt{s}}\sum_{\ell_1=0}^{\infty}f_{\ell_1}(s)f_{\ell_1}^*(s)(2\ell_1+1)P_{\ell_1}(\cos\Theta),$$

where we used the well-known orthogonality relation for spherical functions (Legendre polynomials),

$$\int\frac{d\Omega}{4\pi}P_n(\cos\Theta_1)P_m(\cos\Theta_2) = \frac{\delta_{nm}}{2n+1}P_n(\cos\Theta).$$

Comparing the two sides of the equation, we retrive the unitarity condition for a partial wave with angular momentum ℓ,

$$\text{Im } f_\ell(s) = \tau f_\ell(s)f_\ell^*(s), \qquad (3.7a)$$

$$\tau = \tau(s) \equiv \frac{k_c}{8\pi\sqrt{s}} = \frac{1}{16\pi}\frac{k_c}{\omega_c}, \qquad (3.7b)$$

with k_c the modulus of the intermediate state momentum in the cms:

$$k_c \equiv |\mathbf{k}| = \frac{\sqrt{s^2-2s(m_3^2+m_4^2)+(m_3^2-m_4^2)^2}}{2\sqrt{s}}, \quad \omega_c = \frac{\sqrt{s}}{2}. \qquad (3.8a)$$

For the case of equal masses, $m_3 = m_4 = m$,

$$k_c = \frac{\sqrt{s-4m^2}}{2}. \qquad (3.8b)$$

The solution of the elastic unitarity condition (3.7a) reads

$$f_\ell(s) = \frac{1}{2i\tau(s)}\left[e^{2i\delta_\ell}-1\right] = \frac{\sin\delta_\ell}{\tau}e^{i\delta_\ell}, \qquad (3.9)$$

with $\delta_\ell = \delta_\ell(s)$ the scattering phase in a given angular momentum state.

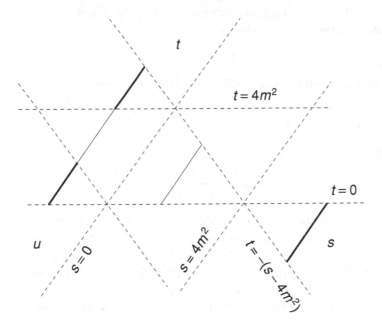

Fig. 3.1   Three integration intervals on the Mandelstam plane. The amplitude is complex at $s > 4m^2$ and $s < 0$ (on the thick lines).

3.3 Analytic properties of partial waves and resonances

We are now going to discuss analytic properties of $f_\ell(s)$. We will need the expression for the partial wave which is complementary to (3.4):

$$f_\ell(s) = \frac{1}{2} \int_{-1}^{1} d\cos\Theta \, P_\ell(\cos\Theta) A(s, t(\cos\Theta)). \qquad (3.10)$$

The cosine of the scattering angle,

$$\cos\Theta = 1 + \frac{2t}{s - 4m^2},$$

varies between -1 and $+1$. On the Mandelstam plane this corresponds to integration over t from $t = -(s - 4m^2)$ up to $t = 0$ (Fig. 3.1). For $s > 0$ the partial wave $f_\ell(s)$ mirrors analytic properties of the amplitude A: it is real for $s < 4m^2$ (since the integration interval then lies inside the triangle where A is real) and is complex above the threshold, $s > 4m^2$. For $s < 0$ $f_\ell(s)$ becomes complex again; this time because of the integration contour hitting t- and u-channel thresholds.

Since $f_\ell(s)$ is real on a finite interval in the s-plane, it assumes complex conjugated values on the sides of the cut:

$$(f(s))^* = f(s^*).$$

Therefore we can represent (3.7a) in terms of discontinuity as

$$\frac{1}{2i}[f_\ell(s+i\epsilon) - f_\ell(s-i\epsilon)] = \tau(s)f_\ell(s+i\epsilon)f_\ell(s-i\epsilon).$$

The formula

$$f_\ell(s+i\epsilon) = \frac{f_\ell(s-i\epsilon)}{1 - 2i\tau(s)f_\ell(s-i\epsilon)} \tag{3.11}$$

solves the problem of analytic continuation of the amplitude (to be precise, of each of its partial wave components) onto the first unphysical sheet related to the two-particle threshold.

So, what are the singularities on the unphysical sheet? Obviously, $f_\ell(+)$ has the same cuts as $f_\ell(-)$. In addition, it acquires new singularities – *poles* – where the denominator in (3.11) vanishes, that is in the points where

$$f_\ell(-) = \frac{1}{2i\tau(s)}. \tag{3.12}$$

These poles on the unphysical sheet(s) are called *resonances*. The position of such a pole depends essentially on the value of the coupling constant. If the coupling is small, so is the physical scattering amplitude. The amplitude can reach a finite value which is required by the resonance condition (3.12) only if interaction is strong enough.

We can reverse the statement:

It is the resonance states that bear additional essential information about interaction that we talked about in the beginning of this lecture.

What is the reason that we have not met singularities more complicated than simple poles?

To answer the question we return to the integral equation in the general form of (3.2). The t_1, t_2 integrations run over finite intervals; moreover, the *kernel* of the equation,

$$\phi(s,t,t_1) = \int dt_2 K(s,t_1,t_2)A(s-i\epsilon,t_2),$$

is a smooth regular function since it is determined by the amplitude on the physical sheet where the amplitude is regular. Therefore, our equation is a standard integral equation of the Fredholm type whose solutions may have only poles (at the points where the Fredholm determinant vanishes).

3.4 Three-particle unitarity condition

Producing only poles is an intrinsic property of the simplest – two-particle – unitarity. The continuation of the three-particle unitarity relation is technically much more involved. We shall outline the main plan of the corresponding analysis. The sketch that follows will suffice for us to grasp the qualitative features of the answer.

Below the four-particle threshold the unitarity condition has the following schematic structure:

$$\frac{1}{i}\left[\;\bigcirc(+)\; - \;\bigcirc(-)\;\right] = \;\bigcirc(+)\bigcirc(-)\; + \;\bigcirc(+)\times\bigcirc(-)\;. \tag{3.13}$$

A new object entered namely, the amplitude $A_{2\to3}$. For the latter we have its proper unitarity equation,

$$\frac{1}{i}\left[\;\bigcirc(+)\; - \;\bigcirc(-)\;\right] = \;\bigcirc(+)\bigcirc(-)\; + \;\bigcirc(+)\times\bigcirc(-)\;. \tag{3.14}$$

The last kernel $A_{3\to3}(-)$ contains an *irreducible* scattering amplitude, $\boxed{}$ but also various *reducible* pieces of the type $\begin{smallmatrix}a\\b\\c\end{smallmatrix}\!\!-\!\!\bigcirc\!-$ so that symbolically

$$\;\bigcirc(+)\times\bigcirc(-)\; - \;\bigcirc(+)\times\boxed{}(-)\; + \;\bigcirc(+)\times\bigcirc(-)\;. \tag{3.15}$$

If the last contribution were not present, the result would have been similar to what we had in the two-particle case. That is, an analytic continuation of $\bigcirc(+)$ beneath the cut would have produced only poles, namely the old two-particle and new three-particle resonances. The new piece, however, is not of the Fredholm type. The corresponding kernel is *singular* as it contains $\delta(p_c - p'_c)$ – one of the particles did not scatter.

How would this complicate the answer? To understand the key features let us keep only the singular term in (3.14) and iterate the equation for $A_{2\to3}(+)$:

$$\bigcirc(+) \;\sim\; \bigcirc(-) \; + \; i\;\bigcirc(+)\times\bigcirc(-)$$

$$= \; i\;\bigcirc(-)\times\bigcirc(-) \; + \; i^2\;\bigcirc(-)\times\bigcirc(-)\times\bigcirc(-) \; + \cdots \tag{3.16}$$

Comparing the chain of the $A_{2\to 2}(-)$ blocks in (3.16) with that of the iterated two-particle unitarity condition,

we observe that

$$\tag{3.17}$$

In the course of the analytic continuation, the $(-)$ amplitudes stay on the physical sheet while the amplitude $A_{2\to 2}(+)$ moves to the first unphysical sheet where, as we know, it has resonance poles in the pair energy s_{ab}.

How will this affect analytic properties of the two-particle scattering amplitude on the l.h.s. of (3.13)? Let us substitute the resonance pole term into the r.h.s. of (3.17) and then into the unitarity condition (3.13):

This is a typical diagram for a threshold branch cut due to the exchange of two poles, one of which is a normal particle and the other – a resonance.

Thus, on the *second* unphysical sheet related to the three-particle cut $9\mu^2 < s < 16\mu^2$ we find, apart from poles with complex masses, also *cuts* – creation thresholds of pairs of a particle with a resonance (from the first unphysical sheet).

In perturbation theory poles have led to the appearance of threshold cuts on the physical sheet. Analogously, on the other sheets there emerge particle–resonance, resonance–resonance, etc. thresholds.

Now we are in a position to formulate the qualitative answer for the analytic structure of the amplitude:

> The full analyticity image is poles – particles and resonances – all other singularities being the unitarity driven consequence of the existence of these poles.

3.5 Properties of resonances

Both theoretically and experimentally resonances are as important as ordinary particles. Therefore we need to learn how to describe them, in the first place.

Let us draw an analogy. A particle is characterized by its mass and spin. A usual pole amplitude we describe in terms of a diagram

$$\text{(3.18)}$$

Does it make sense to draw an analogous diagram for a resonance?

$$\text{(3.19)}$$

If the series (3.4) converges,

$$A(s,t) = \sum_{\ell=0}^{\infty} (2\ell+1)\, f_\ell(s)\, P_\ell(\cos\Theta), \tag{3.20}$$

the full amplitude will but repeat all the poles of the partial waves. The contribution to the amplitude of the resonance in the ℓ-wave,

$$f_\ell(s) = \frac{r}{m_{\text{res}}^2 - s}, \tag{3.21}$$

will be

$$A^{\text{pol}} = (2\ell+1)\frac{r\, P_\ell(\cos\Theta)}{m_{\text{res}}^2 - s}. \tag{3.22}$$

Now we should check that (3.22) coincides indeed with the Born amplitude of the s-channel exchange of a particle with spin $\sigma = \ell$.

3.5.1 Angular dependence

How would we write a Feynman diagram for a particle with a given spin σ? In the case of a *scalar*, $\sigma = 0$, the amplitude

$$A^{\text{pol}} = \quad \raisebox{-1em}{} \quad = \frac{g^2}{m^2 - s}$$

does not depend on angles.

The propagator of a *vector* particle, $\sigma = 1$, contains vector indices and the exchange diagram takes the form

$$A = g^2\, \Gamma_\mu D^{\mu\nu}(k)\Gamma_\nu\,; \tag{3.23}$$

$$D^{\mu\nu}(k) = \frac{-g^{\mu\nu} + b \cdot k^\mu k^\nu}{m^2 - k^2}. \tag{3.24}$$

In QED we dealt with a vector particle – a photon. There we were allowed to omit the term $b \cdot k^\mu k^\nu$. Remember why? Because due to the conservation of the electromagnetic current the interaction was insensitive to this piece in the photon Green function. Not so in a general case, and we ought to determine the coefficient b in (3.24).

A vector particle has three polarization states that are 'propagated' by the propagator

$$D^{\mu\nu}(k) = \sum_{\lambda=1}^{3} \frac{e^\mu_\lambda(k) e^\nu_\lambda(k)}{m^2 - k^2} . \tag{3.25}$$

How to choose three out of four independent vectors? To this end we have to invoke an additional condition

$$k_\mu e^\mu_\lambda = 0 . \tag{3.26}$$

In the rest frame of the particle this condition turns into $me^0_\lambda = 0$,

$$e^\mu_\lambda = (0, \mathbf{e}_\lambda), \qquad \lambda = 1, 2, 3.$$

These are usual space vectors describing three possible spin projections. Taking into account the transversality condition (3.26), the Green function becomes

$$D^{\mu\nu}(k) = \frac{\frac{k^\mu k^\nu}{m^2} - g_{\mu\nu}}{m^2 - k^2} . \tag{3.27}$$

Now, what to write for the vertex function? Possible vector structures are

$$\Gamma^\mu = a p^\mu_1 + b p^\mu_2 = \alpha (p_1 + p_2)^\mu + \beta (p_1 - p_2)^\mu .$$

It is clear that the first term will not contribute to the pole. Using (3.27) we get

$$(p_1 + p_2)_\mu D^{\mu\nu}(k) \equiv k_\mu D^{\mu\nu}(k) = \frac{k^\nu}{m^2} \cdot \frac{k^2 - m^2}{m^2 - k^2} \implies \text{finite.}$$

The pole contribution to the amplitude (3.23) takes the form

$$A^{\text{pol}} = \quad\overset{1}{\underset{2}{\diagdown}}\!\!\underset{}{\overbrace{\sigma=1}}\!\!\overset{3}{\underset{4}{\diagup}}\quad = g'^2 (p_1 - p_2)_\mu D^{\mu\nu}(k)(p_3 - p_4)_\nu. \tag{3.28}$$

To understand the meaning of this expression we go again to the cms of colliding particles (which we take for simplicity to have equal masses). Then the vertex functions

$$(p_1 - p_2)_\mu = (0, 2\mathbf{q}), \qquad (p_3 - p_4)_\nu = (0, 2\mathbf{q}')$$

turn into three-vectors describing relative momenta of initial and final state particles, correspondingly:

$$A^{\text{pol}} = g'^2 \frac{4(\mathbf{q} \cdot \mathbf{q}')}{m^2 - s}.$$

As compared with the case $\sigma = 0$, a scalar product appeared in the numerator,

$$(\mathbf{q} \cdot \mathbf{q}') \propto \cos \Theta = P_1(\cos \Theta). \tag{3.29}$$

It is straightforward to verify that for an arbitrary spin σ the Feynman diagram describing s-channel particle exchange gives an expression proportional to

$$P_\sigma(\cos \Theta_{\text{cms}}),$$

as we have checked for the simplest cases $\sigma = 0, 1$.

So, the angular dependence turned out to be the same for particle and resonance exchanges.

3.5.2 Factorization and unitarity

Now we have to check another very important property of 'particles': factorization. What does it mean? When we draw a Feynman graph, the initial and final states enter as a product, that is, they are *factorized*. If the analogy between resonances and particles is to be preserved, the residue r in (3.21) should split into the product of two constants each belonging to the proper vertex,

$$A^{\text{res}} = a \mathrel{\rangle\!\text{-----}\!\langle} b = g_a \cdot \frac{1}{m^2 - s} \cdot g_b. \tag{3.30}$$

The fact that it is indeed so is not accidental and is tightly related to unitarity.

Up to now we have considered elastic scattering and did not differentiate between initial- and final-state systems (a and b in (3.30)).

It is straightforward to generalize the notion of partial amplitudes for the case of multi-channel scattering problem by introducing the unitary scattering matrix

$$SS^\dagger = I, \qquad \sum_c S_{ac} S_{cb}^\dagger = I_{ab},$$

where a, b, c mark different channels of the reaction. The generalized expression for the partial wave amplitude describing $a \to b$ transition reads*

$$f_{ab} = \frac{1}{2i\sqrt{\tau_a \tau_b}} \left[S_{ab} - 1 \right]. \qquad (3.31)$$

The scattering matrix S can be diagonalized with the help of a unitary tranformation U,

$$S_{ab} = U_{ac}^\dagger S_{cd} U_{db},$$

with S a diagonal unitary matrix

$$S_{cd} = S_c I_{cd}, \quad S_c = \exp(2i\delta_c).$$

Our resonance is a pole in one of its eigenvalues at some complex value

$$s = M^2 = M_1^2 - iM_2^2.$$

Let it be the first element S_1; we may write in the pole approximation

$$S_1 = \frac{M^{*2} - s}{M^2 - s} e^{2i\beta} = \frac{-2i\,\mathrm{Im}\,M^2}{M^2 - s} e^{2i\beta} + \text{ regular}.$$

Here 2β describes the scattering phase away from the resonance:

$$S_1 \simeq e^{2i\beta} \quad \text{for } \left| s - M_1^2 \right| \gg M_2^2.$$

The pole contribution to the full scattering matrix becomes

$$S_{ab} \simeq U_{a1} \left[\frac{-2i\,\mathrm{Im}\,M^2}{M^2 - s} e^{2i\beta} \right] U_{1b}^\dagger = U_{a1} \frac{2iM_2^2}{M^2 - s} e^{2i\beta} U_{b1}^*; \qquad (3.32a)$$

$$f_{ab} \simeq \frac{U_{a1}}{\sqrt{\tau_a}} \frac{M_2^2}{M^2 - s} e^{2i\beta} \frac{U_{b1}^*}{\sqrt{\tau_b}} = \frac{g_a g_b^*}{M^2 - s} e^{2i\beta}, \qquad (3.32b)$$

where we have introduced the constants

$$g_a \equiv U_{a1} \cdot \frac{M_2}{\sqrt{\tau_a}} \qquad (3.32c)$$

that measure coupling of the resonance to different particle states a and play the rôle of (complex) interaction constants. In fact g_a can be taken to be real. This follows from the invariance of the scattering martix with respect to time reversal. Indeed, in this case the S-matrix is symmetric, $S_{ab} = S_{ba}$. This gives $U_{a1}U_{b1}^* = U_{a1}^* U_{b1}$ so that U_{a1} and U_{b1} have equal

---

* We don't write ℓ implying that the total angular momentum is included in the channel indices a, b (ed.).

phases which cancel in (3.32a), (3.32b) and, consequently, we can redefine

$$g_a \implies |U_{a1}| \cdot \frac{M_2}{\sqrt{\tau_a}}. \tag{3.33}$$

The resonance contribution to the full amplitude takes the form very similar to the Born diagram for spin-ℓ particle exchange:

$$A^{\text{res}} = \frac{g_a(2\ell+1)P_\ell(z)\, g_b}{M^2 - s} \, e^{2i\beta}. \tag{3.34}$$

We see that the resonance appears in all channels, $a \to a$, $a \to b$, $b \to b$ etc., and that the transition amplitudes between various channels are factorized. This property followed directly from unitarity.

Thus we have checked that both properties of normal particles – the angular dependence of the amplitude and factorization – hold for complex poles as well.

3.5.3 Width of the resonance

For resonances, however, there is one more statement which makes no sense for ordinary particles, namely, the relation between $\text{Im}\, M^2$ and the coupling constants that also follows from unitarity. Recalling that the matrix U is unitary,

$$\sum_a U_{a1}^* U_{a1} = 1,$$

and expressing U_{a1} via (3.33) one relates the set of g_a with the full *width* of the resonance:

$$\sum_a \tau_a g_a^2 = \sum_a \text{-}\text{-}\!\!\left\langle a \right\rangle\!\!\text{-}\text{-} = M_2^2 \equiv M_1 \Gamma. \tag{3.35}$$

The constant g_a is the transition amplitude between the resonance and ordinary particles with τ_a their phase space. Thus the amount by which the pole is shifted from the real energy axis is determined by the total probability of the resonance decay into real particles.

3.6 A resonance or a particle?

How to establish a generic link between particles and resonances? Let us discuss a simple example of how a resonance emerges.

Imagine that in some reaction a π^0 meson appears as a pole:

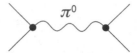

Now we switch on the electromagnetic interaction with a very small interaction constant. Then a new transition will emerge, $\pi^0 \to \gamma\gamma$. How will this small correction to strong dynamics affect the propagation of π^0?

Pion self-energy is determined (in the hadronic language) by a variety of possible intermediate states such as 3π, $N\bar{N}$, etc.:

$$\Sigma_h = \qquad\qquad\qquad\qquad + \qquad\qquad\qquad + \cdots \quad (3.36)$$

The point in k^2 where the denominator of the Green function

$$G(k) = \frac{1}{m_0^2 - k^2 - \Sigma(k^2)}$$

turns into zero determines the renormalized pion mass:

$$m_\pi^2 = m_0^2 - \Sigma(m_\pi^2). \qquad (3.37)$$

For (3.37) to have a real solution, Σ has to be real. We know that the graphs of (3.36) become complex at $k^2 > (3m_\pi)^2$ and $k^2 > (2M_N)^2$, respectively, which scales are significantly higher than m_π^2. That was the case before we turned QED on. Now, however, we shall also have the graph

$$\Sigma_{\text{e.m.}} = \qquad\qquad\qquad, \quad \Sigma(k^2) = \Sigma_h(k^2) + \Sigma_{\text{e.m.}}(k^2), \quad (3.38)$$

which has a threshold (and therefore a branch cut) at $k^2 = (2m_\gamma)^2 = 0$. As a result, (3.37) will have no real solution and the π^0 Green function will acquire a 'resonance' form

$$G(k^2) = \frac{1}{m^2 - k^2 - im_2^2},$$

$$m_2^2 = \text{Im}\,\Sigma_{\text{e.m.}}(m^2) = \qquad\qquad = g_{\pi^0 \to \gamma\gamma}^2 \cdot \tau_{\gamma\gamma}. \quad (3.39)$$

What happened? After the introduction of a small-mass intermediate state the pion pole occurred right on the cut! This, however, would have contradicted the unitarity condition. Therefore the pole (together with all multi-state cuts that it generates) moves under the two-photon cut onto the unphysical sheet:

This analysis tells us that the difference between particles and resonances is elusive. If we believe that electromagnetic interaction has no major effect on the basic properties of pions as strongly interacting particles, then it becomes insignificant whether a π-meson is a particle or a resonance, whether the corresponding pole lies on the real axis or slightly below.

Concluding, we have proved that a resonance as a contribution to the scattering amplitude is identical to a particle with definite quantum numbers and possesses the usual factorization properties that are characteristic for particle exchange. Therefore we can describe resonances with the help of Feynman diagrams as we did for particles, the only difference being a complex mass whose imaginary part is related to the total decay probability of the resonance into stable particles.

3.7 Observation of resonances

This is the last question that we need to address in this lecture. Imagine that a beam hits a target and some particles are produced. The probability amplitude to observe one of these particles a at a given point $\mathbf{r}$ at time t,

$$x = (t, \mathbf{r}),$$

is given by the expression

$$A = \int \frac{d^4 k}{(2\pi)^4 i} \frac{e^{-ik^\mu (x-y)_\mu}}{m^2 - k^2 - i\epsilon} f(y, p, p_i, \ldots), \qquad (3.40)$$

where we have explicitly written the propagator of our particle a.

In a real experiment the observation time t is macroscopically large, $t - y_0 \gg m^{-1}$. Therefore the phase factor in (3.40) is oscillating fast with k, and the integral would be exponentially small if not for the singularity of the propagator function. Since $t - y_0 > 0$, we can close the integration contour in k_0 around the pole at $k_0(\mathbf{k}) = \sqrt{m^2 + \mathbf{k}^2}$, that is to put particle

a on mass shell:

$$A = \int \frac{d^3k}{2k_0(\mathbf{k})(2\pi)^3} \, e^{-ik_0(\mathbf{k})(t-y_0)+i(\mathbf{k}\cdot(\mathbf{r}-\mathbf{y}))} f. \qquad (3.41)$$

At $t \to \infty$ plane waves in this sum cancel each other everywhere but on a classical trajectory where $\mathbf{r}$ is linearly increasing with t:

$$\boldsymbol{\nabla}_\mathbf{k}[\, k_0(\mathbf{k})(t-y_0) - (\mathbf{k}\cdot(\mathbf{r}-\mathbf{y}))\,] = 0 \;\Rightarrow\; \mathbf{r} = \mathbf{v}t + \mathbf{r}_0, \;\; \mathbf{v} \equiv \frac{dk_0(\mathbf{k})}{d\mathbf{k}} = \frac{\mathbf{k}}{k_0}.$$

It is easy to verify that the probability of observing the particle at $\mathbf{r}$ falls with the distance as

$$w \propto |A|^2 \propto \frac{1}{\mathbf{r}^2},$$

which is in accord with the increasing size of the surface of the observation sphere.

If we were to register *two* particles a and b at the same point,

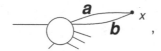

the probability would fall faster with $|\mathbf{r}|$ because two particles prefer to separate at large distances.

Consider now the general case of the observation of two particles.

$$(3.42)$$

Repeating literally the above analysis we will obtain that the observation probability is concentrated along the trajectories

$$\mathbf{r}_1 = \mathbf{v}_1 t + \mathbf{r}_{10}, \quad \mathbf{r}_2 = \mathbf{v}_2 t + \mathbf{r}_{20}.$$

By measuring the directions of particle momenta we can determine, within the Heisenberg uncertainty, classical trajectories, and verify that the particles indeed originate from the interaction region.

Suppose now that a and b may combine into a resonance state:

$$(3.43)$$

How does this graph differ from the usual particle creation in (3.42)? In the case of a complex pole with a small imaginary part $\operatorname{Im} M^2 = M_2^2 \ll M_1^2$

we have

$$k_0 = \sqrt{M^2 + \mathbf{k}^2} \simeq \sqrt{M_1^2 + \mathbf{k}^2} - \frac{iM_2^2}{2\sqrt{M_1^2 + \mathbf{k}^2}} \equiv E - \frac{iM_2^2}{2E}. \quad (3.44)$$

The amplitude then falls exponentially at large times,

$$A \propto e^{-ik_0 t} \propto \exp\left\{-\frac{1}{2}\frac{M_2^2}{E}t\right\} \equiv \exp\left\{-\frac{\Gamma}{2}\cdot\frac{M_1}{E}t\right\}, \quad (3.45)$$

and the probability decreases as

$$|A|^2 \propto \exp\left\{-\Gamma t \cdot \frac{M_1}{E}\right\}. \quad (3.46)$$

Here the ratio M_1/E is the usual Lorentz time dilatation factor. Introducing the proper life-time τ of the resonance,

$$\tau^{-1} \equiv \Gamma = \frac{M_2^2}{M_1}, \quad (3.47)$$

we conclude that for finite times $t \sim \tau \cdot E/M_1$ our resonance state may propagate as a whole. Therefore, restoring trajectories of its decay products we will see that a and b originate not from the target but fly away from a point at some finite distance from the interaction region. This distance will vary event by event. It is clear, however, that the probability of a large displacement is exponentially small.

3.7.1 Non-exponential decay?

Sometimes people talk about the 'non-exponentiality' of a decay. Calculating concrete integrals I will always find contributions that fall as a power of t (most often as $t^{-3/2}$) in addition to the resonance exponent. Does this imply that the decay law (3.46) is inaccurate?

We register a resonance by observing its decay products. Therefore there will always be a background due to production of particles a and b directly off the target. Let us switch on our measuring devices one hour after irradiating the target; some extremely slow particles could have been produced that crawl and hit the detectors after this immense time is elapsed. This power-behaving 'tail' has nothing to do with the decay of the resonance.

3.7.2 Resonance in the invariant mass distribution

Till now we were considering resonances with a small width Γ. For resonances that decay relatively fast (large Γ) a direct visual observation of the resonance path becomes impossible. In this case resonances are extracted by examining energetic characteristics of the process.

$$f(p_1, p_2, p_3) \frac{1}{M^2 - (p_4 + p_5)^2} g(p_4, p_5). \quad (3.48)$$

The resonance propagator in (3.48) introduces a *peak* in the distribution over the invariant mass s_{45} of the pair of particles 4 and 5:

$$\frac{d\sigma}{ds_{45}} \propto |A|^2 \propto \frac{1}{(s_{45} - M_1^2)^2 + M_2^4}. \quad (3.49)$$

Let us remark that instead of observing decay products it suffices to measure the recoil momentum p_3, since due to momentum conservation $s_{45} = (p_4 + p_5)^2 = (p_1 + p_2 - p_3)^2$.

This method works only when the value of the resonance production amplitude f is sufficiently large. It may turn out, however, that in reactions that are available to experimenters the resonance is produced with a small probability.

3.7.3 Phase analysis

In addition to the invariant mass spectrum there is another method (though less unambiguous) of extracting resonances right from elastic scattering – the phase analysis. Recall the general expression for the partial wave amplitude describing elastic ab scattering

$$f_\ell(s) = \frac{1}{2i\tau(s)} \left[\eta(s) \, e^{2i\delta_\ell(s)} - 1 \right]. \quad (3.50)$$

The scattering phase factor near the resonance is

$$e^{2i\delta_\ell} = \frac{M_1^2 - s + iM_2^2}{M_1^2 - s - iM_2^2} \, e^{2i\beta_\ell}. \quad (3.51)$$

If the amplitude away from the resonance is small (as, for example, for e^+e^- scattering), then $\eta \simeq 1$, $\beta_\ell \simeq 0$. In such a case the resonance is impossible to miss since the amplitude at the peak hits the *maximal* value allowed by unitarity,

$$f_\ell = \frac{i}{\tau} = f_\ell^{\max} \qquad \text{for} \quad s = M_1^2 \quad (\eta = 1, \, \beta_\ell = 0). \quad (3.52)$$

What to do if the non-resonant scattering is large and the peak does not stick out from the background?

By measuring the shape of the angular distribution, one can extract a few first partial waves by fitting the cross section with an approximate

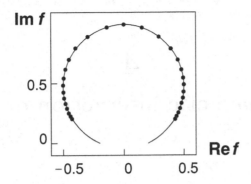

Fig. 3.2   Example of an Argand plot.

expression for the amplitude

$$A(s, \cos\Theta) \simeq \sum_{\ell=0}^{\ell_0} (2\ell + 1) f_\ell(s) P_\ell(\cos\Theta).$$

The following remarkable property of $f_\ell(s)$ then comes onto the stage. The resonant factor in (3.51),

$$\frac{M_1^2 - s + iM_2^2}{M_1^2 - s - iM_2^2} \simeq \frac{(M_{\text{res}} - E) + i\Gamma/2}{(M_{\text{res}} - E) - i\Gamma/2},$$

equals 1 on both sides of the resonance, $|E - M_{\text{res}}| \gg \Gamma$, but changes fast in a relatively small energy interval $|E - M_{\text{res}}| \sim \Gamma$. If we put a point on a complex plane to mark the value of the amplitude f_ℓ at a given energy, this point will make a *full circle* when the energy variable crosses the mass of the resonance (provided η, β, τ do not change essentially inside the interval of the order Γ), see Fig. 3.2.

Combining the three methods that we have described here, a plethora of hadronic resonances has been established.

The studies of the spectrum of hadrons tells us that the fact of stability/ instability of a hadron under consideration, whether it is a particle or a resonance, is not of major importance for strong interaction dynamics. It often looks accidental, depending on an interplay of factors that we rather consider insignificant today.

4

Electromagnetic interaction of hadrons

What do we mean when we talk about electromagnetic interactions of hadrons? The simplest of them is the interaction of hadrons with leptons. As you know there are particles – leptons – that do not participate in the strong interaction. They interact electromagnetically, and this interaction* can be described in the field theoretical (QFT) framework.

We suppose, and this is our main hypothesis, that a charged lepton interacts with hadrons only via a photon field. The second part of our hypothesis must contain something about the interaction of a photon with hadrons. Here we will say, once again within the QFT concept, that it is still a pointlike interaction (as in the usual QED) which has to be treated with account of all possible field-theoretical corrections:

$$\text{(4.1)}$$

Strong interaction being *strong*, the photon will always interact with a virtual particle in the intermediate state that depicts the internal structure of a dressed hadron.

Since QED amplitudes are small, we can restrict ourselves to the one-photon-exchange picture.

4.1 Electron–proton interaction

Consider the electron–proton scattering amplitude in the first order in α_{em}. Both particles have spin-$\frac{1}{2}$, and from general considerations we can

---

* as well as the weak one, with the advent of the Glashow–Weinberg–Salam theory. (ed.)

immediately write

$$= e_1 e_2 \big[\bar{u}(k')\gamma_\mu u(k)\big] \frac{1}{q^2} \big[\bar{U}(p')\Gamma_\mu U(p)\big]. \qquad (4.2)$$

What can we say about the vertex Γ_μ? It has got to be a vector since the photon has spin 1. We have at our disposal the proton momenta p_μ and p'_μ and the Dirac matrix γ_μ:

$$\Gamma_\mu = a(q^2)\gamma_\mu + b(q^2)\sigma_{\mu\nu}q^\nu \qquad \big(\sigma_{\mu\nu} \equiv \tfrac{1}{2}[\gamma_\mu,\gamma_\nu]\big). \qquad (4.3)$$

This is the most general structure that satisfies the on-mass-shell current conservation condition $q^\mu \cdot [\bar{U}(p')\Gamma_\mu U(p)] = 0$ (higher powers of γ-matrices reduce to (4.3) due to the Dirac equation $\bar{U}\hat{p}' = m\bar{U}$, $\hat{p}U = mU$).

Already at this stage we have obtained a strong prediction that derives from the fact that the unknown functions a and b in (4.3) do not contain any s-dependence!

It is worthwhile to remark that in discussing electromagnetic interactions of hadrons we face a situation which is essentially different from that of the strong interaction. In pure hadron interactions we were basically dealing with real particles and on-mass-shell amplitudes. Now, by virtue of the *smallness* of the lepton–hadron interaction, we were able to select a single graph whose amplitude depends on the virtual momentum q^2. Formerly, hadron–hadron scattering amplitudes depended seriously on s and t. Now we have a serious dependence only on $t = q^2$ while the s-dependence turns out to be trivial as it is exclusively due to free particles' spinors.

This observation alone gives rise to the *Rosenblutte formula* checking which one directly verifies that e and p indeed interact only via the electromagnetic force. (The Rosenblutte formula is also being intensively checked experimentally in an attempt to find a possible difference between the leptons – an electron e and a muon μ.)

4.1.1 Electric charge

The photon–proton vertex will be determined by the ensemble of all the diagrams (4.1) which I would have to *calculate*, given a concrete QFT. For example, suppose there were only nucleons and pions. Then I would draw the bare interaction and corrections:

The first thing to do is to calculate the *electric charge* – the value of the amplitude at $q \to 0$.

Bare proton. Suppose the proton possessed a bare electric charge, e_0. Then with account of higher-order radiative corrections it would get modified,

$$e_0 \implies e_0 \, Z_1^{-1} Z_2 \sqrt{Z_3} \equiv e_p,$$

where Z_1^{-1} is the vertex correction and the factor $Z_2 = (\sqrt{Z_2})^2$ comes from initial and final proton wave functions. After renormalization we would have the on-mass-shell condition

$$\Gamma_\mu(q = 0; \, p^2 = p'^2 = m_p^2) = e_p \cdot \gamma_\mu; \quad a(0) \equiv 1. \tag{4.4}$$

The Ward identity taught us that in the theory with a conserved current

$$Z_1 = Z_2, \implies e_p = e_0 \cdot \sqrt{Z_3},$$

that is, charge renormalization is related to the photon only. This is exactly what we call *charge conservation*. Namely, if we plug in equal bare electric charges for the proton and the electron, then the renormalized physical charges will stay equal, irrespective of the nature of the charged particle and interactions it is subject to.

In fact we have only 'half a theory' of this important phenomenon, since (except for a few attempts) we have no pure theoretical reason for ascribing to electron and proton equal (and opposite) bare charges. This is just an experimental fact (and a very solid one in that).

Quarks. What if there is no bare proton at all? We can imagine the proton to be a bound state of some point-like constituents, quarks for example. Then we will have to work with *photon–quark* interaction amplitudes; the proton charge will be simply given by the sum of quark charges,

$$e_p = (e_1 + e_2 + e_3) + \cdots .$$

So essentially we would have an approach rather similar to the previous one but at the level of constituents. One way or another, we cannot move away from the field-theoretical concept of point-like interaction if we intend to keep things under control.

4.1.2 Magnetic moment

Let us consider a small momentum transfer and keep the first power of q in the expression for the vertex (4.3):

$$\Gamma_\mu \simeq e_p(\gamma_\mu + b(0)\sigma_{\mu\nu} q^\nu) . \tag{4.5}$$

Here we have extracted from the vertex the renormalized charge e_p and set $a(0) = 1$.

What is the physical meaning of the linear term in (4.3)? The Dirac vertex γ_μ contains interaction with the charge as well as with the *magnetic moment*. Using the identity

$$\bar{u}(k')(k + k')_\mu u(k) = \bar{u}(k')\left[2m\gamma_\mu + \sigma_{\mu\nu}q^\nu\right]u(k),$$

we may rewrite the amplitude of electron scattering, $A^\mu\,\bar{u}(k')\,\gamma_\mu\,u(k)$, as

$$A^\mu\frac{(k + k')_\mu}{2m} \cdot \bar{u}(k')u(k) \;-\; A^\mu \cdot \bar{u}(k')\frac{\sigma_{\mu\nu}q^\nu}{2m}u(k).$$

With the electron at rest ($\mathbf{k} = 0$), the first term does not contribute to scattering in a magnetic field: $A_0 = 0$, $\mathbf{A} \cdot \mathbf{q} = 0$ ($\mathbf{div\,A}(x) = 0$). The second contribution survives and describes an interaction with the magnetic moment of a spin-$\frac{1}{2}$ charge.

The linear term from the proton (4.5) adds up with the Bohr *magneton* $e/2m$, resulting in the magnetic moment of the proton

$$\mu_p = -\frac{e_p}{2m_p}\left[1 - 2m_p b(0)\right].$$

In QED we have seen that the magnetic moment of the electron acquires an 'anomalous' contribution $\alpha_{em}/2\pi$ (Schwinger, 1962). This was a consequence of the internal structure of the electron that became apparent when radiative corrections had been taken into consideration. The magnetic moment of the proton also changes on account of the interaction, and not only electromagnetically, but of the strong one in the first place.

While the electromagnetic charge renormalizes in a universal way, independently of the proton's nature, the value of its magnetic moment depends crucially on the concrete properties of the strong interaction.

4.2 Form factors

The physical meaning of the functions $a(q^2)$ and $b(q^2)$ becomes apparent from the non-relativistic analogy.

Consider the scattering of a charge off an extended target, e.g. an atom. In quantum mechanics, the scattering amplitude is given by the Fourier integral of the Coulomb potential $V(r)$ which, in turn, one obtains by integrating the charge density $\rho(r')$ over the volume of the target:

$$f \propto \int d^3r\, V(r)\, e^{-i\mathbf{q}\cdot\mathbf{r}}, \qquad V(r) = \int d^3r'\frac{e^2}{|\mathbf{r} - \mathbf{r}'|}\rho(r').$$

Combining the two expressions one arrives at

$$f \propto \frac{e^2}{|\mathbf{q}|^2} \cdot F(q^2), \qquad F(q^2) \equiv \int d^3r' \, \rho(r') \, e^{-i\mathbf{q}\cdot\mathbf{r}'}, \qquad (4.6)$$

with F the *electric form factor* of the target atom. Analogously, our functions a and b are the proton form factors characterizing, correspondingly, the distribution of the *charge* and that of the electric *current* inside the proton.

In the case of a spinless object there is no preferred direction, and the magnetic moment is identically zero. Therefore, if we substitute a π-meson for a proton we will have only one (electric) form factor. Now the only vector at our disposal that satisfies the current conservation condition $q^\mu \Gamma_\mu = 0$ (for $p^2 = p'^2 = \mu^2$) is the sum of the pion momenta, $(p + p')_\mu$; hence, the photon–pion interaction vertex contains a single structure:

$$\Gamma_\mu^\pi = e \cdot a(q^2)(p + p')_\mu.$$

What else can be said about form factor(s)?

Let us look at the analytic properties of a form factor as a function of momentum transfer q^2. It is clear that the form factor is real for $q^2 < 0$ since no real process may occur in the intermediate (t-channel) state. At the same time, for positive virtuality above the two-pion threshold, $q^2 > 4\mu^2$, the form factor becomes complex-valued due to the $\gamma^* \to \pi^+\pi^-$ transition.

The relativistic theory allows us to link the scattering form factor to a completely different physical phenomenon, namely an *annihilation* of a pair of leptons into a pair of hadrons,

$$e^+e^- \;\to\; N\bar{N} \qquad \text{or} \quad e^+e^- \;\to\; \pi^+\pi^-.$$

The very same function that describes an internal electromagnetic structure of a pion in the $e\pi$ scattering process in the region $q^2 < 0$, at $q^2 > (2\mu)^2$ determines the cross section of e^+e^- annihilation into two pions!

4.2.1 Analytic properties of pion form factor

Knowing that the form factor is an analytic function of q^2, I can write

$$a(q^2) = 1 + \frac{q^2}{\pi} \int_{4\mu^2}^{\infty} \frac{dQ^2 \, \mathrm{Im}\, a(Q^2)}{Q^2(Q^2 - q^2)}. \qquad (4.7)$$

I chose to write down the dispersion relation with *one subtraction* in order to exploit the knowledge of the normalization $a(0) = 1$ (which only helps the integral to converge faster).

As we know, the imaginary part is directly related to cross sections of real processes. The latter are subject to various restrictions which then must affect the form factor itself. Let us examine how serious these restrictions actually are by taking the pion form factor $a(q^2)$ as an example:

$$2\,\mathrm{Im}\,\Gamma_\mu = \sum_n \quad \gamma^* \qquad\qquad . \tag{4.8}$$

This equation tells us that the dispersion theory provides us with a system of linear equations for 'form factors' $\gamma^* \to (n * \pi)$, where the rôle of the kernel is played by the pure strong interaction amplitudes $(n * \pi) \to (m * \pi)$.

Consider for simplicity the region $4\mu^2 < q^2 < 16\mu^2$ where the two pion unitarity relation holds:

$$2\,\mathrm{Im}\,a(q^2)(p_1 - p_2)_\mu = \quad \gamma^* \qquad \pi \qquad = \qquad q \qquad p_1' \quad p_1 \qquad p_2' \quad p_2 \qquad . \tag{4.9}$$

The kernel is the $\pi\pi \to \pi\pi$ scattering amplitude. Moreover, π is spinless ($s = 0$), while the total angular momentum of the $\pi\pi$ system, $\mathbf{J} = \boldsymbol{\ell} + \mathbf{s}$, must be equal the photon spin, $J = 1$. Hence, only one partial wave $\ell = 1$ of the $\pi\pi$ amplitude (P-wave) will contribute here.

Let us sketch how this selection occurs. The r.h.s. of (4.9) contains integration over the intermediate-state pion momentum:

$$\int \frac{d^4 p_1'}{(2\pi)^2}\, \delta_+(p_1'^2 - \mu^2)\delta_+(p_2'^2 - \mu^2)\,(p_1' - p_2')_\mu a(q^2) \cdot A^*(p_1', p_2'; p_1, p_2).$$

In the $\pi\pi$ centre-of-mass frame it reduces to the angular integral over the direction of the relative momentum $(p_1' - p_2')_\mu \implies 2\mathbf{p}_c'$ which multiplies the scattering amplitude $A^*(\mathbf{p}_c, \mathbf{p}_c')$. Writing down the partial wave expansion of the latter,

$$A^*(\mathbf{p}_c, \mathbf{p}_c') = \sum_\ell (2\ell + 1)f_\ell^*(q^2)P_\ell(\cos\Theta_{\mathbf{p}_c\mathbf{p}_c'}),$$

we observe that the only term with $\ell = 1$ survives the integration:

$$(2\ell + 1)\int \frac{d\Omega'}{4\pi}\, 2\mathbf{p}_c' \cdot P_\ell(\cos\Theta_{\mathbf{p}_c\mathbf{p}_c'}) = 2\mathbf{p}_c\delta_{\ell,1}.$$

This matches the structure of the l.h.s.,

$$(p_1 - p_2)_\mu \operatorname{Im} a(q^2) \quad \Longrightarrow \quad 2\mathbf{p}_c \operatorname{Im} a(q^2),$$

and the unitarity relation (4.9) takes the form

$$\operatorname{Im} a(q^2) = \tau a(q^2) f_1^*(q^2),$$

with τ the phase space volume (3.7b) of the two-pion state. Substituting the general solution of the two-particle unitarity condition (3.9)

$$f_\ell = \frac{1}{2i\tau} \left(e^{2i\delta_\ell} - 1 \right),$$

we have

$$\frac{a(q^2) - a^*(q^2)}{2i} = \tau \cdot \frac{-1}{2i\tau} \left(e^{-2i\delta_1} - 1 \right) a(q^2) \quad \Longrightarrow \quad \frac{a^*}{a} = e^{-2i\delta_1}. \quad (4.10)$$

The unitarity condition simply tells us that the *phase* of the pion form factor equals that of the $\pi\pi$ scattering amplitude. The origin of the complexity of the form factor lies in *re-interaction* between pions in the final state.

Above the four-pion threshold the situations gets more complicated. Nevertheless, for the sake of simplicity, let us suppose that the relation (4.10) holds for all q^2 values. Then I would be able to calculate the form factor straight away! To this end consider the function $F = \ln a(q^2)$ and write the corresponding dispersion relation,

$$F(q^2) = \frac{1}{\pi} \int \frac{dQ^2}{Q^2 - q^2} \delta_1(Q^2) + \text{'regular'}$$

with 'regular' marking a possible non-singular (analytic) piece. Then,

$$a(q^2) = P(q^2) \times \exp\left\{ \frac{1}{\pi} \int \frac{dQ^2}{Q^2 - q^2} \delta_1(Q^2) \right\},$$

with P a polynomial in q^2. The latter can be replaced by a constant if I suppose a good behaviour at $q^2 \to \infty$. Then, making use of $a(0) = 1$, I would finally predict the form factor from the knowledge of the $\pi\pi$ scattering phase:

$$a(q^2) = \exp\left\{ \frac{q^2}{\pi} \int_{4\mu^2}^{\infty} \frac{dQ^2}{Q^2(Q^2 - q^2)} \delta_1(Q^2) \right\}. \quad (4.11)$$

Unfortunately, literally this formula is incorrect since we have neglected many-particle channels which do essentially contribute at large Q^2. As a semi-quantitative estimate, however, (4.11) works reasonably well.

4.2.2 Pion radius and ρ meson

At large negative q^2 we expect hadron form factors to be falling fast since it is a truly point-like charge that can only give $a(q^2) = \mathcal{O}(1)$ in this limit. On the other hand, at large *positive* $q^2 \gg \mu^2$ it has to decrease too. This time, because a high-virtuality photon can produce many different multi-particle states, so that the probability of a given exclusive channel, $\gamma^* \to \pi\pi$, must fall. Therefore, $a(q^2)$ has to have somewhere a *maximum*.

How could this be? An analytic function exhibiting a maximum makes us think of a nearby singularity. We saw that our form factor has the pion scattering phase as its source (in other words, strong interaction of pions). Suppose there is a *resonance* in the strong $\pi\pi$ interaction amplitude at some $q^2 = M^2$. Then this resonance will drive the behaviour of the form factor. Effectively, we will be looking for the process of the $\gamma^* \to \pi\pi$ transition via a resonance state.

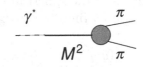

Substituting a resonance for $A_{\pi\pi}$ in the unitarity relation (4.9),

a rough guess for its contribution to the form factor would be

$$a(q^2) = \quad \propto \frac{1}{M^2 - q^2}. \qquad (4.12\text{a})$$

Strictly speaking, to build a realistic model I would have to analyze vertices and take into consideration their q^2 dependence away from the pole position. However, it suffices to invoke, once again, the restriction $a(0) = 1$ in order to reasonably fix the numerator in (4.12a):

$$a(q^2) \simeq \frac{M^2}{M^2 - q^2}. \qquad (4.12\text{b})$$

At small positive q^2, the slope of the q^2 dependence will tell me the *characteristic mass* the annihilation process goes through,

$$a(q^2) = 1 + \frac{q^2}{M^2} + \cdots . \qquad (4.12\text{c})$$

On the other hand, the same expansion can be carried out in the physical region of the scattering channel, $q^2 < 0$. Recall the non-relativistic expression for the charge form factor. Expanding (4.6) in $Q^2 = -q^2 > 0$

we get

$$a(Q^2) = \int d^3r \, \rho(r) \, e^{-i\mathbf{Q}\cdot\mathbf{r}} \simeq a(0) - \int d^3r \frac{(\mathbf{Q}\cdot\mathbf{r})^2}{2}\rho(r)$$

$$= 1 - \frac{Q^2}{2\cdot 3}\langle r^2\rangle.$$

(4.13a)

Here we have introduced the average squared radius of the distribution of the charge inside the pion,

$$\langle r^2 \rangle \equiv \int d^3r \, \rho(r) \cdot r^2,$$

(4.13b)

and used the spherical symmetry: $\langle r_z^2 \rangle = \frac{1}{3}\langle r^2 \rangle$.

Comparing the two expressions,

$$a(q^2) = 1 + \frac{q^2}{M^2} + \cdots \quad \Longleftrightarrow \quad a(q^2) = 1 + \frac{q^2\langle r^2\rangle}{6} + \cdots,$$

(4.14)

we conclude that the charge radius is directly related to the mass of the resonance in the annihilation channel.

In reality the P-wave pion–pion scattering is indeed dominated by the resonance – a vector meson ρ with a mass $m_\rho \simeq 750$ MeV. Firstly, this tells us that the annihilation process $e^+e^- \to \pi^+\pi^-$ should show a prominent peak at $q^2 \simeq m_\rho^2$ (and it does) and, secondly, that the position of this peak determines the electromagnetic radius of the pion:

In the case of nucleons the situation is somewhat more complicated. First of all, the mass of the $N\bar{N}$ state ($q^2 \sim 4\,\text{GeV}^2$) is very large compared to the position of the $\pi\pi$ threshold ($q^2 \sim 0.1\,\text{GeV}^2$). This pushes the physical region of the $e^+e^- \to N\bar{N}$ process far away from the scattering channel and from the first singularity. Besides, unlike pions which 'resonated' in only one meson ρ, a nucleon is also linked to another vector meson ω (with isospin 0). In reality the nucleon form factor behaves rather like $[M^2/(M^2 - q^2)]^2$, which may result from some destructive interference between various meson exchanges.

4.3 Isotopic structure of electromagnetic interaction

It is obvious that the electromagnetic interaction does not respect isotopic invariance:

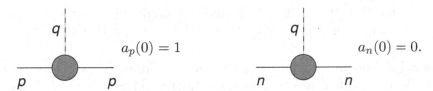

$$a_p(0) = 1 \qquad a_n(0) = 0.$$

Let us generalize the Dirac spinor U describing the proton wave function in (4.2), to represent the isotopic doublet of nucleons,

$$U = \begin{pmatrix} p \\ n \end{pmatrix}. \tag{4.15}$$

Then, with the help of Pauli matrices τ_i ($i = 1, 2, 3$) and of the unit matrix in the 2×2 isotopic space, we can represent the electromagnetic vertex simultaneously for the proton and the neutron as

$$\Gamma_\mu = \bar{U} \left(\Gamma_\mu^{(0)} \cdot \mathbf{I} + \Gamma_\mu^{(1)} \cdot \tau_3 \right) U. \tag{4.16}$$

(We could not use τ_1 and τ_2 here since they would transfer a proton into a neutron, $p + \gamma \to n$, which we rather would not do.) Projecting onto the proton and neutron states, we get

$$\Gamma_\mu^{(0)} + \Gamma_\mu^{(1)} = |p\rangle \Gamma_\mu \langle p|, \quad \Gamma_\mu^{(0)} - \Gamma_\mu^{(1)} = |n\rangle \Gamma_\mu \langle n|.$$

In particular, at $q^2 = 0$ this will give us

$$\Gamma_\mu^{(0)}(0) - \Gamma_\mu^{(1)}(0) = 0.$$

This isotopic beautification does not seem to bring us much profit. Still, from the point of view of the crossing channel, $\Gamma^{(0)}$ and $\Gamma^{(1)}$ may happen to acquire more fundamental meaning than the proton and neutron vertices themselves.

Look at the case of pions. Since pion π_α is a triplet in the $\mathbf{T}$ space ($\alpha, \beta = 1, 2, 3$), we have now *three* independent diagonal matrices that can be used to construct the $\pi\pi\gamma$ interaction vertex:

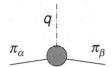

$$\begin{pmatrix} \pi^+ \\ \pi^0 \\ \pi^- \end{pmatrix}^\dagger \left[\Gamma_\mu^{(0)} \cdot \mathbf{I} + \Gamma_\mu^{(1)} \cdot \mathbf{T}_3 + \Gamma_\mu^{(2)} \cdot \mathbf{T}_3^2 \right] \begin{pmatrix} \pi^+ \\ \pi^0 \\ \pi^- \end{pmatrix}. \tag{4.17}$$

The term proportional to $\Gamma_\mu^{(k)}$ corresponds to isospin $\mathbf{T} = k$. For a $\pi\pi$ system the possibilities are: a scalar ($\mathbf{T} = 0$), a vector ($\mathbf{T} = 1$) and a tensor ($\mathbf{T} = 2$) in the isotopic space.

It becomes clear that electromagnetic form factors of a particle belonging to some huge isotopic multiplet *could belong* to high-rank tensor structures in the $\mathbf{T}$ space. In reality, however, photons seem to couple only with the $\mathbf{T} = 0$ and $\mathbf{T} = 1$ channels (though the accuracy of this experimental finding is presently not very high). In particular, in (4.17) for pions $\Gamma_\mu^{(2)} \simeq 0$.

How might one understand this phenomenon? Suppose that all hadrons were built of *isotopic doublets*, $\mathbf{T} = \frac{1}{2}$, like quarks. Then it is *quarks* which participate in the electromagnetic interaction,

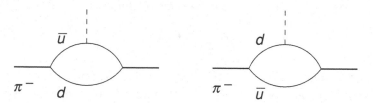

and we have only an iso-scalar and an iso-vector in the photon channel:

$$\Gamma_{\text{em}} \sim \mathbf{I} + \boldsymbol{\tau}_3. \qquad (4.18)$$

It is important to bear in mind that, since strong interactions respect the isospin symmetry, whichever diagrams I include to account for interactions between quarks, coupling of the photon to π-meson will retain the scalar + vector isotopic structure of (4.18) (in the first order in α_{em}).

This is an example of how studying electromagnetic interactions may produce highly non-trivial hints about the nature of hadrons.

4.4 Deep inelastic scattering

At large virtual-photon momentum transfer, $-q^2 \gg 1\,\text{GeV}^2$, electromagnetic proton form factors decrease fast, as a large inverse power of q^2.

This may look surprising at the first sight. Indeed, from the point of view of the dispersion relation,

$$F(q^2) = \frac{1}{\pi} \int \frac{dQ^2\,\text{Im}\,F(Q^2)}{Q^2 - q^2},$$

in order to ensure a fast falloff of F at large negative q^2, the imaginary part must oscillate, and in a very specific way.

4.4.1 Parton concept

Let us look at the problem from a classical perspective instead. It is easy to make $F(q^2)$ fast falling if we take the charge density $\rho(r)$ in the NQM formula (4.6) to be a smooth non-singular function.

There exists another way to explain the smallness
of elastic scattering. Imagine the proton consisting
of some number of *weakly bound* point-like charges.

Then, physically, the scattering of an electron off
such a composite object will *always* be inelastic:
the photon will interact with one of the charges
and kick it out, 'destroying' the target proton.
This picture is similar to what happens to an
atom: when an electron is kicked off from a core

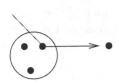

shell, leaving a vacancy, the excited atom 'decays' by emitting photons,
Auger electrons, ...

The elastic proton scattering will only be possible in the configurations
when all the quarks happen to be very close to
each other, at small distances $(\Delta r)^2 \sim 1/|q^2| \ll \langle r^2 \rangle$.
In this picture, the elastic channel suppression is due
to the smallness of the probability of such small-
distance configurations (equivalent to weak binding).

How to determine which answer is closer to reality? Since the q^2 behaviour of the elastic form factor does not help to discriminate the two
pictures, let us look at more complex – and more interesting – *inelastic*
processes. In the first picture the total inelastic cross section will be as
small as the elastic one, since in the scenario of a smooth charge density
ρ, a small-wavelength photon would simply find no-one to interact with,
elastically or otherwise. In the second scenario, an *inelastic* cross section
is not small at all. On the contrary, it is determined by the probability
for the photon to interact with one of the internal point-like constituents
of the proton.

4.4.2 DIS cross section

We will study the process called *deep inelastic lepton–proton scattering*
(DIS); the word 'deep' stresses the fact that the a highly virtual photon
with $Q^2 = -q^2 \gg \langle r^2 \rangle^{-1}$ penetrates *deep* into the proton's interior.

In non-relativistic quantum mechanics, electron scattering off an atom
having N electrons is given by the transition matrix element

$$\rho_{\mathbf{k},\mathbf{k}'}^{0,n} \propto \int d^3\mathbf{r}\, e^{i\mathbf{q}\cdot\mathbf{r}} \int \prod d^3\mathbf{r}_i\, \psi_n^*(\mathbf{r}_i) \sum_{k=1}^{N} \frac{e^2}{|\mathbf{r} - \mathbf{r}_k|}\, \psi_0(\mathbf{r}_i)$$

$$\sim \frac{e^2}{\mathbf{q}^2} \int \prod d^3\mathbf{r}_i\, \psi_n^*(\mathbf{r}_i) \sum_{k=1}^{N} e^{i\mathbf{q}\cdot\mathbf{r}_k}\, \psi_0(\mathbf{r}_i).$$

Here $\mathbf{k}$ and $\mathbf{k}'$ are the initial and final electron momenta, and ψ_0 and ψ_n mark the wave functions of the initial ground state and of the excited final state of the atom, respectively.

We are interested in the cross section summed over all possible final states of the excited atom, $n \leq n_0$. If n_0 is large, the product of the final-state wave functions that enters the expression for the cross section can be simplified using the (*almost*) completeness relation:

$$\sum_{n=0}^{n_0} |\psi_n(\mathbf{r}')\rangle\langle\psi_n(\mathbf{r})| \simeq \sum_{n=0}^{\infty} |\psi_n(\mathbf{r}')\rangle\langle\psi_n(\mathbf{r})| = \delta(\mathbf{r} - \mathbf{r}'). \qquad (4.19)$$

The cross section then reduces to

$$\frac{d\sigma}{d\mathbf{q}^2} \sim \frac{e^4}{\mathbf{q}^4} \int \prod d^3\mathbf{r}_i \, \psi_0^*(\mathbf{r}_i) \sum_{j,k=1}^{N} e^{i\mathbf{q}\cdot(\mathbf{r}_j - \mathbf{r}_k)} \psi_0(\mathbf{r}_i). \qquad (4.20)$$

When the photon wavelength is much smaller that the typical distance between the atomic electrons, $\mathbf{q}^2 \gg \langle \mathbf{r}_{jk}^2 \rangle^{-1}$, the *interference* terms with $j \neq k$ in (4.20) become negligible. We are left with the sum of N diagonal contributions:

$$\frac{d\sigma}{d\mathbf{q}^2} \sim \frac{e^4}{\mathbf{q}^4} \sum_{k=1}^{N} \psi_0^*(\mathbf{r}_i) \, \psi_0(\mathbf{r}_i) = \frac{e^4}{\mathbf{q}^4} \times N,$$

where we have used the normalization condition for the ground state wave function. Thus, the total inelastic eA cross section reduces to the sum of independent interactions of quasi-free individual electrons under two conditions, namely:

(1) the 'resolution' of the photon should be large enough to separate individual electrons inside the target atom, $\mathbf{q}^2 \gg R^{-2}$;

(2) the energy transferred to the atom should be sufficient to have a large enough number of excited states in order to employ the completeness relation (4.19).

Let us turn now to the relativistic theory and learn to write the corresponding cross section. The squared matrix element for the ep scattering process with the production of n particles is shown in Fig. 4.1. To obtain the cross section, we have to square the virtual photon–proton interaction amplitude A_μ bearing the photon index μ and convolute it with the corresponding electron scattering tensor $T_e^{\mu\nu}$. We also write down the phase space volume for $n + 1$ particles (the scattered electron k' and n produced final state hadrons with momenta $\{p_i\}$), and include the square of

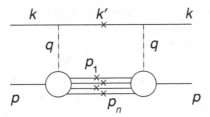

Fig. 4.1 Discontinuity of the forward photon–proton scattering amplitude.

the photon propagator, $(1/q^2)^2$, and the flux factor $J_{ep} \simeq 4(pk)$:

$$d\sigma = \frac{e^2}{J_{ep}} \int \frac{d^4k'}{(2\pi)^3} \delta_+(k'^2 - m_e^2) \sum_n \prod_{i=1}^n \left[\int \frac{d^4p_i}{(2\pi)^3} \delta_+(p_i^2 - m_i^2) \right]$$

$$T_e^{\mu\nu} \frac{1}{q^4} A_\mu(p, q, \{p_i\}) A_\nu^*(p, q, \{p_i\}) (2\pi)^4 \delta\left(p + k - k' - \sum_{i=1}^n p_i \right). \tag{4.21}$$

If we *average* over initial electron polarizations, the electron tensor $T_e'^{\mu\nu}$ is given by the expression

$$T_e^{\mu\nu} = \mathrm{Tr}\left[\frac{(\hat{k} + m_e)}{2} \cdot \gamma^\mu (\hat{k}' + m_e)\gamma^\nu \right] = 2(k^\mu k'^\nu + k'^\mu k^\nu) + g^{\mu\nu} q^2. \tag{4.22}$$

First comes an observation similar to the one we made when discussing the Chew–Low method (see (2.69)): we may replace the sum over produced hadrons by the imaginary part of the amplitude of the forward scattering:

$$\sum_n \underset{p_n}{\overset{p_1}{\bigcirc\!\!\!\bigcirc}} = 2\,\mathrm{Im} \underset{p}{\overset{q}{\bigcirc}}\overset{q}{\underset{p}{}} \equiv e^2 W^{\mu\nu}(pq, q^2). \tag{4.23}$$

The only peculiarity of this 'optical theorem' is that here one of the colliding objects is a virtual photon, so that the tensor $W_{\mu\nu}$ (bearing, once again, vector photon indices) depends on two variables rather than on the energy of the collision only, $W = W(pq, q^2)$. Due to conservation of the electromagnetic current, this tensor must be orthogonal to q^μ (and q^ν) and can be therefore represented as follows:

$$\frac{1}{2\pi} W_{\mu\nu} = \left(-g_{\mu\nu} + \frac{q_\mu q_\nu}{q^2} \right) W_1 + \frac{1}{M^2}\left(p_\mu - \frac{pq}{q^2} q_\mu \right)\left(p_\nu - \frac{pq}{q^2} q_\nu \right) W_2, \tag{4.24}$$

with W_1, W_2 known as the *structure functions*. (If the spin vector s of the initial particle is fixed, in $W_{\mu\nu}$ (as well as in $T_e^{\mu\nu}$) there appears an additional structure proportional to the anti-symmetric tensor $i\epsilon_{\mu\nu\alpha\beta} q^\alpha s^\beta$.)

The differential cross section takes the form

$$d\sigma = \frac{\alpha^2}{4\pi} \frac{1}{q^4} T_e^{\mu\nu} W_{\mu\nu} \frac{d^3 k'}{(pk)\, E'}.$$ (4.25)

Let us calculate the product of the tensors. The electron one, $T_e^{\mu\nu}$, also respects the current conservation; as becomes obvious from its expression equivalent to (4.22):

$$T_e^{\mu\nu} = (k + k')^\mu (k + k')^\nu - q^\mu q^\nu + g^{\mu\nu} q^2; \quad (k + k') \cdot q = k^2 - k'^2 = 0.$$

Therefore, we can drop the terms proportional to q_μ and/or q_ν from the hadron tensor when calculating the convolution:

$$(4.24) \implies -g_{\mu\nu} W_1 + \frac{p_\mu p_\nu}{M^2} W_2.$$

The convolution yields

$$\frac{1}{2\pi} T_e^{\mu\nu} W_{\mu\nu} = -4 \left[(kk') + q^2 \right] W_1 + \left[\frac{4(pk)(pk')}{M^2} + q^2 \right] W_2.$$ (4.26a)

Using $q^2 = (k' - k)^2 = 2m_e^2 - 2(kk')$,

$$\frac{1}{2\pi} T_e^{\mu\nu} W_{\mu\nu} \simeq 4(kk')\, W_1 + 2 \left[\frac{2(pk)(pk')}{M^2} - (kk') \right] W_2,$$ (4.26b)

where we have dropped the electron mass m_e as negligibly small.

In the laboratory system where the target proton is at rest, $p = (M, \mathbf{0})$, we introduce the electron scattering angle Θ and approximate $(kk') \simeq EE'(1 - \cos\Theta)$ in (4.26b) to get

$$\frac{1}{2\pi} T_e^{\mu\nu} W_{\mu\nu} \simeq 4EE' \left[2W_1 \cdot \sin^2\frac{\Theta}{2} + W_2 \cdot \cos^2\frac{\Theta}{2} \right].$$ (4.27)

Substituting (4.27) into (4.25), the cross section becomes

$$\frac{d\sigma}{d\cos\Theta\, dE'} = \frac{4\pi\alpha^2 E'^2}{q^4 M} \left[W_2 \cdot \cos^2\frac{\Theta}{2} + 2W_1 \cdot \sin^2\frac{\Theta}{2} \right].$$ (4.28)

By measuring the scattered electron momentum, one extracts the dependence of the structure functions $W_{1,2}(pq, q^2)$ by measuring the direction of the scattered electron momentum and the energy $\nu = E - E'$ transferred to the hadron system:

$$(pq) = M\nu, \quad q^2 \simeq -2EE'(1 - \cos\Theta).$$

Let us make a comparison with the *elastic* proton scattering. Apart from the form factor in the photon–proton vertex,

$$\Gamma_\mu \sim e_p\, \gamma_\mu \cdot \Gamma_{\text{el}}(q^2),$$

the calculation of the hadron tensor becomes identical to the lepton one:

$$\frac{W_{\mu\nu}^{\text{el}}}{2\pi} = \left[2(p_\mu p_\nu' + p_\mu' p_\nu) + g_{\mu\nu} q^2\right]\Gamma_{\text{el}}^2(q^2) \cdot \delta(2pq + q^2)$$
$$= \left[(p + p')_\mu (p + p')_\nu - q_\mu q_\nu + g_{\mu\nu} q^2\right]\Gamma_{\text{el}}^2(q^2)\,\delta(2pq + q^2). \tag{4.29}$$

The final hadron state consists now of the recoiling proton only, and the delta-function puts it on the mass shell: $p'^2 - M^2 = (p + q)^2 - M^2 = 0$. It is easy to extract the functions W_i corresponding to elastic scattering. Observing that

$$q = p' - p \implies p_\mu - \frac{pq}{q^2} q_\mu = \tfrac{1}{2}(p + p')_\mu;$$

by comparing (4.29) with the general decomposition (4.24) we derive

$$W_2^{\text{el}} = 4M^2\,\Gamma_{\text{el}}^2 \cdot \delta(2pq + q^2), \quad W_1^{\text{el}} = -\frac{q^2}{4M^2} W_2^{\text{el}}. \tag{4.30}$$

Rewriting the phase space element in (4.28) terms of invariants,

$$\frac{E'^2}{M} \cdot d\cos\Theta\, dE' = \frac{E'^2}{M} \cdot \frac{dq^2}{2EE'}\, dq_0 = \frac{pk'}{pk} \cdot dq^2 \frac{d(2pq)}{4M^2},$$

the differential cross section takes the form

$$\frac{d\sigma}{dq^2\, d(2pq)} = \frac{4\pi\alpha^2}{q^4} \cdot \frac{pk'}{pk} \cdot \left[\frac{W_2}{4M^2}\cos^2\frac{\Theta}{2} + \frac{W_1}{2M^2}\sin^2\frac{\Theta}{2}\right]. \tag{4.31}$$

When the energy of the incident electron is large, the scattering angle becomes very small, $\Theta^2 \simeq |q^2|M^2/(pk)^2 \propto |t|/s^2 \to 1$; in this limit

$$\frac{d\sigma^{\text{el}}}{dq^2} \simeq \frac{4\pi\alpha^2}{q^4} \frac{W_2^{\text{el}}}{4M^2}\, d(2pq) = \frac{4\pi\alpha^2}{q^4} \cdot \Gamma_{\text{el}}^2(q^2). \tag{4.32}$$

Thus, W_2^{el} is nothing but the square of the elastic proton form factor.

An *inelasticity* of the interaction in a general case can be characterized by a dimensionless variable ω which measures the invariant mass of the final hadron system in units of the momentum transfer $|q^2|$:

$$W^2 = (p + q^2) - M^2 = -q^2(\omega - 1), \quad \omega \equiv \frac{2(pq)}{-q^2} \geq 1.$$

In the elastic process, the invariant photon–proton energy is determined by the on-mass-shell condition $(p + q)^2 = 2pq + q^2 + M^2 = M^2$ corresponding *exactly* to $\omega = 1$. If we take ω not too close to unity, $\omega = \mathcal{O}(1)$, and keep it fixed while increasing $|q^2|$, the lepton–hadron interaction in this kinematics is called *deep inelastic scattering* (DIS). The inelasticity W^2 of such a process is proportional to the squared transferred momentum q^2 (the latter characterizing the 'hardness' of the process).

The inelastic cross section can be expressed then as

$$\frac{d\sigma}{dq^2\,d\omega} \simeq \frac{4\pi\alpha^2}{q^4} \cdot F_2(q^2,\omega), \tag{4.33}$$

where we have introduced the *scaling function* F_2 – an analogue of the squared form factor in (4.32):

$$F_2(q^2,\omega) = -\frac{q^2}{4M^2} W_2(pq, q^2); \quad F_2^{(\mathrm{el})}(q^2,\omega) = \Gamma_{\mathrm{el}}^2(q^2)\delta(\omega - 1).$$

The SLAC experiment has found that F_2 (and $F_1 = W_1$) becomes *independent* of q^2 starting from $|q^2| \sim 2 - 4\,\mathrm{GeV}^2$. This shows that the picture of a smooth charge distribution inside a proton cannot be correct, hence in such a case the 'inelastic form factor' would be falling with $|q^2|$ together with the elastic one.

On the contrary, the observed *Bjorken scaling* regime $F(q^2,\omega) \simeq f(\omega)$ perfectly fits the second picture: (4.33) tells us that the cross section of an inelastic ep process equals that of the Rutherford elastic scattering off a point-like particle. It is a point-like charge inside the proton – a quasi-free 'parton' – that takes an impact.

The question arises, can we say anything about the parton spin? Let us ask ourselves, why did we get two structure functions in the first place, not ten?

The virtual photon linking the lepton with the hadron block has three polarizations $(e^\lambda q) = 0$, two orthogonal to the scattering plane $\{p, q\}$, and one lying in it:

$$\frac{g_{\mu\nu} - q_\mu q_\nu/q^2}{q^2} = \sum_{\lambda=1}^{3} e_\mu^\lambda(q)e_\nu^{\lambda*}(q); \quad (e^\lambda p) = 0,\ \lambda = 1, 2.$$

The photon does not change its polarization in the cause of scattering:

$$M^{\lambda\lambda'} = e_\mu^\lambda W^{\mu\nu} e_\nu^{\lambda'*} = M^\lambda \delta_{\lambda\lambda'}.$$

Invoking (4.24) we see that the scattering cross section $\sigma^\perp$ of a *transversal* photon ($\lambda = 1, 2$) is determined by the structure containing the $g_{\mu\nu}$ tensor that is, by the function W_1, while the *longitudinal* (in-plane) one, $\sigma^\parallel$ ($\lambda = 3$), – by a definite linear combination of W_1 and W_2.

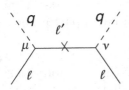

Imagine that some charged parton with momentum ℓ absorbs the virtual photon and scatters elastically. If the parton has spin *zero*, its electromagnetic vertex is proportional to the momentum, $\Gamma_\mu \propto (\ell + \ell')_\mu$, and we get

$$\frac{W_{\mu\nu}}{2\pi} \sim (\ell + \ell')_\mu \frac{\mathrm{Im}}{\pi} \left[\frac{1}{m^2 - \ell'^2 - i\epsilon}\right] (\ell + \ell')_\nu \sim \frac{4\ell_\mu\ell_\nu}{q^2} \delta\left(\frac{2\ell q}{q^2} + 1\right).$$

In this case the longitudinal photon interacts with a normal cross section, while the transverse polarizations would be power suppressed (provided the parton inside the proton has a limited transverse momentum):

$$\frac{\sigma^\parallel}{\sigma_{\text{point}}} = \mathcal{O}(1), \qquad \frac{\sigma^\perp}{\sigma_{\text{point}}} \sim \frac{\langle \ell_\perp^2 \rangle}{|q^2|} \ll 1.$$

On the other hand, for a spin-$\frac{1}{2}$ parton, the structure of $W_{\mu\nu}$ reproduces that of the electron tensor (4.22):

$$W_{\mu\nu} \propto \left(2\frac{\ell_\mu\ell'_\nu + \ell'_\mu\ell_\nu}{q^2} + g_{\mu\nu}\right) \delta\left(\frac{2\ell q}{q^2} + 1\right).$$

Here, on the contrary, the *longitudinal* polarization gets suppressed,

$$\ell^\mu \cdot \left(\ell_\mu\ell'_\nu + \ell'_\mu\ell_\nu - g_{\mu\nu}(\ell\ell')\right) \propto \ell^2 = m^2,$$

and we have a situation just *opposite* to the previous case

$$\frac{\sigma^\perp}{\sigma_{\text{point}}} = \mathcal{O}(1), \qquad \frac{\sigma^\parallel}{\sigma_{\text{point}}} \sim \frac{\langle \ell_\perp^2 \rangle + m^2}{|q^2|} \ll 1.$$

The experiment shows

$$\sigma^\perp \gg \sigma^\parallel$$

($\sigma^\parallel/\sigma^\perp \sim 1/5$), hinting at spin-$\frac{1}{2}$ partons (quarks?).

The overall impression is that the picture with quarks in the rôle of partons stands up to scrutiny.

Two phenomena have to be understood:

(1) the fact that the inelastic cross section is not small, $F = \mathcal{O}(1)$; and

(2) the Bjorken scaling, $F(q^2, \omega) \simeq f(\omega)$.

If the Bjorken scaling phenomenon is verified by future more detailed and more accurate experiments[†] we will be facing a serious puzzle! Actually,

---

† To hope this would *not* is a sin, because it is *beautiful*, in the first place.

the existence of this scaling challenges all we knew in 'the past'. While the observed scaling is apparently well explained by a naive parton model, it cannot hold from the field-theoretical point of view.

Indeed, imagine a proton consisting of 'points', subject to some QFT interaction. The Bjorken scaling emerges indeed in the toy model of scalar fields with the $\lambda\varphi^3$ interaction. Why? Because all the integrals in this theory converge in the ultraviolet momentum region, and at large $|q^2|$ all the corrections vanish leaving us with a point-like particle without form factor. Unfortunately, this is true not only for inelastic but for the elastic scattering as well.

In principle, we could get a falling elastic form factor back if we suppose that the fields φ may form bound states, in which case a falloff of $F_{\rm el}^2(q^2)$ at large q^2 would be explained by the decay of a bound state into its point-like constituents. In spite of such an 'improvement', this model still does not suit us, for two reasons. Firstly, one cannot build real hadrons out of spinless particles, and, secondly, we would have $\sigma^\parallel \gg \sigma^\perp$, in contradiction with experiment.

As soon as we introduce *fermions*, the theory seizes to be super-convergent and becomes (at best) renormalizable. Ultraviolet logarithmic divergences appear that have to be renormalized, etc. But this means that interaction corrections are *never small* for any, whichever large, q^2.

In a logarithmic quantum field theory, inside a physical particle there are always virtual exchanges with arbitrarily large momenta, exceeding the DIS momentum transfer: $|q_{\rm virt}^2| \gg |q^2|$. Under these circumstances, when probing a hadron with higher and higher 'resolution', we encounter more and more 'constituents'; hence, the exact Bjorken scaling regime simply *cannot hold*.

5

Strong interactions at high energies

5.1 The rôle of cross-channels

In this lecture we return to the study of the four-point amplitude $A(s,t)$. As we already know, it describes three different crossing reactions in the corresponding channels on the Mandelstam plane $s = (p_a + p_b)^2$, $t = (p_a - p_c)^2$, $u = (p_a - p_d)^2$:

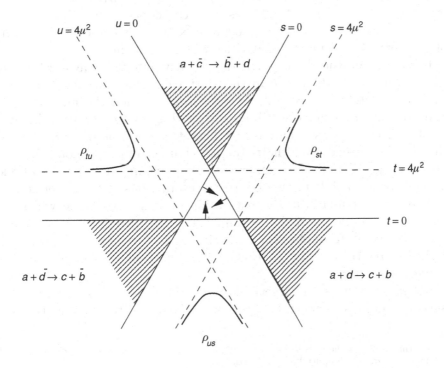

In Lecture 3 we examined analytic properties of the amplitude at small energies when, in fact, a small number of partial waves contributed. We discussed that the behaviour of $A(s,t)$ was governed by singularities of these partial waves. We have shown that unitarity led to the appearance of poles on unphysical sheets of the scattering amplitude – resonances – as well as to threshold branch cuts related to these poles. We have also discussed the physical origin of these singularities. Moreover, that was all we could possibly say about the strong interaction since perturbation theory is not applicable in principle (in terms of hadrons, $g^2/4\pi \sim 14$) and the Lagrangian is not known. Even if we knew the underlying Lagrangian, that would not have helped us in calculating amplitudes we were interested in.*

The method we have employed in Lecture 3 fails when we increase the interaction energy because the unitarity condition becomes prohibitively complicated due to the opening of new multi-particle production channels. In spite of this, at *very high* energies s, when the number of inelastic channels becomes large, an interesting new simplification arises in certain regions on the Mandelstam plane.

Take finite $|t|$ and large $s \gg m^2$ and consider *near-to-forward* two-particle scattering at small angles $\Theta_s \sim \sqrt{-t/s} \ll 1$. For $|t| \sim m^2$ it is the nearest singularities in the *t-channel* that will determine the behaviour of the amplitude, while u-channel singularities are irrelevant as they lie far away: $|u| \simeq s \gg m^2$. Analogously, we can expect a similar for *near-to-backward* scattering: $|u| \sim m^2$ fixed, $s \to \infty$, in which case a finite number of the nearest u-channel singularities will be relevant.

It is clear that the idea of extracting the asymptotic behaviour by means of the *analytic continuation* in the scattering angle variable (either t or u) is bound to be successful. At very large s when the number of the opened inelastic channels is immense, adding one or two would by no means affect the elastic amplitude. Therefore it seems natural to expect that in the $s \to \infty$ limit the t- (or u-) channel unitarity conditions will play a major rôle. This, together with the s-channel unitarity, will allow us to draw a possible picture of strong interactions at high energies.

The key instrument in the realization of this programme will be the *analytic continuation* of the t-channel unitarity condition valid at finite positive $t \sim m^2$ to unphysical scattering angles $\cos \Theta_t \sim s/t \to \infty$ which region is relatively close to that of the physical s-channel scattering, namely, $-t \sim m^2$.

Recall that it is the information coming from *cross-channels* that makes a major difference between the relativistic theory and the usual

---

* Now that we know the Lagrangian – that of QCD – this does not help us much in calculating hadron interaction amplitudes either. (ed.)

non-relativistic quantum mechanics. Still, to better understand the physical peculiarities of strong interactions it is often advisable to turn to non-relativistic theory. Therefore, prior to addressing the specific consequences of the relativistic nature of the theory we shall start our discourse by looking, at a qualitative level, at the main characteristics of an interaction using familiar quantum-mechanical notions. Namely, we will look at the possible behaviour of the interaction *strength* and the interaction *radius* in the asymptotic high-energy regime.

5.2 Qualitative picture of elastic scattering

Consider a four-point amplitude at $s \to \infty$. For the sake of simplicity we will restrict ourselves to the elastic scattering of spinless particles with equal masses μ. From general considerations it is clear that the scattering is concentrated mainly at small angles, in the near-to-forward direction.

At high energy many inelastic channels contribute to the unitarity condition,

$$\text{Im}_s A(s,t) \equiv A_1(s,t) = \frac{1}{2} \sum_n \quad \overset{A(s+i\varepsilon) \qquad A(s-i\varepsilon)}{\text{(diagram)}} \tag{5.1}$$

so that at $t = 0$ ($p_1 = p_1'$, $p_2 = p_2'$) the imaginary part of the forward amplitude is given by a sum of a large number of positive contributions. According to the optical theorem,

$$\text{Im}\, A(s,0) \simeq s\sigma_{\text{tot}}, \qquad s \gg \mu^2. \tag{5.2}$$

Therefore, if σ_{tot} is constant (or changing slowly) at high energies, which is what experiment tells us, then the forward amplitude increases like s. Let us see what such a behaviour corresponds to in the language of partial waves:

$$\text{Im}\, A(s,t) = \sum_{\ell=0}^{\infty} (2\ell+1)\, \text{Im}\, f_\ell(s) P_\ell(z), \tag{5.3a}$$

$$z \equiv \cos\Theta_s = 1 + \frac{2t}{s-4\mu^2}. \tag{5.3b}$$

At $t = 0$ we have $P_\ell(1) = 1$. Since from the unitarity condition (3.7) it follows that

$$0 \le \text{Im}\, f_\ell(s) \le 16\pi, \tag{5.4}$$

see (3.9), each term in (5.3a) is positive and bounded from above. This means that the growth of the amplitude $A(s,0) \propto s$ that we are looking for may come only from the increase of the number of terms $\ell_0(s)$ that contribute significantly to the series, $\ell < \ell_0(s)$.

From quantum mechanics we know that high-energy scattering off a finite-range potential with radius ρ_0 is of quasi-classical nature. Therefore we can introduce an impact parameter ρ by identifying

$$\ell = k_c\rho, \tag{5.5}$$

with k_c the cms momentum (3.8),

$$k_c = \frac{\sqrt{s - 4\mu^2}}{2} \simeq \frac{1}{2}\sqrt{s}.$$

Now, to define an interaction radius we equate

$$\ell_0(s) = k_c\rho_0. \tag{5.6a}$$

Since our interaction is *strong* it is natural to expect the partial waves with $\ell < \ell_0$ (that is $\rho < \rho_0$) to be *saturated*,

$$\mathrm{Im}\, f_\ell(s) = \mathcal{O}(1) \quad \text{for } \ell < \ell_0, \tag{5.6b}$$

and to be negligibly small for $\ell > \ell_0$ when, in the classical language, the projectile misses the target, $\rho > \rho_0$. Now we estimate the size of the imaginary part of the forward elastic amplitude by simply truncating the sum at $\ell \sim \ell_0 \gg 1$:

$$\mathrm{Im}\, A(s,0) \sim \ell_0^2 \sim s \cdot \rho_0^2. \tag{5.7}$$

Invoking (5.2) gives then a natural formula for the cross section,

$$\sigma_{\mathrm{tot}} \sim \rho_0^2. \tag{5.8}$$

In the first lecture we discussed the strong interaction characterized by a finite radius which is determined by hadron masses. In a relativistic theory (in marked contrast with non-relativistic quantum mechanics) the 'potential' depends in general on particle velocities. Therefore the interaction radius *may* vary with energy,

$$\rho_0 = \rho_0(s). \tag{5.9}$$

Strictly speaking, the notion of the interaction potential is inapplicable in relativistic theory. Therefore (5.6) in fact serves as a *definition* of the interaction radius through the number of saturated partial waves.

5.2.1 Forward scattering

In the case of $\Theta_s \neq 0$ ($t < 0$) terms on the r.h.s. of (5.1) enter with different phases and the resulting amplitude decreases fast with $|t|$ increasing: the elastic cross section has a sharp peak in the forward direction.

How could we estimate the range of angles the scattering amplitude is concentrated in? Since essential partial waves have $\ell \sim \ell_0 \gg 1$, characteristic scattering angles can be estimated as

$$\Theta < \Theta_0 \sim \frac{1}{\ell_0} \sim \frac{1}{k_c \rho_0} \ll 1. \tag{5.10}$$

The cms momentum transfer at high energies is

$$|\mathbf{q}| = \left| \mathbf{p}'_1 - \mathbf{p}_1 \right| \sim k_c \Theta_0 \sim \frac{1}{\rho_0}. \tag{5.11}$$

This means that in terms of Mandelstam variables the amplitude is concentrated in the region of finite momentum transfers, $|t| \sim 1/\rho_0^2$.

5.2.2 Backward peak

The qualitative expectation is that the scattering at *finite* angles is suppressed in the $s \to \infty$ limit holds provided the factor Im f_ℓ in the expansion (5.3a) is a smooth function of ℓ. That was the case in non-relativistic quantum mechanics (where ℓ enters analytically the centrifugal term of the Hamiltonian). In the relativistic theory, on the contrary, Im f_ℓ *oscillates* with ℓ and this leads to a new, interesting phenomenon – the appearance of the second narrow peak in the *backward* direction ($\pi - \Theta \ll 1$).

To single out this oscillating behaviour of Im f_ℓ we need to employ analytic properties of $A(s,t)$. To this end it is convenient to express the partial wave $f_\ell(s)$ defined by (3.10) in terms of the function

$$Q_\ell(z) \equiv \frac{1}{2} \int_{-1}^{1} \frac{dz' \, P_\ell(z')}{z - z'}, \tag{5.12}$$

representing the second solution of the Legendre equation, regular at infinity:

$$Q_\ell(z) \overset{|z| \to \infty}{\sim} z^{-\ell-1}.$$

From (5.12) it immediately follows that $Q_\ell(z)$ has a logarithmic branch cut between -1 to $+1$. Discontinuity over the cut returns the P_ℓ function:

$$Q_\ell(z + \imath\epsilon) - Q_\ell(z - \imath\epsilon) = -\imath\pi P_\ell(z), \quad -1 < z < 1. \tag{5.13}$$

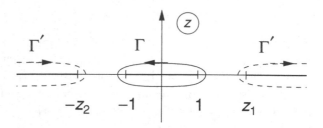

Fig. 5.1   Change of the integration contour $\Gamma \to \Gamma'$ in the representation (5.14) for the partial-wave amplitude.

Using this fact we can replace integration over real z by a contour integral

$$\int_{-1}^{1} dz\, P_\ell(z)\cdots = \frac{1}{\pi i} \int_\Gamma dz\, Q_\ell(z)\cdots,$$

where Γ runs clock-wise embracing the interval $[-1,1]$ as displayed in Fig. 5.1. This allows us to represent the partial-wave amplitude as

$$f_\ell(s) = \int_{-1}^{1} dz\, P_\ell(z) A(s,z) = \frac{1}{2\pi i} \int_\Gamma dz\, Q_\ell(z) A(s,z). \tag{5.14}$$

We took the liberty of replacing the second argument of the amplitude t by the cosine of the scattering angle (5.3b); the two variables, when s is kept fixed, are simply proportional. Now we look at the analytic features of the amplitude. As a functions of t it has two unitarity cuts corresponding to the opening of t- and u-channel thresholds, $t \geq t_{\min}$ and $u \geq u_{\min}$, respectively. In the z plane they translate into the cuts

$$z_1 < z < +\infty; \qquad z_1 = 1 + \frac{2 \cdot t_{\min}}{s - 4\mu^2}; \tag{5.15a}$$

$$-\infty < z < -z_2; \qquad z_2 = 1 + \frac{2 \cdot u_{\min}}{s - 4\mu^2}. \tag{5.15b}$$

(In our toy model where all particles are identical, $t_{\min} = u_{\min} = (2\mu)^2$.) Since Q_ℓ falls on the large circle, $|z| \to \infty$, we can deform and replace Γ by another contour Γ' which runs around the left and right cuts as shown in Fig. 5.1. This gives us

$$f_\ell(s) = \frac{1}{\pi} \int_{z_1}^{\infty} dz\, Q_\ell(z) A_3(s,z) + \frac{1}{\pi} \int_{-z_2}^{-\infty} dz\, Q_\ell(z) A_2(s,z), \tag{5.16}$$

where $A_3 = \operatorname{Im}_t A$ and $A_2 = \operatorname{Im}_u A$ denote discontinuities ('imaginary parts') of the amplitude $A(s,t)$ in the t and u channels. Using the relation

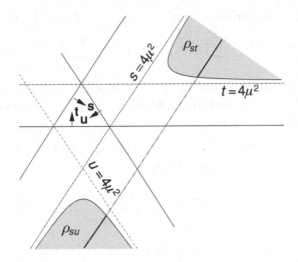

Fig. 5.2 Integration over double spectral functions ρ_{st} and ρ_{su} determining the imaginary part of the partial wave, $\operatorname{Im} f_\ell(s)$, in (5.20).

(valid for integer ℓ)

$$Q_\ell(-z) = (-1)^{\ell+1} Q_\ell(z), \tag{5.17}$$

we may rewrite (5.16) in the following form ($z_u = -z$):

$$f_\ell(s) = \frac{1}{\pi} \int_{z_1}^\infty dz\, Q_\ell(z) A_3(s,z) + \frac{(-1)^\ell}{\pi} \int_{z_2}^\infty dz_u\, Q_\ell(z_u) A_2(s,-z_u). \tag{5.18}$$

Evaluating the imaginary part of the partial wave, we obtain

$$\operatorname{Im} f_\ell = \operatorname{Im} f_\ell^{\text{right}} + (-1)^\ell \operatorname{Im} f_\ell^{\text{left}}, \tag{5.19}$$

where

$$\operatorname{Im} f_\ell^{\text{right}}(s) = \frac{1}{\pi} \int_{z_1}^\infty dz\, Q_\ell(z) \rho_{st}(s,t(z)), \tag{5.20a}$$

$$\operatorname{Im} f_\ell^{\text{left}}(s) = \frac{1}{\pi} \int_{z_2} dz\, Q_\ell(z) \rho_{su}(s,u(-z)). \tag{5.20b}$$

It is worth noticing that the imaginary part of the partial wave is determined by discontinuities across the Landau singularities – spectral functions ρ_{st} and ρ_{su} as shown in Fig. 5.2. Contributions of the right and left cuts in the t-plane (5.20) are smooth functions of ℓ. The oscillating factor $(-1)^\ell$ is explicitly written in (5.19) so now we are ready to return to the partial-wave expansion (5.3a) and scan through all angles to see what happens to the amplitude.

$\theta \simeq 0$. In the forward direction we have $z \simeq 1$ and $P_\ell(1) = 1$.

$$A_1(s, z=1) = \sum_\ell (2\ell + 1) \operatorname{Im} f_\ell^{\text{right}} + \sum_\ell (-1)^\ell \cdot (2\ell + 1) \operatorname{Im} f_\ell^{\text{left}}.$$

The contribution of the left cut is small (of the order of a single partial wave) and the large contribution $\propto s$ is coming from the right cut.

$\theta \simeq \frac{\pi}{2}$. For large angles $z \simeq 0$ so that $P_{2n+1}(0) \simeq 0$ and, for large n, $P_{2n}(0) \simeq \frac{2(-1)^n}{\sqrt{\pi n}}$. We obtain

$$A_1(s, z=0) \simeq \sum_{\ell=2n} (-1)^n \cdot \frac{2(4n+1)}{\sqrt{\pi n}} \left(\operatorname{Im} f_{2n}^{\text{right}} + \operatorname{Im} f_{2n}^{\text{left}} \right).$$

The series is oscillating, resulting in $A_1 = \mathcal{O}(1) \ll s$.

$\theta \simeq \pi$. For backward scattering, $z \simeq -1$, $P_\ell(-1) = (-1)^\ell$. Here

$$A_1(s, z=-1) = \sum_\ell (-1)^\ell \cdot (2\ell + 1) \operatorname{Im} f_\ell^{\text{right}} + \sum_\ell (2\ell + 1) \operatorname{Im} f_\ell^{\text{left}}$$

and, contrary to NQM, we have again a same-sign series which originates this time from the *left* cut of the relativistic amplitude.

The qualitative behaviour of the amplitude as a function of the scattering angle is shown in Fig. 5.3.

If particles are *identical* then 'forward' and 'backward' directions are indistinguishable so that the scattering amplitude becomes obviously symmetric, $A(\Theta) = A(\pi - \Theta)$, both in relativistic and non-relativistic theories. However, if participating particles are different, then from the point of view of NQM this situation looks totally bizarre.

Why would a 180° scattering – the full reflection of particle momenta – be profitable? In NQM, to encourage backward scattering one would have

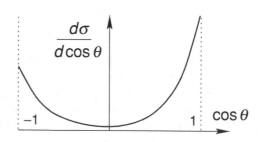

Fig. 5.3 Angular dependence of relativistic two particle scattering.

to organize a head-on collision whose probability is falling with energy as

$$\sigma \propto \lambda^2 \sim 1/s \qquad (5.21)$$

(with λ the cms wavelength $\lambda \sim 1/k_c$). The answer lies in the following consideration.

In a relativistic theory the 'potential' may carry quantum numbers. Therefore there is another possibility as an alternative to the large momentum transfer: rather then exchanging *momenta*, colliding particles may swap their *identities* instead. This means that by nature the backward scattering phenomenon is not different from diffractive scattering which is characterized by finite momentum transfer, but accompanied by an exchange of quantum numbers!

Have we ever met such a phenomenon before? In fact we have. Recall the Compton scattering in QED. In the leading order of perturbation theory we had two contributing Feynman diagrams,

In the $s \to \infty$ limit the first amplitude becomes negligible. It corresponds to interaction in a single partial wave – a head-on collision – so that the qualitative estimate (5.21) applies. At the same time the second amplitude describes a peripheral interaction with finite momentum transfer $|u| \sim m_e^2$ so that high-energy electron–photon scattering occurs mostly backwards ($s \simeq |t| \gg |u|$).

5.3 Analyticity of elastic amplitude and interaction radius

Let us study the ℓ-dependence of partial waves $f_\ell(s)$. We know that in the physical region of the s-channel $A(s,t)$ has no singularities. Therefore the partial-wave expansion series

$$A(s,t) = \sum_{\ell=0}^{\infty}(2\ell+1)f_\ell(s)P_\ell(z) \qquad (5.22)$$

must be absolutely converging, together with all its derivatives. To ensure such a regularity, f_ℓ must decrease fast with ℓ in the $\ell \to \infty$ limit.

5.3.1 f_ℓ at large ℓ

At the first sight a power falloff would seem to be sufficient since at large ℓ

$$P_\ell(z) \simeq J_0(\ell\Theta) \simeq \sqrt{\frac{2}{\pi\ell}} \cos\left(\ell\Theta - \frac{\pi}{4}\right); \tag{5.23}$$

$$\ell \gg 1, \ \Theta \ll 1, \quad \ell\Theta = \mathcal{O}(1).$$

We know, however, that the series must stay convergent in the unphysical region of positive t as well, up to $t = t_0$ where the first t-channel singularity is positioned (for example, $t_0 = 4m_\pi^2$ for $\pi\pi \to \pi\pi$ scattering). According to (5.3b), $t > 0$ corresponds to $z > 1$ where the scattering angle is imaginary,

$$z = \cos\Theta = 1 + \frac{2t}{s - 4m_\pi^2} = \cosh\chi, \quad \Theta = i\chi, \tag{5.24}$$

and the Legendre polynomials start to grow exponentially with ℓ:

$$P_\ell(z) \sim e^{i\ell\Theta} + e^{-i\ell\Theta} \sim e^{\ell\chi(t,s)}. \tag{5.25}$$

Up to $t \leq t_0$, this increase has to be damped by the fall-off of partial waves. Consequently,

$$f_\ell(s) \stackrel{\ell \gg 1}{=} C(\ell, s) e^{-\ell\chi_0}, \tag{5.26a}$$

where the factor C is non-exponential in ℓ and

$$\cosh\chi_0 \equiv 1 + \frac{2 \cdot t_0}{s - 4m_\pi^2}. \tag{5.26b}$$

In the $s \to \infty$ limit we have

$$\cosh\chi_0 \simeq 1 + \frac{\chi_0^2}{2} \to 1, \quad \chi_0 \simeq \sqrt{\frac{4t_0}{s}} \simeq \frac{\sqrt{t_0}}{k_c}. \tag{5.27}$$

In terms of the impact parameter (5.5), $\rho = \ell/k_c$, the large-ℓ asymptotic regime (5.26) takes the form

$$f_\ell(s) \implies f(\rho, s) = C(\rho, s) e^{-\rho/r_0}, \qquad r_0 \equiv 1/\sqrt{t_0}. \tag{5.28}$$

In particular, for $\pi\pi$ scattering ($t_0 = 4m_\pi^2$) the condition (5.27) gives

$$\chi_0 \simeq \frac{2m_\pi}{k_c} \quad \text{and} \quad r_0 = \frac{1}{2m_\pi}.$$

We conclude that partial waves with large ℓ that correspond to impact parameters ρ exceeding the interaction radius, $\rho \gg \rho_0$, fall exponentially as $\exp(-\rho/r_0)$.

It is important to stress that the nature of the two parameters ρ_0 and r_0 is essentially different: the radius ρ_0 we have introduced in (5.6a) as a

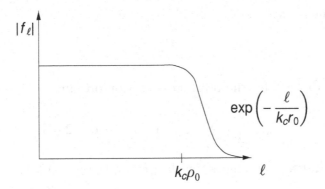

Fig. 5.4   Qualitative picture of the magnitude of partial waves.

measure of the of the *value* of characteristic angular momentum ℓ_0 below which partial waves are saturated, while r_0 determines the rate of the falloff of small partial waves with $\ell \gg \ell_0$, see Fig. 5.4.

5.3.2 Black disc

What can be said about partial-wave amplitudes below ℓ_0? It is impossible to answer this question without knowing the details of strong interaction dynamics. Nevertheless, it is straightforward to give an estimate *from above* for elastic, inelastic and total cross sections at high energies.

In Lecture 3 we discussed elastic unitarity for partial waves. Above the two-particle threshold the elastic unitarity condition (3.7) generalizes as

$$\operatorname{Im} f_\ell = \tau |f_\ell|^2 + \Delta_\ell, \tag{5.29}$$

where $\Delta_\ell \geq 0$ accounts for contribution of inelastic scattering channels: $2 \to 3$, $2 \to 4$, etc. The general solution for partial waves reads

$$f_\ell(s) = \frac{1}{2i\tau(s)} \left[\eta_\ell(s)\, e^{2i\delta_\ell(s)} - 1 \right], \tag{5.30a}$$

where η_ℓ is the so-called elasticity parameter,

$$\eta_\ell^2 = 1 - 4\tau\Delta_\ell, \qquad 0 \leq \eta_\ell \leq 1. \tag{5.30b}$$

Elastic unitarity corresponds to

$$\eta_\ell = 1, \qquad \Delta_\ell \equiv 0. \tag{5.31a}$$

On the contrary, at very high energies when inelastic channels dominate the unitarity condition it is natural to expect

$$\eta_\ell \simeq 0, \qquad \Delta_\ell \simeq \Delta_{\max} \simeq 4\pi. \tag{5.31b}$$

Invoking the optical theorem (5.2) we have

$$\sigma_{\text{tot}} = \frac{1}{s} \sum_\ell (2\ell + 1) \, \text{Im} \, f_\ell \tag{5.32}$$

and, using (5.30a), for the total cross section obtain

$$\sigma_{\text{tot}} \simeq \frac{1}{2\tau s} \sum_\ell^{\ell_0} (2\ell + 1) \left[1 - \eta_\ell \cos(2\delta_\ell) \right]. \tag{5.33a}$$

Then, integrating over angles the elastic amplitude squared, it is straight-forward to derive

$$\sigma_{\text{el}} \simeq \frac{1}{4\tau s} \sum_\ell^{\ell_0} (2\ell + 1) \left[1 - 2\eta_\ell \cos(2\delta_\ell) + \eta_\ell^2 \right]. \tag{5.33b}$$

Finally, from $\sigma_{\text{tot}} = \sigma_{\text{el}} + \sigma_{\text{in}}$ we get the total *inelastic* cross section,

$$\sigma_{\text{in}} \simeq \frac{1}{4\tau s} \sum_\ell^{\ell_0} (2\ell + 1) \left[1 - \eta_\ell^2 \right]. \tag{5.33c}$$

The maximal value of $\text{Im} \, f_\ell$ in (5.32) can be estimated as

$$[\text{Im} \, f_\ell(s)]^{\text{max}} = \frac{1}{2\tau(s)} \simeq 8\pi, \tag{5.34}$$

where we dropped the exponential term in (5.30a): it is unrealistic to expect that all scattering phases with large ℓ are artificially adjusted so that this oscillating contribution will matter. This gives the upper boundary

$$\sigma_{\text{tot}} \leq [\sigma_{\text{tot}}]^{\text{max}} \simeq \frac{8\pi}{s} \cdot \ell_0^2 \simeq 2\pi\rho_0^2. \tag{5.35}$$

If we accept the natural *maximal inelasticity* hypothesis (5.31b), then the relations (5.33) give

$$\sigma_{\text{el}} = \sigma_{\text{in}} = \tfrac{1}{2}\sigma_{\text{tot}} = \pi\rho_0^2. \tag{5.36}$$

This corresponds to the well-known quantum-mechanical picture of scattering off a black disc of radius ρ_0: half of σ_{tot} is inelastic and corresponds to the geometrical cross section of a fully absorbing disc (impact parameters $\rho \leq \rho_0$) while the other half is due to the elastic diffraction of the plane wave (at $\rho > \rho_0$) off the sharp edge of the disc.

Let us make a few additional remarks.

(1) The exponential falloff of partial waves at $\ell \gg 1$ is related to the absence of *massless particles* in the theory.

(2) The concrete value of r_0 is different for different reactions. For $\pi\pi$ scattering we had $r_0 = 1/2m_\pi$. For nucleon scattering we would have $t_0 = m_\pi^2$ and $r_0 = 1/m_\pi$ instead since there is a pion pole in the amplitude of the t-channel reaction $N\bar{N} \to \pi \to N\bar{N}$.

5.3.3 Behaviour of Im f_ℓ at large ℓ

With the help of (5.26) and the unitarity relation (3.7) we can verify that, as we have seen in Lecture 2, the behaviour of Im $A(s,t)$ is governed not by physical t-channel poles and/or thresholds but by the *Landau–Mandelstam singularities*.

Indeed, in the finite energy region of $\pi\pi$ scattering $4m_\pi^2 < s < 16m_\pi^2$ where inelastic channels are 'closed' ($\Delta_\ell \equiv 0$, $\eta_\ell \equiv 1$),

$$\text{Im} f_\ell = \tau |f_\ell|^2 \sim e^{-2\cdot\ell\chi_0}, \quad \ell \gg 1. \tag{5.37}$$

As a result, the series (5.3) for Im $A(s,t)$ will stay convergent *above* $t_0 = 4m_\pi^2$. The critical value of t at which Im A will develop a singularity will be determined not by $\chi(t,s) = \chi_0$ as for the amplitude A itself but by the condition $\chi(t,s) = 2 \cdot \chi_0$:

$$\cosh\chi = 2\cosh^2\chi_0 - 1; \ 1 + \frac{2t}{s - 4m_\pi^2} = 2\left(1 + \frac{2\cdot 4m_\pi^2}{s - 4m_\pi^2}\right)^2 - 1. \tag{5.38}$$

From (5.38) follows the relation

$$(s - 4m_\pi^2)(t - 16m_\pi^2) = 64m_\pi^2. \tag{5.39}$$

This is nothing but the Landau–Mandelstam equation describing the position of the singularity – one of the Karplus curves shown in Fig. 5.5 that corresponds to the double discontinuity ρ_{st} of the graph (5.40). Let me remind you that the 'box' graph that we studied in Lecture 2 (with two instead of four lines in the intermediate t-channel state) is absent for true pions since a pion cannot transfer into a $\pi\pi$ system due to G-parity conservation. (We

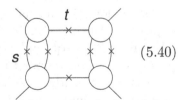

$$(5.40)$$

excluded another graph similar to (5.40) but with two vertical and four horizontal lines by having chosen $s < 16m_\pi^2$.)

Thus the behaviour of the *imaginary part* of the scattering amplitude, Im A, is determined by the first Landau–Mandelstam singularity $t_i(s)$ given by (5.39) rather than by t_0 as the amplitude A itself. At high energies the difference between $t_i(s)$ and t_0 is, however, washed away. Indeed, already for $s > (4m_\pi)^2$ a two-π state in the t channel becomes allowed and the corresponding Karplus curve on Fig. 5.5 appears. Since *all* Karplus

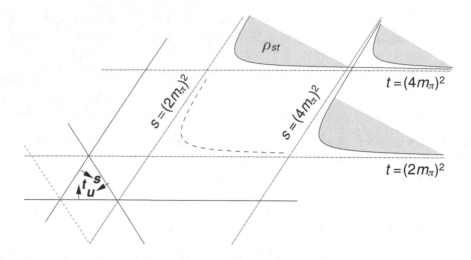

Fig. 5.5　Karplus curves for two- and four-pion states in s- and t-channel. The curve marked ρ_{st} corresponds to (5.39). The Landau–Mandelstam singularity shown by dashed line is absent for $\pi\pi$ scattering.

curves corresponding to two-pion exchange in the t-channel (and any number of particles in the s-channel) tend to the same value $t_i(s) \to (2m_\pi)^2 = t_0$ in the high-energy limit, we get

$$\text{Im}\, f_\ell \sim f_\ell \sim \mathrm{e}^{-\ell\chi_0} \simeq \mathrm{e}^{-2m_\pi\rho}, \quad s \to \infty.$$

5.4 Impact parameter representation

We will find it convenient to use the impact parameter representation in what follows. So let us look into the physical meaning of partial-wave amplitudes $f_\ell(s) = f(\rho, s)$ in the ρ-space.

For fixed $t = -\mathbf{q}^2$ and large s-scattering occurs at small angles $\Theta \simeq \sqrt{-t/s} \ll 1$. The momentum transfer $\mathbf{q}$ is then orthogonal to the cms momentum of incident pions, $\mathbf{q} \perp \mathbf{k}_c$, $q \simeq k_c\Theta$. For small angles the following approximation applies:

$$P_\ell(\cos\Theta) \simeq J_0(\ell\Theta).$$

Since in the partial-wave expansion large ℓ are essential, we can replace the sum in (5.22) by the integral,

$$A(s, q^2) \simeq \int 2\ell d\ell f_\ell(s) P_\ell(\cos\Theta) \simeq k_c^2 \int 2\rho d\rho f(\rho, s) J_0(q\rho), \qquad (5.41)$$

and make use of the integral representation

$$J_0(q\rho) = \int \frac{d\phi}{2\pi} e^{i(\mathbf{q}\cdot\boldsymbol{\rho})}, \qquad (\mathbf{q}\cdot\boldsymbol{\rho}) \equiv q\rho \cdot \cos\phi \qquad (5.42)$$

to arrive at a well-known quantum-mechanical formula

$$A(s, q^2) = \frac{k_c^2}{\pi} \int d^2\boldsymbol{\rho} \, e^{i(\mathbf{q}\cdot\boldsymbol{\rho})} f(\boldsymbol{\rho}, s). \qquad (5.43)$$

The amplitude (5.43) in NQM describes the following scattering process. Consider a plane wave propagating along the z-axis and hitting the target at $z = 0$. Right behind the target the wave function will be modified by the factor $\lambda(\rho)$ describing the absorption of the incident wave at a given impact parameter,

$$\psi_{\text{out}}(z, \boldsymbol{\rho})|_{z=+0} = \psi_{\text{in}}(z, \boldsymbol{\rho}) \cdot \lambda(\rho), \quad \psi_{\text{in}}(z, \boldsymbol{\rho}) = e^{ikz}. \qquad (5.44)$$

For example, for a totally absorbing sharp-edge target (a 'black disc' of radius R_0)

$$\lambda(\rho) = \vartheta(\rho - R_0).$$

On the other hand, the plane wave expansion of the final field that we observe at large distance from the target is given by

$$\psi_{\text{out}} = \psi_{\text{in}} \cdot \int d^2\mathbf{q} \, e^{i(\mathbf{q}\cdot\boldsymbol{\rho})} a(q), \qquad (5.45)$$

where $a(q)$ is the scattering amplitude. Comparing (5.45) at $z = +0$ with (5.44) and inverting the Fourier representation (5.43) we conclude that $f(\boldsymbol{\rho}, s)$ has the meaning of the impact parameter distribution of the field right behind the target.

5.5 Constant interaction radius hypothesis

We found above that the partial-wave amplitude at large ρ has a structure

$$f(\rho, s) = C(\rho, s) e^{-2m_\pi\rho} \qquad (5.46)$$

with C a non-exponential function of ρ. This factor may depend on its two variables in a rather complicated way so that the interaction radius may turn out to be energy-dependent, $\rho_0 = \rho_0(s)$. This is what happens in reality. However, before discussing seriously this phenomenon, it is natural to look at the consequences of the hypothesis of a constant radius:

$$\rho_0(s) = \text{const.}$$

5.5.1 Consequences of the hypothesis $\rho_0(s) = const$

It is easy to convince oneself that this hypothesis actually means that the C-factor in (5.46) *factorizes*:

$$C(\rho, s) = c(\rho\mu) \cdot h(s/\mu^2). \tag{5.47}$$

Indeed, if the functional dependence on ρ and s *did not* separate, the impact parameter distribution of the field would be changing with energy.

What are the main consequences of the hypothesis (5.47)? They are

$$A_{\mathrm{el}}(s, t) = ish(s) \cdot F(t), \tag{5.48a}$$

$$\sigma_{\mathrm{tot}} = \frac{\mathrm{Im}\, A(s, 0)}{s} = h(s)F(0), \tag{5.48b}$$

$$\frac{d\sigma_{\mathrm{el}}}{dt} = \frac{1}{16\pi} \left| \frac{A_{\mathrm{el}}(s, t)}{s} \right|^2 = \frac{\sigma_{\mathrm{tot}}^2}{16\pi} \cdot \left| \frac{F(t)}{F(0)} \right|^2. \tag{5.48c}$$

From (5.48c) we conclude that the constant radius implies an energy-independent *shape* of the differential elastic scattering cross section:

$$\frac{1}{\sigma_{\mathrm{tot}}^2(s)} \frac{d\sigma_{\mathrm{el}}(s, t)}{dt} = \mathrm{const}(s).$$

A constant radius is also consistent with the *Pomeranchuk theorem* (Pomeranchuk, 1958).

5.5.2 Pomeranchuk theorem

Consider the scattering of a particle a and its antiparticle $\bar{a}$ on the same target b. If the total cross sections are asympotically constant,

$$\lim_{s \to \infty} \sigma_{\mathrm{tot}}^{ab}(s) = \sigma_a \quad \text{and} \quad \lim_{s \to \infty} \sigma_{\mathrm{tot}}^{\bar{a}b}(s) = \sigma_{\bar{a}},$$

then

$$\sigma_a = \sigma_{\bar{a}}.$$

This is a non-trivial statement as it has nothing to do with charge conjugation invariance: $\sigma_{\mathrm{tot}}^{ab} = \sigma_{\mathrm{tot}}^{\bar{a}\bar{b}}$. The isotopic structure of ab- and $\bar{a}b$-scattering amplitudes may be absolutely different as, for example, is the case of pp and $p\bar{p}$ interactions. Experimentally, cross sections of particle and antiparticle interactions with a given target hadron differ remarkably at moderate energies but become practically equal already at energies s above few hundred GeV^2 (see below, Fig. 14.3 on page 378).

Let us sketch the idea of the proof. Discontinuities of the amplitude on the left and right cuts cannot be different if we want $A(s, t)$ to be "regular" at infinity. Indeed, let

$$\operatorname{Im} A_{ab}(s,0) \stackrel{s\to\infty}{=} \sigma_a s,$$

$$\operatorname{Im} A_{\bar{a}b}(u,0) \stackrel{u\to\infty}{=} \sigma_{\bar{a}} u.$$

(5.49)

If the cross sections are different, $\sigma_a \neq \sigma_{\bar{a}}$, then the *real parts* $\operatorname{Re} A$ corresponding to discontinuities (5.49),

$$A_{\text{right}}(s) \simeq \frac{\sigma_a s}{\pi} \ln(-s), \qquad A_{\text{left}}(s) \simeq \frac{\sigma_{\bar{a}} u}{\pi} \ln(-u), \qquad (5.50)$$

will not compensate one another:

$$A_{ab}(s) \stackrel{s\to\infty}{=} A_{\text{right}}(s) + A_{\text{left}}(-s) = i\sigma_a s + \frac{\sigma_a - \sigma_{\bar{a}}}{\pi} \cdot s \ln s,$$

$$A_{\bar{a}b}(u) \stackrel{u\to\infty}{=} A_{\text{right}}(-u) + A_{\text{left}}(u) = i\sigma_{\bar{a}} u + \frac{\sigma_a - \sigma_{\bar{a}}}{\pi} \cdot u \ln u.$$

(5.51)

The amplitudes (5.51) contradict the initial assumption of the theorem (namely, $\sigma_{\text{tot}} = \text{const.}$) as they produce

$$\sigma_{\text{el}}(s) \propto (\sigma_a - \sigma_{\bar{a}})^2 \ln^2 |s| \stackrel{|s|\to\infty}{\gg} \sigma_{\text{tot}} = \text{const.}$$

Does the Pomeranchuk theorem teach us anything about $f(\rho, s)$ and, specifically, about the interaction radius? In other words, to what extent does the asymptotic equality $\sigma_a = \sigma_{\bar{a}}$ depend on the hypothesis of the constant radius ρ_0?

Let us see how the logarithmic growth of $\operatorname{Re} A$, which is necessary for the theorem to be broken, could appear in principle:

$$\operatorname{Re} A(s,0) = \frac{s}{4\pi} \int d^2\boldsymbol{\rho} \, \operatorname{Re} f(\rho, s). \qquad (5.52)$$

From the unitarity condition (5.29) ($\tau \simeq 1/16\pi$ for $s \gg \mu^2$) we have

$$\operatorname{Im} f = \frac{1}{16\pi} |f|^2 + \Delta > \frac{1}{16\pi} |\operatorname{Re} f|^2,$$

which gives

$$\operatorname{Re} f(\rho, s) < \sqrt{16\pi \operatorname{Im} f(\rho, s)}. \qquad (5.53)$$

On the other hand, to violate the Pomeranchuk theorem we have to have

$$\operatorname{Re} f(\rho, s) \sim \operatorname{Im} f(\rho, s) \cdot \ln s \qquad (5.54)$$

in the essential integration region $\rho \lesssim \rho_0(s)$. Combining (5.53) and (5.54) produces

$$\operatorname{Im} f < \frac{\text{const}}{\ln^2 s}. \qquad (5.55)$$

Wishing to preserve the constancy of the total cross section,

$$\sigma_{\text{tot}} = \frac{1}{4\pi} \int d^2\rho \, \text{Im} \, f(\rho, s) = \rho_0^2 \cdot \frac{\langle \text{Im} \, f \rangle}{4}, \qquad (5.56)$$

and taking into consideration the inequality (5.55), we arrive at the necessity to abandon the constant radius, $\rho_0(s) \gtrsim c \ln s$. We shall learn soon that the radius *cannot grow faster* than $\ln s$. This means that only in the extreme case of the fastest possible growth of the interaction radius, $\rho_0(s) \propto \ln s$, the Pomeranchuk theorem may fail.

5.6 Possibility of a growing interaction radius

We discussed the case when the interaction radius, and thus the shape of the elastic peak, does not change with energy, the amplitude factorizes and the Pomeranchuk theorem holds. In the next lecture we shall formally demonstrate that the $\rho_0 = $ const regime is actually forbidden as it contradicts t-channel unitarity.

Let us ask ourselves, whether we can force the radius to grow with energy? What sort of physical processes might be responsible for that, at a qualitative level? Strangely enough, such a possibility does exist.

Recall that in perturbative language we have obtained the constant radius $\rho_0 \sim 1/2\mu$ by considering the nearest singularity in t of the amplitude; two-meson t-channel exchange gave us

$$\implies \quad A(s, \rho) \propto e^{-2\mu\rho}. \qquad (5.57)$$

5.6.1 Long-living fluctuations and the growing radius

Let us study the space–time structure of the simplest perturbative diagram corresponding to the processes (5.57):

$$x_1^0 < x_2^0 < y_2^0 < y_1^0. \qquad (5.58)$$

The process (5.58) can be 'spelled out' in time as follows. First, the projectile particle a experienced a virtual decay at the point x_1; one of its offspring at x_2 hit and excited the target b which, in its turn, decayed in

y_2, ... How far can a virtual particle migrate in the transverse plane? From the uncertainty relation, during the lifetime of the fluctuation,

$$\Delta t \sim \Delta E^{-1} \sim \mu^{-1}, \qquad (5.59a)$$

a virtual particle k_1 may shift in the transverse plane at a distance

$$|\mathbf{x}_{2\perp} - \mathbf{x}_{1\perp}| \sim \mu^{-1}. \qquad (5.59b)$$

Could we allow it to move farther than that so as to make the interaction radius growing? At the first sight this seems to be an easy thing to do: by minimizing the energy uncertainty we may increase the lifetime t in (5.59a) significantly. In the laboratory frame where the particle a is fast and has a very large momentum $p \equiv p_z \simeq s/2\mu \gg \mu$, virtual splitting $a \to 1 + 2$ introduces energy uncertainty

$$
\begin{aligned}
\Delta E &= E_{\text{interm}} - E_{\text{init}} = \sqrt{\mathbf{k}_1^2 + \mu^2} + \sqrt{\mathbf{k}_2^2 + \mu^2} - \sqrt{\mathbf{p}^2 + \mu^2} \\
&\simeq \frac{\mathbf{k}_{1\perp}^2 + \mu^2}{2k_{1z}} + \frac{\mathbf{k}_{2\perp}^2 + \mu^2}{2k_{2z}} - \frac{\mu^2}{2p} \simeq \frac{1}{2p}\left[\frac{\mu^2 + \mathbf{k}_\perp^2}{x(1-x)} - \mu^2 \right], \quad (5.60)
\end{aligned}
$$

where we used $|\mathbf{k}_\perp| \sim \mu \ll p$ and have introduced the decay momentum fraction x,

$$k_{1z} \equiv xp, \quad k_{1z} = (1-x)p \quad (x \sim 1 - x \sim 1).$$

The energy difference

$$\Delta E \sim \frac{\mu_\perp^2}{x(1-x)\,p}, \qquad \mu_\perp^2 \equiv \mathbf{k}_\perp^2 + \mu^2 = \mathcal{O}(\mu^2).$$

is minimal for $x \sim \frac{1}{2}$ and can be made extremely small at high energy, $\Delta E \propto \mu^2/p$. The corresponding fluctuation time gets Lorentz dilated:

$$\Delta t \sim \frac{x(1-x)p}{\mu_\perp^2} \gg \frac{1}{\mu}. \qquad (5.61)$$

Unfortunately this does not help us to achieve our goal: at high energies the decay angle decreases in the same proportion as the lifetime increases, so that the transversal displacement of the offspring remains finite,

$$|\Delta\boldsymbol{\rho}| \sim \Delta t\,|\mathbf{v}_\perp| \sim \frac{x(1-x)p}{\mu^2} \cdot \left| \frac{\mathbf{k}_\perp}{xp} \right| \sim \frac{1}{\mu},$$

the same as in (5.59b).

However, our exercise was not completely useless as we learned that at high energies long-living fluctuations may be constructed.

There is another problem with the process of (5.58). If our particle k_1 is point-like and interacts with the target 'head-on', its cross section is

proportional to the wavelength squared,

$$\sigma \sim \pi \lambda_1^2 \sim \frac{1}{k_{1z}^2} \simeq \frac{1}{(x\,p)^2} \propto \frac{1}{(x\,s)^2},$$

and falls very fast with s, unless we chose $x \sim \mu/p \ll 1$, which would take us back to the small lifetime $\Delta t \sim 1/\mu$ in (5.61)!

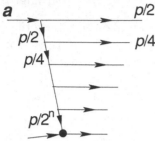

However, there is a way to reconcile a normal cross section with longevity. To this end we have to allow the virtual particle k_1 to decay further in order to sequentially degrade its energy. Now the particle that hits the target b has a momentum

$$p_n \sim \frac{p}{2^n}, \tag{5.62}$$

with n the number of decays of the projectile particle a. Here we supposed, for the sake of simplicity, that in each decay the longitudinal momentum of the parent is shared equally, $x \sim \frac{1}{2}$. Now, if we reach $p_n \sim \mu$, the interaction with the target will have a normal cross section $\sigma \sim \mu^{-2}$ typical for interaction of particles with small collision energy. To get there we will have to emit

$$n \simeq \frac{\ln(p/\mu)}{\ln 2} \tag{5.63}$$

particles. This would not have been easy if the interaction constant were small. If, on the contrary, we accept that the probability of $1 \to 2$ splittings is $\mathcal{O}(1)$ (which is not unnatural for strong dynamics), this would provide us with a realistic model of the interaction radius growing with energy.

Indeed, in the course of $n \sim \ln(s/\mu^2)$ decays a virtual particle experiences n moves in the transverse plane of the typical size $|\Delta\rho|$ each. If emission processes are strongly correlated as shown in Fig. 5.6(a), we can get the growth as fast as

$$\rho_0(s) \sim n(s) \cdot |\Delta\rho| \sim \frac{1}{\mu} \ln \frac{s}{\mu^2}. \tag{5.64a}$$

Fig. 5.6   Correlated (a) and uncorrelated (b) motion in the ρ-space.

If, on the other hand, sequential emissions are independent, as in Fig. 5.6(b), then our particle experiences a *Brownian motion* in the impact parameter space and, on the average, moves away from the origin by

$$\rho_0(s) \sim \sqrt{n(s) \cdot |\Delta\rho|^2} \sim \frac{1}{\mu}\sqrt{\ln\frac{s}{\mu^2}}. \tag{5.64b}$$

Thus we have constructed a viable picture of how the interaction radius may become energy-dependent from the point of view of interaction dynamics.

5.6.2 Growing radius and causality

How the possibility of a growing interaction radius can be envisaged from analytic properties of the amplitude? Suppose that the factor $C(\rho, s)$ in the expression for the asymptotic of partial waves (5.46) grew as a power of energy, $C(\rho, s) \propto s^N$, so that

$$f(\rho, s) \simeq \text{const}\, s^N \cdot e^{-2\mu\rho}, \qquad \rho \gg \mu^{-1}. \tag{5.65}$$

Recall now how the notion of the interaction radius was introduced in (5.6). Partial waves $f(\rho, s)$ are exponentially small at very large ρ. With ρ decreasing, the partial wave grows and eventually hits the saturation limit. The radius ρ_0 was defined as the value of ρ where it happens:

$$f(\rho_0, s) \simeq 8\pi.$$

Applying this definition to (5.65), we obtain

$$\rho_0(s) \simeq \frac{N}{2\mu}\ln s. \tag{5.66}$$

Formally speaking, we could have forced the radius to increase with s even faster. For example, if instead of a power we chose $C(\rho, s) \propto \exp(a\sqrt{s})$ this would have led to $\rho_0(s) \propto \sqrt{s}$.

However, such a steep growth of partial waves looks intrinsically dangerous: we know that this type of growth of the *full amplitude* $A(s, t)$ may result in the violation of causality.

It is time to reverse the logic. Let us *impose* the usual inequality that suffices to ensure causality,

$$|A(s, t)| \leq s^{N(t)} \qquad \text{for } s \to \infty, \tag{5.67}$$

and derive a possible growth of $\rho_0(s)$ that would dutifully respect it. Here $N(t)$ is limited in a finite interval of t (the number of necessary *subtractions* in the dispersion relation in s for a given t, see Lecture 2).

First we will take a rough model in which all partial waves with $\ell \leq \ell_0(s)$ are *saturated* while those with $\ell > \ell_0(s)$ are negligible. Then for the amplitude we have an estimate

$$|A(s,t)| \leq 16\pi \sum_{\ell=0}^{\ell_0(s)} (2\ell+1)|P_\ell(z)|\,. \qquad (5.68)$$

In the $t > 0$ region we have $z = \cosh\chi > 1$, Legendre polynomials increase exponentially with ℓ, see (5.25), so that (5.68) is dominated by the last term of the sum:

$$|A(s,t)| \sim \ell_0\, e^{\ell_0 \chi_0(s)}, \qquad (5.69a)$$

where, according to (5.27),

$$\chi_0(s) = \chi(t,s)|_{t=4\mu^2} \simeq \frac{2\mu}{k_c}, \qquad \cosh\chi(t,s) = 1 + \frac{2t}{s-4\mu^2}\,. \qquad (5.69b)$$

By comparing (5.69) with (5.67) we obtain the maximal growth of the characteristic angular momentum,

$$\ell_0(s) \leq \frac{N_1}{\chi_0(s)}\ln s = k_c \cdot \frac{N_1}{2\mu}\ln s,$$

where $N_1 \equiv N(t)|_{t=4\mu^2}$ is the maximal number of subtractions that we need for positive t up to the first t-channel singularity at $t = 4\mu^2$.

Thus from the boundary (5.67) motivated by the causality consideration we derive two remarkable results:

$$\rho_0(s) \leq \frac{N_1}{2\mu}\ln s, \qquad (5.70a)$$

$$\sigma_{\text{tot}} \leq c\ln^2 s, \qquad c = \left(\frac{N_1}{2\mu}\right)^2 \cdot \frac{\langle \text{Im}\, f\rangle}{4}, \qquad (5.70b)$$

where we have used the impact parameter representation (5.56) in order to derive the upper bound for the total cross section (5.70b).

Inequalities (5.70) constitute the essence of the *Froissart theorem*.
Two remarks are in order concerning these results.

(1) Since $\text{Im}\, f$ does not exceed the unitarity limit, the coefficient c in (5.70b) is restricted by

$$c \leq \left(\frac{N_1}{2\mu}\right)^2 \cdot 4\pi\,.$$

(2) Moreover, we can claim that $N_1 < 2$ since otherwise we would have had among hadrons, as we shall see later, an elementary particle with spin $\sigma = 2$ and mass smaller that $2m_\pi$.

Combining these two observations we obtain

$$c \le \frac{4\pi}{m_\pi^2} \simeq 240\,\mathrm{mb}.$$

5.6.3 Froissart theorem

The previous consideration was not very accurate as we arrived at (5.70) using a rough truncation of the partial-wave expansion in (5.68). Now we are ready to give a more rigorous proof of the Froissart theorem (Froissart, 1961).

We will exploit analytic properties of the elastic amplitude $A(s, z)$ as a function of the cosine of the scattering angle Θ,

$$z \equiv \cos\Theta = 1 + \frac{2t}{s - 4\mu^2}.$$

To prove the theorem not much is needed. It suffices to state that, as in the perturbation theory:

(1) singularities of $A(s, z)$ in z lie outside the physical region of the s-channel, $-1 \le z \le +1$; and in addition that

(2) for finite $|z|$ $A(s, z)$ is polynomially bounded,

$$|A(s, z)| < cs^N.$$

Then the energy growth of the interaction radius and of the total cross section is limited by (5.70). Move the integration contour C in the Cauchy representation,

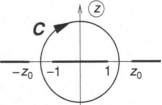

$$A(s, z) = \frac{1}{2\pi i} \int_C \frac{A(s, z')}{z' - z}\, dz',$$

away from the s-channel cut $-1 \le z \le 1$ into the unphysical region. This can always be done if there are no massless particles in the theory (in which case the t-channel singularities at $z = \pm z_0$ collide with the tips of the physical interval, ± 1). Then for the partial-wave amplitude (3.10),

$$f_\ell(s) = \frac{1}{2} \int_{-1}^{1} dz\, A(s, z) P_\ell(z) = \frac{1}{2\pi i} \int_C dz'\, A(s, z') Q_\ell(z'), \qquad (5.71)$$

we have a simple estimate

$$|f_\ell(s)| < \frac{L_\mathcal{C}}{2\pi} |A(s,z)|_{\max} \cdot |Q_\ell(z)|_{\max}, \qquad (5.72)$$

with subscripts max denoting maximal values of the functions on the contour, and $L_\mathcal{C}$ the length of the latter.

For $|z| > 1$ the exact boundary for the Legendre function on the contour has the form

$$|Q_\ell(z)| < c'_\ell \exp(-\ell\chi_{\min}), \quad \chi_{\min} \equiv \min_{z\in\mathcal{C}}\{\cosh^{-1} z\} \stackrel{s\to\infty}{\simeq} \frac{\sqrt{t}}{k_c}, \quad (5.73)$$

where factor c'_ℓ is non-exponential in ℓ at $\ell \to \infty$. Then (5.72) gives

$$|f_\ell(s)| < c_\ell s^N \exp\left\{-\frac{\ell}{k_c}\sqrt{t_{\min}}\right\}, \qquad (5.74)$$

which estimate is valid for arbitrary ℓ. So, $t_{\min}$ in (5.74) is the minimal value of t along the integration path $\mathcal{C}$. But the contour can be moved! Were it not for the cross-channel cuts $[-\infty, -z_0]$, $[z_0, +\infty]$, we could have kept 'inflating' the contour. By so doing we would *increase* $t_{\min}$ and thus strengthen the upper bound (5.74). Therefore the strongest boundary for $|f_\ell|$ that we may get is determined by the condition (5.74) with $t_{\min}$ equated with the position of the nearest singularity $t_0 = 4\mu^2$:

$$\min_\mathcal{C}\min_z\{\chi(t,s)\} = \cosh^{-1}\left(1 + \frac{2\,t_0}{s - 4\mu^2}\right) \implies \sqrt{t_{\min}} = \sqrt{t_0} = 2\mu.$$

The rest of the proof proceeds as above. Namely, we define $\ell_0(s)$ from the saturation condition $|f_{\ell_0}| = \text{const}$ in (5.74) and immediately obtain the maximal growth of the radius,

$$\rho_0(s) \equiv \frac{\ell_0(s)}{k_c} \leq \frac{N}{2\mu}\ln s \equiv \rho_F(s), \qquad (5.75)$$

and of the total cross section,

$$\sigma_{\text{tot}} \simeq \frac{1}{s}\sum_{\ell=0}^{\ell_0(s)}(2\ell+1)\,\text{Im}\,f_\ell \propto \frac{\ell_0^2(s)}{s} \leq \tilde{c}\ln^2 s.$$

Let us note that the extreme Froissart regime $\rho_0 = \rho_F$ corresponds to a clear physical picture of a disc which grows fast with energy, changing neither its transparency nor the sharpness of the edge (the latter being determined by the parameter $r_0 = 1/2\mu$, see Section 5.3).

5.6.4 Pomeranchuk theorem for the case of growing radius

Now that we know that both ρ_0 and σ_{tot} may grow logarithmically, we return to the question of what can be said about the asymptotic behaviour of particle and antiparticle total interaction cross sections with a given target. Let

$$\sigma_{tot}^a = C_1 \ln^\gamma s, \quad \sigma_{tot}^{\bar{a}} = C_2 \ln^\gamma s, \quad \gamma \leq 2.$$

We can easily construct the corresponding amplitudes in analogy with the case of constant cross sections considered in Section 5.5, cf. (5.51):

$$A_{ab}(s,0) \stackrel{s\to\infty}{=} A_{\text{right}}(s) + A_{\text{left}}(-s)$$

$$\simeq \frac{sC_1}{\pi(\gamma+1)} \ln^{\gamma+1}(-s) - \frac{sC_2}{\pi(\gamma+1)} \ln^{\gamma+1} s \qquad (5.76)$$

$$= i s \sigma_{tot}^a + \frac{s \ln s}{\pi(\gamma+1)} \cdot \Delta\sigma, \qquad \Delta\sigma \equiv \sigma_{tot}^a - \sigma_{tot}^{\bar{a}}.$$

Suppose that $C_1 \neq C_2$ ($\Delta\sigma \neq 0$). Then the imaginary part of the amplitude is relatively small and can be neglected, and not only for $t = 0$ but for finite $t < 0$ as well. The generalization of (5.76) will read

$$A_{ab}(s,t) \simeq \frac{s \ln s}{\pi(\gamma+1)} \Delta\sigma \cdot F(t,s) \qquad (5.77)$$

with the factor F such that for forward scattering $F(0,s) \equiv 1$. Now we construct the differential elastic scattering cross section,

$$\frac{d\sigma_{el}}{dq^2} = \frac{1}{16\pi} \left| \frac{A(s,t)}{s} \right|^2 \simeq \frac{|F(t,s)|^2}{16\pi} \left(\frac{\Delta\sigma \ln s}{\pi(\gamma+1)} \right)^2 . \qquad (5.78)$$

Integration over momentum transfer gives the total elastic cross section:

$$\sigma_{el} = \int dq^2 \frac{d\sigma_{el}}{dq^2} \simeq \left(\frac{\Delta\sigma \ln s}{\pi(\gamma+1)} \right)^2 \frac{1}{16\pi} \int dq^2 |F(q^2,s)|^2 . \qquad (5.79)$$

The integral in (5.79) is determined by the interaction radius (see Fig. 5.4 on page 121):

$$\int dq^2 |F(q^2,s)|^2 = \frac{1}{\rho_0^2(s)}.$$

To avoid contradiction we need to impose the restriction

$$\sigma_{el} = \left(\frac{\Delta\sigma \ln s}{\pi(\gamma+1)} \right)^2 \frac{1}{16\pi \rho_0^2(s)} \leq \sigma_{tot}^a.$$

This inequality can be translated into

$$\frac{\Delta\sigma(s)}{\sigma(s)} \leq \text{const} \sqrt{\frac{\rho_0^2(s)}{\sigma(s)\ln^2 s}} = \frac{\text{const}}{\sqrt{\sigma(s)}} \cdot \frac{\rho_0(s)}{\rho_F(s)}, \quad (5.80)$$

with ρ_F the radius corresponding to the Froissart regime (5.75).

There are two possibilities.

$\gamma = 0$. Constant total cross sections σ_{tot}^{ab}, $\sigma_{\text{tot}}^{\bar{a}b}$. From (5.80) then follows that the asymptotic inequality $\sigma_{\text{tot}}^{ab} \neq \sigma_{\text{tot}}^{\bar{a}b}$ is possible only when $\rho_0(s) = \rho_F(s)$, that is in the case of the extreme energy growth of the radius.

$0 < \gamma \leq 2$. Logarithmically growing cross sections:

$$\sigma = C_1 \ln^\gamma s, \qquad \Delta\sigma \leq \text{const} [\ln s]^{\gamma/2} \cdot \frac{\rho_0(s)}{\rho_F(s)}. \quad (5.81)$$

In this case $C_1 = C_2$, that is the particle and antiparticle cross sections have the same asymptotic behaviour; their difference may grow with s as well though slower, with (at least) a twice smaller exponent.

In fact, it is unclear how the interaction radius $\rho_0(s)$ behaves with energy. Formally we only proved that it cannot increase faster than $\ln s$. More physical information is needed in order to choose between different regimes, e.g. $\rho \sim \sqrt{\ln s}$ and $\rho \sim \ln s$, which possibilities were offered by the picture with the number of interactions increasing with energy that we have discussed above, see (5.64).

6

t-channel unitarity and growing interaction radius

Until now we have been exploiting analyticity and unitarity in the *s*-channel. We saw, in particular, how the *s*-channel unitarity put restrictions on the picture of strong interactions in the impact parameter plane and gave rise to the Pomeranchuk and Froissart theorems. As you remember, analyticity is related to causality and unitarity means that the sum of probabilities of all possible channels of particle *creation* equals one.

There is, however, one more condition, the one that is not easy to formulate. Namely, the probability that colliding particles *exchange* something,

$$A(s,t) \sim$$

also *cannot be bigger than one.* But in what sense?

The problem one faces trying to formulate such a restriction lies in the fact that it is *real* (on-mass-shell) particles that we can measure and 'count' while the exchange particles are *virtual*. Talking about virtual particles we would have to abandon our general picture in which all what matters are particle masses and on-mass-shell amplitudes ('imaginary parts').

We could make exchange particles real (and thus 'countable') if we chose positive *t* above corresponding thresholds,

(6.1)

137

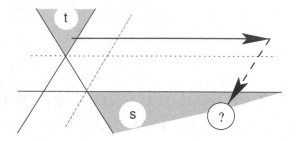

Fig. 6.1   Analytic continuation from t-channel to large imaginary scattering angles, $z_t \propto s \to \infty$ is close to the high energy s-channel scattering region.

This chain reminds unitarity relation written for t-channel scattering. Where does the t-channel unitarity operate? It holds for $t > 4\mu^2$ and negative s (momentum transfer) while we are interested in s positive, and large. This region is unphysical from the point of view of the s-channel so that (6.1) seems to be of little relevance.

On second thought, our amplitudes are analytic functions. We may try to formulate the new restriction we are looking for, by starting from $t > 4\mu^2$ and then *continuing* the t-channel unitarity condition to large s. In so doing we will find ourselves not far from an important physical region of the s-channel which describes high-energy processes with finite momentum transfer $|t| \ll s$, see Fig. 6.1.

By expressing the s-channel amplitude via its imaginary parts (discontinuities) in t- and u-channels,

$$A_3(s,t) \equiv \mathrm{Im}_t\, A = \frac{1}{2}\sum_n \underbrace{}_{t}\,, \qquad A_2(s,t) \equiv \mathrm{Im}_u\, A = \frac{1}{2}\sum_n \underbrace{}_{u}\,,$$

we would obtain *specific for the relativistic theory* consequences of the fact that the s-channel interaction is not arbitrary but occurs via exchange of particles in cross-channels. It is interesting to understand, what sort of new restrictions upon $f(\rho, s)$ the t-channel unitarity will impose.

Regretfully, the programme of analytic continuation of t-channel unitarity conditions was not fully completed. We only know how to carry out such continuation in simple cases, the simplest of which is the region $4\mu^2 < t < 16\mu^2$ (for pions) where the two-particle unitarity holds. This was done for the first time by Mandelstam.

6.1 Analytic continuation of two-particle unitarity

In Lecture 3 we have discussed the s-channel two-particle unitarity condition (3.1). Let us rewrite it for the t-channel scattering, that is treating $t = (p_1 + [-p_3])^2$ as energy, $t > 4\mu^2$, and $s < 0$ (and $u < 0$) as momentum transfer(s).

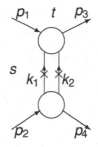

The unitarity condition takes the form

$$\mathrm{Im}_t\, A(t,s) = \frac{1}{2i}\left[A(t+i\epsilon, s) - A(t-i\epsilon, s)\right]$$

$$\equiv A_3(t, z) = \frac{1}{2} \cdot \text{⬡} = \tau \int \frac{d\Omega}{4\pi} A(\mathbf{p}_1, \mathbf{k}) A^*(\mathbf{k}, \mathbf{p}_2),$$

(6.2a)

where

$$z = \cos\Theta_{12} = 1 + \frac{2\,s}{t - 4\mu^2}$$

(6.2b)

and τ is now the t-channel phase-space volume factor

$$\tau = \tau(t) = \frac{k_c(t)}{8\pi\sqrt{t}} = \frac{1}{16\pi}\sqrt{\frac{t - 4\mu^2}{t}}.$$

(6.2c)

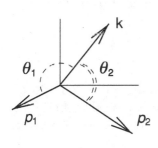

The internal amplitudes $A(t, z_1)$ and $A^*(t, z_2)$ depend on the energy t and on the corresponding scattering angles,

$$z_i \equiv \cos\Theta_i = 1 + \frac{2\,s_i}{t - 4\mu^2}, \quad (i = 1, 2)$$

where $\Theta_{1(2)}$ is the angle between the initial (final) cms momentum $\mathbf{p}_{1(2)}$ and the intermediate-state momentum $\mathbf{k}$.

In order to continue (6.2) to large $z \propto s$, we are going to analyse analytic properties of the t-channel imaginary part A_3 in z. As a first step it is convenient to trade the angular integration for symmetric integrals over z_1 and z_2. Choosing the polar axis $\mathbf{z}$ along $\mathbf{p}_1$, we write

$$d\Omega = d(\cos\Theta_1) \cdot d\phi = dz_1 \cdot dz_2 \times J^{-1}.$$

The trigonometric relation

$$z_2 = zz_1 + \sqrt{(1 - z^2)(1 - z_1)^2} \cdot \cos\phi$$

gives us the dependence $z_2 = z_2(\phi)$ and we derive

$$\left(\frac{dz_2}{d\phi}\right)^2 = (1-z^2)(1-z_1^2)\sin^2\phi = (1-z^2)(1-z_1^2) - (z_2-zz_1)^2.$$

The Jacobian is proportional to $|\sin\phi|$ and equals $J = \frac{1}{2}\sqrt{-K}$, where

$$K \equiv (z_2 - zz_1)^2 - (1-z^2)(1-z_1^2) = (z^2 + z_1^2 + z_2^2) - 1 - 2zz_1z_2$$
$$= (z - z_1z_2)^2 - (1-z_1^2)(1-z_2^2). \tag{6.3}$$

(The symmetry of the Jacobian $z_1 \leftrightarrow z_2$ should have been expected. The fact that it turned out to be symmetric with respect to *all three* cosines is less obvious, though true. We will exploit it in what follows.) We arrive at

$$A_3(t, z) = \frac{\tau}{2\pi} \int\int \frac{dz_1\, dz_2}{\sqrt{-K(z, z_1, z_2)}} A(t, z_1) A^*(t, z_2), \tag{6.4}$$

where integration limits are determined by the condition $-K \geq 0$. Examining the integral in new variables we note that the dependence on z is localized in K so that we have a kind of an integral representation with the kernel $(-K)^{-1/2}$.

6.1.1 z_2 integration: pinch

What sort of integral is this? Let us move step by step and study first the integration over z_2 while keeping z_1 fixed. The z_2 integral runs from z_2^- to z_2^+, that is, between two zeros of K (where $\sin\phi = 0$):

$$z_2^\pm = zz_1 \pm \sqrt{(1-z^2)(1-z_1^2)}. \tag{6.5}$$

After that we will have to integrate over z_1 in the interval $[-1, 1]$. Now that the double integral has been explicitly written, we need to find out what happens to it when we move outside the physical t-channel region $(-1 \leq z \leq 1)$ and keep increasing z.

Quite an exercise in the 'theory of functions of complex variable' is awaiting us. The task of continuing our integral would have been hopeless if we did not possess the knowledge of analytic properties of the amplitude. What should we expect? We know that in unphysical regions of the Mandelstam plane there are 'spectral domains' where A_3 becomes complex. Increasing s, I will inevitably hit these domains. It makes sense therefore to prepare ourselves to this eventuality.

A clever thing to do is to replace the integral over a fixed interval by path integration in the complex z_2-plane along the contour embracing the cut of the function $1/\sqrt{K}$:

$$\int_{z_2^-}^{z_2^+} \frac{dz_2}{\sqrt{-K}} = \frac{i}{2} \int_C \frac{dz_2}{\sqrt{K}}.$$

In addition to the square-root branch cut $[z_2^-, z_2^+]$, our integrand as a function of z_2 has physical singularities of the amplitude $A(t, z_2)$ on the z_2-plane. These are s- and u-thresholds that start at $z_2 = \pm z_{20}$ and run to $\pm \infty$, correspondingly:

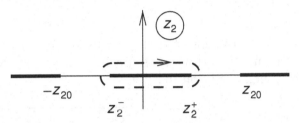

Now we increase z and pass through $z = 1$. When $z = 1$, the endpoints of the cut (6.5) *collide* at $z_2^- = z_2^+ = z_1$ and for $z > 1$ they become complex conjugate. At this point nothing dramatic happens to the answer since I will keep deforming calmly the integration contour by following the metamorphosis of the cut. How might our integral develop a singularity? Only if the tip of the cut would collide with one of the threshold singularities $\pm z_{20}$, pinching the contour.

At which value of the external variable z does the *pinch* occur? One needs to solve the equation, for example, $z_2^+ = z_{20}$. It is easy to realize that this equation has the same structure as (6.5) namely,

$$z = z^{\mathrm{pinch}}(z_1) = z_1 z_{20} \pm \sqrt{(1 - z_1^2)(1 - z_{20}^2)}. \qquad (6.6)$$

Since $z_{20} > 1$, the position of the pinch point corresponds to a complex z.

6.1.2 z_1 integration: contour trapping

The time has come to look for singularities of the integrand as a function of z_1. Introducing

$$f(t, z, z_1) = \frac{i\tau}{4\pi} \int_C \frac{dz_2}{\sqrt{K(z, z_1, z_2)}} A^*(t, z_2),$$

we have

$$A_3 = \int_{-1}^{1} dz_1 \, A(t, z_1) f(t, z, z_1).$$

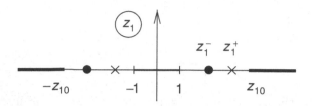

Fig. 6.2   Integration contour $[-1, 1]$ and singularities in the z_1 plane: right (A_1) and left cuts (A_2) of the amplitude $A(t, z_1)$ and pinch points z_1^- (circle) and z_1^+ (cross). The left pair of points solves an alternative pinch condition $z^+ = -z_{20}$.

First of all, there are thresholds $z_1^\pm$ of the amplitude $A(t, z_1)$. Secondly, the singularities of the function $f(z_1)$ whose position $z_1 = z_1(z)$ we determine by inverting the pinch condition (6.6):

$$z_1^{\text{pinch}}(z) \equiv z_1^\pm = z z_{20} \pm \sqrt{(z^2 - 1)(z_{20}^2 - 1)}. \qquad (6.7)$$

Mark that $z_1^\pm$ are real since $z_{20} > 1$ and $z \geq 1$. The structure of the z_1-plane will look as shown in Fig. 6.2 A symmetric pair of singularities $\tilde{z}_1^\pm$ on the left side of the z_1-plane solves the pinch equation $z_2^+ = -z_{20}$ complementary to (6.6). We will follow those on the right side of the plane, $z_1^\pm$.

With z increasing, the two singular points start off from $z_1^\pm = z_{20}$ at $z = 1$ and separate, z_1^- moving to the left and z_1^+ to the right. Can z_1^- collide with the integration interval $[-1, 1]$? From (6.7) it is clear that $z_1^- = +1$ indeed takes place at $z = z_{20}$. This is, however, the *absolute minimum* of $z_1^-(z)$ for a real z. (It becomes obvious if we parameterize $z = \cosh \eta$, $z_{20} = \cosh \eta_{20}$ resulting in $z_1^- = \cosh(\eta - \eta_{20}) \geq 1$.) This means that with z moving above z_{20}, the position of singularity reflects from $+1$ and increases indefinitely. A peculiar situation: z_1^- barely touches the integration interval and bounces off.

So is there a singularity or not? Let us show that the point $z = z_{02}$ is in fact *not singular*. We face here a curious phenomenon (I wonder if you have met anything of this sort in your maths course.) Imagine that while changing some external parameter, a singularity of the integrand touches the tip of the integration contour. To determine whether the answer for the *integral* will be singular at this value of the parameter we have to compare two ways of passing by this point, from above and from below:

In order to have everything smooth and well defined, we will have to deform the contour correspondingly, and differently in two cases:

The analytic continuation of the integral that was initially defined on $[-1, 1]$ acquired an additional piece running from $+1$ to the new position of the singularity and back, the only difference between the two ways being the *direction* of the loop. If the two paths led to different results then we found a singularity of the integral. There exists, however, a trivial case when the two expressions coincide: when the singularity is a square root, so that the values on the sides of the cut are just opposite in sign. This being our case, we conclude $z = z_{20}$ to be a regular point.

However, something did happen. Namely, in spite of the fact that the function is non-singular, its explicit representation in terms of a contour integral has changed. The phenome-

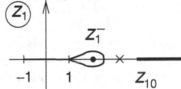

non we encountered is called 'contour trapping'. Now that we have the added loop that follows the movement of the point $z_1^-(z)$, a real possibility to develop a singularity finally emerges. The integral for A_3 becomes singular at the value of z when point $z_1^-(z)$ bumps on the threshold of the amplitude $A(t, z_1)$ at z_{10} and pinches the contour that it trapped and dragged along.

So, would there have been no singularity if not for z_{10}? Sure. In this case we would have $A_3(z) = $ const (or a polynomial in z at most). This was in fact implicit from the beginning: if the integrand $A(z_1)$ did not depend on the scattering angle Θ_1, the l.h.s. of (6.2a) would have been independent of the angle Θ as well; in other words, it would not have singularities in z.

Finally, solving the 'collision' equation $z_1^- = z_{10}$ for the position of singularity, we obtain

$$z = z_{10}z_{20} + \sqrt{(z_{10}^2 - 1)(z_{20}^2 - 1)}. \qquad (6.8)$$

Substituting explicit expressions for z_{i0} we derive again the familiar equation describing the *Karplus curve* – the boundary of the double spectral function ρ_{st}.

6.1.3 Imaginary part of the imaginary part

Let us take z above the singularity, $z_1^- > z_{10}$, and calculate the imaginary part of A_3 that is the discontinuity in s (in z):

$$\rho_{st} = \mathrm{Im}_s\, A_3(t, z) = \int_{z_{10}}^{z_1^-} dz_1\, \mathrm{Im}_s\, A(t, z_1) \cdot \Delta f(t, z, z_1). \qquad (6.9a)$$

What is Δf in this expression? Recall that we had the contour pinched in the z_2-integration as well; Δf stands for the corresponding discontinuity over the cut of the amplitude in the z_2-plane:

$$\Delta f(t, z, z_1) \sim \int_{z_{20}}^{z_2^+} \frac{dz_2}{\sqrt{K}}\, \mathrm{Im}_s\, A(t, z_2). \qquad (6.9b)$$

We arrive at the expression of the same structure as (6.4) for the A_3 itself but integrated over a different region,

$$\rho_{st} \sim \iint \frac{dz_1\, dz_2}{\sqrt{K(z, z_1, z_2)}}\, \mathrm{Im}_s\, A(t, z_1)\, \mathrm{Im}_s\, A^*(t, z_2). \qquad (6.10a)$$

We don't need to worry about the lower limits of the integrals since the factors $\mathrm{Im}\, A(t, z_i)$ themselves know about z_{10}, z_{20}. As for the upper limits, they are given by the inequality

$$z_1 z_2 + \sqrt{(z_1^2 - 1)(z_2^2 - 1)} \leq z \qquad (6.10b)$$

which is equivalent to $K > 0$, see (6.3).

Anything else? Until now we have been studying only positive z_1, z_2. Considering analogously left-side singularities on Fig. 6.2 we will restore the u-channel contribution. Using our old notation for imaginary parts of the amplitude in s and u channels, $A_1 \equiv \mathrm{Im}_s\, A$ and $A_2 \equiv \mathrm{Im}_u\, A$, the final formula reads

$$\rho_{st} = \frac{\tau}{\pi} \iint \frac{dz_1\, dz_2}{\sqrt{K(z, z_1, z_2)}} [\, A_1(t, z_1) A_1^*(t, z_2) + A_2(t, z_1) A_2^*(t, z_2)\,], \qquad (6.11a)$$

where the integration is performed over the region (6.10b); $z_1 > 1$, $z_2 > 1$.

If I chose to continue analytically the t-channel unitarity condition to $s \to -\infty$ (instead of $+\infty$), I would obtain a similar integral expression for another double spectral function,

$$\rho_{ut} = \frac{\tau}{\pi} \iint \frac{dz_1\, dz_2}{\sqrt{K(z, z_1, z_2)}} [\, A_1(t, z_1) A_2^*(t, z_2) + A_2(t, z_1) A_1^*(t, z_2)\,]. \qquad (6.11b)$$

Mandelstam equations (6.11) solve the problem of analytic continuation of the t-channel unitarity condition. Thus we learned how to express 'imaginary parts of the imaginary parts' ρ_{ij} via the imaginary parts A_i of the amplitude themselves!

6.1.4 Mandelstam representation

We have obtained the double discontinuity ρ_{st} in the following order: we were sitting in the t channel at $t > 4\mu^2$, took $A_3 = \text{Im}_t A$, then, by continuing A_3 to $|z_t| > 1$, moved to the s-channel and there evaluated $\text{Im}_s A_3$. We could have done it in the opposite order, namely start from $A_1 = \text{Im}_s A$ in the s-channel, $t < 0$, and then increase t to access $\text{Im}_t A_1$ at $t > 4\mu^2$. It is natural to expect that this way we would have got the same expression (6.11a) for ρ_{st},

$$\rho_{st}(s,t) = \text{Im}_s A_3(s,t) = \text{Im}_t A_1(s,t).$$

Although a formal proof does not exist, this statement would be definitely correct if the amplitude admitted the double integral representation

$$A(s,t) = \frac{1}{\pi^2} \iint \frac{\rho_{st}(s',t')\, ds'\, dt'}{(s'-s)(t'-t)} + [s \to u] + [t \to u], \qquad (6.12)$$

where the integration region is restricted by the Karplus curve in the s'–t' plane. Since 1958, when Mandelstam suggested the representation (6.12) for the invariant amplitude Mandelstam (1958), no Feynman graph has been found which would violate it (provided all participating particles are stable, $m_a < m_b + m_c + m_d$).

The spectral density ρ_{st} corresponds to simultaneously evaluating discontinuities over s in t and bears information about unitarity in both channels. This object is therefore well suited to support our expectation that probabilities of particle *creation* and particle *exchange* are not independent. Such inter-dependence is a specific feature of the relativistic theory, in marked difference to non-relativistic quantum mechanics.

6.2 $\rho_0 = $ const, $\sigma_{\text{tot}} = $ const contradicts *t*-channel unitarity

The Froissart theorem provided us only with *upper bounds* for growth rates of $\rho_0(s)$ and $\sigma_{\text{tot}}(s)$. Now we will show that in a relativistic theory the radius $\rho_0(s)$ *must* virtually always *grow* with s. (To be precise, it is allowed *not to grow* only if the total cross section falls faster than $1/\ln s$ at asymptotically high energies.)

As we have discussed above, the hypothesis $\rho(s) \to $ const implies that the s- and t-dependence of the scattering amplitude factorize,

$$A(s,t) \overset{s \to \infty}{=} s \cdot F(t), \qquad (6.13)$$

where we choose $A \propto s$ to ensure asymptotically constant σ_{tot}. Once the amplitude has such a form in the physical region of the s channel, then, by virtue of analyticity, it has to have the same structure at positive t as well, it seems. We will suppose that (6.13) holds for finite t (of any sign), but then it has to satisfy the equations (6.11) that we have derived for moderate positive t (in the interval $4\mu^2 < t < 16\mu^2$). Let us see if it really does. From (6.13) we get

$$A_1(s,t) \simeq s \operatorname{Im} F \equiv s \cdot F_1(t); \quad \rho_{st} \simeq s \cdot \operatorname{Im} F_1(t) \quad \text{for } t > 4\mu^2. \quad (6.14a)$$

Analogously for the *antiparticle* scattering amplitude, in the crossing channel, $u \to \infty$,

$$A_2(u,t) \simeq u \cdot F_2(t); \qquad \rho_{ut} \simeq u \cdot \operatorname{Im} F_2(t) \quad \text{for } t > 4\mu^2. \quad (6.14b)$$

Thus, we wrote down explicitly all the ingredients of the Mandelstam relations (6.11) for the double spectral densities ρ_{st} and ρ_{ut}. This means that we can verify our model (6.13) provided the dominant contribution to the integral comes from the region of large internal energies. Let us start calculating the integral (6.11a) *supposing* that $z_1, z_2 \gg 1$ and then verify that this is indeed true.

Approximating the Jacobian

$$-K = \left[z - z_1 z_2 + \sqrt{(z_1^2 - 1)(z_1^2 - 1)}\right]\left[z - z_1 z_2 - \sqrt{(z_1^2 - 1)(z_1^2 - 1)}\right]$$

$$\simeq z(z - 2z_1 z_2),$$

and substituting the asymptotic approximation (6.14) for the block amplitudes A_1 and A_2 we obtain

$$\rho_{st} \simeq \frac{\tau}{\pi} \int \frac{dz_1\, dz_2 \cdot z_1 z_2}{\sqrt{z(z - 2z_1 z_2)}} \cdot \left[\frac{t - 4\mu^2}{2}\right]^2 [F_1 F_1^* + F_2 F_2^*].$$

The integrand depends only on the product $z_1 z_2 = x$, therefore

$$\rho_{st} \propto \int_{z_{10}}^{z/z_{20}} \frac{dz_1}{z_1} \int_{z_1 z_{20}}^{z/2} \frac{x\, dx}{\sqrt{z(z - 2x)}} \simeq z \int_{z_{10}}^{z/z_{20}} \frac{dz_1}{z_1} \int_0^{1/2} \frac{y\, dy}{\sqrt{1 - 2y}} \quad (6.15)$$

$$\propto z \ln \frac{z}{z_{10} z_{20}} \propto s \ln s.$$

The inconsistency of our calculation with (6.14a) is apparent:

$$\operatorname{Im} F_1(t) = \lim_{s \to \infty} \frac{\rho(s,t)}{s} \overset{?}{\simeq} \frac{c}{\mu^2} \ln s + \frac{1}{\mu^2} \cdot \mathcal{O}(1). \quad (6.16)$$

The unwanted dominant contribution $\mathcal{O}(\ln s)$ came from the specific integration region $z_1 z_2 \sim z$, $z_0 \ll z_1, z_2 \ll z$ (which, by the way, confirms our initial decision to use asymptotic formulae for the internal blocks). To scrutinize other regions won't help since the integrand is positively definite so that there can be no cancellation.

6.2.1 $\rho_0(s)$ for arbitrary $\sigma_{\text{tot}}(s)$

Let us release the $\sigma_{\text{tot}} = $ const condition and look whether the radius can stay asymptotically constant in the general case. The generalization reads

$$A_1(s,t) \simeq s \cdot h(s) \cdot F_1(t),$$

$$\rho_{st} \simeq s \cdot h(s) \cdot \operatorname{Im} F_1(t), \qquad (\rho \equiv 0 \text{ for } t < 4\mu^2). \tag{6.17}$$

Then (cf. (6.15))

$$\rho(s,t) \propto \int \frac{dz_1\, dz_2\, z_1 z_2 \cdot h(z_1) h(z_2)}{\sqrt{z(z - 2z_1 z_2)}} \simeq z \int_{z_{10}}^{z/z_{20}} \frac{dz_1}{z_1} h(z_1) \int_0^{1/2} \frac{y\, dy}{\sqrt{1-2y}} h\left(\frac{zy}{z_1}\right).$$

Since the y-integral converges, we can substitute a constant $c = \langle y \rangle = \mathcal{O}(1)$ for y in the argument of the second h-function to obtain

$$\operatorname{Im} F_1(t) \propto \frac{\rho(s,t)}{zh(z)} \sim \frac{1}{h(z)} \int_{z_{10}}^{z/z_{20}} \frac{dz_1}{z_1} h(z_1) h\left(\frac{z}{z_1} c\right). \tag{6.18}$$

To avoid contradiction, the r.h.s. of (6.18) has to have a finite $z \to \infty$ limit. It is easy to see that this is possible only if

$$h(z) < \frac{\text{const}}{\ln z}, \qquad z \to \infty. \tag{6.19}$$

Only in this case which corresponds to a falling total cross section,

$$\sigma_{\text{tot}}(s) < \frac{\text{const}}{\ln s}, \qquad s \to \infty,$$

the constant interaction radius would not contradict t-channel unitarity.

We wrote the unitarity condition valid for $4\mu^2 < t < 16\mu^2$, made use of the concrete form of the amplitude at $s \to \infty$ and finite t and came to a contradiction with the hypothesis $\sigma_{\text{tot}} = $ const.

What is the reason for that?

The picture that caused us trouble is that of Fig. 6.3(a). It is related to the production process of two showers of particles in a high energy $\pi\pi$ collision with the exchange of a pion, Fig. 6.3(b).

From the very beginning we supposed that the total $\pi\pi$ interaction cross section is constant at high energy. But it is this $\sigma_{\pi\pi}$ that twice

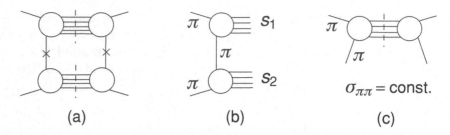

Fig. 6.3   On the 'black disc' ansatz (6.13) versus *t*-channel unitarity.

enters the graph we have selected. Now, however, we can vary the total energy partitioning between showers, $s\mu^2 \sim s_1 s_2$, adding contributions with different s_1 and s_2. Since for each of the two pion–pion interaction sub-processes Fig. 6.3(c) $\sigma_{\text{tot}} \to$ const, we obtained an additional $\ln s$ enhancement due to integration over shower masses.

Does our contradiction mean that $\sigma_{\pi\pi}$ cannot be constant in the high energy limit? No. This only tells us that it is wrong to think that the t dependence of the amplitude is determined exclusively by the nearest singularity due to one-pion exchange: $\rho_0 \simeq (2\mu)^{-1} =$ const. Multi-meson exchanges must be important, interfering with one pion; the higher the energy s, the more the amplitude has to 'remember' about the faraway singularities in t. In other words, we can no longer consider the interaction radius to be energy independent, unless σ_{tot} falls with s.

Thus, trying to preserve asymptotic constancy of the total cross section, we have to abandon the factorization ansatz (6.13) and look for a more complicated structure of the amplitude; we have to have ρ_0 changing with energy.

There were times when the constancy of the interaction radius was held in deep respect, people thought that it had a deep physical meaning. Later it transpired that the truth is just the opposite: it is *practically impossible* to have it not growing with energy.

6.2.2 Numerical estimate

How 'serious' is the contradiction with unitarity that we have faced? Look more attentively at our relation:

$$\text{Im}\, F_1 \gtrsim \frac{\tau}{4\pi} \cdot \left[\frac{t - 4\mu^2}{2} \right] \cdot 2\,|F_1|^2 \cdot \ln s.$$

For $t \sim \mu^2$ we can take

$$\text{Im}\, F_1(t) \sim F_1(t) \sim F_1(0) = \sigma_{\text{tot}} \sim \frac{1}{\mu^2}$$

as a rough estimate. Stepping away from the t threshold by $(t - 4\mu^2) \sim \mu^2$ then gives for the numerical coefficient c of the logarithmic term in (6.16)

$$c = \frac{\tau}{4\pi} \cdot \left[(t - 4\mu^2)F_1(t)\right] \sim \frac{\tau}{4\pi} \simeq \frac{1}{4\pi}\frac{1}{16\pi} \sim \frac{1}{600}.$$

This means that though the radius *has* to grow, it may do so very slowly: the formal contradiction starts to be really important only at fantastically high energies, $\ln s \sim 600$.

6.2.3 Modelling a growing radius

Let us attempt to model a growing radius. We wrote $A_1 = s\,F(t)$ for the black-disc picture and failed. Try

$$A_1(s, t) = s^{\alpha(t)} F(t), \tag{6.20}$$

such that

$$\alpha(0) = 1, \qquad \alpha(t) < 1 \quad \text{for } t < 0.$$

For a finite t we then have approximately

$$A_1(s, t) \simeq s\,e^{\alpha' t \ln s} F(t) = s\,e^{-\alpha' \mathbf{q}^2 \ln s} F(-\mathbf{q}^2). \tag{6.21}$$

The essential momentum transfer $\mathbf{q}$ in (6.21) is

$$|\mathbf{q}| \simeq \frac{1}{\sqrt{\alpha' \ln s}},$$

which immediately translates into the energy-dependent radius

$$\rho_0(s) \simeq \sqrt{\alpha' \ln s}. \tag{6.22}$$

What will change in the t-channel unitarity condition? Examine the r.h.s. of (6.11a):

$$A_1(z_1, t)A_1^*(z_2, t) \propto z_1^{\alpha(t)} z_2^{\alpha^*(t)}.$$

Above the threshold, $t > 4\mu^2$, both $F(t)$ and $\alpha(t)$ in (6.20) will become complex in general:

$$\implies z_1^{\alpha_1 + i\alpha_2} z_2^{\alpha_1 - i\alpha_2} = (z_1 z_2)^{\alpha_1} \exp\left\{i\alpha_2 \ln\frac{z_1}{z_2}\right\}, \qquad \alpha_2 = \text{Im}\,\alpha(t).$$

Recall that the logarithmic s-dependence occurred due to fact that the integrand depended solely on the product $z_1 z_2$. Now, on the contrary, we

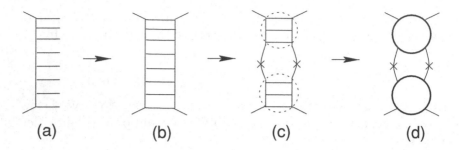

Fig. 6.4   Relation of ladder-like inelastic processes to t-channel unitarity.

have an oscillating function of the ratio z_1/z_2 which will force the integral to converge and produce

$$\int d\ln\frac{z_1}{z_2}\,\exp\left\{i\alpha_2\ln\frac{z_1}{z_2}\right\}\sim\frac{1}{\alpha_2}$$

in place of $\ln s$. Our consistency condition (6.16) will turn into

$$\operatorname{Im}F_1(t)\simeq\frac{c}{\mu^2}\cdot\frac{1}{\alpha_2}+\frac{1}{\mu^2}\cdot\mathcal{O}(1)\,. \tag{6.23}$$

The formal contradiction is gone. Moreover since c is numerically small, we may have $\alpha't\ll 1$ and $\alpha(t)\simeq 1$ in a broad region of momentum transfer t.

A comment is in order. In Section 5.6 we saw how to 'construct' a growing radius. To do so we allowed a fast incident particle to slow down before hitting the target, by emitting a whole 'comb' of virtual particles on the way.

Imagine that inelastic processes have indeed the structure of a 'comb' as shown in Fig. 6.4(a).

Then, by s-channel unitarity, squaring the amplitude (a) we get the forward scattering amplitude as a 'ladder' of Fig. 6.4(b). A remarkable thing about this picture, based on *repetitions* in the t-channel, is that it directly solves t-channel unitarity!

Indeed, by taking discontinuity in t somewhere along the graph, Fig. 6.4(c), the upper and the lower parts of the 'ladder' will sum up into the full interaction amplitudes as shown by blocks in Fig. 6.4(d). Hence, the *necessity* (as well as *opportunity*) of having the radius grow with energy is related to the possibility of *repetitions* in the t-channel which are the key to the t-channel unitarity.

To move further we need to investigate which solutions are reasonable, what are realistic strong interaction amplitudes in the deep asymptotic regime. One could continue along the lines of this lecture and study the restrictions imposed by cross-channel unitarity conditions.

It turns out, however, that there is a more elegant way to find asymptotics of relativistic amplitudes by establishing a transparent link with the old non-relativistic theory.

7

Theory of complex angular momenta

This theory will allow us to keep track of analyticity and unitarity in the t-channel when analysing s-channel phenomena.

The first step is to write down the amplitude expansion in terms of t-channel partial waves $f_\ell(t)$ rather than $f_\ell(s)$ as we did before to derive the Froissart theorem. We write

$$A(s,t) = \sum_{n=0}^{\infty}(2n+1)f_n(t)P_n(z), \qquad (7.1\text{a})$$

where now

$$z \equiv z_t = \cos\Theta_t = \frac{2\,s}{t - 4\mu^2} \qquad (7.1\text{b})$$

stands for the cosine of the scattering angle in the t-channel process.

The unitarity condition in the t-channel limits the size of each partial amplitude $|f_n(t)| = \mathcal{O}(1)$. In the s-channel, we have obtained the growing amplitude $A(s,t)$ by summing up a large number of terms $\ell \lesssim \ell_0(s)$ in the partial-wave expansion in $f_\ell(s)$. Now we keep t finite and a finite number of partial waves $f_n(t)$ with $n \lesssim n_0(t) = \mathcal{O}(1)$ will contribute.

How will the series (7.1a) behave in the $s \to +\infty$ limit? From the t-channel point of view this region on the Mandelstam plane is absolutely unphysical as it corresponds to large *imaginary* scattering angles (7.1b) $z \gg 1$. We have mentioned before more than once that this unphysical region bears information about high energies in the s-channel. So let us try to imagine what sort of behaviour of the series at large z we could expect.

The partial wave expansion (7.1a) was written in the physical region of the t-channel. Since partial-wave amplitudes are falling fast at large n, we can split the sum into two pieces,

$$A(s,t) = \sum_{n=0}^{n_0(t)} (2n+1) f_n(t) P_n(z) + \sum_{n_0(t)}^{\infty} \cdots ,$$

and drop the infinite series term. (Formally speaking, this 'tail' will diverge for large z but there is no special reason for it to be large and, more importantly, to change significantly when we will move from positive $t = \mathcal{O}(\mu^2)$ down to $t < 0$ as to reach the s-channel domain.) Then for $z \to \infty$ we will have a qualitative estimate

$$A(s,t) \sim z^{n_0(t)} \propto s^{n_0(t)}. \tag{7.2}$$

Being rather brutal, this estimate nevertheless tells us what we could expect. Namely, that the large-s asymptote of the s-channel amplitude is governed by the *characteristic angular momentum* $n_0(t)$ in the cross-channel. What remains is to learn how to determine this characteristic momentum.

7.1 Sommerfeld–Watson representation

This representation was applied by T. Regge to the problem of analytic continuation of the partial-wave expansion.

The quest is, how to invent, in a more or less unique way, a function $f_\ell(t)$ that would be *analytic* in ℓ and would coincide with the partial waves in (7.1a) in every integer point,

$$f_\ell(t)|_{\ell=n} = f_n(t), \quad n = 0, 1, 2, \ldots, \infty.$$

If we succeeded, the problem of analytic continuation of the series (7.1a) to large z would have been relatively easy to solve. Indeed, suppose we knew how to construct such a function. Then I would write a simple formula (the Sommerfeld–Watson integral)

$$A(s,t) = \frac{1}{2i} \int_C \frac{d\ell}{\sin \pi \ell} f_\ell(t) P_\ell(-z), \tag{7.3}$$

where the contour $\mathcal{C}$ encircles all integer points $n \geq 0$ anti-clockwise:

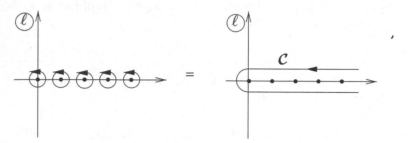

The function P_ℓ is regular in ℓ. Evaluating the residues of the singular factor $1/\sin \pi\ell$ at $\ell = n$ and bearing in mind that $P_n(-z) = (-1)^n P_n(z)$, we recover the original sum (7.1a).

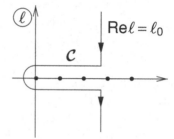

The idea is as follows. Imagine that we managed to choose f_ℓ 'good enough' so as the contour can be *deformed* as shown here on the right. Then everywhere along $\mathcal{C}$

$$\mathrm{Re}\,\ell \leq \ell_0 = \text{const.}$$

This gives us an upper bound

$$P_\ell(z) \propto z^\ell = z^{\mathrm{Re}\,\ell + i\,\mathrm{Im}\,\ell}, \quad A(s,t) \propto |P_\ell(z)| \lesssim z^{\ell_0}, \quad z \to \infty.$$

Such an inequality does not make much sense since it depends on the choice of the integration contour: the boundary gets stronger as we move the contour to the *left*. What prevents us from strengthening the upper bound indefinitely? The function f_ℓ has singularities somewhere in the ℓ plane. Shifting the contour in (7.3) is possible until we hit such a singularity at some point $\ell = \alpha(t)$. Thus it is the position of the *rightmost singularity* of the partial wave $f_\ell(t)$ that will determine the asymptotic behaviour of the amplitude,

$$A(s,t) \sim z^{\alpha(t)} \propto s^{\alpha(t)}, \quad s \to \infty. \tag{7.4}$$

Our qualitative expectation has been made precise: we gave definite meaning to the 'characteristic angular momentum' n_0 in (7.2) by having linked it with analytic properties of the partial-wave amplitude considered as a function of a complex variable ℓ.

This is the key idea of the theory of complex angular momenta.

Two things were needed for this programme to succeed namely, that

(1) f_ℓ is analytic in the right half-plane, $\mathrm{Re}\,\ell > N$; and   (7.5a)

(2) f_ℓ falls along each beam $|\ell| \to \infty$ in this half-plane.   (7.5b)

Let us see how fast f_ℓ has to decrease with $|\ell|$. In the physical region $z = \cos\Theta \in [-1, +1]$ where (7.1a) was written,

$$P_\ell(-z) \sim J_0(\ell[\pi - \Theta]) \sim \frac{1}{\sqrt{\ell}} \left[e^{i\ell(\pi-\Theta)} + e^{-i\ell(\pi-\Theta)} \right].$$

This means that the ratio

$$\frac{P_\ell(-z)}{\sin \pi\ell} \tag{7.6}$$

in (7.3) falls *exponentially* for all angles but $\Theta = 0$ where

$$\left| \frac{P_\ell(-z)}{\sin \pi\ell} \right| \propto \frac{1}{\sqrt{\ell}}, \qquad |\ell| \to \infty.$$

So it suffices to have

$$|f_\ell| < \ell^{-3/2}$$

to ensure the convergence of the integral and the possibility of the contour deformation.

It is easy to show that the problem of extrapolating the function from its values in integer points onto the entire complex plane has no more than one solution under a much weaker condition, namely,

$$|f_\ell| < e^{|\ell|\pi}, \qquad |\ell| \to \infty. \tag{7.7}$$

This is known as the *Carlson theorem*. We will not prove it. Let us remark, however, that the statement of the theorem is essentially trivial. Suppose we found a solution $f_\ell^{(1)}(t)$. To construct a different one we would have to add a function that vanished in all integers $\ell = n$, that is something of the form

$$f_\ell^{(2)}(t) = f_\ell^{(1)}(t) + \delta f_\ell(t); \qquad \delta f_\ell(t) = g_\ell(t) \cdot \sin \pi\ell.$$

But the factor $\sin \pi\ell$ grows exponentially along the imaginary axis, $\propto e^{|\ell|\pi}$, just violating the condition (7.7).

7.2 Non-relativistic theory

One may ask the same question of the behaviour of the amplitude at $\cos\Theta \gg 1$ in the framework of the non-relativistic scattering theory, though it makes not much sense here. Nevertheless, the programme that we have outlined above can be carried out literally and rigorously.

In the non-relativistic theory we have the Schrödinger equation at our disposal,

$$\left[-\frac{\hbar^2}{2m}\boldsymbol{\nabla}^2 + U(r) \right] \psi(\mathbf{r}) = E\psi(\mathbf{r}).$$

The radial part, $F_n(r)$, of the wave function $\psi(\mathbf{r}) = F_n(r)Y_{n,m}(\Theta, \phi)$ satisfies the equation

$$\left[-\frac{\hbar^2}{2m}\boldsymbol{\nabla}_r^2 + \frac{n(n+1)}{r^2} + U(r) \right] F_n = EF_n(r). \tag{7.8}$$

Among the two solutions of (7.8),

$$F_n^{(1)}(r) \propto r^n, \qquad F_n^{(2)}(r) \propto r^{-n-1}, \qquad r \to 0, \tag{7.9}$$

we choose the first one that is regular at $r = 0$. Having fixed the wave function at the origin, at $r \to \infty$ it behaves as

$$F_n(r) \propto a_n(k)\frac{e^{-ikr}}{r} + b_n(k)\frac{e^{ikr}}{r}. \tag{7.10}$$

This corresponds to the S-matrix element $S_n(k) = b_n/a_n$ and to the scattering amplitude

$$f_n(k) = \frac{1}{2i}\left(\frac{b_n(k)}{a_n(k)} - 1 \right). \tag{7.11}$$

Why did we keep the angular momentum n to be an integer? In order to have a non-singular angular dependence of the wave function. However, as long as we are interested in the non-physical region $\cos\Theta > 1$, no-one would forbid us to look upon n in the radial Schrödinger equation (7.8) as an arbitrary continuous parameter.

So we substitute $n \to \ell$ and treat ℓ as a complex number. Then we will solve (7.8), choose $F_\ell(r) \propto r^\ell$ as before and find the functions a and b from the large-r behaviour (7.10). The ratio b_ℓ/a_ℓ does not depend on the normalization of the wave function and describes the 'scattering amplitude' as a function of ℓ.

Have we satisfied the necessary conditions (7.5) for deforming the contour in the Sommerfeld–Watson integral? The condition (7.5a) is fulfilled with $N = \frac{1}{2}$. Indeed, any solution of (7.8) will be an analytic function of ℓ since the parameter ℓ enters the equation *analytically*. Where does the restriction $\text{Re}\,\ell > \frac{1}{2}$ come from in the first place? It emerges from the choice of the solution proper in (7.9): for the prescription to be *unique* we must impose

$$\text{Re}\,\ell > \text{Re}(-\ell - 1) \qquad \Longrightarrow \qquad \text{Re}\,\ell > \tfrac{1}{2}.$$

The second condition (7.5b) is satisfied as well. It is clear that the scattering amplitude falls fast in the $|\ell| \to \infty$ limit: due to the repulsive centrifugal potential in (7.8), with ℓ increasing, the wave function $F_\ell(r)$ gets more and more suppressed at finite distances $r < r_0$ where the scattering potential $U(r)$ is concentrated.

Thus a_ℓ and b_ℓ are regular analytic functions of ℓ, together with the wave function $F_\ell(r)$, in the right half-plane $\mathrm{Re}\,\ell > \frac{1}{2}$. The amplitude f_ℓ in (7.11) is then regular everywhere but the points $a_\ell(k) = 0$ where it acquires *poles*.

We know already that the poles of the amplitude correspond to *resonances*. The position of such a pole, a complex energy $\overline{E}$, is determined by the equation

$$f_n(\overline{E}) = \infty$$

and depends, obviously, on n as a parameter of the Schrödinger equation, $\overline{E} = E(n)$. The equation

$$f_\ell(\overline{E}) = \infty$$

will have solutions for non-integer values of ℓ as well. In Lecture 3 we were considering the position of the resonance in *energy*, $\overline{E} = E(\ell)$, keeping $\ell = n$ fixed. Now we are studying the same object but from a different angle: we fix E and look at the position of the pole in the ℓ plane, $\bar{\ell} = \ell(E)$.

Having an analytic amplitude which falls properly with ℓ in the right half-plane and has no singularities but poles, we perform the Sommerfeld-Watson trick. Closing the contour around the rightmost pole at $\ell = \alpha(t)$ we will obtain

$$A^{\mathrm{pole}}(s,t) \;\; = \;\; \frac{-r(t)}{\sin \pi \alpha(t)} (2\alpha(t) + 1) P_{\alpha(t)}(-z), \qquad (7.12)$$

$$A(s,t) \;\; \simeq \;\; A^{\mathrm{pole}}(s,t) \propto z^{\alpha(t)}, \qquad z \to \infty. \qquad (7.13)$$

as was foreseen in (7.4).

This is a remarkable result.

We have two particles interacting via potential U. In an attractive potential, $U < 0$, there may be bound states or resonance states in various partial waves. Their energies depend on the angular momentum n of the state, $t = t(n)$. By simply *inverting* this dependence, $\ell = \ell(t)$, we get information about the large-z asymptotics: at a given energy t, the behaviour of the amplitude in the $z \to \infty$ limit is determined by the resonance that has *maximal angular momentum* $\ell(t)$ corresponding to this energy.

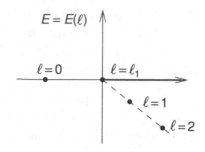

Fig. 7.1 Energy level in a shallow potential well changes continuously with orbital momentum ℓ.

Take not too deep a potential well, such that there exists a level for $\ell = 0$ but not for $\ell = 1$ when the centrifugal repulsion switches on. Imagine that we change ℓ continuously. With ℓ increasing, the energy level will be pushed up until at some $\ell = \ell_1 < 1$ it will cross $E = \overline{E}(\ell_1) = 0$ and move into the continuum, $E > 0$. For $\ell > \ell_1$ there is no *bound state* (a discrete energy level) in our potential any more. However, the level *as a solution of the Schrödinger equation* will not vanish in thin air. What will be its fate then? It can neither belong to the continuous spectrum (by unitarity), nor have a complex energy on the physical sheet (which is forbidden by causality). The only option for the level is to dive onto the *unphysical* sheet and acquire a complex mass there. That is, to become a resonance as displayed in Fig. 7.1.

Redraw the picture now. Let us change the energy and see what will happen to the angular momentum of the level.

At $E = \overline{E}(0)$ we have a pole in the partial wave $\ell = 0$. Increasing the energy, we will find the corresponding value of ℓ. At $E = 0$ we will have $\ell = \ell_1$. If we want to continue keeping ℓ real, we would have to lead E into the complex plane. If instead we continue to keep the *energy* real and increasing, then the pole in the ℓ-plane will move onto the upper half-plane as shown in Fig. 7.2.

It suffices to draw this curve in order to determine the asymptotics of the amplitude at large z. And this was first realized in the framework of a non-relativistic theory. T. Regge (1959) found the way to quantify the value of the characteristic angular momentum n_0 in (7.2),

$$A(s,t) \propto P_{n_0(t)}(z).$$

We may say that in the quantum-mechanical context n_0 measures the strength of the potential. It tells us, what the maximal value of the angular

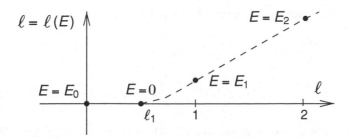

Fig. 7.2   Movement of the pole in the ℓ-plane in non-relativistic theory.

momentum is for which the attraction is still stronger than the centrifugal repulsion and the wave function is still concentrated at small distances so that the partial waves with $n \leq n_0(t)$ are large and contribute significantly to the partial-wave expansion (7.1a).

7.3 Complex ℓ in relativistic theory

It was not at all clear whether this programme could be carried out in a relativistic theory where the potential (if any) depends on particle velocities. Nevertheless, it turned out that the results that we have obtained for potential scattering are *almost* correct in the relativistic framework.

7.3.1 u-channel and a problem with analytic continuation

Why 'almost'? As we have discussed above in Section 7.1, the very supposition that an analytic function f_ℓ existed, immediately allowed me to analytically continue the series (7.1a), originally defined for $z \in [-1, +1]$, to arbitrary $|z| > 1$. We saw that for any complex value of z (but real positive z corresponding to $\Theta = 0$) the ratio (7.6) of $P_\ell(-z)$ and $\sin \pi \ell$ falls *exponentially* along the integration contour. In particular, for $z < 0$ ($\Theta = \pi$) the Legendre function $P_\ell(-z)$ does not increase at all along the imaginary ℓ-axis. The Sommerfeld–Watson integral for A then converges (and so do integrals for its derivatives over z). But this contradicts the fact that $A(z)$ must have singularities at $z < -1$ since we know that our relativistic amplitude has a cut at $s < 0$ corresponding to the u-channel scattering!

Nevertheless, let us try to approach the problem constructively to see where the problem lies and how we might overcome it. So, we start again from two complementary formulae for integer n:

$$A(s,t) \equiv A(z,t) \;=\; \sum_n (2n+1) f_n(t) P_n(z), \qquad (7.14a)$$

$$f_n(t) \;=\; \frac{1}{2} \int_{-1}^{1} dz\, P_n(z) A(z,t), \qquad (7.14b)$$

and search for a way of continuing (7.14) to complex angular momenta ℓ.

The task of defining f_ℓ seems easy at the first glance. A straightforward generalization of (7.14b) by a simple substitution $n \to \ell$ will, however, not satisfy us. Indeed, since the z-integral involves all points in the interval $[-1,+1]$, including $z = -1$, it is this end-point corresponding to $\Theta = \pi$ that will make f_ℓ, so defined, behave as $\exp(i\pi\ell)$. But this behaviour violates Carlson's theorem (uniqueness of the analytic continuation) thus forcing us to abandon this bold attempt.

To find a smarter way let us make use of the knowledge of the analytic structure of the amplitude:

$$A(z,t) \;=\; \frac{1}{\pi} \int_{z_0}^{\infty} dz'\, \frac{A_1(z',t)}{z'-z} + \frac{1}{\pi} \int_{z_0}^{\infty} dz_u\, \frac{A_2(-z_u,t)}{z_u+z}, \qquad (7.15)$$

where we represented the contribution of the left cut in terms of a positive integration variable $z_u = -z$. I wrote the dispersion relation without subtractions. They are necessary in principle to have the integrals convergent. 'Subtraction' in the dispersion integral means extracting a polynomial in z of some degree N. However, I am now going to study partial waves with sufficiently large n. If I take $n > N$, then, due to the orthogonality of P_ns, the subtracted polynomial will not affect my partial waves f_n. So we may substitute our 'analytic wisdom' (7.15) into (7.14b) to obtain

$$f_n(t) = \frac{1}{\pi} \int_{z_0}^{\infty} dz\, A_1(z,t) Q_n(z) - \frac{1}{\pi} \int_{z_0}^{\infty} dz_u\, A_2(-z_u,t) Q_n(-z_u), \quad (7.16)$$

where

$$Q_n(z) \equiv \frac{1}{2} \int_{-1}^{1} dz'\, \frac{P_n(z')}{z-z'}. \qquad (7.17)$$

The new expression (7.16) better suits our purpose; it is tempting to try

$$Q_n(z) \;\Longrightarrow\; Q_\ell(z)$$

with $Q_\ell(z)$ the second solution of the Legendre equation that is regular at infinity:

$$Q_\ell(z) \propto z^{-\ell-1}, \qquad |z| \to \infty.$$

Such a behaviour is perfectly satisfactory for continuing the first term in (7.16). In the second term, however, we get

$$Q_\ell(-z) \propto (-z)^{-\ell-1} = -e^{-i\pi\ell}|z|^{-\ell-1},$$

and the second part of the partial wave again acquires too fast an exponential increase with $\mathrm{Im}\,\ell$.

So what's the way out? Using the relation

$$Q_n(-z) = (-1)^{n+1}Q_n(z)$$

valid for integer n, we can rewrite (7.16) as

$$f_n(t) = \frac{1}{\pi} \int_{z_0}^\infty dz\, A_1(z,t)Q_n(z) + \frac{(-1)^n}{\pi} \int_{z_0}^\infty dz\, A_2(-z,t)Q_n(z). \quad (7.18)$$

This way we localize the problem before attempting the analytic continuation to complex n. Actually, we have already derived an analogous formula for s-channel partial-wave amplitudes in Lecture 5 when we discussed the relativistic phenomenon of the appearance of the *backward peak* in the differential angular cross section, cf. (5.18).

7.3.2 Continuing separately even and odd angular momenta

We have to abandon the idea of constructing an analytic continuation of f_n from *all* integer points anyway: as we already know such an attempt is bound to fail because of the existence of the u-channel singularities. We are led to try to continue even and odd angular momenta separately,

$$f_\ell^{(+)}\Big|_{\ell=2n} = f_{2n}, \qquad f_\ell^{(-)}\Big|_{\ell=2n+1} = f_{2n+1}.$$

By so doing we get rid of the oscillating factor in (7.18) and obtain two functions,* both behaving nicely at large $|\ell|$:

$$f_\ell^{(\pm)}(t) = \frac{1}{\pi} \int_{z_0}^\infty dz\, Q_\ell(z)A_1(z,t) \pm \frac{1}{\pi} \int_{z_0}^\infty dz\, Q_\ell(z)A_2(-z,t). \quad (7.19)$$

Now for the analytic continuation to be unique, a stronger condition than (7.7) must be imposed:

$$\left|f_\ell^{(\pm)}\right| < \exp\left(\tfrac{1}{2}|\ell|\pi\right).$$

It is easy to verify that the functions defined by (7.19) do satisfy this condition easily (along the imaginary axis they don't increase at all).

---

* Equation (7.19) is known as the *Gribov–Froissart projection* (ed.).

Thus the price we pay for solving the problem of continuation is the introduction of *two* analytic functions in place of one. What is the reason for that?

In a relativistic theory one 'potential' is not enough. There is always another diagram corresponding to what is known in nuclear physics under the name of *exchange potential* V_{exch}:

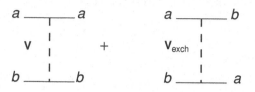

The two graphs differ by the transposition $a \leftrightarrow b$ which, for spinless particles, introduces the factor $(-1)^\ell$. Therefore for *even* and *odd* orbital momenta we have

$$V \Longrightarrow V + V_{\text{exch}} \; (\ell = 2n) \qquad \text{and} \qquad V \Longrightarrow V - V_{\text{exch}} \; (\ell = 2n+1),$$

correspondingly. Having two different full potentials means that there isn't any analytic relation between partial waves with even and odd angular momenta.

7.3.3 Sommerfeld–Watson representation for $f_\ell^{(\pm)}$

Let us split the amplitude into symmetric and anti-symmetric parts with respect to $s \leftrightarrow u$,

$$A(z,t) = A^+(z,t) + A^-(z,t), \quad A^\pm(z,t) \equiv \frac{A(z,t) \pm A(-z,t)}{2}, \quad (7.20)$$

and treat these two amplitudes separately:

$$A^\pm(z,t) = \sum_{n=\text{even/odd}} (2n+1) f_n^{(\pm)}(t) P_n(z). \qquad (7.21)$$

Recall that above we wrote the dispersion relation without subtractions, which was fine for the purpose of analysing partial waves with $n > N$. The integrals (7.19) for $f_\ell^{(\pm)}$ are defined also for sufficiently large angular momenta. This is necessary to ensure convergence: if the amplitude behaves at large z as $|A(z)| = \mathcal{O}(z^N)$, then

$$f_\ell^{(\pm)} \sim \int^\infty dz \, Q_\ell(z) A(z) \sim \int^\infty dz \cdot z^{-\ell-1} \cdot z^N < \infty \quad \Longrightarrow \quad \operatorname{Re}\ell > N.$$

This means that the series (7.21) for the amplitude $A^\pm$ can be represented by the Sommerfeld–Watson integral only *partially*. Namely, it is the infinite series of partial waves with $n \geq N$ in

$$A = \sum_{n=0}^{N}(2n+1)f_n P_n + \sum_{n=N+1}^{\infty}(2n+1)f_n P_n$$

that can be combined into an analytic function of ℓ, while a few first partial waves remain, generally speaking, arbitrary.

Why did a few partial waves remain unaccounted for? In a non-relativistic language, we may imagine adding to the potential a singular term $\delta_0 V(\mathbf{r}) = \mathrm{const} \cdot \delta(r)$. It contributes to the S-wave scattering only and as a result f_0 would fall out of the family. Introducing $\delta_1 V(\mathbf{r}) \propto \boldsymbol{\nabla}\delta(r)$ we would analogously spoil the P-wave f_1. If we continue these singular series by summing up an infinite number of derivatives we may violate causality (polynomial boundary $|A| \lesssim s^N$). As we shall see soon from the s-channel unitarity condition, the number of 'special' partial waves that remain unaccounted for in fact cannot be larger than two: $\ell = 0$ and $\ell = 1$.

Now that we have removed the oscillating factor $(-1)^\ell$ that used to cause too fast an exponential increase, we can deform the contour $\mathcal{C}$ in
the Sommerfeld–Watson representation embrac-
ing the points $n \geq N+1$ by straightening it and
sending along the imaginary axis at $\mathrm{Re}\,\ell = \ell_0$ so
that $N < \ell_0 < N+1$:

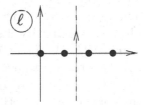

$$A^{\pm}(s,t) = \sum_{n=0}^{N}(2n+1)f_n(t)P_n(z) \cdot \tfrac{1}{2}[1 \pm (-1)^n]$$

$$+ \frac{i}{4}\int_{\ell_0-i\infty}^{\ell_0+i\infty}\frac{d\ell\,(2\ell+1)}{\sin\pi\ell}f_\ell^{(\pm)}(t)[P_\ell(-z) \pm P_\ell(z)]. \tag{7.22}$$

When the contour transformation is done, we are ready to leave the physical region of the t-channel and to study the large-s regime. Taking $z \to \infty$ will affect only the oscillating factor $\exp(i\,\mathrm{Im}\,\ell \cdot \ln z)$ but not the convergence of the integral.

Let us verify that the formula (7.22) is what we have been looking for. To this end, examine the analytic properties of $A^{\pm}$. They obviously must be those that we put in. Namely at positive $s > 4\mu^2$ we should encounter a non-zero *absorptive part* which appears with the opening of the first s-channel threshold:

$$A_1^{\pm} = \frac{A(s+i\epsilon) - A(s-i\epsilon)}{2i} = \tfrac{1}{2}[A_1(s) \pm A_2(s)], \tag{7.23a}$$

as it follows directly from the definition of the amplitudes (7.20). Also, if we decrease s, starting from $u > 4\mu^2$ we should see the cross-channel absorptive part

$$A_2^\pm \equiv \frac{A(u + i\epsilon) - A(u - i\epsilon)}{2i} = \tfrac{1}{2}[A_2(u) \pm A_1(u)];$$

$$A_2^\pm(z, t) = \pm A_1^\pm(-z, t). \tag{7.23b}$$

Taking $s > 0$ we have $z > 1$, and the first Legendre function under the integral in (7.22) becomes complex since $P_\ell(z)$ with non-integer ℓ has a *logarithmic cut* running from -1 to $-\infty$. The phase of the argument in $P_\ell(-z)$ starts to matter. Comparing two ways of defining $(-z)$,

$$\frac{1}{2i}[P_\ell(e^{i\pi}z) - P_\ell(e^{-i\pi}z)] = \sin \pi\ell \cdot P_\ell(z),$$

and substituting into (7.22) we derive the absorptive part

$$A_1^\pm(s, t) = \frac{1}{4i}\int_C d\ell(2\ell + 1)f_\ell^{(\pm)}(t)P_\ell(z). \tag{7.24}$$

The complexity of $P_\ell(-z)$ has, however, nothing to do with physics. Therefore at $s > 0$ the amplitude must stay real until we meet its first physical singularity. Let us see how it happens in our formula.

In (7.24) the poles in the integer points have disappeared inviting us to move the contour back to the right and close it at $+\infty$. If $|z| < 1$ we can always do so to obtain $A_{\text{abs}}^\pm \equiv 0$ as expected. As for $z > 1$, closing the contour will be still possible as long as the integrand falls in the right half-plane.

We have studied the large-ℓ asymptote of f_ℓ in Section 5.3. Applying (5.26) to the t-channel partial waves,

$$f_\ell(t) \propto \exp(-\ell\chi_0), \qquad \cosh \chi_0 \equiv z_0 = 1 + \frac{2 \cdot 4\mu^2}{t - 4\mu^2}, \tag{7.25a}$$

and comparing with the asymptote of the Legendre functions (5.25),

$$P_\ell(z) \propto \exp(\ell\Theta), \qquad \cosh \Theta = z, \tag{7.25b}$$

we immediately see that the absorptive part $A_1 = 0$ as long as $\Theta < \chi_0$, that is up to $s = 4\mu^2$ when we hit the s-channel threshold singularity and the partial-wave expansion (7.14a) diverges.

Considering analogously $z < 0$, it is easy to obtain the *u-channel* absorptive part independently,

$$A_2^\pm(z, t) = \pm\frac{1}{4i}\int_C d\ell(2\ell + 1)f_\ell^{(\pm)}(t)P_\ell(-z) = \pm A_1^\pm(-z, t), \tag{7.26}$$

in accord with the expectation (7.23b).

Thus we have derived two formulae relating partial waves with a definite *signature* to the absorptive part of the amplitude with a definite $s \leftrightarrow u$

symmetry:

$$A_{\text{abs}}^{\pm}(s,t) = \frac{1}{4i} \int_C d\ell (2\ell+1) f_\ell^{(\pm)}(t) P_\ell(z); \qquad (7.27a)$$

$$f_\ell^{(\pm)}(t) = \frac{2}{\pi} \int_{z_0}^{\infty} dz \, Q_\ell(z) A_{\text{abs}}^{\pm}(z,t), \qquad (7.27b)$$

with z_0 defined in (7.25a). Remember, z in these formulae is a cosine of the *t-channel* scattering angle,

$$z \equiv z_t(s,t) = 1 + \frac{2s}{t - 4\mu^2} = \frac{s-u}{t - 4\mu^2}. \qquad (7.27c)$$

Let us note an attractive feature of (7.27a): the expression for A_{abs} is free from an undetermined sum of few 'non-analytic' terms present in the Sommerfeld–Watson representation for the amplitude A itself. Moreover, A_{abs} is a valuable thing: continuing to $t < 0$, we will get hold of the imaginary part of the s-channel amplitude which interests us much.

On its own, the expression (7.27a) for the absorptive part is sort of trivial, something resembling the Mellin transformation. It is complementary to (7.27b) for $f_\ell^{(\pm)}$ which expression is slightly less obvious as it exploits analytic properties of the amplitude. Still, if not for the unitarity condition, the translation of A_{abs} into f_ℓ, and back again, would have had not much value (although performing Mellin transform may be sometimes useful). The essence of the issue lies in that the singularities in ℓ of $f_\ell^{(\pm)}$ are determined by the *physical spectrum* of particles and resonances.

7.4 Analytic properties of partial waves and unitarity

In order to determine the character of the large-s asymptotics of the scattering amplitude, we need to learn what singularities $f_\ell^{(\pm)}(t)$ has in the ℓ-plane. Moreover, till now we were sitting at $t > 4\mu^2$ while it is the physical region of the s-channel, $t < 0$, that really interests us. In non-relativistic quantum mechanics we saw how the unitarity condition has translated the poles of the amplitude on the unphysical sheet (resonances) into singularities in ℓ of the partial wave f_ℓ – the Regge trajectories $\ell = \ell(t)$. We are about to try the same path in the relativistic theory.

7.4.1 Redefining partial waves

So we will keep $\text{Re}\,\ell > N$ and discuss the properties of $f_\ell(t)$ defined by (7.27b). We have introduced partial waves with complex ℓ at $t > 4\mu^2$. There $f_\ell^{(\pm)}$ are complex, because so are A_1 and A_2 (see the path 1 in

Fig. 7.3). Below the t-channel threshold the absorptive parts are real and it would have been nice if the partial waves at $t < 4\mu^2$ were real too. However when the sign of $t - 4\mu^2$ in the expression (7.27c) changes, z becomes negative. We have then to watch for $Q_\ell(z) = z^{-\ell-1} F(z)$, with F a regular even function of z, which acquires an ℓ-dependent phase. This phase will provide f_ℓ with a 'kinematical' complexity which has nothing to do with analyticity (since $A_{\rm abs}$ is real!). Let us have a look at the vicinity of the threshold, $0 < t - 4\mu^2 \ll \mu^2$. Here we have $z \simeq 2\,s/(t - 4\mu^2) \gg 1$, and (7.27b) gives

$$ f_\ell^{(\pm)} \propto (t - 4\mu^2)^\ell \cdot \int_{4\mu^2}^{\infty} \frac{ds}{s^{\ell+1}} [A_1 \pm A_2]. $$

For integer n this is nothing but the usual threshold behaviour, $f_n \propto k_c^{2n}$. To get rid of the trivial phase factor it is convenient to redefine partial waves by introducing

$$ f_\ell^{(\pm)}(t) \equiv (t - 4\mu^2)^\ell \cdot \phi_\ell^{(\pm)}(t). \tag{7.28} $$

The new partial wave ϕ_ℓ is given by the integral

$$ \phi_\ell^{(\pm)}(t) = \frac{2}{\pi} \int_{4\mu^2}^{\infty} \frac{2\,ds}{(t - 4\mu^2)^{\ell+1}} Q_\ell(z) A_{\rm abs}^{\pm}, $$

where we choose to integrate over s rather than z as in the original formula (7.19). Moving to $t < 4\mu^2$, we reflect the argument of $Q_\ell(z \to -z)$ and write down a more convenient expression,

$$ \phi_\ell^{(\pm)}(t) = \frac{4}{\pi} \int_{4\mu^2}^{\infty} \frac{ds}{(4\mu^2 - t)^{\ell+1}} Q_\ell \left(\frac{2s}{4\mu^2 - t} - 1 \right) A_{\rm abs}^{\pm}(s, t), \tag{7.29} $$

which makes it clear that ϕ_ℓ stays real in the interval $0 < t < 4\mu^2$. If we decrease t further, at $t < 0$ the Legendre function $Q_\ell(z)$ becomes complex ($|z| < 1$) when $4\mu^2 < s < 4\mu^2 - t$ (interval $[a, b]$ on the path #2 in Fig. 7.3). A *physical* singularity will emerge later, when the integration line on the Mandelstam plane crosses the Karplus curve where $\rho_{su} \neq 0$ and the *absorptive part* becomes complex (interval $[c, d]$ on the line #3 in Fig. 7.3).

7.4.2 Two-particle unitarity condition for ϕ_ℓ

Recall how we have used the two-particle unitarity condition to find the discontinuity on the right cut of the partial wave with integer n:

$$ {\rm Im}\, f_n(t) = \tau f_n(t) f_n^*(t), \qquad 4\mu^2 < t < 16\mu^2, $$

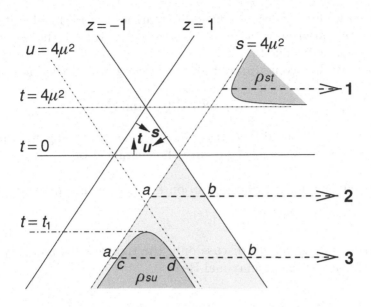

Fig. 7.3   Integration paths in the representation (7.29) for $\phi_\ell(t)$. For $t < 0$, on the interval $[a, b]$ the Legendre function $Q_\ell(z)$ is complex. When $t < t_1$, the third spectral function contributes to complexity of ϕ_ℓ along the $[c, d]$ interval.

or, in terms of ϕ,

$$\frac{1}{2i}[\phi_n(t + i\epsilon) - \phi_n(t - i\epsilon)] = C_n \phi_n(t + i\epsilon) \phi_n(t - i\epsilon),$$

$$C_n \equiv \tau \cdot (t - 4\mu^2)^n.$$

Now we can state that the same relation holds for arbitrary complex ℓ:

$$\frac{1}{2i}[\phi_\ell(t + i\epsilon) - \phi_\ell(t - i\epsilon)] = C_\ell \phi_\ell(t + i\epsilon) \phi_\ell(t - i\epsilon),$$

$$C_\ell \equiv \tau \cdot (t - 4\mu^2)^\ell. \tag{7.30}$$

Why? Thanks to the Carlson theorem. Indeed, as it is easy to see from the properties of ϕ_ℓ, the difference between the l.h.s. and the r.h.s. of (7.30) does not increase at infinity faster than $\exp(\frac{1}{2}\pi|\ell|)$ and equals zero in all even (odd) points. Therefore it is zero on the entire ℓ-plane.

The unitarity condition (7.30) applies to $4\mu^2 < t < 16\mu^2$. Above the four-pion threshold in the unitarity condition there appear *integral terms* the continuation of which to complex angular momenta is not so easy.

Now that we briefly described the singularities in t for fixed ℓ, it is time to turn to the question which really interests us – that of the structure of singularities in ℓ for fixed t.

Where may they come from? From the divergence of the integral

$$\phi_\ell \sim \int^\infty dz\, Q_\ell(z) A_{\mathrm{abs}}(z)$$

at some sufficiently small $\mathrm{Re}\,\ell$. It is difficult to say anything starting from nothing. Therefore we will begin with the classification.

(1) 'Fixed singularity' whose position $\ell = \ell_0$ does not depend on t.

(2) 'Moving singularity', $\ell = \ell_0(t)$.

For the time being we will discuss only the rightmost singularity in the ℓ-plane, the one with the maximal $\mathrm{Re}\,\ell_0$.

7.4.3 Fixed singularities in the ℓ plane

The first statement we can make about the fixed singularity (should it happen to be the rightmost one) is that it may occur only on the *real axis*. Indeed, whatever the nature of the singular point, its contribution to the asymptote of $A_{\mathrm{abs}}(z)$ is proportional to z^{ℓ_0}. If $\mathrm{Im}\,\ell_0 \neq 0$, this factor *oscillates* fast at large z. But this would contradict the positivity of the cross section, $\sigma_{\mathrm{tot}} \propto A_{\mathrm{abs}}$ (in the next lecture we will verify that the t-*derivative* of $A_{\mathrm{abs}}(s,t)$ must also be positive in the interval $0 < t < 4\mu^2$).

Now, since $\ell_0 \neq \ell_0(t)$, we can choose $4\mu^2 < t < 16\mu^2$ to see what the unitarity condition would tell us. The r.h.s. of (7.30), for *real* ℓ_0, reduces to $|\phi_\ell|^2$ and we have

$$\mathrm{Im}\,\phi_\ell = C_\ell \cdot |\phi_\ell|^2 \quad \Longrightarrow \quad |\phi_\ell| < C_\ell^{-1}. \qquad (7.31)$$

Therefore the singularity may be only a *weak* one, namely such that in the singular point the partial-wave amplitude stays finite like, for example, $\phi_\ell \sim \sqrt{\ell - \ell_0}$. But this is exactly the case when, as we have discussed in the previous lecture, the cross section *must fall* at large s.

Let us check that, indeed, $|\phi_{\ell_0}| < \infty$ implies a falling cross section. Dropping irrelevant factors we write

$$A_{\mathrm{abs}}(s) \sim \int_C d\ell\, \phi_\ell \cdot s^\ell$$

and shift the contour to the left in search for singularity. If ϕ_ℓ had a pole we would have taken the residue and obtained a power asymptote

$$A_{\mathrm{abs}}^{\mathrm{pole}} = \mathrm{const} \cdot s^{\ell_0}.$$

The pole is, however, forbidden by the t-channel unitarity restriction (7.31). (By the way, a particular case of this veto is the familiar classical diffraction picture, $A_{\text{abs}}(s,t) = sF(t)$, which corresponds to the fixed pole singularity at $\ell_0 = 1$.) Therefore ϕ_ℓ may only have a *branch cut* starting at ℓ_0 and running to the left. Integrating the discontinuity $\Delta\phi_\ell$ along the cut we get

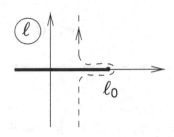

$$A_{\text{abs}} \simeq s^{\ell_0} \cdot \int_0^\infty dx \, \Delta\phi(x) \, e^{-x \ln s},$$

where we have introduced $x = \ell_0 - \ell > 0$ as an integration variable. When s is large, the integral converges at $\langle x \rangle \sim 1/\ln s \ll 1$ so that only the very tip of the cut matters. Parametrizing the discontinuity of the partial wave as $\Delta\phi(x) \propto x^\gamma$ and evaluating the integral we obtain

$$A_{\text{abs}} \propto \frac{s^{\ell_0}}{(\ln s)^{\gamma+1}}. \tag{7.32}$$

The *finiteness* condition (7.31) tells us that $\gamma \geq 0$. As a result

$$\sigma \propto s^{-1} A_{\text{abs}} \sim \frac{s^{\ell_0 - 1}}{(\ln s)^{\gamma+1}} < \frac{1}{\ln s}, \tag{7.33}$$

where we have used the maximal power value $\ell_0 = 1$ allowed by the Froissart theorem.

Fixed singularities in NQM. Do fixed singularities exist in quantum mechanics? Yes, and they are related to the '*falling on the centre*' phenomenon, with the behaviour of the potential at small distances. As we have already discussed, in non-relativistic quantum mechanics a singularity appears in ℓ when the choice between the two solutions of the Schrödinger equation at the origin, $r = 0$, becomes ambiguous. When discussing NQM scattering, we have tacitly implied that the interaction potential was less singular than the centrifugal barrier, $r^2 \cdot V(r) \to 0$ at $r \to 0$, in which case the singularity was at $\ell_0 = -\frac{1}{2}$ (when $r^{-\ell} \sim r^{\ell+1}$). In the opposite case $V(r)$ itself will govern the $r \to 0$ asymptotic behaviour of the wave function $\psi_\ell(r)$, and a fixed singularity may emerge at any ℓ_0.

It is important to stress that such fixed singularities correspond to definite physics, namely a super-singular behaviour of the interaction at small distances. It seems they are very unlikely to have any relation with the approximate constancy of the total cross section.

Strictly speaking, the question of a possible rôle of fixed singularities remained unsolved.

7.4.4 Moving singularities

More difficult to analyse are moving singularities. At the same time they are much more interesting. Given the function $\ell = \ell_0(t)$, we might invert it and consider $t = t_0(\ell)$ as a singularity on the t-plane about which we have already learned a thing or two!

Above we have described *all* singularities of the partial wave amplitude on the t-plane: there is nothing but the right cut, $t > 4\mu^2$ and the left cut, $t < 0$. So where are then these new ℓ-dependent singularities?

$$\phi_\ell^{(\pm)}(t)$$

$$\underset{t_0}{\rule{2cm}{0.4pt}} \quad \underset{0}{\Big|} \quad \underset{4\mu^2}{\rule{1.5cm}{0.4pt}}$$

Recall that our analysis of $\phi_\ell^{(\pm)}(t)$ was carried out for $\operatorname{Re}\ell > N$. New singularities show up at smaller ℓ. How can this happen? It is clear that a singularity cannot just 'pop up' suddenly with ℓ decreasing. Our amplitude has cuts on the complex t-plane, and the possibility arises for the singularity to move from beneath a cut and appear on the physical sheet at some $\ell < N$. This means that such a moving singularity is always present but 'hidden' on the *unphysical sheets* of the amplitude at large $\operatorname{Re}\ell$. Therefore, in order to learn which singularities the partial wave with $\operatorname{Re}\ell < N$ may have on the t-plane, it suffices to find out (as we did before when we studied resonances in Lecture 3) what the singularities are on the unphysical sheets at $\operatorname{Re}\ell > N$.

Left cut. The first important statement: No moving singularities emerge from the sheets linked to the *left cut*.

We take $t < 0$, write

$$\phi_\ell^{(\pm)}(t + i\epsilon) = \phi_\ell^{(\pm)}(t - i\epsilon) + \Delta\phi_\ell^{(\pm)}(t)$$

and move the argument $t + i\epsilon$ down under the cut to explore the corresponding sheet. The amplitude $\phi_\ell^{(\pm)}(t - i\epsilon)$ on the r.h.s. of the equation stays on the physical sheet. Since it is regular there, the partial wave on the l.h.s. diving under the cut will exhibit singularities of the discontinuity over the left cut, $\Delta\phi_\ell^{(\pm)}(t)$. The latter, however, cannot have any (moving) singularities. This is a consequence of a simple fact that, as we have repeatedly stressed before, it is given by integrals over *finite* intervals. Indeed, one contribution to $\Delta_t\phi$ comes from the complexity of Q_ℓ at $t < 0$: $\Delta_t Q_\ell(-z) = -\frac{\pi}{2}P_\ell(-z)$ (for $-1 < z < 1$); the other one appears at $t < t_1$ due to the discontinuity of $A_{\text{abs}}^\pm$: $\operatorname{Im}_t \operatorname{Im}_s A^\pm \equiv \rho_{su}^\pm$. From (7.29) we obtain

$$\Delta\phi_\ell^{(\pm)}(t) = -2 \int_{s_a}^{s_b} \frac{ds}{(4\mu^2 - t)^{\ell+1}} P_\ell\left(\frac{2s}{4\mu^2 - t} - 1\right) A_{\text{abs}}^\pm(s, t + i\epsilon)$$

$$+ \frac{4}{\pi} \int_{s_c}^{s_d} \frac{ds}{(4\mu^2 - t)^{\ell+1}} Q_\ell\left(\frac{2s}{4\mu^2 - t + i\epsilon} - 1\right) \rho_{su}^\pm(s, t). \tag{7.34}$$

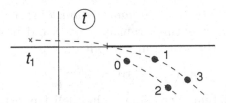

Fig. 7.4   Resonances on Regge trajectories $\ell^{\pm}(t)$ and the movement of a Regge pole onto the physical sheet.

The integrals run over finite regions; they converge and cannot produce singularities. As for an explicit ℓ-dependence of the integrands, P_ℓ is regular on the entire ℓ plane; Q_ℓ is 'almost regular': strictly speaking, it has *poles* in negative integers, $\ell = -1, -2, \ldots$, but these do not concern us here as they are not related to moving singularities.

Right cut. Exactly as it was the case of integer angular momenta that we have explored in Lecture 3, on the unphysical sheet linked to the two-particle cut there may be only poles:

$$\phi_\ell^{(\pm)}(+) = \frac{\phi_\ell^{(\pm)}(-)}{1 - 2iC_\ell\,\phi_\ell^{(\pm)}(-)}, \quad \phi_{\ell_0}^{(\pm)}(t) = \frac{1}{2iC_{\ell_0}(t)} \quad \Longrightarrow \quad \ell_0 = \ell^{\pm}(t).$$

A remarkable thing! We knew that resonances with different spins n live on the unphysical sheet. Now not only have we got the statement about the large-s behaviour but also about the resonances themselves. In Lecture 3 we had independent equations for resonance masses,

$$\phi_n^{(\pm)}(t) = [\,2iC_n(t)\,]^{-1} \quad \Longrightarrow \quad m_n^2 = t(n).$$

Now we see that all these resonances are analytically linked to each other as shown in Fig. 7.4. This discovery laid the basis for the classification of all hadrons according to 'Regge trajectories' they belong to. Real and imaginary parts of the position of the pole on the t-plane give the squared mass and the width of the resonance.

Moreover, two (generally speaking different) analytic curves $\ell^{\pm}(t)$ that combine together resonances with even spins and those with odd spins, are those very same curves that determine the asymptotic behaviour of the symmetric and anti-symmetric parts of the scattering amplitude, correspondingly.

What sort of information may this give us in practice? By studying t-channel particle scattering at relatively small energies t, experimenters find resonances, measure their masses, decay widths and determine their spins. Imagine that we put 'many points' on a Regge trajectory and in so doing approximately found $\operatorname{Re}\ell(t)$. This is a mere classification at this point. Now, let us extrapolate the curve to $t = 0$, and below. This will tell us the *characteristic angular momentum*, $\ell^{\pm}(t_1)$, corresponding to a given value of the momentum transfer $t_1 < 0$ and immediately give the energy behaviour of the scattering amplitude,

$$A^{\pm}(s, t_1 \leq 0) \propto s^{\operatorname{Re}\ell^{\pm}(t_1)},$$

in a completely different – crossing – channel! Understanding this cross-channel relation constitutes the main achievement of the theory of complex angular momenta.

8

Reggeon exchange

We have discussed analytic properties of the partial-wave amplitude $f_\ell^\pm(t)$. Further, having realized that moving singularities $\ell = \alpha(t)$ in the complex angular momentum plane can be investigated also as singularities in the energy plane, $t = t(\ell)$, we discussed how they appear from the unphysical sheets, connected with the right (unitary) cuts of the amplitude. We found that on the first unphysical sheet partial waves can have only poles, i.e. the situation here turned out to be the same as for integer angular momenta, $\ell = n$.

Recall that the picture we had for integer n was dominated by poles. We put in particles (poles on the physical sheet), and they generated singularities on other sheets linked to production thresholds of two or more particles (or resonances).

It is clear that first the pole singularities must be studied. In non-relativistic quantum mechanics, by increasing the interaction strength, I can turn a resonance (a virtual state) into a real particle (a bound state). In the theory of complex angular momenta the same phenomenon takes place with the decrease of ℓ: a resonance pole moves from the unphysical sheet, through the tip of the unitary cut, onto the physical sheet, see Fig. 7.4 above. In this sense ℓ is akin to an interaction constant of the non-relativistic theory. So, in this lecture we are going to discuss the properties of Regge poles and the picture of the strong interactions in the pole approximation.

As we shall see later, Regge poles generate *branch cuts* in the ℓ plane, in a manner similar to the generation of threshold branchings via unitarity conditions by poles (particles) in the case of integer angular momenta. The existence of these new branching singularities will seriously affect some of the results of the present lecture.

8.1 Properties of the Regge poles. Factorization

In the previous lecture we have introduced two analytic functions $f_\ell^\pm(t)$. What sort of amplitude corresponds to a single pole in the partial wave of definite signature? Substituting the pole expression

$$f_\ell^\pm(t) = \frac{r^\pm(t)}{\ell - \alpha^\pm(t)}$$

into the Sommerfeld–Watson integral (7.22) results in the following contribution to the scattering amplitude in the s channel:

$$A^\pm(s,t) = -\frac{\pi}{2} r \frac{2\alpha + 1}{\sin \pi\alpha} [P_\alpha(-z) \pm P_\alpha(z)], \quad z = 1 + \frac{2s}{t - 4\mu^2}, \quad (8.1)$$

where we have suppressed the signature label $\pm$ for the residue, $r = r^\pm(t)$ and the trajectory, $\alpha = \alpha^\pm(t)$.

Let us look at the singularities of $A^+(s,t)$. The amplitude with a positive signature has poles at those t values for which $\alpha^+(t) = 2n$ is an even number. Assume, e.g. that $\alpha(t) = 0$ at $t = m_0^2$. This means that if $t \to m_0^2$, the physical partial wave $f_0(t)$ tends to infinity. Indeed, near the pole we have

$$f_\ell^+(t) \simeq \frac{r(t)}{\ell - \alpha(m_0^2) - \alpha'(m_0^2)(t - m_0^2)},$$

and, since $\alpha(m_0^2) = 0$,

$$f_0(t) \simeq \frac{r(m_0^2)}{\alpha'(m_0^2) \cdot (m_0^2 - t)}.$$

Then the pole term of $A^+(s,t)$,

$$A_0^+(s,t) = \frac{r}{\alpha'} \frac{1}{m_0^2 - t} = \quad \sigma = 0$$

coincides with the exchange diagram for a spin $\sigma = 0$ particle (with residue $g^2 = r/\alpha'$).

The next singularity of A^+ appears at some $t = m_2^2$ where $\alpha(m_2^2) = 2$:

$$A_2^+(s,t) = \frac{r(m_2^2)}{\alpha'(m_2^2)} \frac{2\alpha(m_2^2) + 1}{m_2^2 - t} P_2(z).$$

This is, however, just the contribution of the diagram for a spin $\sigma = 2$ particle exchange:

$$= \Gamma_{\mu_1\mu_2} \frac{d_{\mu_1\mu_2,\mu_3\mu_4}}{m_2^2 - t} \Gamma_{\mu_3\mu_4} = g^2 \frac{P_2(\cos\Theta_t)}{m_2^2 - t}.$$

Hence, in the physical points of its proper signature, $\alpha^+ = 2n$, the trajectory $\ell = \alpha^+(t)$ reproduces the particle (resonance) exchange with even spin values for the amplitude (8.1). Similarly, the contribution of the Regge pole of negative signature $A^-(s,t)$ contains the exchange of odd spin particles.

Let us note that a trivial generalization of a Feynman diagram

$$\sigma = g^2 \frac{P_\sigma(z)}{m_\sigma^2 - t} \tag{8.2}$$

to the case of non-integer spin σ, based on the assumption that the mass is a continuous function of the spin, $m_\sigma^2 \to m^2(\ell)$, does not give a reggeon amplitude (8.1). Indeed, the amplitude (8.2) has a pole for any ℓ value, whereas in the expression (8.1) the poles emerge in integer points only. In addition, the Regge amplitude (8.1) bears a non-trivial complexity due to the factor $[P_\alpha(-z) \pm P_\alpha(z)]$, which cannot be obtained by the generalization (8.2) either.

The expression (8.1) can be considered as the contribution to the scattering amplitude coming from the exchange of a 'particle' of variable spin – a *reggeon* – with a propagator

$$\frac{[P_{\alpha(t)}(-z) \pm P_{\alpha(t)}(z)]}{\sin\pi\alpha(t)} \qquad \text{instead of the usual} \qquad \frac{P_\sigma(z)}{m_\sigma^2 - t}.$$

The analogy between the reggeon and the particles (resonances) will be complete, if we find that the reggeon residue is factorizable. In this case we will be able to speak not only about 'propagation' of the Regge pole but also about the vertices of its 'emission' and 'absorption'.

As in the case of resonances, the factorization is, essentially, a consequence of unitarity. We presented a formal proof of factorization for resonances by diagonalizing the S-matrix in Lecture 3. Let us show in a more transparent way how factorization appears. Consider, e.g. three different reactions

$$\pi\pi \to \pi\pi, \qquad \pi K \to \pi K, \qquad KK \to KK. \tag{8.3a}$$

For each amplitude (8.3a) its own partial waves can be introduced in the t-channel:

$$\varphi_\ell(\pi\pi \to \pi\pi), \qquad g_\ell(\pi\pi \to K\bar{K}), \qquad h_\ell(K\bar{K} \to K\bar{K}). \qquad (8.3b)$$

In the region $4\mu^2 < t < 16\mu^2 < 4m_K^2$ there exists only one intermediate state and the unitarity condition takes the simple form

$$\text{Im}\,\varphi_\ell(t) = C_\ell\,\varphi_\ell(t)\varphi_\ell^*(t); \qquad (8.4a)$$

$$\text{Im}\,g_\ell(t) = C_\ell\,\varphi_\ell(t)g_\ell^*(t); \qquad (8.4b)$$

$$\text{Im}\,h_\ell(t) = C_\ell\,g_\ell(t)g_\ell^*(t). \qquad (8.4c)$$

Here $C_\ell = \tau(t - 4\mu^2)^\ell$ with $\tau = \omega_c/16\pi k_c$ the invariant phase-space volume (3.7) of the $\pi\pi$ state, and the factor $(t - 4\mu^2)^\ell$ appears owing to the definition of the partial waves φ_ℓ, g_ℓ, h_ℓ, see (7.30). We have seen already that the unitarity conditions (8.4) can be continued onto arbitrary complex ℓ values by rewriting them in terms of discontinuities on the cuts:

$$\frac{1}{2i}[\varphi_\ell(+) - \varphi_\ell(-)] = C_\ell\varphi_\ell(-)\varphi_\ell(+), \qquad (8.5a)$$

$$\frac{1}{2i}[g_\ell(+) - g_\ell(-)] = C_\ell\varphi_\ell(-)g_\ell(+), \qquad (8.5b)$$

$$\frac{1}{2i}[h_\ell(+) - h_\ell(-)] = C_\ell g_\ell(-)g_\ell(+), \qquad (8.5c)$$

where $(+) = t + i\varepsilon$, $(-) = t - i\varepsilon$. Equations (8.5) allow us to move to the first unphysical sheet and check there the singularities of the partial amplitudes (see Lecture 7):

$$\varphi_\ell(+) = \frac{\varphi_\ell(-)}{1 - 2iC_\ell\,\varphi_\ell(-)}, \tag{8.6a}$$

$$g_\ell(+) = \frac{g_\ell(-)}{1 - 2iC_\ell\,\varphi_\ell(-)}, \tag{8.6b}$$

$$h_\ell(+) = h_\ell(-) + \frac{2iC_\ell\,g_\ell^2(-)}{1 - 2iC_\ell\,\varphi_\ell(-)}. \tag{8.6c}$$

For a certain $t = t(\ell)$ (or equivalently $\ell = \alpha(t)$) where the $\pi\pi$ partial wave $\varphi(-)$ on the unphysical sheet equals $\varphi_\ell(-) = 1/2iC_\ell$, every partial amplitude (8.6) acquires the same pole:

$$\varphi_\ell(t) \simeq \frac{r_{\pi\pi}(t)}{\ell - \alpha(t)}; \qquad g_\ell(t) \simeq \frac{r_{\pi K}(t)}{\ell - \alpha(t)}; \qquad h_\ell(t) \simeq \frac{r_{KK}(t)}{\ell - \alpha(t)}.$$

It easily follows from equations (8.6) that the residues of the amplitudes in this pole satisfy the relation

$$r_{\pi\pi} \cdot r_{KK} = r_{\pi K}^2. \tag{8.7a}$$

It is just this last expression (8.7a) which verifies the factorization of the reggeon residue:

$$r_{\pi\pi} = \tilde{g}_\pi^2, \qquad r_{\pi K} = \tilde{g}_\pi \tilde{g}_K, \qquad r_{KK} = \tilde{g}_K^2, \tag{8.7b}$$

with $\tilde{g}_\pi$ and $\tilde{g}_K$ the coupling constants of the reggeon with particles.

Hence, the contribution of the Regge pole to our amplitudes is

$$A_{\pi\pi}^\pm(s, t) = -\frac{\pi}{2}(2\alpha + 1)[\tilde{g}_\pi(t)]^2 \cdot \frac{[P_\alpha(-z_{\pi\pi}) \pm P_\alpha(z_{\pi\pi})]}{\sin(\pi\alpha)}$$

$$A_{\pi K}^\pm(s, t) = -\frac{\pi}{2}(2\alpha + 1)\tilde{g}_\pi(t)\tilde{g}_K(t) \cdot \frac{[P_\alpha(-z_{\pi K}) \pm P_\alpha(z_{\pi K})]}{\sin(\pi\alpha)}$$

$$A_{KK}^\pm(s, t) = -\frac{\pi}{2}(2\alpha + 1)[\tilde{g}_K(t)]^2 \cdot \frac{[P_\alpha(-z_{KK}) \pm P_\alpha(z_{KK})]}{\sin(\pi\alpha)}.$$

Here $z_{\pi\pi}$, $z_{\pi K}$, and z_{KK} are cosines of the t-channel scattering angles of the corresponding reactions. Taking into account that at large s the cosines also factorize,

$$z_{\pi\pi} \simeq \frac{s}{2|\mathbf{p}_\pi^{(t)}|^2}, \qquad z_{\pi K} \simeq \frac{s}{2|\mathbf{p}_\pi^{(t)}||\mathbf{p}_K^{(t)}|}, \qquad z_{KK} \simeq \frac{s}{2|\mathbf{p}_K^{(t)}|^2},$$

where

$$\left|\mathbf{p}_\pi^{(t)}\right| = \tfrac{1}{2}\sqrt{t - 4\mu^2}, \quad \left|\mathbf{p}_K^{(t)}\right| = \tfrac{1}{2}\sqrt{t - 4m_K^2},$$

and that

$$P_\alpha(z_{ab}) \propto z_{ab}^\alpha \sim \frac{s^\alpha}{\left|\mathbf{p}_a^{(t)}\right|^\alpha \cdot \left|\mathbf{p}_b^{(t)}\right|^\alpha},$$

we come to the main conclusion: Regge pole exchange can be described by the diagram

$$= A_{ab}^\pm(s,t) \overset{s \to \infty}{=} g_a(t) g_b(t) D^\pm(s,t). \qquad (8.8\text{a})$$

Here

$$D^\pm(s,t) = -\frac{(-s)^{\alpha^\pm(t)} \pm s^{\alpha^\pm(t)}}{\sin \pi \alpha^\pm(t)} = s^{\alpha^\pm(t)} \cdot \xi_\alpha^\pm \qquad (8.8\text{b})$$

is the reggeon propagator,

$$\xi_\alpha^\pm = -\frac{e^{-i\pi\alpha^\pm(t)} \pm 1}{\sin \pi \alpha^\pm(t)} \qquad (8.8\text{c})$$

is called the signature factor, and $g_a(t)$ and $g_b(t)$ are the vertices of the reggeon 'emission' and 'absorption'.

For t values close to the mass squared of a real particle, the amplitude (8.8) transforms into the corresponding Feynman diagram. Essentially, we learned how to write the exchange for a particle whose spin depends continuously on its 'virtual mass'.

A characteristic feature of the reggeon propagator is its complexity. The vertex functions $g(t)$ and the trajectory $\alpha(t)$ are real for $t < 4\mu^2$. The signature factor $\xi_{\alpha(t)}$ is, however, always complex:

$$\xi_{\alpha(t)}^+ = i - \cot\frac{\pi\alpha(t)}{2}, \qquad \xi_{\alpha(t)}^- = i + \tan\frac{\pi\alpha(t)}{2}. \qquad (8.9)$$

Near those integer ℓ values where ξ_α has poles ($\alpha^+ = 2n$, $\alpha^- = 2n + 1$) it is almost real (as it should be for a particle exchange). For integer ℓ values where ξ_α has no poles ($\alpha^+ = 2n + 1$, $\alpha^- = 2n$), i.e. in physical points of an 'alien' signature, $\xi_\alpha = i$ and the amplitude is purely imaginary (as, for example, in classical diffraction off a black target).

Hence, although the reggeon has properties of the usual particles, it differs from those essentially. Drawing a reggeon exchange, we have to understand that it corresponds to some real states in the s-channel. This means that it is not an elementary object but a complex one (an ensemble of diagrams).

What these diagrams look like, what is an s-channel image of a reggeon exchange, we will investigate in Lecture 9 in detail.

8.2 Quantum numbers of reggeons. The Pomeranchuk pole

In the previous section we began to draw an analogy between a reggeon and a particle, the spin of which depends on its mass continuously. However, usual particles possess also other characteristics namely, internal quantum numbers such as parity P, charge conjugation C, isotopic spin and its projection, I and I_3, strangeness S, baryon number B, etc. Do these exist for reggeons?

The answer is simple, and it is contained in the very origin of the Regge poles. We have come to Regge poles by analytically continuing the t-channel unitarity condition. This conditions is, however, diagonal not only in the total angular momentum j but also with respect to arbitrary conserved quantum numbers (that commute with j). This means that, without noticing it, in fact we have obtained a Regge trajectory from an amplitude with definite baryon charge, isospin, strangeness etc. in the t-channel. Since the unitarity conditions for amplitudes with different quantum numbers are continued independently, it is natural to expect that the corresponding trajectories $\alpha(t)$ will also be different.

Consequently, definite quantum numbers can be assigned to every Regge trajectory. That is, unless there is a special degeneracy, either accidental or following from some symmetry. (The isotopic invariance of strong interactions may serve as an example of such a symmetry, which leads to degenerate trajectories with different electric charges but belonging to one isotopic multiplet.) For $t > 0$ on each trajectory there are only resonances with the same quantum numbers, differing only by their spins.

Let us note that a particle situated on a Regge trajectory can be considered as a composite one, formed of two particles in the t-channel reaction. One may ask whether all particles are placed on trajectories or if there are also *non-reggeized*, elementary ones. We postpone the theoretical investigation of this question to Lecture 11. Here we just sketch the experimental situation.

Virtually all the well established resonances, both bosonic and fermionic ones (fermionic Regge poles will be considered in Section 8.7), fit the

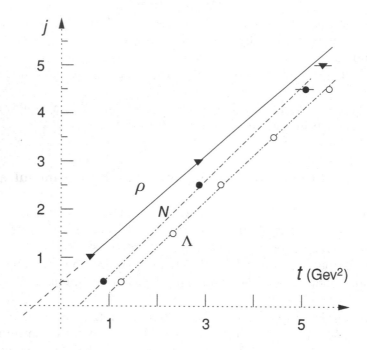

Fig. 8.1   Chew–Frautschi plots. Dot-dashed lines sample baryon trajectories $N_{I=1/2}$ and $\Lambda_{I=0}$, the latter demonstrating degeneracy in signature.

Regge trajectories as demonstrated by Fig. 8.1. With a good accuracy, the trajectories turn out to be linear, having approximately equal slopes (these are the so-called Chew–Frautschi diagrams).

Extrapolating the trajectory to $t = 0$ on Fig. 8.1, we obtain a prediction for the asymptotics of the scattering amplitude in the crossing channel. Hence, an interesting possibility appears to verify directly the theory of complex angular momenta.

As an example, let us consider the pion–nucleon charge exchange reaction $\pi^- p \to \pi^0 n$. At high energies s the forward amplitude ($t = 0$) shows a power growth:

$$A(s, 0) \propto s^{\alpha(0)} = s^{0.57 \pm 0.02}. \tag{8.10}$$

In a system of two pions with isospin $I = 1$ the vector and tensor resonances $\rho_{\sigma=1}(770)$ and $\rho_{\sigma=3}(1690)$ are well established. A straight line going through these points in Fig. 8.1 crosses the $t=0$ axis at $j=\alpha(0) \simeq 0.5$, in a good agreement with (8.10).

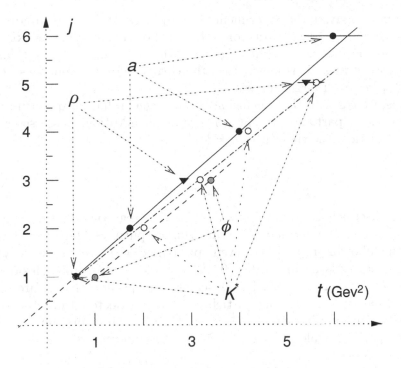

Fig. 8.2   Examples of meson Regge trajectories.

Some meson trajectories are displayed in Fig. 8.2. Series of ρ and a mesons show *signature degeneracy*. The masses of *isoscalars* ω, f practically coincide with the masses of their respective *isovector* partners, ρ and a, and obviously lie on the same line. Strange $I = \frac{1}{2}$ mesons K^* with spins from 1 to 5 seem to be approximately degenerate in signature too. The situation with mass spectra of *unnatural parity* mesons, like *pseudoscalars* π and η, *axial vector* resonances, etc. is less clear.

The theory of complex angular momenta finds a striking confirmation in the so-called 'exotic' reactions, when the quantum numbers of the t-channel reaction are such that neither of the resonances can contribute. For example, the reaction

$$A \sim s^{-1.5} \qquad (8.11)$$

requires strangeness exchange $S = 2$. Mesons with strangeness 2 were not observed. (Note that their existence would contradict the quark model.)

Correspondingly, in the experiment the amplitude (8.11) falls rapidly with energy. Even if a Regge trajectory with such quantum numbers exists, on the Chew–Frautschi diagram Fig. 8.2 it must lie much lower than the usual trajectories. This means that the corresponding resonances, if any, would have much larger masses.

Thus, we came to the conclusion that a reggeon has all quantum numbers of usual particles (except spin), and, in addition, possesses a new characteristic – the signature $P_j = (-1)^j$.

8.2.1 'Naturality'

We will discuss in Section 8.5 how to determine reggeon quantum numbers by considering the πN and NN scattering amplitudes. Here we will make only the following remark. For even-spin particles ($\ell = \alpha^+$ trajectory) the positive parity is natural: $J^P = 0^+, 2^+, 4^+$, i.e. scalars and tensors. The states $J^P = 0^-, 2^-, 4^-$, i.e. pseudoscalars and pseudotensors, are of unnatural parity. For odd-spin particles ($\ell = \alpha^-$ trajectory) we observe the opposite situation: the states $J^P = 1^-, 3^-, 5^-, \ldots$ (vectors, tensors) have natural parities while $J^P = 1^+, 3^+, 5^+, \ldots$ (pseudovectors, pseudotensors) unnatural ones.

To make the notion of 'naturality' independent of the signature, it is convenient to introduce, instead of the usual spatial parity P, a new quantum number P_r

$$P_r \equiv P \cdot P_j = P(-1)^j \qquad (8.12a)$$

which characterizes the 'pseudity' of particles lying on a given Regge trajectory:

$$
\begin{array}{ll}
P_r = +1 & \left\{ \begin{array}{l} \alpha^+ : 0^+, 2^+, 4^+, \ldots \\ \alpha^- : 1^-, 3^-, 5^-, \ldots \end{array} \right. \\[2ex]
\text{'natural parity'} & \\[2ex]
P_r = -1 & \left\{ \begin{array}{l} \alpha^+ : 0^-, 2^-, 4^-, \ldots \\ \alpha^- : 1^+, 3^+, 5^+, \ldots \end{array} \right. \\[1ex]
\text{'unnatural parity'} &
\end{array}
$$

The same procedure is carried out for the charge (C) parity. Let me remind you that the C-parity can be introduced only if the charge conjugation of the system of particles leads to the same system (we are interested in t-channel states). For example, for the $\pi^+\pi^-$-pair we have $C = (-1)^\ell$, with ℓ the orbital moment. To characterize the charge parity independently of the signature, we introduce the quantum number C_r:

$$C_r \equiv C(-1)^j, \qquad (8.12b)$$

which in our case $(j = \ell)$ gives $P_r = (-1)^{\ell+j} = +1$. If so, on trajectories with 'normal' charge and space parities, $P_r = C_r = +1$, lie mesons with

$$J^{PC} : 0^{++}, 1^{--}, 2^{++}, 3^{--}, \ldots .$$

A t-channel state with a non-zero isospin, $I \neq 0$, does not have a definite C-parity since the charge conjugation does not commute with isospin. Then one introduces instead the G-parity which is determined via C-parity of the neutral component of the multiplet,

$$G = C(-1)^I, \qquad G_r \equiv C_r(-1)^I, \tag{8.12c}$$

and analogously to (8.12b) we have a signature-independent quantum number G_r.

8.2.2 High energy symmetry

One of the most important consequences of the reggeized exchange is the appearance of a new symmetry (absent at low energies) in the high energy asymptotics of a two-particle amplitude. Let us demonstrate this by considering pion–nucleon scattering.

Since $I_\pi = 1$ and $I_N = \frac{1}{2}$, for this process two t-channel states are possible, with isospins $I = 0, 1$:

$$A(s,t) = C_0 A_0^{(t)} + C_1 A_1^{(t)}, \tag{8.13a}$$

while in the s-channel there can be states with isospins $I = \frac{1}{2}$ and $\frac{3}{2}$:

$$A(s,t) = C_{\frac{1}{2}} A_{\frac{1}{2}}^{(s)} + C_{\frac{3}{2}} A_{\frac{3}{2}}^{(s)}. \tag{8.13b}$$

The amplitudes A_0, A_1, $A_{\frac{1}{2}}$ and $A_{\frac{3}{2}}$ are not independent; $A_{\frac{1}{2}}$ and $A_{\frac{3}{2}}$ can be expressed in terms of A_0, A_1 and vice versa.

Asymptotically $A(s,t)$ can be determined by the contribution of the leading Regge pole. Consider, e.g. the case $\alpha_0(t) > \alpha_1(t)$. Then the reggeon exchange with an isospin $I = 0$ dominates and

$$A(s,t) \overset{s\to\infty}{\simeq} C_0 A_0. \tag{8.14}$$

The equality (8.14) leads to a definite relation between the s-channel amplitudes $A_{\frac{1}{2}}$ and $A_{\frac{3}{2}}$: both contributions are expressed in terms of A_0 and have the same energy dependence. Note that there is nothing of this kind in the low-energy region. For example, in the amplitude $A_{\frac{3}{2}}$ there is a magnificent resonance $\Delta(1236)$. Having isospin $I = \frac{3}{2}$, it is obviously absent in $A_{\frac{1}{2}}$; as a result, the energy profiles of $A_{\frac{1}{2}}$ and $A_{\frac{3}{2}}$ are essentially different.

Moreover, let us examine contributions of Regge poles with opposite signatures to, e.g. the $\pi^+ N \to \pi^+ N$ amplitude,

$$A_{\pi^+ N}(s,t) = A_{\pi^+ N}^+(s,t) + A_{\pi^+ N}^-(s,t). \qquad (8.15a)$$

If A^- is negligible in the $s \to \infty$ limit (which is the case as we shall see shortly), the full amplitude (8.15a) becomes $s \leftrightarrow u$ symmetric and we obtain

$$A_{\pi^+ N}(s,t) \simeq A_{\pi^- N}(s,t). \qquad (8.15b)$$

It is important to stress that this statement is *stronger* than that of the Pomeranchuk theorem about the equality of total particle and antiparticle *cross sections*, since (8.15b) holds for *amplitudes*, at arbitrary t.

So, the very existence of quantum numbers for Regge poles introduces an additional symmetry that manifests itself at high energies when among all possible t-channel quantum numbers only those survive that ensure the maximal energy exponent – the leading Regge trajectory.

8.2.3 Vacuum pole

Continuing the discussion of reggeon quantum numbers, let us ask whether we could say which of the poles is the rightmost? The answer is:

in the interval $0 \leq t \leq 4\mu^2$ (of which $t = 0$ is the point of the most interest to us) the *rightmost pole* in the j-plane has to have positive signature, $P_j = +1$, and all the quantum numbers of the *vacuum*.

First we look at the signature and show that the leading pole is not allowed to have $P_j = -1$. Indeed, owing to the optical theorem

$$\sigma_{ab} \propto \operatorname{Im} A_{ab}(s,t) = g_a g_b s^\alpha \cdot \operatorname{Im} \xi_\alpha > 0.$$

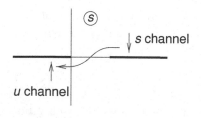

If the signature were negative, that is the amplitude was *odd* with respect to $s \to -s$, this would result in a *negative cross section* for the crossing u-channel reaction, $\sigma_{\bar{a}b} \propto \operatorname{Im} A_{\bar{a}b} \simeq \operatorname{Im} A_{ab}(-s) < 0$, violating the s-channel unitarity.

Physically, t-channel unitarity taught us only that the trajectories $\alpha^+(t)$ and $\alpha^-(t)$ were different. To acquire a more subtle information, namely $\alpha^+(t) > \alpha^-(t)$, we need to turn to the s-channel unitarity (which, as you remember, is a substitute for 'potential' for the t-channel).

Further, it is clear that the leading pole cannot have non-zero strangeness, baryon charge and alike, since in such a case the partial cross section of the corresponding non-diagonal transition would have been asymptotically larger than the total one, $\sigma_{ab\to a'b'} > \sigma_{ab}^{\text{tot}}$.

A bit more tricky is the situation with quantum numbers like isospin when a diagonal transition $ab \to ab$ can still correspond to a $I \neq 0$ state in the t-channel. We shall clarify the logic of the proof by an example for which we take $\pi\pi$ scattering. The pion being an isovector ($I_\pi = 1$), three two-pion states are possible, with t-channel isospin values $I = 0, 1, 2$.

α β

γ δ

These states are described, correspondingly, by the scalar product of two isovectors ($I=0$), their vector product ($I=1$) and the symmetric irreducible (zero trace) tensor:

$$A^{(0)}_{\alpha\beta\gamma\delta} = \delta_{\alpha\beta}\delta_{\gamma\delta} \cdot A_0,$$

$$A^{(1)}_{\alpha\beta\gamma\delta} = \sum_\sigma \varepsilon_{\alpha\beta\sigma}\varepsilon_{\gamma\delta\sigma} \cdot A_1,$$

$$A^{(2)}_{\alpha\beta\gamma\delta} = \sum_{\rho_1,\rho_2} M^{\rho_1\rho_2}_{\alpha\beta} M^{\rho_1\rho_2}_{\gamma\delta} \cdot A_2; \quad M^{\rho_1\rho_2}_{\alpha\beta} = \delta^{\rho_1}_\alpha \delta^{\rho_2}_\beta + \delta^{\rho_2}_\alpha \delta^{\rho_1}_\beta - \frac{2}{3}\delta_{\alpha\beta}\delta^{\rho_1\rho_2}.$$

Now we have to prove that the rightmost pole must be contained by A_0. Obviously, it cannot belong to A_1, since $\pi\pi$ is a symmetric (Bose) system and therefore the isospin-asymmetric states, like $I=1$, correspond to *odd* signature.

So only A_2 remains under suspicion. Let us consider a diagonal transition – an elastic zero-angle scattering process: $\alpha = \beta$, $\gamma = \delta$. Fix the charge (I_3) of one of the π-mesons, say, $\gamma = \delta = 2$ and examine one by one the total cross sections of the three isotopic states of the other one: $\alpha = \beta = 1, 2, 3$. If the leading pole is part of A_2, one of the cross sections turns out to be negative, since, due to the irreducibility of the tensor $I = 2$,

$$\sigma_{12\to12} + \sigma_{22\to22} + \sigma_{32\to32} \propto \text{Tr}\, M = \sum_{\alpha=1}^{3} M^{\rho_1\rho_2}_{\alpha\alpha} = 0.$$

This example shows what happens with any symmetry. The idea is essentially as follows. If there is some internal symmetry, the two-particle–reggeon vertex has the structure of an irreducible tensor. However, elastic amplitudes, and thus total cross sections, are given by *diagonal elements* which sum up to zero due to the irreducibility condition, and, whichever the detailed situation, one can always find a state whose cross section will turn out to be negative.

We conclude: if there is no degeneracy, then the rightmost pole has quantum numbers of the vacuum and signature $P_j = +1$. This pole was named by Gell-Mann the 'Pomeranchuk pole' (the name later drifted to 'pomeron'), since it satisfies the Pomeranchuk theorem automatically. To be more precise, the 'Pomeranchuk pole' is called the vacuum pole with an additional condition imposed, namely, that its 'intercept' is

$$\alpha_{\mathbf{P}}(0) = 1, \tag{8.16}$$

the condition that, due to the optical theorem, guarantees asymptotic constancy of the total interaction cross sections.

Strictly speaking, the hypothesis (8.16) does not follow from the theory. You may ask, why would we need this hypothesis in the first place? What is so attractive about asymptotically constant cross sections?

First of all, our interaction is *strong* and we see no reason why cross sections should fall with the increase of energy. But if σ_{tot} does not fall, then the constancy remains the only option, since any positive power would violate the Froissart theorem.

A more serious argument is the experimental situation. With the increase of the incident energy $p_{\text{lab}} = s/2m$ by hundred times, from 10^{10} to 10^{12} eV where the measurements were carried out, the total proton–proton and pion–proton cross sections change only by ten percent.[*] Consequently, it is natural to assume that the cross sections are basically constant, while the observed relatively small deviations are driven by correction effects that are slower than a power of s (e.g. logarithmic).

8.3 Properties of the Pomeranchuk pole

Thus, the asymptotic behaviour of the elastic amplitude at $s \to \infty$ can be described by the vacuum pole (pomeron $\mathbf{P}$) exchange as

$$A_{ab}(s,t) = g_a(t)g_b(t)\xi_\alpha s^{\alpha(t)}, \quad \xi_\alpha = i - \cot\frac{\pi\alpha(t)}{2}, \tag{8.17}$$

where for small values of t we can approximate the pomeron trajectory as $\alpha(t) \simeq \alpha_{\mathbf{P}}(0) + \alpha'_{\mathbf{P}} \cdot t = 1 + \alpha'_{\mathbf{P}} \cdot t$.

Let us discuss the characteristic features of the vacuum pole exchange.

---

[*] Thirty years and three orders of magnitude in energy later, the cross sections $\sigma_{pp} \simeq \sigma_{p\bar{p}}$ have grown by about 50%, see below Fig. 14.3, page 378. (ed.)

8.3.1 Pomeranchuk theorem

The Pomeranchuk theorem,

$$\sigma_{ab}^{tot} = \sigma_{\bar{a}b}^{tot}, \tag{8.18}$$

is a consequence of the fact that the signature of the $\alpha_{\mathbf{P}}(t)$ trajectory is positive. In addition, $\mathrm{Re}\, A(s,0) = 0$, since $j = \alpha_{\mathbf{P}}(0) = 1$ is a point of 'alien' signature and the signature factor in (8.17) is purely imaginary. This means that the pomeron exchange is analogous, in the NQM language, to diffraction off absorbing target.

8.3.2 Factorization of total cross sections

Factorization relation for total cross sections,

$$\left(\sigma_{ab}^{tot}\right)^2 = \sigma_{aa}^{tot} \cdot \sigma_{bb}^{tot}, \tag{8.19}$$

follows from the factorisation of the reggeon exchange (8.8). The knowledge of the πN and NN cross sections enables us to predict

$$\sigma_{\pi\pi} = \frac{(\sigma_{\pi N})^2}{\sigma_{NN}} \simeq \frac{(25\,\mathrm{mb})^2}{40\,\mathrm{mb}} \simeq 16\,\mathrm{mb}.$$

Unfortunately, so far there is no experimental verification for the total cross sections. For *inelastic* cross sections such expectations have been verified and they agree reasonably well.

8.3.3 Factorization of differential cross sections

Consider reactions of excitation of a target hadron $(b \to c)$ by different projectiles (a). For example, proton (nucleon) excitation, $N \to N^*$, by pion and kaon beams. Relations like

$$\frac{\sigma(\pi N \to \pi N^*)}{\sigma(KN \to KN^*)} = \frac{\sigma(\pi N \to \pi N)}{\sigma(KN \to KN)}$$

were checked many times. As it turns out, the factorization of differential cross sections at high energies generally holds within 10–20%.

8.3.4 Falloff of charge-exchange reactions

Another consequence of the pomeron picture is the disappearance of all sorts of 'charge-exchange' $2 \to 2$ reactions in the high-energy limit. Indeed, any such process corresponds to a non-diagonal transition (e.g. $\pi^- p \to \pi^0 n$) with non-vacuum quantum numbers in the t-channel. Being devoid of the vacuum-pole exchange, its amplitude must therefore be *suppressed* as a power of s relative to the elastic amplitude.

8.3.5 Growing radius and shrinkage of diffractive cone

Substituting the elastic scattering amplitude

$$A_{ab} \;=\; s \cdot g_a g_b \, e^{(\alpha_{\mathbf{P}}(t)-1)\ln s} \left(i - \cot \frac{\pi \alpha_{\mathbf{P}}(t)}{2} \right) \tag{8.20}$$

in the expression for the differential cross section, at small $t \simeq -\mathbf{q}_\perp^2$ we obtain

$$\frac{d\sigma}{dq^2} = \frac{1}{16\pi} \left| \frac{A}{s} \right|^2 \simeq \frac{g_a^2 g_b^2}{16\pi} \, e^{-2\alpha_{\mathbf{P}}' \ln s \cdot q_\perp^2}. \tag{8.21a}$$

Hence, the diffractive cone *shrinks* logarithmically with the increase of s:

$$q_{\text{char}}^2 \simeq \frac{1}{2\alpha_{\mathbf{P}}' \ln s}, \tag{8.21b}$$

which corresponds to the growth of the radius of interaction

$$\rho_0 \sim \sqrt{\alpha_{\mathbf{P}}' \ln s}. \tag{8.21c}$$

8.3.6 Impact parameter diffusion

What is the picture in the impact parameter plane corresponding to the pomeron exchange? Let us invert the Fourier representation (5.43) for the impact parameter function $f(\boldsymbol{\rho}, s)$,

$$f(\boldsymbol{\rho}, s) = \frac{\pi}{k_c^2} \int \frac{d^2 \mathbf{q}_\perp}{(2\pi)^2} \, e^{-i(\mathbf{q}_\perp \cdot \boldsymbol{\rho})} A(s, -q_\perp^2).$$

Substituting for A the pomeron amplitude (8.20) we obtain

$$f(\boldsymbol{\rho}, s) \;\simeq\; 4\pi i g_a(0) g_b(0) \int \frac{d^2 \mathbf{q}_\perp}{(2\pi)^2} \, e^{-i(\mathbf{q}_\perp \cdot \boldsymbol{\rho}) - \alpha_{\mathbf{P}}' q_\perp^2 \ln s}$$

$$= \; \frac{i g_a g_b}{\alpha_{\mathbf{P}}' \ln s} \exp \left(-\frac{\rho^2}{4\alpha_{\mathbf{P}}' \ln s} \right). \tag{8.22}$$

We see that the partial amplitude is not *saturated*; on the contrary, its magnitude is small at large values of $\ln s$: the target is not at all 'black' but rather 'grey' and gets increasingly transparent with the growth of the energy. At the same time the interaction radius also grows since the exponential falloff of $f(\boldsymbol{\rho})$ starts at larger and larger impact parameters as shown in Fig. 8.3.

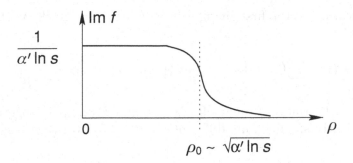

Fig. 8.3   Impact parameter profile of the partial wave corresponding to pomeron exchange.

The total cross section remains constant,

$$\int d^2\rho \, \text{Im} \, f(\rho, s) \;=\; g_a g_b \;=\; \sigma^{\text{tot}} = \text{const},$$

but this is achieved in a rather peculiar way.

By the way, as we have already seen in Section 5.6, the growth of the radius $\rho_0(s) \propto \sqrt{\ln s}$ corresponds to a 'random walk' of a point in which the interaction of the projectile with the target takes place in the impact parameter plane. This fact can be also seen directly from (8.22) for $f(\rho, s)$ which very expression is nothing but the Green function of the two-dimensional diffusion process,

$$\frac{\partial}{\partial \xi} f(\rho, \xi) - \alpha' \nabla_\rho^2 f(\rho, \xi) = \delta(\rho)\delta(\xi), \qquad (8.23)$$

with $\xi = \ln s$ in the rôle of the diffusion time.

8.3.7 Properties of the pomeron trajectory

Let us discuss the properties of $\alpha_{\mathbf{P}}(t)$. In the interval $0 \le t \le 4\mu^2$ we can make two statements, namely, that:

(1) the pomeron trajectory $\alpha_{\mathbf{P}}(t)$ is real and

(2) monotonically increasing, $\alpha'_{\mathbf{P}}(t) > 0$.

Indeed, the imaginary part of the amplitude

$$\text{Im} \, A(s, t) = r(t)s^{\alpha(t)}, \quad r(t) \equiv g_a(t)g_b(t), \qquad (8.24)$$

remains real up to the first singularity at $t = 4\mu^2$ where the partial wave expansion

$$\text{Im } A(s,t) = \sum_\ell (2\ell+1) \text{ Im } f_\ell(s) P_\ell(z_s), \quad z_s = 1 + \frac{2t}{s - 4\mu^2}, \qquad (8.25)$$

diverges. Furthermore, we have $\text{Im } f_\ell \geq 0$, and $P_\ell(z_s) > 0$, together with all its derivatives. Therefore, differentiating (8.25) over t we have

$$\frac{d}{dt} \text{Im } A(s,t) > 0. \qquad (8.26)$$

Substituting the pomeron amplitude (8.24),

$$\frac{d}{dt}\left(r(t)s^{\alpha(t)}\right) = r'(t)s^{\alpha(t)} + r\alpha'(t)\ln s \cdot s^{\alpha(t)} > 0. \qquad (8.27)$$

Since $s \to \infty$, the main contribution comes from the last term in (8.27), resulting in $\alpha'_\mathbf{P}(t) > 0$. (If by any chance $\alpha'_\mathbf{P}(t) = 0$, then the second derivative will be positive at this point, etc.)

Is there a complexity in $\alpha(t)$ for $t < 0$?

Let us demonstrate that in spite of the partial wave $\varphi_\ell(t)$ having the left cut, the trajectory remains regular at $t \leq 0$. Indeed, if $\alpha(t)$ were complex, then $\varphi_\ell(t)$ would have acquired at $t = 0$ additional singularities in ℓ:

$$\Delta\varphi_\ell(t) = \frac{\Delta r(t)}{\ell - \alpha(t)} + \frac{r(t)}{(\ell - \alpha(t))^2}\Delta\alpha(t).$$

We know, however, that moving singularities in ℓ of $\varphi_\ell(t)$ can come only from under the right cut[†] at $t \geq 4\mu^2$, see Lecture 3. Thus, $\Delta r = \Delta \alpha \equiv 0$. As a consequence, $\alpha_\mathbf{P}(t)$ remains real also when $t < 0$.

For $t > 4\mu^2$ the pomeron trajectory becomes complex. By explicitly solving the two-particle unitarity condition for the partial wave near the threshold, $t \simeq 4\mu^2$, it is straightforward to show that the trajectory moves onto the *upper* plane, $\text{Im } \alpha_\mathbf{P}(t) > 0$. Whether it stays there for arbitrary large t remains an open question. This is true in NQM but cannot be rigorously proved in the relativistic theory, although it appears the most natural hypothesis. If this is the case and if we could write the dispersion relation

$$\alpha_\mathbf{P}(t) = 1 + \frac{t}{\pi}\int_{4\mu^2}^\infty \frac{\text{Im } \alpha_\mathbf{P}(t')}{t'(t'-t)} \, dt', \qquad (8.28)$$

[†] As we shall see below in Section 8.6, a branch-point singularity in the trajectory may appear at $t=0$ if *two* trajectories collide in this point.

then differentiating (8.28) over t an arbitrary number of times we would have *all derivatives* of $\alpha_{\mathbf{P}}$ to be *positive* in the interval $0 \leq t \leq 4\mu^2$. Once again, this property holds in the NQM where the monotonic increase of $\alpha(t)$ can be directly linked to the natural movement of the energy level (decrease of the binding) with the increase of the centrifugal barrier (angular momentum ℓ).

In the relativistic theory the representation (8.28) was not proved and so we cannot make a strong statement of the positivity of all derivatives. Nevertheless, using the t-channel unitarity condition we have demonstrated that at least its *first derivative* is positive. This is a remarkable example of how the cross-channel unitarity allows one to *partially* recover the NQM results, in particular the natural behaviour of the characteristic t-channel angular momentum.

8.3.8 Mesons on the pomeron trajectory?

Let us note, finally, that the Pomeranchuk pole with $\alpha_{\mathbf{P}}(0) = 1$ differs essentially from the non-vacuum ones.

All other Regge trajectories (both the trajectories with non-vacuum quantum numbers and the subleading vacuum pole, the so-called $\mathbf{P}'$ with $\alpha_{\mathbf{P}'} \simeq 0.5$) appear to have the same slope, $\alpha' \simeq 1\,\mathrm{GeV}^{-2} \sim 1/m_N^2$, while the slope of the pomeron trajectory turns out to be about four times smaller,

$$\alpha'_{\mathbf{P}} \simeq 0.27\,\mathrm{GeV}^{-2} \sim 1/4m_N^2. \tag{8.29}$$

Moreover, so far it is not clear whether any resonances lie on the pomeron trajectory. Given the slope (8.29), we should expect a resonance with spin 2 and $m_2^2 \sim 4\,\mathrm{GeV}^2$ belonging to the pomeron trajectory. The experimental situation remains uncertain: there seem to be two to three candidates for such a tensor meson.

In the linear approximation for $\alpha_{\mathbf{P}}(t)$, the trajectory would cross the line $\alpha(t) = 0$ at some negative $t = m_0^2 < 0$ thus giving rise to a scalar meson with an *imaginary mass*! M. Gell-Mann advanced arguments in favour of *vanishing of the pomeron residue* at this point, $g_{\alpha(m_0^2)} = g_0 = 0$. Nowadays this problem is no longer considered acute, since as we will see later, at negative t the dominant rôle is played by ℓ-plane singularities other than poles (Regge cuts).

8.4 Structure of the reggeon residue

8.4.1 Scattering of particles with spins

Up to now we have investigated the scattering of spinless particles (like pions) or, more precisely, the processes in which spins of participating

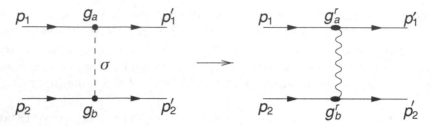

Fig. 8.4   Particle-exchange diagram (a) and the Regge-pole exchange (b).

particles were inessential. Scattering of particles with spin is an interesting problem on its own. We saw that in the spinless case a Regge amplitude differs from the amplitude of a virtual particle exchange only by the modification of the propagator ($\sin \pi \alpha(t)$ in the denominator) and by the redefinition of the vertex g_a (Fig. 8.4). It is clear that the case of particles with spin will not differ in this respect. We could formally solve the problem by introducing particle states with definite helicity (projection of spin on the direction of three-momentum), play with spherical functions and continue the corresponding amplitudes in j. The same goal can be achieved, however, by following, essentially, the perturbation theory. We shall choose the latter path which is more rewarding.

The main problem here is how to write the *vertices*. This is easy to resolve for the case of particle exchange (Feynman graphs); a generalization to reggeons will be carried out simply by substituting the reggeon propagator for that of the exchanged particle.

The first question we have to ask is, where the *factorization* did come from? Even in the case of spinless scattering particles, the amplitude of an exchange of a particle with spin σ (Fig. 8.4(a)) does not factorize into a product but is given by a *sum of products*,

$$
A(s,t) = \Gamma_{\{\mu_1,\ldots,\mu_\sigma\}}(p_1,p_1') D^{[\sigma]}_{\{\mu\}\{\nu\}}(q) \Gamma_{\{\nu_1,\ldots,\nu_\sigma\}}(p_2,p_2') \cdot \frac{1}{m^2 - q^2},
$$

$$
D^{[\sigma]}_{\{\mu\}\{\nu\}}(q) = \sum_{m=1}^{2\sigma+1} e^m_{\{\mu\}}(q) e^m_{\{\nu\}}(q),
$$

(8.30)

which sum runs over all polarization states. The answer is simple: of the whole sum only one polarization survives at high energies, the one that gives the contribution with the highest power s.

Each function $\Gamma_{\{\}}$ in (8.30) depends only on the momenta of particles entering the corresponding vertex and knows nothing about the large s. The momenta p_1 and p_2 get connected when we sum over the polarization of the exchange particle. To see how this happens, let us consider the example of a vector particle exchange:

$$D^{[1]}_{\mu\nu}(q) = -\sum_{m=1}^{3} e^m_\mu e^m_\nu \qquad \left(D^{[1]}_{\mu\mu} = 3\right), \tag{8.31}$$

where $\{e^m\}$ are three vectors determining a space, orthogonal to the momentum q:

$$(e^m q) = 0, \qquad m = 1, 2, 3. \tag{8.32}$$

Since q is space-like, we can find two independent light-like vectors $e^\pm$ among (8.32):

$$e^{+2} = e^{-2} = 0, \quad (e^+ e^-) = 1; \quad (e^\pm q) = 0.$$

These light-like polarization vectors can be constructed using linear combinations of the initial particle momenta, p_1 and p_2. To ensure the orthogonality condition (8.32), we would have to have

$$p^\mu_1 q_\mu \simeq p^\mu_2 q_\mu \simeq 0. \tag{8.33}$$

Being an important characteristic feature of high-energy processes, let us check this fact in detail.

Sudakov kinematics of high-energy $2 \to 2$ *processes.* Before going into calculations, a simple observation first. When two particles scatter elastically, in their cms the energy is not transferred, $q_0 = 0$, and $-t = \mathbf{q}^2 = 2p_c^2(1 - \cos\Theta_s) \simeq p_c^2 \cdot \Theta_s^2$. So, for large collision energies, $s \simeq 4p_c^2 \gg |t|$, the scattering angle is small, $\Theta_s \sim \sqrt{-t/s} \ll 1$. This makes the *longitudinal* component of the momentum transfer, $q_{\|} \sim \frac{1}{2}p_c\Theta_s^2$, much smaller that the transverse one, $q_\perp \propto p_c\Theta_s$; in other words, $\mathbf{q}$ is approximately orthogonal to the common direction of the colliding particles, $\mathbf{z}$. In Lorentz invariant terms, this statement is equivalent to (8.33) that we are about to verify formally.

Let us consider the kinematics of the $2 \to 2$ scattering process in the most general case, $p_1, p_2 \to p'_1 p'_2$, using the Sudakov decomposition.

$$p_1 = p^+ + \gamma_1 p^-, \quad p_2 = p^- + \gamma_2 p^+; \tag{8.34a}$$

$$\gamma_{1,2} = \frac{m_{1,2}}{s} \tag{8.34b}$$

(where we have approximated $2p^+p^- \simeq s$). In the linear approximation in $\gamma_i = \mathcal{O}(1/s)$, the inverse relations read

$$p^+ \simeq p_1 - \gamma_1 p_2, \quad p^- \simeq p_2 - \gamma_2 p_1;$$

actually, for our purposes it often suffices to equate

$$p^+ \simeq p_1, \quad p^- \simeq p_2,$$

which approximation holds, component by component, when both p_0^+ and p_0^- are large (which is true for virtually all reference frames but the laboratory frame where one of the incoming particles is at rest).

We write the transferred four-momentum as

$$q_\mu = \alpha p_\mu^+ + \beta_q p_\mu^- + q_{\perp\mu}; \quad -(q_{\perp\mu})^2 = \mathbf{q}_\perp^2 \geq 0, \qquad (8.35a)$$

$$q^2 = \alpha_q \beta_q s - \mathbf{q}_\perp^2, \qquad (8.35b)$$

where $q_\perp^\mu$ lies in the $(\mathbf{x}, \mathbf{y})$ plane.

Making use of (8.34) and (8.35) we evaluate square momenta of the outgoing particles,

$$(p_1')^2 = (p_1 + q)^2 = (1 + \alpha_q)(\gamma_1 + \beta_q)s - \mathbf{q}_\perp^2 = {m_1'}^2,$$

$$(p_2')^2 = (p_2 + q)^2 = (\gamma_2 - \alpha_q)(1 - \beta_q)s - \mathbf{q}_\perp^2 = {m_2'}^2,$$

to obtain

$$-\alpha_q s - \beta_q m_2^2 = {m_2'}^2 - m_2^2 - q^2, \qquad (8.36a)$$

$$\beta_q s + \alpha_q m_1^2 = {m_1'}^2 - m_1^2 - q^2. \qquad (8.36b)$$

In the high energy limit, $|q^2| \sim m^2 \ll s$, we have $-\alpha_q \sim \beta_q \sim m^2/s \ll 1$ and (8.36) simplifies

$$-\alpha_q s \simeq {m_2'}^2 - m_2^2 - q^2; \qquad \cdot (8.37a)$$

$$\beta_q s \simeq {m_1'}^2 - m_1^2 - q^2. \qquad (8.37b)$$

We conclude that the 'longitudinal' contribution to the invariant t becomes negligible,

$$q^2 = q_\|^2 - \mathbf{q}_\perp^2, \quad q_\|^2 \equiv -\alpha_q \beta_q s \sim \frac{q^4}{s} \ll |q^2|,$$

so that the momentum transfer becomes essentially *transversal*,

$$q^2 = -\mathbf{q}_\perp^2 \cdot \left(1 + \mathcal{O}\left(\frac{m^2}{s}\right)\right). \qquad (8.38)$$

This statement can be enforced and applied to all components of the four-vector,

$$q_\mu = \alpha_q s \cdot \frac{p_\mu^+}{s} + \beta_q s \cdot \frac{p_\mu^-}{s} + q_{\perp\mu} \simeq q_{\perp\mu},$$

once again, in any but the laboratory frame of reference (where one of the normalized momentum vectors $p^\pm/s$ is of the order of unity).

Polarization vectors. We return to the construction of the polarization vectors. Now that we have verified the condition (8.33) of the approximate orthogonality of the momentum transfer q to the (p_1, p_2) plane, we can treat q_μ as having, say, only a **y** component: $q_\mu = (q_0; q_x, q_y, q_z) \simeq (0; 0, \sqrt{-q^2}, 0)$.

In the cms of the t-channel where $q_\mu = (\sqrt{q^2}; \mathbf{0})$, the polarization vectors are all space-like:

$$
\begin{pmatrix} t \\ x \\ y \\ z \end{pmatrix} : \quad e^1_{(t)} = \begin{pmatrix} 0 \\ 0 \\ 0 \\ 1 \end{pmatrix}, \quad e^2_{(t)} = \begin{pmatrix} 0 \\ 0 \\ 1 \\ 0 \end{pmatrix}, \quad e^3_{(t)} = \begin{pmatrix} 0 \\ 1 \\ 0 \\ 0 \end{pmatrix}. \tag{8.39}
$$

The complex Lorentz boost that takes us to the s-channel, transforms the momentum transfer as follows

$$
q^\mu_{(t)} = \begin{pmatrix} -i\sqrt{-q^2} \\ 0 \\ 0 \\ 0 \end{pmatrix} \implies q^\mu = \begin{pmatrix} 0 \\ 0 \\ \sqrt{-q^2} \\ 0 \end{pmatrix} ;
$$

it swaps the time- and y-components of a four-vector, $t \to iy$, $y \to it$, while leaving the x and z components unchanged. Under this transformation, the complex linear combinations of the first two polarizations in (8.39), in the s-channel turn into two light-like vectors:

$$
e^+ = \frac{e^1 - ie^2}{\sqrt{2}} = \frac{1}{\sqrt{2}} \begin{pmatrix} 1 \\ 0 \\ 0 \\ 1 \end{pmatrix}, \quad e^- = -\frac{e^1 + ie^2}{\sqrt{2}} = \frac{1}{\sqrt{2}} \begin{pmatrix} 1 \\ 0 \\ 0 \\ -1 \end{pmatrix}. \tag{8.40}
$$

From the t-channel point of view, the combinations $e^1 \pm ie^2$ correspond to *circular polarizations* describing states with a spin projection ± 1 onto the **x**-axis, i.e. in our case, onto the normal to the scattering plane. The light-like vectors (8.40) are nothing but the Sudakov vectors we have introduced above to represent the (t, z) interaction plane:

$$
e^\pm \equiv \sqrt{\frac{2}{s}} p^\pm; \quad e^+ \simeq \sqrt{\frac{2}{s}} p_1, \ e^- \simeq \sqrt{\frac{2}{s}} p_2. \tag{8.41a}
$$

In terms of these vectors the sum over polarizations (8.31) takes the form

$$-\sum_{m=1}^{3} e_\mu^m e_\nu^m = e_\mu^+ e_\nu^- + e_\mu^- e_\nu^+ - e_\mu^\perp e_\nu^\perp,$$
(8.41b)

where $e^\perp = e^3 = e_{(t)}^3$ of (8.39) is a space-like unit vector normal to the scattering plane $(\mathbf{x})$.

We have thus introduced convenient basis vectors. Let us see the contributions of various polarizations to the scattering amplitude:

$$= \Gamma_\mu^a(1) \cdot D_{\mu\nu}^{[1]} \cdot \Gamma_\nu^b(2) \times \frac{1}{m^2 - q^2}$$

$$\propto (\Gamma^a e^+)(e^- \Gamma^b) + (\Gamma^a e^-)(e^+ \Gamma^b) - (\Gamma^a e^\perp)(e^\perp \Gamma^b).$$
(8.42)

Which polarization leads to the leading contribution $s^\sigma = s$ in (8.42)? The general form of the vertex for a scalar particle is

$$\Gamma_\mu^a(1) = \Gamma_\mu^a(p_1, q) = a(q^2)p_{1\mu} + b(q^2)q_\mu.$$
(8.43)

The term q_μ can be dropped, see (8.32); the transverse polarization does not contribute, $e^\perp p_1 = e^\perp p_2 = 0$. We have

$$e_\mu^+ p_1^\mu = \frac{m_1^2}{\sqrt{2s}}, \qquad e_\mu^- p_2^\mu = \frac{m_2^2}{\sqrt{2s}};$$

$$e_\mu^- p_1^\mu = \sqrt{\frac{s}{2}}, \qquad e_\mu^+ p_2^\mu = \sqrt{\frac{s}{2}}.$$

Consequently, at $s \to \infty$ of the whole sum (8.42) only one term survives,

$$\left(\Gamma_\mu^a(1)e_\mu^-\right)\left(\Gamma_\mu^b(2)e_\mu^+\right) \sim (\sqrt{s})^2 = s,$$

in which each vertex is multiplied by the polarization vector directed along the momentum p_i of the *opposite* vertex. Hence, we can factorize the asymptotic amplitude as follows:

$$A(s,t) = \left(\Gamma_\mu^a(1)\frac{e_\mu^-}{\sqrt{s}}\right)\left(\Gamma_\nu^b(2)\frac{e_\nu^+}{\sqrt{s}}\right) \cdot s \cdot D(q^2).$$
(8.44)

The analogue of the reggeon residue,

$$\left(\Gamma_\mu^a(1)\frac{e_\mu^-}{\sqrt{s}}\right) \sim \left(\Gamma_\mu^a(1)\frac{p_{2\mu}}{s}\right) \equiv g_a^r(1),$$

is an invariant; in the case of spinless particles it can depend only on q^2 via $a(q^2)$, see (8.43). If the colliding particles have non-zero spins, the vertex function Γ_μ^a changes: it will depend also on the polarizations of the incoming and outgoing particles (1), (1′):

$$\Gamma_\mu^a(1) = \Gamma_\mu^a\left(p_1, q; \epsilon_i^\lambda, \epsilon_f^\rho\right).$$

However, the general structure of the Regge residue remains unchanged; a large contribution will still be coming only from multiplication of the vertex (1) by the polarization $e^- \propto p_2$. The residue g_a^r though will now depend not only on q^2, but also on the spin projections of the initial and final particles in the upper vertex (1) on the momentum direction of the particle (2). Thus for $\sigma = 1$ we have obtained

$$g_a^r(1) \equiv \left(\Gamma_\mu^a(1)\frac{p_{2\mu}}{s}\right), \quad g_b^r(2) \equiv \left(\Gamma_\mu^b(2)\frac{p_{1\mu}}{s}\right). \tag{8.45}$$

Let us see now, what the residue for a $\sigma = 2$ particle exchange looks like. A spin $\sigma = 2$ wave function can be constructed as a symmetric product of two vector states,

$$e_{\mu_1\mu_2}^\lambda = e_{\mu_1}^{\lambda_1}e_{\mu_2}^{\lambda_2} + e_{\mu_2}^{\lambda_1}e_{\mu_1}^{\lambda_2}. \tag{8.46}$$

Now each vertex bears two vector indices and the amplitude has the structure

$$A(s,t) \quad \propto \quad \Gamma_{\mu_1\mu_2}^a(1)D_{\mu_1\mu_2,\nu_1\nu_2}^{[2]}\Gamma_{\nu_1\nu_2}^b(2) \times \frac{1}{m^2 - q^2}$$

$$D_{\mu_1\mu_2,\nu_1\nu_2} = \sum_\lambda e_{\mu_1\mu_2}^\lambda e_{\nu_1\nu_2}^\lambda. \tag{8.47}$$

Strictly speaking, we have to subtract from (8.46) an admixture of a $\sigma = 0$ state to make our symmetric tensor traceless. We, however, can ignore this fact since the corresponding subtraction term contains $g_{\mu_1\mu_2}$ which, acting on one vertex, yields no large s-dependent contribution.

Expressing (8.46) in terms of the polarizations $e^\pm$, the propagator $D_{\mu_1\mu_2,\nu_1\nu_2}$ in (8.47) will contain various products

$$e_{\mu_1}^+ e_{\mu_2}^+ e_{\nu_1}^- e_{\nu_2}^-, \quad e_{\mu_1}^+ e_{\mu_2}^- e_{\nu_1}^- e_{\nu_2}^+, \quad \ldots$$

It is easy to see that similarly to the case of $\sigma = 1$, only the term $e_{\mu_1}^- e_{\mu_2}^- e_{\nu_1}^+ e_{\nu_2}^+$ produces a contribution $\sim s^2$. (By the way, this term is, on its own, a symmetric irreducible tensor since $(e^-)^2 = 0$.) So, the upper vertex produces

$$\left(\Gamma^a_{\mu_1 \mu_2}(1) \cdot \frac{e^-_{\mu_1}}{s} \frac{e^-_{\mu_2}}{s} \right),$$

and the analogue of the Regge residue for a spin-2 particle exchange acquires the form

$$g^r_a(1) \equiv \left(\Gamma^a_{\mu_1 \mu_2}(1) \frac{p_{2\mu_1} p_{2\mu_2}}{s^2} \right), \quad g^r_b(2) \equiv \left(\Gamma^b_{\nu_1 \nu_2}(2) \frac{p_{1\nu_1} p_{1\nu_2}}{s^2} \right). \quad (8.48)$$

How do (8.45) and (8.48) generalize to the case of non-integer angular momentum exchange? For particle exchange with an integer spin σ the residues $g^r_a(1)$ turned out to be polynomials of p_2/s of the order σ. In the general case (which corresponds to a sum of various t-channel exchanges) the s-dependence of the amplitude becomes more complicated:

$$A(s,t) = \Gamma_a \left(p_1, q; \epsilon_i^{\lambda_1}, \epsilon_f^{\rho_1}; \frac{p_2}{s} \right) \Gamma_b \left(p_2, q; \epsilon_i^{\lambda_2}, \epsilon_f^{\rho_2}; \frac{p_1}{s} \right) \cdot s^\alpha \cdot \xi_\alpha. \quad (8.49)$$

However, the structure of the residues resembles that of simple particle exchange: each vertex depends on its internal variables (momenta and polarizations) *and* on the direction of the momentum in the opposite vertex.

In other words, the Regge factorization *remembers* the reaction plane (p_1, p_2) via the polarization vectors $e^\pm$ which ensure the maximal contribution to the scattering amplitude.

8.4.2 Quasi-elastic processes

Up to now we have been discussing $2 \to 2$ scattering processes. Do Regge poles contribute to multi-particle production? It is easy to imagine a high-energy collision in which, say, one of the incident particles becomes excited and subsequently decays giving rise to a three-particle final state. A question can be raised, for example, for which lifetimes of the intermediate state in Fig. 8.5(a) our previous analysis will remain valid. It is possible to rigorously deduce contribution of Regge poles to such processes from the corresponding unitarity conditions for multi-particle amplitudes of the type shown in Fig. 8.5(b). I will not do that but suggest you guess the answer by looking at the diagrams describing the t-channel exchange of a particle with spin instead.

Recall the result of our analysis of reggeon amplitudes for the scattering of particles with spin: the reggeon vertex Γ acquired an additional dependence on the polarizations of participating particles. Similarly, here the vertex describing the $1 \to 2$ particle conversion of Fig. 8.5(b) will depend

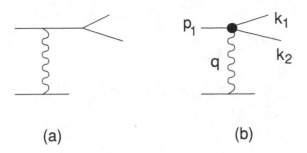

(a)   (b)

Fig. 8.5   Production of an unstable particle (a) and new reggeon vertex (b).

not only on the total momentum transfer, $q = k_1 + k_2 - p_1$, but also on the relative momentum $k_1 - k_2$; in other words, on the two final particle momenta separately,

$$\Gamma(1) = \Gamma\left(p_1, k_1, k_2; \frac{p_2}{s}\right). \tag{8.50}$$

It is clear that some condition must be imposed on the final state momenta in order to preserve the dominance of one polarization as before: if 'wrong' polarizations multiplying momenta k_1, k_2 produced large contributions, the factorization would be lost. First of all, we have to have, as always, the momentum transfer q^2 to be finite; otherwise the very reggeon approach would fail. But this is not enough. We need not only the *total* momentum $k_1 + k_2$ (which determines q^2) but also *each* k_i to be 'almost parallel' to the incoming p_1. The final particles must *fly together*, in a dense 'bunch'.

To give a precise meaning to 'flying together' let us consider a more general case of a quasi-elastic process when both incident particles produce bunches of particles which move close to the directions of the incoming hadrons as shown in Fig. 8.6. A process of this kind is called diffractive dissociation. If the relative energies of the particles in such a bunch are

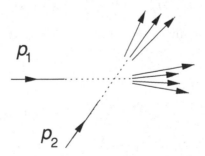

Fig. 8.6   Quasi-elastic process of (double) diffractive dissociation.

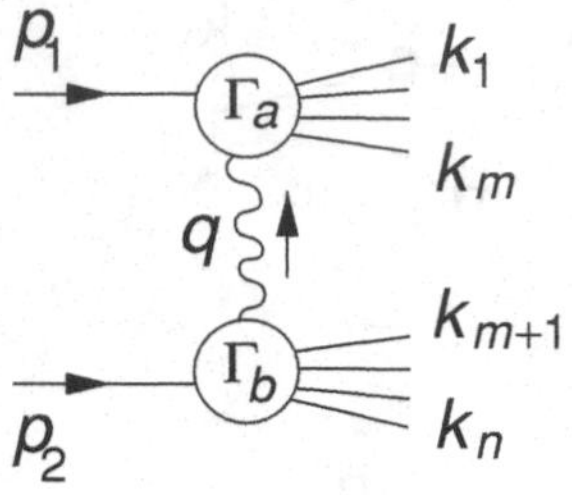

$$P_1 = \sum_{i=1}^{m} k_i, \qquad M_1^2 = P_1^2;$$

$$P_2 = \sum_{i=m+1}^{n} k_i, \qquad M_2 = P_2^2.$$

Fig. 8.7   Kinematics of high-energy diffractive dissociation.

small, it may be treated as a composite particle, and at large s the process can be effectively described as a two-particle scattering.

Let us express all momenta in terms of the Sudakov variables:

$$k_i = \alpha_i p^+ + \beta_i p^- + k_{i\perp}, \qquad q = \alpha_q p^+ + \beta_q p^- + q_\perp; \qquad (8.51)$$

$$p_1 = p^+ + \gamma_1 p^-, \qquad p_2 = p^- + \gamma_2 p^+.$$

Recall that $p^\pm$ are zero-norm vectors ($p^{\pm 2} = 0$, $2p^+p^- = s$), and $\gamma_{1,2} \simeq m_{1,2}^2/s$. In a frame where both scattering particles are fast (e.g. the cms of s-channel), p^+ almost coincides with p_1, as p^- does with p_2. Then, in (8.51), $-\alpha_q$ is the fraction of the momentum p_1, transferred by the reggeon to the lower vertex Γ_b, while β_q is the fraction of p_2 transferred up to Γ_a, see Fig. 8.7. How large are the longitudinal Sudakov components of the momentum transfer q? We have calculated them above in (8.37):

$$-\alpha_q = \frac{\Delta M_2^2 + \mathbf{q}_\perp^2}{s}, \qquad \beta_q = \frac{\Delta M_1^2 + \mathbf{q}_\perp^2}{s}, \qquad (8.52)$$

where $\Delta M_1^2 = \left(\sum_{i=1}^{m} k_i\right)^2 - m_1^2$ and $\Delta M_2^2 = \left(\sum_{i=m+1}^{n} k_i\right)^2 - m_2^2$ are the 'excitation energies' of the upper and lower bunches of particles.

If we keep invariant masses of the bunches *finite* in the high-energy limit,

$$M_1^2, M_2^2 = \mathcal{O}(m^2), \qquad s \to \infty, \qquad (8.53)$$

our previous logic remains intact and we arrive at a reggeon exchange amplitude describing the diagram of Fig. 8.7,

$$A_\alpha(s, q^2; s_{ij}, t_{1i}, t_{2i}) \qquad (8.54)$$

$$= \Gamma_a\left(p_1, \epsilon_1; \{k_i, \epsilon_i\}_1^m; \frac{p_2}{s}\right) \Gamma_b\left(p_2, \epsilon_2; \{k_i, \epsilon_i\}_{m+1}^n; \frac{p_1}{s}\right) s^\alpha \xi_\alpha,$$

where $s_{ij} = (k_i + k_j)^2$, $t_{1i} = (p_1 - k_i)^2$, $t_{2i} = (p_2 - k_i)^2$. This is just the answer that one derives as a result of a rather cumbersome procedure of analytic continuation of partial waves for multi-particle amplitudes.

The expression (8.54) is, essentially, the same as we have written for non-zero spin particle scattering. It differs from (8.49) only by the structure of the vertex functions: now Γ includes also the variables $\{k_i, \epsilon_i\}$ (momenta and polarizations of the offspring) describing 'internal movement' of the particles in each 'bunch'.

Under the condition (8.53), the longitudinal momentum transfer is vanishingly small at large s (in any reference frame where colliding particles are fast). For example, in the s-channel cms, $p^{\pm} = \frac{1}{2}\sqrt{s}(1; 0, 0, \pm 1)$,

$$
\begin{aligned}
q_{\|} &\equiv \alpha_q p^+ + \beta_q p^- = \frac{\sqrt{s}}{2}(\alpha_q + \beta_q, 0, 0, \alpha_q - \beta_q) \\
&= \frac{1}{\sqrt{s}}\left(\frac{\Delta M_1^2 - \Delta M_2^2}{2}; 0, 0, -\frac{\Delta M_1^2 + \Delta M_2^2}{2} - \mathbf{q}_{\perp}^2\right).
\end{aligned} \tag{8.55a}
$$

As for its contribution to the squared momentum transfer, $q^2 = q_{\|}^2 - \mathbf{q}_{\perp}^2$,

$$
\left|q_{\|}^2\right| = -\alpha_q \beta_q s \simeq \frac{\Delta M_1^2 \cdot \Delta M_2^2}{s} \ll \mathbf{q}_{\perp}^2 \simeq -q^2. \tag{8.55b}
$$

Finally, let us address the question of the applicability of the reggeon inelastic diffraction amplitude (8.54), which we have vaguely formulated above as 'to fly together' for particles in the two bunches.

Momentum conservation in the upper and lower blocks gives us

$$
-\alpha_q = 1 - \sum_{i=1}^{m} \alpha_i, \qquad \beta_q = \sum_{i=1}^{m} \beta_i - \gamma_1; \tag{8.56a}
$$

$$
-\alpha_q = \sum_{i=m+1}^{n} \alpha_i - \gamma_2, \qquad \beta_q = 1 - \sum_{i=m+1}^{n} \beta_i; \tag{8.56b}
$$

in addition to

$$
\mathbf{q}_{\perp} = \sum_{i=1}^{m} \mathbf{k}_{i\perp} = -\sum_{i=m+1}^{n} \mathbf{k}_{i\perp}.
$$

To reconcile (8.56) with the condition $-\alpha_q \sim \beta_q = \mathcal{O}(s^{-1})$ following from (8.52) and (8.53), particles in the upper bunch must have $\alpha_i = \mathcal{O}(1)$ and small β_i components, and those in the lower bunch, on the contrary, $\beta_j = \mathcal{O}(1)$ and $\alpha_j = \mathcal{O}(s^{-1})$. Making use of the on-mass-shell relations for the produced particles,

$$
k_i^2 = \alpha_i \beta_i s - k_{i\perp}^2 = m_i^2 \qquad (\alpha_i > 0, \beta_i > 0),
$$

the requirements on particle momenta in the two beams read

$$\alpha_i \sim 1, \qquad \beta_i = \frac{m_i^2 + \mathbf{k}_{i\perp}^2}{\alpha_i s} \ll 1, \qquad i = 1 \div m; \qquad (8.57a)$$

$$\beta_i \sim 1, \qquad \alpha_i = \frac{m_i^2 + \mathbf{k}_{i\perp}^2}{\beta_i s} \ll 1, \qquad i = (m+1) \div n. \quad (8.57b)$$

The conditions (8.57) are just expressions of the fact that the particles fly in 'bunches' in both beams. For particles belonging to one beam, e.g. the upper one, we have

$$
\begin{aligned}
s_{ij}^{aa} &= (\alpha_i + \alpha_j)(\beta_i + \beta_j)s - (\mathbf{k}_{i\perp} + \mathbf{k}_{j\perp})^2 \\
&\sim (\alpha_i^a \beta_j^a + \beta_i^a \alpha_j^a)\, s \sim \left(1 \cdot \frac{m^2}{s} + \frac{m^2}{s} \cdot 1\right) s = \mathcal{O}(m^2), \quad (8.58a)
\end{aligned}
$$

while for particles from different beams,

$$s_{ij}^{ab} \sim \left(\alpha_i^a \beta_j^b + \beta_i^a \alpha_j^b\right) s \sim \left(1 \cdot 1 + \frac{m^2}{s} \cdot \frac{m^2}{s}\right) s = \mathcal{O}(s). \qquad (8.58b)$$

In the kinematical configuration described by (8.57), the first term in (8.58b) for the invariant energy between the two particles from opposite beams is *much larger* than the second one. In fact, this is the *necessary condition* for the validity of the reggeon expression (8.54) we are looking for. It becomes clear if we rewrite the r.h.s. of (8.58b) in the following terms,

$$\alpha_i^a \beta_j^b s = 2(k_i^a \cdot e^-)(e^+ \cdot k_j^b); \qquad \beta_i^a \alpha_j^b s = 2(k_i^a \cdot e^+)(e^- \cdot k_j^b).$$

We immediately see that the strong inequality

$$\alpha_i^a \beta_j^b \gg \beta_i^a \alpha_j^b \quad \Longrightarrow \quad \left(\frac{\alpha_i}{\beta_i}\right)_{\text{bunch } a} \gg \left(\frac{\alpha_j}{\beta_j}\right)_{\text{bunch } b} \quad \Longrightarrow \quad \frac{(\alpha_i)_a}{(\alpha_j)_b} \gg 1$$

guarantees the dominance of that one polarization that gives rise to the factorized reggeon amplitude.

At the same time, for two particles from *the same bunch* the ratio of the Sudakov variables is not too large. Indeed, if we take two particles having relatively small energy in their cms, $\mathbf{p}_1 + \mathbf{p}_2 = 0$; $p_{10}, p_{20} = \mathcal{O}(m)$,

$$s_{12} = (p_1 + p_2)^2 \sim 4m^2.$$

Then in the reference frame where both particles have large velocity along the **z**-axis we have

$$p_{10} \simeq p_{1z} + \frac{m^2 + \mathbf{p}_{1\perp}^2}{2p_{10}}, \qquad p_{20} \simeq p_{2z} + \frac{m^2 + \mathbf{p}_{2\perp}^2}{2p_{20}},$$

and the invariant energy reads

$$s_{12} \simeq 2(p_{10}p_{20} - p_{1z}p_{2z}) \simeq (m^2 + \mathbf{p}_{1\perp}^2)\frac{p_{20}}{p_{10}} + (m^2 + \mathbf{p}_{2\perp}^2)\frac{p_{10}}{p_{20}}$$

$$\sim m_\perp^2 \left(\frac{p_{20}}{p_{10}} + \frac{p_{10}}{p_{20}}\right).$$

Comparing the two expressions for s_{12} we conclude that to 'fly together' means for fast particles that the *ratio* of their energies is neither too small nor too large:

$$\frac{p_{20}}{p_{10}} \sim \frac{p_{10}}{p_{20}} \sim 1.$$

Meantime the *difference* of energies may be very large.

8.5 Elastic scatterings of π and N off the nucleon

Now we will illustrate the results of the previous section by studying the pion–nucleon and nucleon–nucleon reactions, $\pi N \to \pi N$ and $NN \to NN$. To understand what Regge poles can contribute to the asymptotics of these processes at all, we have to consider first in detail the $\pi\pi$ and NN vertices in Fig. 8.8(a).

8.5.1 The pion vertex

Of all invariants that one can construct from the three four-vectors entering the pion vertex, $\Gamma_\pi(p_1, q; p_2/s)$, only one variable, q^2, survives at $s \to \infty$, so we have

$$g_\pi = \Gamma_\pi\left(p_1, q; \frac{p_2}{s}\right) = g_\pi(q^2).$$

Let us clarify which reggeon quantum numbers can be emitted in the $\pi\pi$ vertex. To do that we have to look at the system of two pions from the

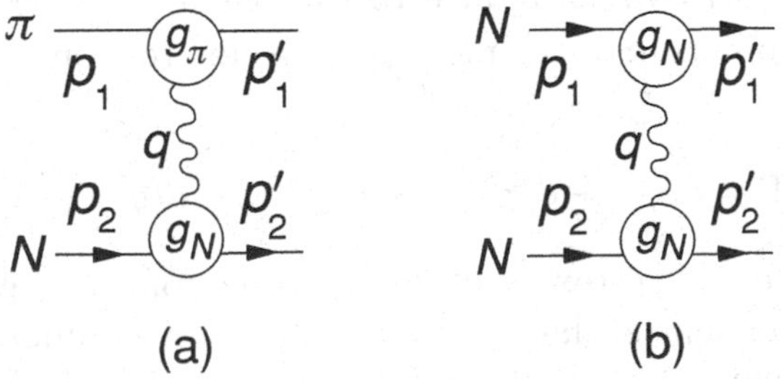

Fig. 8.8   Reggeon exchange amplitudes for πN (a) and NN scattering (b).

Table 8.1   *Reggeons coupled to pion → pion. All have $P_r = C_r = +1$.*

I	G_r	sgn	Reggeon	$\alpha(0)$	α'	Resonance
0	+	+	**P**	1	0.25	?
			f (**P′**)	0.5	1	$f_2(1270)$, $f_4(2050)$, $f_6(2510)$
1	−	−	ρ	0.5	1	$\rho(770)$, $\rho_3(1690)$, $\rho_5(2350)$
2	+	+	?			none

t-channel. Since the pions are spinless, the spatial parity is

$$P = (-1)^\ell = (-1)^j \quad \Longrightarrow \quad P_r = +1.$$

The pions are identical Bose-particles, therefore the vertex must be symmetric under the transmutation of two particles, that is of their positions and isotopic indices: $+1 = (-1)^\ell(-1)^I = (-1)^j(-1)^I$. Consequently, $\pi\pi$ states with a total isospin $I = 0, 2$ in the t-channel are linked to reggeons with positive signature and those with $I = 1$ to ones with negative signature. The $\pi\pi$ charge parity is also unique: $C = (-1)^\ell \Longrightarrow C_r = +1$.

Possible quantum numbers of the $\pi\pi$ system are listed in Table 8.1.

8.5.2 The nucleon vertex

Now we turn to the lower vertex in Fig. 8.8(a), that for the nucleon. In the construction of the nucleon vertex, Γ_N, we may employ Lorentz invariants and Dirac matrices,

$$\hat{\Gamma}_N = a \cdot \mathbf{I} + b \cdot \gamma_\mu \times (\text{vectors})^\mu + c\gamma_5 + d \cdot \gamma_5\gamma_\mu \times (\text{vectors})^\mu,$$

to be sandwiched between final and initial nucleon spinors,

$$g_N = \bar{u}(p'_2)\hat{\Gamma}_N \left(p_2, q; \frac{p_1}{s}\right) u(p_2).$$

As we have already learned, $\hat{\Gamma}$ depends on the momentum in the 'alien' vertex through the four-vector $\hat{p}_1/s$. Convoluting γ_μ with vectors p_2 and q produce $\hat{p}_2$ and $\hat{p}'_2$ which reduce to the scalar m_N when acting on the neighbouring spinors. So the most general expression for $\hat{\Gamma}$ we are left with reads

$$\hat{\Gamma} = a(q^2) + b(q^2)\frac{\hat{p}_1}{s} + c(q^2)\gamma_5 + d(q^2)\gamma_5\frac{\hat{p}_1}{s}. \tag{8.59}$$

Contrary to the $\pi\pi$ vertex, the nucleon one contains, generally speaking, four independent scalar functions. We need to understand whether all these structures are really independent or some of them can mix with each other. To this end we have to look at the $\bar{N}N$ system in the

t-channel to see what this system can transfer to; in other words, what are the quantum numbers of each of the four terms in (8.59).

Given the total angular momentum j, we have four possible states:

$$
\begin{aligned}
(1,2): \quad & \sigma = 1, \quad \ell = j \pm 1, \\
(3): \quad & \sigma = 1, \quad \ell = j, \\
(4): \quad & \sigma = 0, \quad \ell = j.
\end{aligned} \tag{8.60}
$$

Here ℓ is the orbital momentum of the $\bar{N}N$ pair and σ its full spin.

Bearing in mind that the *internal parity* of a fermion and its antifermion is opposite, spatial parity of the $\bar{N}N$ pair is

$$
P = (-1) \times (-1)^\ell = \begin{cases} (-1)^j & \text{for } \ell = j \pm 1, \\ (-1)^{j+1} & \text{for } \ell = j. \end{cases}
$$

This separates the first two and the last two states in (8.60):

$$
P_r(1,2) = +1, \quad P_r(3,4) = -1.
$$

Consider now the charge parity C (or G-parity which generalizes it onto full isotopic multiplets). Under charge conjugation of a neutral system, say, $p\bar{p}$, the particle becomes an antiparticle and vice versa,

$$
p(x_1, \sigma_1) + \bar{p}(x_2, \sigma_2) \quad \rightarrow \quad \bar{p}(x_1, \sigma_1) + p(x_2, \sigma_2).
$$

To get back to the initial state one has to exchange space coordinates and spin variables. This operation results in the phase factor

$$
C = (-1)^\ell \times (-1)^\sigma = \begin{cases} (-1)^j & \Longrightarrow C_r = +1 \quad \text{for } (1,2) \text{ and } (4), \\ (-1)^{j+1} & \Longrightarrow C_r = -1 \quad \text{for } (3). \end{cases}
$$

Now we know quantum numbers of the four states listed in (8.60):

$$
\begin{aligned}
(1,2): \quad & \sigma = 1, \quad \ell = j \pm 1, \quad P_r = +1, \quad C_r = +1, \\
(3): \quad & \sigma = 1, \quad \ell = j, \quad P_r = -1, \quad C_r = -1, \\
(4): \quad & \sigma = 0, \quad \ell = j, \quad P_r = -1, \quad C_r = +1.
\end{aligned} \tag{8.61}
$$

We conclude that the first two states can actually mix, so that we can expect three (rather than four) distinct sets of reggeons that couple to a nucleon.

It is interesting to notice that the list (8.61) does not contain one combination, namely $(P_r, C_r) = (+, -)$: such a state cannot be constructed from a fermion and an antifermion. Remarkably, particles with this specific combination of quantum numbers were not observed experimentally! This *may mean* that all mesons are indeed built of a fermion–antifermion pair (as in the quark model).

Let us learn how to establish spatial and charge parities for each term of the reggeon vertex (8.59). To do that we have to go into the t-channel cms and perform the symmetry operation in one vertex (obviously, under the reflection of *both* vertices the amplitude stays invariant). We start with spatial reflection, that is, changing signs of three-momenta without touching spins. Under this operation,

$$p_2^{(t)} \to (p_{20}^{(t)}; -\mathbf{p}_2^{(t)}), \quad p_2'^{(t)} \to (p_{20}'^{(t)}; -\mathbf{p}_2'^{(t)}),$$

the spinors transform as follows, $u \to \gamma_0 u$, $\bar{u} \to \bar{u}\gamma_0$, so that the vertex gets wrapped by γ_0 matrices,

$$\hat{\Gamma}_N \to \gamma_0 \hat{\Gamma}_N \gamma_0 \qquad (\gamma_0^2 = \gamma_0, \ \gamma_0\gamma_i\gamma_0 = -\gamma_i, \ \gamma_0\gamma_5\gamma_0 = -\gamma_5).$$

Then we have to recall that $p_1 \simeq p^+$ represents in the t-channel a circular polarization vector which has no energy component, $p_1^{(t)} \simeq (0, \mathbf{p}_1^{(t)})$; as a result, $\gamma_0 \hat{p}_1^{(t)} \gamma_0 = -\hat{p}_1^{(t)}$. Finally, under the reflection, the t-channel scattering angle undergoes $\Theta^{(t)} \to \Theta^{(t)} + \pi$, so that the sign of s also has to be changed since $s \propto \cos\Theta^{(t)}$. All this said, we obtain

$$(a)\colon 1 \to 1, \quad (b)\colon \frac{\hat{p}_1}{s} \to \frac{-\hat{p}_1}{-s}, \quad (c)\colon \gamma_5 \to -\gamma_5, \quad (d)\colon \gamma_5\frac{\hat{p}_1}{s} \to -\gamma_5\frac{-\hat{p}_1}{-s},$$

which gives us P_r parity of each of the four structures in the vertex (8.59):

$$g_{(1,2)} = \bar{u}\left[a + b\frac{\hat{p}_1}{s}\right]u \to +g_{(1,2)},$$

$$g_{(3)} = \bar{u}\left[c\gamma_5\right]u \to -g_{(3)},$$

$$g_{(4)} = \bar{u}\left[d\gamma_5\frac{\hat{p}_1}{s}\right]u \to -g_{(4)}.$$

Charge conjugation is a bit more complicated. The crossing amounts to

$$s \to u \simeq -s, \tag{8.62a}$$

and transforming the nucleon wave functions as follows,

$$u \to C^{-1}\bar{v}^T, \qquad \bar{u} \to v^T C, \tag{8.62b}$$

where v is a spinor describing the antiparticle (superscript T stands for transposition), and C is the charge conjugation matrix:

$$C^{-1}\gamma_\mu C = -\gamma_\mu^T, \qquad C^{-1}\gamma_5 C = \gamma_5^T. \tag{8.62c}$$

Making use of the rules (8.62) it is straightforward to derive that, differently from three other vertices, the vertex g_4 has an odd charge parity.

Both spatial and charge reflections include the $s \to -s$ operation because of which the reggeon propagator $s^\alpha \xi_\alpha$ produces an additional factor $(-1)^\alpha \equiv (-1)^j$. This factor is included in the definition of P_r and C_r.

Table 8.2 *Reggeons coupled with nucleon → nucleon.*

vertex	P_r	C_r	I	G_r	sgn	Reggeon
	$+$	$+$	0	$+$	$+$	$\mathbf{P}$, f $(\mathbf{P'})$
$a + b\dfrac{\hat{p}_1}{s}$			0	$+$	$-$	ω
			1	$-$	$+$	A_2
			1	$-$	$-$	ρ
	$-$	$+$	0	$+$	$+$	η
$c\,\gamma_5$			1	$-$	$+$	π
	$-$	$-$	0	$+$	$-$	f_1
$d\,\gamma_5\dfrac{\hat{p}_1}{s}$			1	$-$	$-$	a_1

In Table 8.2 the corresponding Regge trajectories are presented.

8.5.3 Spin phenomena in πN and NN scattering

We start from the $\pi N \to \pi N$ scattering process of Fig. 8.8(a). Examining Tables 8.1 and 8.2 we conclude that reggeon exchanges with $P_r = C_r = +1$ are possible, which include reggeons with vacuum quantum numbers, $\mathbf{P}$, $f(\mathbf{P'})$, etc. as well as negative signature trajectories like ρ. All such reggeons contribute, but at $s \to \infty$ only the rightmost one, the pomeron $\mathbf{P}$, survives. In this limit elastic scattering dominates, and the amplitude can be written in the following form,

$$A_{\pi p}^{\mathrm{el}}(s, q^2) \simeq g_\pi(q^2)\bar{u}^{\lambda'}(p_2')\left[a(q^2) + 2mb(q^2)\frac{\hat{p}_1}{s}\right]u^\lambda(p_2)\cdot s^\alpha \xi_\alpha, \qquad (8.63)$$

where the pion residue g_π and the nucleon vertex functions a, b are all isospin-diagonal and refer to the pomeron, $\alpha = \alpha_{\mathbf{P}}(q^2)$. Indices λ and λ' marking the Dirac four-spinors stand for polarizations of the incoming and the outgoing protons, correspondingly. Let us see whether there is anything new in (8.63) compared to the case of spinless particles. We observe that the amplitude still depends on spins so that even at tremendously large energies, proton polarization changes (whereas isospin does not: charge exchange reactions, like $\pi^- p \to \pi^0 n$, have died out).

What may happen to spin observables in principle? There are different possibilities: a polarized nucleon may flip its spin, or lose its polarization altogether; an unpolarized one may acquire non-zero polarization, etc. What happens in reality? From the reggeon expression (8.63) it follows that only one of the whole variety of spin phenomena survives in the high energy limit, namely the nucleon spin turns. Moreover, this occurs in a rather unusual form.

What would we expect in a simple diffraction picture? If a particle in NQM scatters at small angle, we could imagine that its spin either does not change as in Fig. 8.9(a) or turns, following the change in the particle-momentum direction (if spin and momentum are firmly tied), Fig. 8.9(b). In the relativistic case, in spite of a miniscule scattering an-

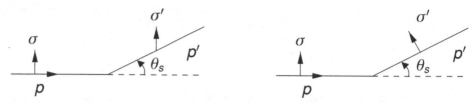

Fig. 8.9

gle $\theta_s \simeq |\mathbf{p}'_\perp|/|\mathbf{p}'| \ll 1$, the spin turns to a *large* angle, $\theta = \mathcal{O}(1)$, which depends on the momentum transfer but not on the collision energy. This shows that the reggeon interaction acts in a non-trivial way, independently on the momentum and the spin of the participating particle.

We start with the first term in the nucleon vertex (8.63),

$$a(q^2)\bar{u}^{\lambda'}(p'_2)u^\lambda(p_2),$$

and treat it in a frame where the proton is fast, $|\mathbf{p}'_2| \simeq |\mathbf{p}_2| \gg |\mathbf{q}|$, and scatters at a small angle

$$\theta_s \simeq \sin\theta_s; \quad \boldsymbol{\theta}_s = \frac{[\mathbf{p}_2 \times \mathbf{p}'_2]}{|\mathbf{p}_2|\,|\mathbf{p}'_2|} \simeq \frac{[\mathbf{e}_z \times \mathbf{q}]}{|p_{02}|}.$$

To evaluate the product of Dirac spinors we can express the final state wave function $u(p'_2)$ via the initial state one, $u(p'_2) = \hat{R}(\boldsymbol{\theta}_s)u(p_2)$, using the rotation matrix $\hat{R}$ which we expand to the first order in θ_s:

$$\hat{R}(\boldsymbol{\theta}_s) = \begin{pmatrix} \exp\!\left(\frac{i}{2}\boldsymbol{\sigma}\cdot\boldsymbol{\theta}_s\right) & 0 \\ 0 & \exp\!\left(-\frac{i}{2}\boldsymbol{\sigma}\cdot\boldsymbol{\theta}_s\right) \end{pmatrix} \simeq \mathrm{I} + \gamma_0 \frac{i\boldsymbol{\sigma}\cdot\boldsymbol{\theta}_s}{2}.$$

This gives

$$\bar{u}^{\lambda'}(p'_2)u^\lambda(p_2) \simeq \bar{u}^{\lambda'}(p_2)\left(1 - \gamma_0 \frac{i\boldsymbol{\sigma}\cdot\boldsymbol{\theta}_s}{2}\right)u^\lambda(p_2)$$

$$= \varphi^{\lambda'*}\left(2m - 2p_{20}\frac{i\boldsymbol{\sigma}\cdot\boldsymbol{\theta}_s}{2}\right)\varphi^\lambda = \varphi^{\lambda'*}\!\left(2m + i[\boldsymbol{\sigma}\times\mathbf{q}]_z\right)\varphi^\lambda.$$

Here we have represented the Dirac wave functions in terms of 2×2 Weyl spinors φ,

$$u^\lambda(p) = \begin{pmatrix} \sqrt{p_0 + m}\,\varphi^\lambda \\ \dfrac{\boldsymbol{\sigma} \cdot \mathbf{p}}{\sqrt{p_0 + m}}\varphi^\lambda \end{pmatrix},$$

and used the well known relations

$$\begin{aligned}
\bar{u}^{\lambda'}(p)u^\lambda(p) &= 2m\varphi^{\lambda'*}\varphi^\lambda, \\
\bar{u}^{\lambda'}(p)\gamma_\mu u^\lambda(p) &= 2p_\mu\varphi^{\lambda'*}\varphi^\lambda.
\end{aligned}$$

In the second term of the nucleon vertex we can set $p_2' = p_2$, up to negligible corrections $\mathcal{O}(s^{-1})$, and use (8.64) to get

$$2mb\frac{p_{1\mu}}{s}\bar{u}(p_2')\gamma^\mu u(p_2) \simeq 2mb\varphi^*\varphi\left[\frac{p_{1\mu}}{s} \cdot 2p_2^\mu + \cdots\right] \simeq 2mb(q^2)\varphi^*\varphi.$$

Combining the two terms, for the nucleon vertex we obtain the expression

$$\begin{aligned}
&\bar{u}^{\lambda'}(p_2')\left[a(q^2) + 2mb(q^2)\frac{\hat{p}_1}{s}\right]u^\lambda(p_2) \\
&\simeq \varphi^{\lambda'*}\left(2m[a(q^2) + b(q^2)] + ia(q^2)[\boldsymbol{\sigma} \times \mathbf{q}]_z\right)\varphi^\lambda.
\end{aligned} \tag{8.64}$$

The matrix structure of the nucleon Regge residue (8.64) tells us that the spin of the proton turns around the normal to the scattering plane, $[\mathbf{n} \times \mathbf{n}']$, to a finite angle θ:

$$\tan\frac{\theta}{2} = \frac{|\mathbf{q}_\perp|}{2m}\frac{a(-\mathbf{q}_\perp^2)}{a(-\mathbf{q}_\perp^2) + b(-\mathbf{q}_\perp^2)} = \mathcal{O}(1). \tag{8.65}$$

Here a few comments are due. First of all, we see that θ is determined by the functions a and b which describe pomeron attachment to target nucleon and do not depend on the type of the projectile. Therefore, for a given q, the nucleon spin will rotate by the same angle (8.65) in πN and NN scattering. Moreover, since in the single-pole approximation all spin amplitudes have the same phase (all complexity is embodied in the universal signature factor ξ_α), polarization can neither emerge nor disappear as a result of collision since these effects are proportional to an interference between amplitudes with and without 'spin–flip':

$$\mathcal{P} \propto \mathrm{Im}\left(A_{\uparrow\uparrow}^* A_{\uparrow\downarrow}\right).$$

For the same reason at asymptotically high energies these is no other more subtle effects such as correlation between two spins in the NN scattering.

8.6 Conspiracy

Concerning NN scattering, one interesting phenomenon remains to be investigated known as 'conspiracy'.

As we have established already, this amplitude contains three separate contributions,

$$A_{NN \to NN} = A_{(1)} \begin{pmatrix} P_r = +1 \\ C_r = +1 \end{pmatrix} + A_{(2)} \begin{pmatrix} P_r = -1 \\ C_r = +1 \end{pmatrix} + A_{(3)} \begin{pmatrix} P_r = -1 \\ C_r = -1 \end{pmatrix},$$

whose energy dependence is described by three reggeons with different, generally speaking, Regge trajectories:

$$A_{(1)} = \left(\bar{u}_1[a(q^2) + b(q^2)\frac{\hat{p}_2}{s}]u_1 \right)\left(\bar{u}_2[a(q^2) + b(q^2)\frac{\hat{p}_1}{s}]u_2 \right) s^{\alpha_1} \xi_{\alpha_1}, \quad (8.66a)$$

$$A_{(2)} = \left(\bar{u}_1[c(q^2)\gamma_5]u_1 \right)\left(\bar{u}_2[c(q^2)\gamma_5]u_2 \right) s^{\alpha_2} \xi_{\alpha_2}, \quad (8.66b)$$

$$A_{(3)} = \left(\bar{u}_1[d(q^2)\gamma_5\frac{\hat{p}_2}{s}]u_1 \right)\left(\bar{u}_2[d(q^2)\gamma_5\frac{\hat{p}_1}{s}]u_2 \right) s^{\alpha_3} \xi_{\alpha_3}. \quad (8.66c)$$

Let us take a moderately large energy where all the poles are essential, add their contributions and... we will be surprised! Look at the value of the amplitude at $\mathbf{q} = 0$, i.e. at the exactly forward scattering. Then the first amplitude (8.66a) trivializes and becomes diagonal with respect to the spin,

$$\bar{u}^{\lambda'}(p_1)[\frac{\hat{p}_2}{s}]u^\lambda(p_1) = \frac{p_2^\mu}{s} \cdot \bar{u}\gamma_\mu u = \frac{p_2^\mu}{s} \cdot 2p_{1\mu} \cdot \delta_{\lambda\lambda'} = \delta_{\lambda\lambda'}.$$

As for the rest, $A_{(2)}$ vanishes and $A_{(3)}$ stays finite. To see this it suffices to look at one vertex in the rest frame, say, $p_1 = (m, \mathbf{0})$, where the four-spinor has only an upper component,

$$u^\lambda(p_1) = \begin{pmatrix} \varphi^\lambda \\ 0 \end{pmatrix},$$

and of three matrices,

$$\gamma_5 = \begin{pmatrix} 0 & 1 \\ -1 & 0 \end{pmatrix}, \quad \gamma_5\gamma_0 = -\begin{pmatrix} 0 & 1 \\ 1 & 0 \end{pmatrix}, \quad \gamma_5\gamma_i = -\begin{pmatrix} \sigma_i & 0 \\ 0 & \sigma_i \end{pmatrix},$$

only the last one (diagonal) survives,

$$\bar{u}(p_1)\gamma_5\gamma u(p_1) = \varphi_1^* \sigma \varphi_1 \equiv \sigma^{(1)},$$

yielding the matrix element $\sigma_z^{(1)}$ upon multiplication by $p_{2\mu}$ in (8.66c). Thus from $A_{(1)}$ and $A_{(3)}$ we have two contributions to the forward amplitude:

$$A(q = 0) = c + c'\sigma_z^{(1)}\sigma_z^{(2)}; \tag{8.67}$$

$\mathbf{z}$ is the collision axis as before.

This expression is formally legitimate (the $\mathbf{z}$-direction is special, while rotational symmetry in the transverse plane is respected). Still it looks strange: from general considerations we could expect the presence of another term in (8.67)

$$\cdots + c''(\boldsymbol{\sigma}_\perp^{(1)} \cdot \boldsymbol{\sigma}_\perp^{(2)}) = c''(\sigma_x^{(1)}\sigma_x^{(2)} + \sigma_y^{(1)}\sigma_y^{(2)}). \tag{8.68}$$

Why have we lost this invariant? It is not a question of the asymptotic behaviour since we have added all Regge pole contributions; so, what has happened?

Let us see whether we could restore the term (8.68) in the forward scattering amplitude. We have to be a bit more accurate. Taking the $q = 0$ limit in (8.66a), we have dropped the second, linear in $\mathbf{q}$, term in the full expression (8.64) for the first amplitude,

$$\propto ia(-\mathbf{q}_\perp^2)[\boldsymbol{\sigma} \times \mathbf{q}_\perp]_z \qquad \left(\propto \sigma_x \quad \text{for } \mathbf{q}_\perp \| \mathbf{y}\right). \tag{8.69}$$

This would have been a bad idea if the coefficient was *singular* at $q_\perp = 0$. Imagine that we force this contribution to be *finite* in the $q_\perp \to 0$ limit. The result is bizarre: the amplitude would remember about the direction of the vector $\mathbf{q}_\perp$ before it vanished! Indeed, if $\mathbf{q}$ is directed along the $\mathbf{y}$-axis, the rescued term contains the matrix σ_x, as envisaged in (8.69). So, we have found the $\sigma_x \otimes \sigma_x$ term rather than the full $\sigma_\perp \otimes \sigma_\perp$ of (8.68).

Let us search for the missing $\sigma_y \otimes \sigma_y$. The vertex in $A_{(2)}$ is a *pseudoscalar* and should therefore contain $\varphi^*(\boldsymbol{\sigma} \cdot \mathbf{q}_\perp)\varphi \propto \sigma_y$ (and the coefficient $c(q^2)$ in (8.66b) could also be singular). Indeed, rewriting (8.66) in the two-component form at small momentum transfer, we obtain

$$A_{(1)} = \left\{f_1 + f_2[\boldsymbol{\sigma}^{(1)} \times \mathbf{q}_\perp]_z\right\}\left\{f_1 + f_2[\boldsymbol{\sigma}^{(2)} \times \mathbf{q}_\perp]_z\right\}s^{\alpha_1}\xi_{\alpha_1}, \tag{8.70a}$$

$$A_{(2)} = f_3^2\left(\boldsymbol{\sigma}^{(1)} \cdot \mathbf{q}_\perp\right)\left(\boldsymbol{\sigma}^{(2)} \cdot \mathbf{q}_\perp\right)s^{\alpha_2}\xi_{\alpha_2}, \tag{8.70b}$$

$$A_{(3)} = f_4^2\,\sigma_z^{(1)}\sigma_z^{(2)}s^{\alpha_3}\xi_{\alpha_3}, \tag{8.70c}$$

where $f_i = f_i(q^2)$ and $\boldsymbol{\sigma}^{(i)} \equiv \varphi_i^*\boldsymbol{\sigma}\varphi_i$.

The strategy of restoring the lost contribution (8.68) is now clear: f_2 and f_3 must be both singular and, moreover, have to have the same $q \to 0$

limit,

$$f_2^2(q_\perp^2) \simeq f_3^2(q_\perp^2) \simeq \frac{c''}{\mathbf{q}_\perp^2} \qquad \text{at } \mathbf{q}_\perp \to 0. \qquad (8.71a)$$

But the two pieces, $\sigma_x \sigma_x$ and $\sigma_y \sigma_y$, belong to different trajectories! In order to preserve the rotational invariance in the $\{x, y\}$ plane at *all energies*, the vacuum trajectory $(A_{(1)})$ must *cross* with a trajectory with π-meson quantum numbers $(A_{(2)})$ at $t = 0$. This explains the name *conspiracy*: two reggeons must 'make a deal' concerning both their trajectories and residues,

$$\alpha_1(0) = \alpha_2(0), \quad r_1(0) = r_2(0). \qquad (8.71b)$$

Is this a miracle? Not entirely. We have lost, in the first place, (8.68) because $(\boldsymbol{\sigma}_\perp \cdot \boldsymbol{\sigma}_\perp)$ does not correspond to a state with definite quantum numbers in the t-channel: it has a mixture of two states with different *parity*.

Recall what was the reason for Regge trajectories with differing quantum numbers to be different? It was the *unitarity* that separated them. Satisfying different unitarity conditions, two reggeons have no reason to be related *unless* there is a specific symmetry in the works. Is there not an additional symmetry at $t = 0$? Obviously, there is.

What is the *parity* of a given object? We go to its rest frame and see which sign the wave function acquires under the spatial reflection. However, if a particle has $m = 0$, one cannot stop it and the notion of parity loses sense. But the point $t = 0$ corresponds exactly to a zero mass object in the t-channel, and our states $A_{(1)}$ and $A_{(2)}$, which differed only by their parity, $P_r = \pm$, become physically indistinguishable. Certainly, this argument does not *prove* the conspiracy phenomenon but makes its possible existence less mysterious.

From the point of view of the interaction in the t-channel, the conspiracy means that the 'potential' possesses an additional symmetry at $t = 0$ which prevents the unitarity condition from separating the amplitudes with different parities. A closer examination of various reactions shows that the conditions (8.71) are not sufficient. Having made f_i singular, we did something serious: inserted a singularity that the total amplitude should not have. As a remedy, another Regge trajectory must enter the conspiracy plot, with an intercept smaller by one unity. (Actually, a whole series of shifted trajectories – 'daughters' – appear.)

To conclude, if forward-scattering amplitudes contain the contribution proportional to $(\boldsymbol{\sigma}_\perp \cdot \boldsymbol{\sigma}_\perp)$ then, within the Regge-pole approach, this is an evidence for conspiracy. Experimentally such a phenomenon is not

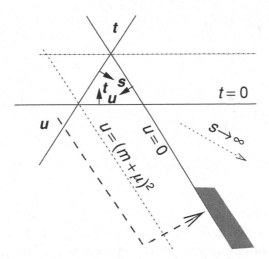

Fig. 8.10   Backward πN scattering on the Mandelstam plane.

observed so far. In any case, an experimental proof is difficult because of the influence of the *reggeon branchings*.

8.7 Fermion Regge poles

Up to now we have studied the behaviour of two-particle scattering amplitudes in the region $|t| \sim m^2$, $s \to \infty$. However, since we formally allowed an incident particle to change identity, this means that we have already treated the backward scattering as well! Look, for example, at the region $|u| \sim m^2$, $s \to \infty$ in the $\pi N \to \pi N$ scattering amplitude shown on the Mandelstam plane in Fig. 8.10.

What is so remarkable about this region? In Lecture 5 we have seen that in the relativistic theory there exists, in addition to a forward peak, also one in the backward direction. The magnitude of the backward peak depends on how willingly the particle is ready to change its individuality (recall the Compton effect).

With the help of the theory of complex angular momenta we have learned how to write the asymptotics for the scattering of spinless particles at a small angle $\theta_s = \sqrt{-t/s}$ and found that it was determined by quantum numbers in the t-channel. Later, we have generalized the obtained results to the case of non-zero spin particles.

Obviously, the same programme can be carried out also for backward scattering, $\theta_u = \sqrt{-u/s} \ll 1$. In the case of spinless particles, or particles with integer spins, the whole story will be a mere repetition, the

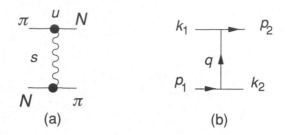

Fig. 8.11   (a) Backward πN scattering; (b) u-channel fermion exchange.

only difference being that the asymptotics will now be determined by the quantum numbers of the *u-channel exchange*. In a backward scattering of a fermion, e.g. backward πN scattering shown in Fig. 8.11(a), an interesting new phenomenon appears owing to the fact that the exchange is now carried out by a reggeon with a *half-integer* angular momentum. Let us investigate this case in detail.

How to write an amplitude corresponding to a fermion exchange? Let us turn to the perturbation theory. The exchange amplitude for a particle with $\sigma = \frac{1}{2}$ shown in Fig. 8.11(b) has the form

$$A(s, u) = \bar{u}(p_2) g(q^2) \frac{1}{m - \hat{q}} g(q^2) u(p_1), \quad q = p_1 - k_2. \tag{8.72}$$

Given the spinor normalization $u(p) \sim \sqrt{p_0}$, at large s this amplitude behaves as $A[\sigma = \frac{1}{2}] \propto \sqrt{s} = s^\sigma$, as expected.

If we have a $\sigma = \frac{3}{2}$ particle exchange, the longitudinal polarization tensor, $e_\mu^- e_\nu^+ \propto p_{1\mu} k_{1\nu}/s$, convolutes with the momenta in the vertices, $k_1^\mu p_1^\nu$ and produces an extra s, yielding $A[\sigma = \frac{3}{2}] \sim s^\sigma$, etc.

Generalizing to the case of an arbitrary spin as before, we shall write the reggeon exchange amplitude as (cf. (8.63))

$$A(s, u) = \bar{u}(p_2) \big[a(q^2) + b(q^2)\hat{q} \big] u(p_1) s^{\alpha(q^2) - \frac{1}{2}} \xi_{\alpha(q^2) - \frac{1}{2}}. \tag{8.73}$$

Here we put $\alpha - \frac{1}{2}$ in the exponent since spinors provide an additional factor $\sqrt{s}$, as they already did before in (8.72).

Following our logic, let us determine the quantum numbers in the u channel. If we assume that there is no degeneracy so that unitarity does separate trajectories with different quantum numbers, we need to learn to write amplitudes with definite parity which correspond to two opposite parity u-channel states with $\ell = j \pm \frac{1}{2}$.

To do so, we move, as usual, to the u-channel cms and carry out spatial reflection in one of the vertices, e.g. in the lower one in Fig. 8.11b. The matrix $\hat{q}$, transversal in the s channel, in the u-channel centre-of-mass frame has only a time component:

$$\hat{q}_{(u)} = \gamma_0 q_{0(u)} = \gamma_0 \sqrt{q^2}.$$

The spinor $u(p_1)$ under reflection produces

$$u(\mathbf{p_1}) \implies i\gamma_0 u(\mathbf{p_1}).$$

Then, the cosine of the u-channel scattering angle changes sign, $z_u = \cos\theta_u \to -z_u$, which corresponds to the reflection $s \to -s$, yielding the factor $(-1)^{\alpha - \frac{1}{2}}$. Altogether, the parity operation acts on the amplitude as follows:

$$A \propto \left[a + b q_{0(u)} \cdot \gamma_0\right] \implies \left[a \cdot \gamma_0 + b q_{0(u)}\right](-1)^\alpha.$$

To have a definite parity $P_r = P \cdot (-1)^\alpha$, one of the conditions

$$a\gamma_0 + b q_{0(u)} = \pm(a + b q_{0(u)}\gamma_0)$$

has to be satisfied, that is, returning to the Lorentz invariant form,

$$a(q^2) = \pm b(q^2)\sqrt{q^2}.$$

Thus, the contribution of a fermionic Regge pole with a definite parity P_r reads

$$A_\pm = r^\pm(q^2)\bar{u}(p_2)\left[\hat{q} \pm \sqrt{q^2}\right]u(p_1) \cdot s^{\alpha_\pm - 1/2}\xi_{\alpha_\pm - 1/2}. \tag{8.74}$$

We see right away that $A_\pm(s, u)$ have rather unusual features.

(1) The existence of a singularity at $q^2 = 0$ contradicts the analytic structure of the Mandelstam plane where the first u-channel (threshold) singularity appears at $q^2 = u = (m + \mu)^2 > 0$.

(2) In the physical region of the s-channel ($q^2 \le 0$) the complexity of $A_\pm$ is not limited to the complexity of the signature factor.

Hence, we cannot assume that the high-energy behaviour of the backward scattering is determined by a single pole of definite parity without contradicting analyticity of the scattering amplitude in u. The only way to avoid this contradiction is to say that the asymptotics is determined by both poles simultaneously:

$$A = A_+ + A_- = r^+ \bar{u}_2(\hat{q} + \sqrt{q^2})u_1 s^{\alpha_+ - \frac{1}{2}}\xi_{\alpha_+ - \frac{1}{2}}$$
$$+ r^- \bar{u}_2(\hat{q} - \sqrt{q^2})u_1 s^{\alpha_- - \frac{1}{2}}\xi_{\alpha_- - \frac{1}{2}}, \tag{8.75a}$$

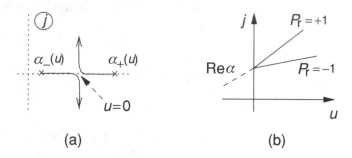

Fig. 8.12   (a) Movement of fermionic reggeons from positive to negative u on the j plane; (b) conspiracy of fermionic trajectories on the Chew–Frautschi plot.

where the two trajectories derive from one function, as do the residues:

$$\alpha^+ \equiv \alpha(\sqrt{u}), \qquad \alpha^- = \alpha(-\sqrt{u}); \tag{8.75b}$$

$$r^+ \equiv r(\sqrt{u}), \qquad r^- = r(-\sqrt{u}). \tag{8.75c}$$

Let us see what are the consequences of this picture.

Conspiracy. First of all, how do the Regge trajectories $\alpha^\pm(u)$ behave?

$u > 0$: the physical region of the u-channel. The trajectories α_+ and α_- are values of the same function of different real arguments.

$u < 0$: the physical region of the s-channel; the arguments are imaginary, complex conjugate, and we have

$$\alpha_+ = \alpha(i\sqrt{-u}) = \alpha^*(-i\sqrt{-u}) = \alpha_-^*.$$

The possible behaviour of the poles in the j-plane while moving from one region to the other is shown in Fig. 8.12(a). At $u > 0$ the trajectories are real and, generally speaking, different, see Fig. 8.12(b). They cross at $u = 0$ ('conspiracy'!); at $u < 0$ they become complex conjugate and diverge in the j plane.

So our attempt to separate fermionic trajectories with different parities, without violating the analyticity, has led us to unavoidable conspiracy. As we have already discussed above in Section 8.6, this phenomenon is due to the uncertainty in the parity of a massless particle.

Let us demonstrate this in perturbation theory. Suppose that an exchange with a Fermi-particle ν takes place described by the diagram in

Fig. 8.11(b) (one can imagine, e.g., backward $\pi^- e^-$ scattering via neutrino exchange). If $m_\nu \neq 0$, we can write two different amplitudes:

$$A_1 = \bar{u}_2 \cdot (i\gamma_5)\frac{m_\nu + \hat{q}}{m_\nu^2 - q^2} \cdot (i\gamma_5)u_1 = \bar{u}_2\frac{m_\nu - \hat{q}}{m_\nu^2 - q^2}u_1,$$

$$A_2 = \bar{u}_2 \cdot \frac{m_\nu + \hat{q}}{m_\nu^2 - q^2} \cdot u_1.$$

On the mass shell, $q^2 = m_\nu^2$, the parity of the amplitude A_1 is $P_1 = -i$ (here we, again, have to carry out a reflection in one of the vertices in the cms of the u-channel); the parity of the second amplitude (A_2) is $P_2 = i$. Taking now $m_\nu = 0$, the two amplitudes become indistinguishable,

$$A_1 = \bar{u}_2\frac{\hat{q}}{q^2}u_1 = -A_2,$$

and we arrive at a degeneracy in parity.

Oscillations. Since the backward scattering amplitude is described by a pair of *complex conjugate* fermionic Regge poles, this should lead to oscillations, both in s and u. Parameterizing $\alpha_\pm(u) = \alpha_1(u) \pm i\alpha_2(u)$, the amplitude can be represented as follows:

$$A(s,u) = |r(u)|s^{\alpha_1 - \frac{1}{2}} \cdot \bar{u}(p_2)\left[f_1\hat{\mathbf{q}}_\perp + f_2\sqrt{\mathbf{q}_\perp^2}\right]u(p_1); \tag{8.76}$$

$$f_1 = f\,\cos[\alpha_2(u)\ln s + \varphi(u)]\,, \quad f_2 = f\,\sin[\alpha_2(u)\ln s + \varphi(u)]\,.$$

To observe these oscillations turns out to be not so simple. Indeed, if we consider the cross section

$$\frac{d\sigma}{d\Omega} \propto \frac{1}{s}AA^\dagger$$

summed (averaged) over the polarization of the final (initial) nucleon,

$$\frac{d\sigma}{d\Omega} \propto \tfrac{1}{2}\mathrm{Tr}\big(A \cdot (\hat{p}_1 + m) \cdot \bar{A} \cdot (\hat{p}_2 + m)\big)\,, \qquad \bar{A} = \gamma_0 A\gamma_0,$$

the oscillations will not manifest themselves, since here the sum $|f_1|^2 + |f_2|^2$ enters, and we obtain

$$\frac{d\sigma}{d\Omega} \propto |r(u)|^2 s^{2(\alpha_1(u)-1)} \cdot f^2(u).$$

The cross section will not oscillate even if we measure the polarization of *one* of the nucleons; only the *spin correlation* between the initial and final fermions will be an oscillating function of energy.

Thus, the theory, in the pole approximation, predicts that:

(1) the baryon resonances lie on Regge trajectories;

(2) cross sections with baryon number exchange decrease as powers of s;

(3) trajectories with different parities conspire at $u = 0$; and

(4) in the asymptotics of the amplitude there are unusual oscillations.

Indeed, many fermionic resonances, up to very high spins, are observed experimentally. Surprisingly, their trajectories look linear in the mass *squared*, $u = m^2$, similarly to the bosonic case (see Fig. 8.1 on page 180), in spite of the fact that the fermion propagator depends on the first power of the mass, $G \propto 1/(m - \hat{q})$, rather than on m^2 as boson propagators do. This means that in the series expansion for the analytic function α which determines the nucleon trajectories according to (8.75b), the square-root term is very small, if not altogether absent:

$$\alpha(\sqrt{u}) \simeq \alpha_0 + \alpha_1 \sqrt{u} + \alpha_2 u + \cdots, \qquad \alpha_1 \simeq 0.$$

But if this is the case then according to (8.75b) there must be *parity degeneracy* for all values of u: $\alpha^+(u) = \alpha^-(u) \simeq \alpha_0 + \alpha_2 u$. Hence, there must exist resonances with opposite parity and *equal masses*, lying on the same trajectory. This is, however, not observed experimentally.

In order to reconcile the theory with the experiment, one can try to 'conceal' this degeneracy by putting the residue of one of the two trajectories equal zero in the position of an unwanted resonance. This may look weird (indeed, how to force a resonance not to interact with anything in the theory of strong interactions?), but in some *dual models* (see Lecture 16) such a possibility is considered.

So, at the moment we have the following situation. On the one hand, we do see fermionic Regge poles. On the other hand, there must be an additional trick in Nature that makes the apparent linearity of baryon trajectories and the absence of parity degeneracy in the spectrum of baryon resonances consistent.

9
Regge poles in perturbation theory

9.1 Reggeons, ladder graphs, and multiparticle production

We have studied two-particle reactions and introduced objects of 'variable spin' – *reggeons* as a generalization of usual particles. The reggeon amplitude,

$$A^{\pm}_{\text{regg}}(s, q^2) \propto \xi_\alpha s^\alpha, \quad \xi_\alpha = \frac{e^{-i\pi\alpha} \pm 1}{-\sin\pi\alpha}, \quad \alpha = \alpha(q^2),$$

differs essentially from the particle-exchange amplitude,

$$A_\sigma(q^2, s) \propto \frac{s^\sigma}{\mu^2 - q^2 - i\varepsilon},$$

by a non-trivial complexity, even at $q^2 < 0$ where particle exchange is real. As the s-channel unitarity tells us, the imaginary part of the elastic amplitude is determined by real processes, mostly by many-particle production since s is very large:

$$\operatorname{Im} A_{\text{el}}(s, q^2) \simeq s\sigma_{\text{tot}} = \tfrac{1}{2} \sum_n \ \text{\raisebox{-0.5ex}{}} \ .$$

This means that the reggeon, having a large imaginary part,

$$\operatorname{Im} A^{\pm}_{\text{regg}} \propto \operatorname{Im} \left[i - \frac{\cos\pi\alpha \pm 1}{\sin\pi\alpha} \right] s^\alpha = s^\alpha,$$

is not an elementary object but is 'composed' of certain inelastic s-channel processes. We have to understand what these processes are.

As we have discussed before, from the point of view of t-channel dynamics, the Regge pole is a bound state of non-relativistic particles. But the interaction of slow particles can be described in terms of a potential:

$$\text{(9.1)}$$

The potential acts without retardation, therefore the dashed lines do not cross. If it turns out that indeed, the reggeon corresponds to potential scattering in the t-channel, this would answer the question which inelastic processes are important at high energies. Cutting through the diagrams on the r.h.s. of (9.1) we obtain *ladders* as an image of inelastic processes 'describing' the reggeon.

$$\text{Im}\, A_{\text{regg}} \sim \sum \quad \boxed{} \qquad \text{(9.2)}$$

These are processes of production of a large number of particles, those very particles that play the rôle of the binding potential in the crossing channel.

9.2 Reggeization in $g\phi^3$ theory

In order to verify that the sum of the diagrams (9.1) gives rise to a Regge behaviour,

$$A = b(t)\, s^{\alpha(t)}, \qquad (9.3)$$

we will address the problem perturbatively and employ the simplest $g\phi^3$ theory with a small coupling. Although this quantum field theory is far from realistic, this academic exercise will teach us important lessons about the phenomenon of *reggeization*. Later we will try to generalize the results beyond the perturbation theory.

9.2.1 Qualitative analysis of higher order diagrams

We are going to construct perturbation theory for the elastic amplitude in the relevant region of the Mandelstam plane:

$$s \simeq -u \to \infty, \quad |t| \sim m^2.$$

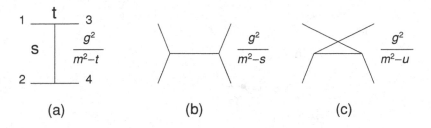

Fig. 9.1   Born diagrams for $A(1, 2 \to 3, 4)$ in $g\phi^3$ theory.

We start from the Born approximation. Of the Born graphs Fig. 9.1, the first one is $\mathcal{O}(g^2/m^2)$ and dominates the asymptotics, while the graphs Fig. 9.1(b) and (c) are much smaller at high energies, $\mathcal{O}(g^2/s)$. Since the coupling has a dimension of mass, in higher orders of the perturbative expansion each extra power of g^2 is accompanied by m^2, or $t \sim m^2$, or $s \simeq -u$ in the denominator. To build up the characteristic behaviour (9.3), we need to search for *large* perturbative corrections of the relative size

$$A^{(n+1)}/A^{(n)} \propto \frac{g^2}{m^2} \ln s. \tag{9.4}$$

It is clear that self-energy and vertex insertions into the Born graphs cannot lead to the Regge structure (9.3), since corrections of this type are functions of only one invariant, either t, or s.

$$\delta G = \quad \delta\Gamma =$$

The theory is convergent in the ultraviolet region, and no corrections may come from the region of large virtual momenta. Therefore only specific diagrams (and for a special reason) may contain the large logarithmic factor (9.4).

In the next order we have three topologically new diagrams of Fig. 9.2. In the last graph, Fig. 9.2(c), all four propagators are large, $|k_i^2| \sim s$, making its contributions negligibly small, $\mathcal{O}(g^4/s^2)$. The first two diagrams

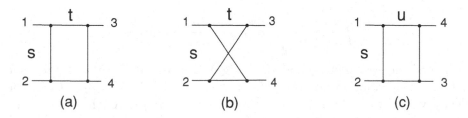

Fig. 9.2   Second-order diagrams.

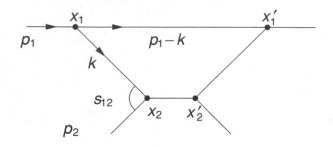

Fig. 9.3   Long-living fluctuation in the rest frame of p_2.

are much larger and can be considered as g^2/m^2 corrections to the Born graphs Fig. 9.1(b) and (c). It is these two which will interest us. Let us show that they contain the $\ln s$ enhancement indeed.

9.2.2 Dominance of ladder graphs: space–time picture

In fact we have already considered the diagram Fig. 9.2(a) when we discussed how to make the interaction radius increasing with energy. In Section 5.6 we observed that in the rest frame of the particle p_2 the projectile p_1, having a very large energy $p_{10} \simeq s/2m$, may fluctuate into a pair of particles. We saw that if virtualities of the offspring are limited, $|k^2| \sim |(p_1 - k)^2| \sim m^2$, the energy uncertainty turns out to be very small, see (5.60):

$$\Delta E \sim \frac{m^2}{x(1-x)p_{10}} \sim \frac{m^3}{s} \ll m, \qquad x = \frac{k_0}{p_{10}}.$$

At high energies large longitudinal distances become important,

$$|x_1' - x_1| \lesssim \Delta t \sim \frac{1}{\Delta E} \sim \frac{x(1-x)s}{m^3} \gg m^{-1}, \qquad (9.5)$$

and the fluctuation may occur long before the projectile hits the target. The origin of the logarithmic growth of our diagram is precisely the integration over large longitudinal distances:

$$\int \frac{e^{i\Delta E(x_1' - x_1)}}{|x_1' - x_1|} \, dx_1' \sim \ln \frac{m}{\Delta E} \sim \ln \frac{s}{m^2}.$$

(The distance between x_2 and x_2' in Fig. 9.3 stays small.) In order to have a long-living fluctuation, see (9.5), both particles k and $(p_1 - k)$ in the decay vertex must be relativistic: $x, (1-x) \gg m^2/s$. At the same time, the target prefers to interact with a *slower* particle, since in the lower vertex in Fig. 9.3 two point-like particles interact in the S-wave state with the cross section

$$\sigma \sim \pi \lambda_c^2 \sim \frac{1}{s_{12}} \simeq \frac{1}{2mk_0} = \frac{1}{xs}.$$

As a result of the interplay of these two tendencies, the amplitude remains small, $A \sim s^{-1}$, but acquires a logarithmic enhancement in the next order,

$$A^{(0)}[\text{Fig. 9.1b}] \simeq \frac{g^2}{-s}, \quad A^{(1)}[\text{Fig. 9.2a}] \sim \frac{g^2}{-s} \times \frac{g^2}{m^2} \ln \frac{s}{m^2}.$$

We have a rather curious situation here. We chose a superconvergent theory with a small coupling constant $g^2/m^2 \ll 1$ and expected that we could rely on the perturbation theory. However, with s increasing, the incident particle gets more and more time to decay; eventually it will *always* do so, even if the interaction constant is small. But this means that the perturbation theory must fail! And this is exactly what happens: with the increase of s the *true expansion parameter* (9.4) sooner or later becomes of the order of unity so that *all orders* of the perturbative expansion become equally important.

The virtual particle k in Fig. 9.3 will decay in its turn, and the process will continue until a relatively slow particle with a momentum $k_n \sim m$ appears, which interacts with the target with a 'normal' cross section $\sigma \sim m^{-2}$.

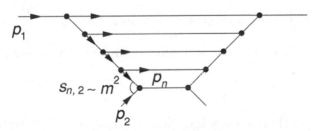

In such a process with n virtual particles along the decay chain, there are n independent integrations over the longitudinal distances. Hence, it is natural to expect that the nth-order correction to the Born graph will be enhanced as $(\ln s)^n$. Our hope is that summing up all ladder graphs we will obtain just the Regge amplitude (9.3):

9.2.3 Calculation of the box diagram

Let us demonstrate how this happens. We begin with the calculation of the behaviour of the diagram of Fig. 9.3 in the kinematical region $s \gg -q^2 \sim m^2$. This is the shortest ladder, with two rungs only, so we call it J_2:

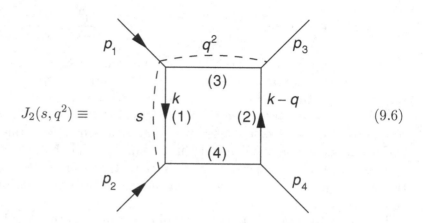

$$J_2(s, q^2) \equiv \qquad\qquad\qquad\qquad\qquad\qquad\qquad (9.6)$$

We define the light-like Sudakov momenta

$$(p_1')^2 = (p_2')^2 = 0; \quad p_1' \simeq p_1 - \gamma p_2, \; p_2' \simeq p_2 - \gamma p_1; \quad \gamma = \frac{m^2}{s}; \; s = 2(p_1' p_2'),$$

to cast the four-vector of the momentum transfer $q = p_1 - p_3$ as

$$q = \beta_q p_1' + \alpha_q p_2' + q_\perp; \quad \alpha_q = \frac{q^2}{s}, \; \beta_q = -\frac{q^2}{s}. \qquad (9.7)$$

Recall that in our high-energy kinematics momentum transfer is 'transversal' in the sense of

$$q^2 \simeq (q_\perp)^2 \cdot \left(1 + \mathcal{O}(s^{-1})\right).$$

By construction, the four-vector $q_\perp^\mu$ is orthogonal to p_1^μ and p_2^μ; in the reference frames where $\mathbf{p}_1$ and $\mathbf{p}_2$ lie on the same line, q_μ becomes the usual three-vector perpendicular to this line (collision axis), $q_\perp = (0; 0, \mathbf{q}_\perp)$,

$$q^2 \simeq -\mathbf{q}_\perp^2.$$

Finite-state particles have finite transverse components, $\mathbf{p}_{4\perp} = -\mathbf{p}_{3\perp} = \mathbf{q}_\perp$, i.e. particles scatter at a small angle. Then we write

$$k = \beta p_1' + \alpha p_2' + k_\perp.$$

Transverse momentum integrals in the $g\phi^3$ theory converge, $|k_\perp| \sim m$, so we will have to look for the logarithmic enhancement in the α–β sector.

Our amplitude reads

$$J_2(s, q^2) = g^4 \int \frac{d^4k}{(2\pi)^4 i} \frac{1}{(1)(2)(3)(4)}, \tag{9.8}$$

where the Feynman denominators are

$$(1) \equiv m^2 - k^2 - i\varepsilon, \qquad (2) \equiv m^2 - (k-q)^2 - i\varepsilon,$$
$$(3) \equiv m^2 - (k - p_1)^2 - i\varepsilon, \qquad (4) \equiv m^2 - (k + p_2)^2 - i\varepsilon.$$

We obtain

$$(1) = -\alpha\beta s + m^2 + \mathbf{k}_\perp^2 - i\varepsilon, \tag{9.9a}$$
$$(2) = -(\alpha - \alpha_q)(\beta - \beta_q)s + m^2 + (\mathbf{k}-\mathbf{q})_\perp^2 - i\varepsilon, \tag{9.9b}$$
$$(3) = -(\alpha - \gamma)(\beta - 1)s + m^2 + \mathbf{k}_\perp^2 - i\varepsilon, \tag{9.9c}$$
$$(4) = -(\alpha + 1)(\beta + \gamma)s + m^2 + \mathbf{k}_\perp^2 - i\varepsilon. \tag{9.9d}$$

Loop integration in terms of Sudakov variables has the structure

$$\int d^4k = \frac{s}{2} \int_{-\infty}^{\infty} d\alpha \int_{-\infty}^{\infty} d\beta \int d^2\mathbf{k}_\perp.$$

Let us start from a rough estimate of the magnitude of the answer to get a feeling what we should expect. Dimension-wise, the answer may turn out to be very small if all the virtualities happen to be of the order of s, the biggest invariant: $d^4k/(k^2)^4 \sim s^{-2}$. So we better try to keep all the denominators as small as possible. Given $k_\perp = \mathcal{O}(m)$, from (9.9c, d) follows an estimate for the virtualities of the 'horizontal' lines,

$$(3) \sim \alpha s, \quad (4) \sim -\beta s. \tag{9.10}$$

β is the fraction of the incoming momentum p_1 transferred to the bottom of the diagram (9.6) and, vice versa, $(-\alpha)$ measures the fraction of p_2 that flows into the top. It looks natural to have $\beta, (-\alpha) \ll 1$; in this case an incident particle passes its large momentum almost entirely to the neighbouring 'horizontal' line, $(p_1 - k) \simeq p_1$, $(p_2 + k) \simeq p_2$.

Wanting to keep virtualities (3) and (4) finite, we demand

$$\beta \sim \frac{m^2}{s}, \quad -\alpha \sim \frac{m^2}{s}, \tag{9.11a}$$

in which case the 'longitudinal parts' of the 'vertical' propagators (9.9a, b) are negligibly small, $\alpha\beta s \sim m^2/s$, and (1) and (2) become purely transversal. Then from the kinematical region (9.11a) we get

$$J_2 \sim \frac{g^4}{s} \int \frac{d^2\mathbf{k}_\perp}{(1)(2)} \iint \frac{d\alpha \, d\beta}{(3)(4)} \sim \frac{-g^4}{sm^2} \int \frac{d(\alpha s)}{\alpha s} \int \frac{d(\beta s)}{\beta s} \sim \frac{-g^4}{sm^2}. \quad (9.11b)$$

This is *almost* the correct answer. We can get more if we release the strong restrictions (9.11a) and allow α and β to vary broader:

$$\frac{m^2}{s} \ll \beta \ll 1, \quad \frac{m^2}{s} \ll -\alpha \ll 1. \quad (9.11c)$$

The virtual propagators (9.10) then become relatively large but we gain a logarithmic enhancement by integrating over the β/α ratio along the hyperbola $\alpha\beta s = \text{const}$.

We are now ready to calculate the amplitude exactly. Take first, e.g. the integral over α. Since it converges at infinity (as $d\alpha/\alpha^4$), the contour can be closed either in the upper or the lower half-plane, whichever we find more convenient. This simplifies the calculation of the integral by residues. The poles in α of the integrand are due to four denominators:

$$\alpha_{(1)} = \frac{m^2 + \mathbf{k}_\perp^2 - i\varepsilon}{\beta s}, \qquad \alpha_{(2)} = \alpha_q + \frac{m^2 + (\mathbf{q} - \mathbf{k})_\perp^2 - i\varepsilon}{(\beta - \beta_q) s},$$

$$\alpha_{(3)} = \gamma - \frac{m^2 + \mathbf{k}_\perp^2 - i\varepsilon}{(1 - \beta) s}, \qquad \alpha_{(4)} = -1 + \frac{m^2 + \mathbf{k}_\perp^2 - i\varepsilon}{(\beta + \gamma) s}. \qquad (9.12)$$

In order to choose the best strategy, we have to look at the *imaginary parts* of the poles $\alpha_{(i)}$ due to Feynman's $i\epsilon$. The configuration of the poles depends on the value of β. First we observe that if $\beta < -\gamma$ all four poles are situated *above* the real axis. In this case we close the contour in the *lower* half-plane to get zero. Analogously, if $\beta > 1$ we get the same result by closing the contour *upwards* and leaving all the poles outside the loop, *below* the real axis.

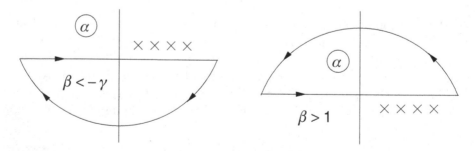

Next, we have two intervals of β where we close the contour on the lower half-plane around one pole, $\alpha = \alpha_{(4)}$, and two poles, (2) and (4), correspondingly:

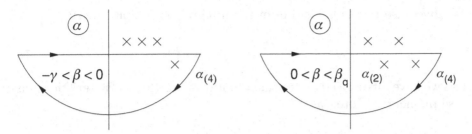

Since both $\gamma = m^2/s$ and $\beta_q = -q^2/s$ are very small, these integration intervals in β are tiny and do not produce the $\ln s$ enhancement we are looking for (cf. (9.11b)). We are left with the large interval $\beta_q < \beta < 1$:

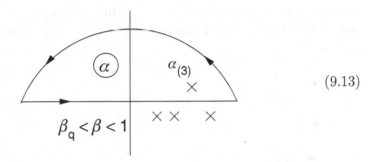

$$(9.13)$$

This interval gives the main contribution due to the logarithmic integration over β under the conditions

$$\frac{m^2}{s} \ll \beta \ll 1. \qquad (9.14)$$

Given these strong inequalities, we estimate $\alpha = \alpha_{(3)} = \mathcal{O}(m^2/s)$, the remaining denominators simplify,

$$(1) \quad \simeq \quad m^2 + \mathbf{k}_\perp^2, \qquad (9.15a)$$

$$(2) \quad \simeq \quad m^2 + (\mathbf{k} - \mathbf{q})_\perp^2, \qquad (9.15b)$$

$$(4) \quad \simeq \quad m^2 + \mathbf{k}_\perp^2 - \beta s - i\varepsilon, \qquad (9.15c)$$

and we obtain

$$J_2 \simeq \int \frac{d^2\mathbf{k}_\perp}{2(2\pi)^3} \frac{g^4}{[m^2 + \mathbf{k}_\perp^2][m^2 + (\mathbf{k} - \mathbf{q})_\perp^2]} \int_{m^2/s}^1 \frac{d\beta}{m^2 + \mathbf{k}_\perp^2 - \beta s - i\varepsilon}. \qquad (9.16)$$

It is the last denominator,

$$\frac{d\beta}{m^2 + \mathbf{k}_\perp^2 - \beta s - i\varepsilon} \simeq -\frac{1}{s} \frac{d\beta}{\beta},$$

that gives rise to the logarithm in the integration region (9.14):

$$J_2 \simeq -\frac{g^2}{s} \ln\frac{s}{m^2} \times \beta(q_\perp^2). \qquad (9.17a)$$

Here we have introduced a convenient notation for the characteristic transverse momentum integral,

$$\beta(q_\perp) \equiv \int \frac{d^2\mathbf{k}_\perp}{2(2\pi)^3} \frac{g^2}{[m^2 + \mathbf{k}_\perp^2][m^2 + (\mathbf{k}-\mathbf{q})_\perp^2]}. \qquad (9.17b)$$

The horizontal lines (3), (4) with large virtual momenta dropped out in the calculation of the asymptotic behaviour of the amplitude (9.6) and we have obtained a reduced diagram which contains only transverse momenta, and reminds of a Feynman diagram of a two-dimensional quantum field theory:

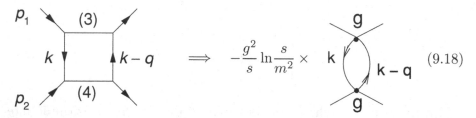

$$\qquad (9.18)$$

Here comes a final refinement before we move to higher orders. The r.h.s. of (9.18) is real, while we know that the box diagram on the l.h.s. has an imaginary part when (and only when) s is positive. Can we find $\mathrm{Im}\, J_2$ without performing any calculations? It is actually very simple. Once we know $\mathrm{Re}\, J_2$, in order to restore $\mathrm{Im}\, J_2$ it suffices to replace in (9.18)

$$\ln s \to \ln(-[s - \imath\varepsilon]) = \ln s - i\pi. \qquad (9.19)$$

This does not invalidate our asymptotic analysis since we have kept only the leading contribution $\propto \ln s$ and systematically omitted all constant corrections, $|i\pi| \ll \ln s$ being one of them. On the contrary, to keep this 'special constant' is legitimate, since it promotes our approximate expression to a true amplitude with proper analyticity.

It is time to guess the full answer. Let us try to add up the Born amplitude Fig. 9.1(b) and the next order correction (9.17) we have just derived:

$$\rightthreetimes\!\!+\!\!\square\!\!+\cdots \;=\; -\frac{g^2}{s} - \frac{g^2}{s}\beta(q_\perp)\cdot\ln(-s) + \cdots$$

$$\overset{?}{=} -\frac{g^2}{s} \cdot e^{\beta(q_\perp)\ln(-s)} = g^2(-s)^{-1+\beta(q_\perp)}. \qquad (9.20)$$

If this guess is right, our two terms would be simply the first terms of the *perturbative series expansion* in $\beta(q_\perp) = \mathcal{O}(g^2/m^2) \ll 1$ for the Regge pole amplitude.

9.2.4 Ladders in the leading logarithmic approximation

What will happen in higher orders? Take the third perturbative order, $A \propto g^6$, and imagine that we found a contribution $A = \mathcal{O}(g^6 \ln s)$. In spite of looking enhanced, this one is *insignificant* since when $g^2 \ln s \sim 1$, it constitutes but a small correction to the previous order,

$$A = A_{\text{Born}} \left(1 + g^2 \ln s + \underline{g^2 \cdot g^2 \ln s} + \cdots \right). \tag{9.21}$$

Now we have to look for $A = \mathcal{O}(g^6 \ln^2 s)$. In each order of the perturbation theory, with adding new internal momentum integration, we need to pick up an additional $\ln s$ enhancement factor. In spite of an immense number of Feynman graphs in high orders, each can be analysed (and if necessary evaluated) approximately, in a search for $(g^2 \ln s)^n$ terms. It is clear that not many diagrams will yield such a strong enhancement. Extracting and assembling such contributions in all orders constitutes the so-called 'leading logarithmic approximation',

$$A_{\text{LLA}} = A_{\text{Born}} \left(1 + \sum_{n=1}^{\infty} f_n \cdot (g^2 \ln s)^n \right); \quad g^2 \ll 1, \ g^2 \ln s \sim 1. \tag{9.22}$$

There is something important to stress. If we want the approximate amplitude A_{LLA} to represent the high-energy behaviour of the true scattering amplitude, the condition $g^2 \ll 1$ is absolutely crucial. Only under this condition may we ignore a plethora of *subleading* corrections, like the one underlined in (9.21).

Let us draw the diagrams that do contribute in the LL approximation; we will check later that others do not. These are the n-particle ladders.

Note that a ladder with a given number of rungs can be constructed by adding the Born amplitude on top of the ladder of the previous order,

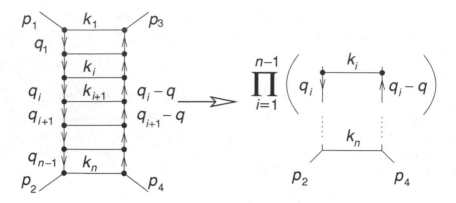

Fig. 9.4   Structure of the ladder diagram.

One can use this t-channel iterative nature of ladder graphs to derive the high energy behaviour of the amplitude $A_{2\text{ton}\to 2}$ *by induction*. However, to gain a better understanding of the kinematics of ladder-type processes we shall proceed with a direct analysis of the structure of ladder amplitudes.

Consider an n-particle ladder shown in Fig. 9.4.

We have $n-1$ internal momentum integrals, and the amplitude $J_n(s, q^2)$ can be written as

$$J_2 \;=\; g^{2n} \prod_{i=1}^{n-1} \left(\int \frac{d^4 q_i}{(2\pi)^4 i} \frac{1}{[m^2 - q_i^2][m^2 - (q_i - q)^2][m^2 - (q_{i-1} - q_i)^2]} \right)$$
$$\times \frac{1}{m^2 - (p_2 + q_{n-1})^2} \,, \tag{9.23}$$

where in all propagators the Feynman shift is implied, $m^2 \to m^2 - i\varepsilon$. We have $(n-1)$ 'blocks', each containing two vertical, q_i and $(q_i - q)$, and one horizontal propagator, $k_i = q_{i-1} - q_i$ ($q_0 \equiv p_1$). The bottom rung, $k_n = p_2 + q_{n-1}$, closes the chain.

For the sake of convenience we will slightly modify the definition of the Sudakov variables as compared to (9.7) and write

$$q = \beta_q p_1' - \alpha_q p_2' + q_\perp, \quad \beta_q = \alpha_q = -\frac{q^2}{s} = \frac{\mathbf{q}_\perp^2}{s}\left(1 + \mathcal{O}\left(\frac{m^2}{s}\right)\right); \tag{9.24}$$
$$q_i = \beta_i p_1' - \alpha_i p_2' + q_{i\perp}, \quad q_i^2 = -\alpha_i \beta_i s - \mathbf{q}_{i\perp}^2.$$

Thus α_i and β_i enter in a more symmetric way: both positive, they describe the fraction of p_2 transferred *up* the ladder and the fraction of p_1 descending the ladder, respectively.

Have a look at the propagators of the horizontal lines, the ladder rungs:

$$k_1^2 = (p_1 - q_1)^2 = (\alpha_1 + \gamma)(1 - \beta_1)s + \cdots, \tag{9.25a}$$

$$k_2^2 = (q_1 - q_2)^2 = (\alpha_2 - \alpha_1)(\beta_1 - \beta_2)s + \cdots, \tag{9.25b}$$

$$k_i^2 = (q_{i-1} - q_i)^2 = (\alpha_i - \alpha_{i-1})(\beta_{i-1} - \beta_i)s + \cdots, \tag{9.25c}$$

where we singled out the α dependence. Suppose now that we evaluate the α_i integrals by putting the rungs, one by one, on the mass shell as we did before in the $n = 2$ case when we closed the contour around the pole (3), see (9.13). The residue in α_1 equals $1/(1 - \beta_1)s$; then α_2 gives $1/(\beta_1 - \beta_2)s$, etc. The bottom rung will produce one more factor,

$$m^2 - k_n^2 = m^2 - (q_{n-1} + p_2)^2 = m^2 + \mathbf{k}_{n\perp}^2 - (1 - \alpha_{n-1})(\beta_{n-1} + \gamma)s, \tag{9.25d}$$

in the denominator. Combining all β-dependent factors, the following structure of the β_i integrals emerges:

$$J_n \sim \frac{1}{1 - \beta_1} \frac{d\beta_1}{\beta_1 - \beta_2} \frac{d\beta_2}{\beta_2 - \beta_3} \cdots \frac{d\beta_{n-2}}{\beta_{n-2} - \beta_{n-1}} \frac{d\beta_{n-1}}{\beta_{n-1} + \gamma}. \tag{9.26}$$

How can we get many logarithms? We need to gain one logarithm per *each* integration. One can easily prove that it is necessary to arrange successive β_is as follows:

$$1 \gg \beta_1 \gg \beta_2 \gg \cdots \gg \beta_i \gg \cdots \gg \beta_{n-1} \gg \gamma = \frac{m^2}{s}. \tag{9.27a}$$

What about the αs? They, too, turn out to be strongly ordered. Indeed, applying (9.27a) to (9.25) leads to the following pattern:

$$\alpha_1 \sim \frac{m_{1\perp}^2}{s}, \quad \alpha_2 \sim \frac{m_\perp^2}{\beta_1 s}, \quad \cdots \quad \alpha_i \sim \frac{m_{i\perp}^2}{\beta_{i-1} s}, \quad \cdots \quad \alpha_{n-1} \sim \frac{m_{n-1,\perp}^2}{\beta_{n-2} s},$$

where $m_{i\perp}^2 \equiv m^2 + \mathbf{k}_{i\perp}^2 = \mathcal{O}(m^2)$. Combining with (9.27a), we get

$$\frac{m_{1\perp}^2}{s} \sim \alpha_1 \ll \alpha_2 \ll \cdots \ll \alpha_i \ll \cdots \ll \alpha_{n-1} \ll 1; \tag{9.27b}$$

$$\alpha_i \beta_{i-1} s \sim m_{i\perp}^2, \quad i = 1, \ldots, n-1 \quad (\beta_0 \equiv 1). \tag{9.27c}$$

Inequalities (9.27) show that the flows of p_1 and p_2 momenta along the ladder are opposite and strongly ordered: ascending the ladder, the fractions of p_2 (α_i) successively *decrease*, in accord with the strong *increase* of β_i (fraction of p_1).

Given the strong ordering of βs, (9.26) reduces to

$$J_n \sim \int^1 \frac{d\beta_1}{\beta_1} \int^{\beta_1} \frac{d\beta_2}{\beta_2} \cdots \int^{\beta_{n-3}} \frac{d\beta_{n-2}}{\beta_{n-2}} \int_\gamma^{\beta_{n-2}} \frac{d\beta_{n-1}}{\beta_{n-1}},$$

giving

$$J_n \sim \frac{1}{(n-1)!} \left(\int_\gamma^1 \frac{d\beta}{\beta} \right)^{n-1}.$$

Now we can complete the calculation of J_n. Due to the estimate (9.27c), a crucial simplification emerges: the longitudinal variables drop out from all *vertical lines*. Indeed,

$$m^2 - q_i^2 = m^2 + \mathbf{q}_{i\perp}^2 + \alpha_i \beta_i s = m_{i\perp}^2 + \alpha_i \beta_{i-1} s \cdot \frac{\beta_i}{\beta_{i-1}} \simeq m_{i\perp}^2,$$

where we have used (9.27c) and the strong β-ordering, $\beta_i/\beta_{i-1} \ll 1$. Thus in the logarithmic region (9.27) the dependences on $\mathbf{q}_{i\perp}$ and β_i fully separate and leave us with $n-1$ identical transverse momentum loop integrals (9.17b) resulting in

$$J_n = -\frac{g^2}{s} \frac{1}{(n-1)!} \ln^{n-1} \left(\frac{-s}{m^2} \right) \cdot [\beta(q_\perp)]^{n-1}. \qquad (9.28)$$

We changed the sign of s under the logarithm to restore analyticity of the amplitude properly, as we have discussed before, when we analysed J_2. Summing over n we obtain the Regge-like expression (9.20) that we have guessed above.

Before discussing the final result, let us verify that the ladder diagrams in the kinematical region (9.27) are indeed the only ones to contribute.

First of all, it is clear from (9.26) that a decision to *swap* βs in the ordering condition (9.27a) immediately results in a loss of (at least) one logarithmic factor. To have βs *strongly* ordered is also essential. If we keep two neighbouring momenta of the same order, $\beta_k \sim \beta_{k+1}$, such a pair can be treated as a single rung with the effective 'mean' momentum $\bar{\beta} = \sqrt{\beta_k \beta_{k+1}}$, effectively reducing the ladder J_n to J_{n-1} and thus losing one $\ln s$,

$$\iint \frac{d\beta_k \, d\beta_{k+1}}{\beta_k \beta_{k+1}} = \int \frac{d\bar{\beta}}{\bar{\beta}} \cdot \int d\ln \left(\frac{\beta_k}{\beta_{k+1}} \right) \implies \ln s \cdot \mathcal{O}(1).$$

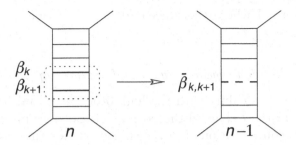

The same thing happens if we allow any rungs to cross.

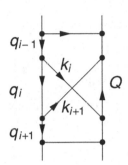

The momentum Q, which used to be $Q = q_i - q$ in the plain ladder, becomes in the crossed configuration

$$Q = [q_{i+1} + q_{i-1} - q_i] - q. \qquad (9.29)$$

It inherits the largest α-component from q_{i+1}, and the β-component from q_{i-1}, so that the new virtual denominator becomes

$$m^2 - Q^2 \simeq \alpha_{i+1}\beta_{i-1}s + m^2 + \mathbf{Q}_\perp^2.$$

Invoking (9.27c) we see that the 'longitudinal' part of the virtuality is no longer negligible; on the contrary, the new denominator is very large,

$$m^2 - Q^2 \sim m_\perp^2 \cdot \frac{\beta_{i-1}}{\beta_i} \gg m_{Q\perp}^2;$$

its presence spoils the logarithmic integration over β_i,

$$\int \frac{d\beta_{i-1}}{\beta_{i-1}} \int^{\beta_{i-1}} \frac{d\beta_i}{\beta_i} \cdot \left(\frac{\beta_{i-1}}{\beta_i}\right)^{-1} \sim \int \frac{d\beta_{i-1}}{\beta_{i-1}} \times 1,$$

and we have $\beta_{i+1} \sim \beta_i$, i.e. the situation that we have just discussed. The appearance of a large virtuality has a transparent physical explanation. Momenta of the ladder rungs, k_i, are very different in scale. For example, in the rest frame of p_2 we have

$$k_i \simeq (\beta_{i-1} - \beta_i)p_1 \approx \beta_{i-1}p_1, \quad k_{i+1} \simeq (\beta_i - \beta_{i+1})p_1 \approx \beta_i p_1,$$

so that the successive particle is much *softer* than its predecessor,

$$\frac{k_{i+1}}{k_i} \simeq \frac{\beta_i}{\beta_{i-1}} \ll 1.$$

This is favourable for the *production* of particles k_i, k_{i+1} (the left side of the graph). However, on the *absorption* side (the right half of the graph), the natural ordering is just opposite. In the lower vertex a hard particle has to be absorbed where there used to be a soft one, causing a large recoil.

9.2.5 Reggeon as a sum of ladder graphs

Up to now we were dealing with the high-order amplitudes that one obtains iterating in the t-channel the 's-channel' Born graph Fig. 9.1b. Exactly the same considerations can be carried out for the 'u-channel' amplitude (9.1c), which differs from the graph that we have iterated by the crossing transformation $s \leftrightarrow u$.

Under the crossing the second-order amplitude of Fig. 9.2 turns into the familiar box,

etc. Thus, in order to obtain the second series of contributions, we simply have to substitute $s \to u \simeq -s$ in our answer (9.28).

The final answer for the high-energy behaviour of the amplitude reads

$$A(s,t) \simeq \sum_{n=1}^{\infty} J_n = \frac{g^2}{m^2}\left[\left(\frac{-s}{m^2}\right)^{-1+\beta(t)} + \left(\frac{s}{m^2}\right)^{-1+\beta(t)}\right], \qquad (9.30a)$$

$$\beta(t) = \int \frac{d^2\mathbf{k}_\perp}{2(2\pi)^3} \frac{g^2}{[m^2+\mathbf{k}_\perp^2][m^2+(\mathbf{k}-\mathbf{q})_\perp^2]}; \quad t \simeq -\mathbf{q}_\perp^2. \quad (9.30b)$$

Strictly speaking, we cannot claim that (9.30) describes the true *asymptotics* $s \to \infty$, since our leading logarithmic approximation (9.22) applies to very large but finite energies,

$$s \lesssim m^2 \exp\left\{\frac{m^2}{g^2}\right\}.$$

Nevertheless, let us compare our result (9.30a) with the Regge pole amplitude,

$$A_{\text{pole}}^{\pm}(s,t) = -\frac{r(t)}{\sin \pi\alpha(t)}\left[\left(\frac{-s}{m^2}\right)^{\alpha(t)} \pm \left(\frac{s}{m^2}\right)^{\alpha(t)}\right].$$

We conclude that we have, indeed, obtained a Regge pole with a positive signature,

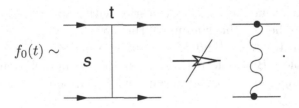

the trajectory

$$\alpha(t) = -1 + \beta(t), \tag{9.31a}$$

and the residue

$$r(t) = -\frac{g^2}{m^2} \sin \pi \alpha(t) \simeq \frac{g^2}{m^2} \beta(t). \tag{9.31b}$$

We have expanded the sinus near -1 since, perturbatively, β is small, $\beta(t) = \mathcal{O}(g^2/m^2) \ll 1$. The residue is therefore of the second order in the coupling, $r = \mathcal{O}\big((g^2/m^2)^2\big)$.

Did we get a bound state? To get a spin zero particle on the Regge trajectory, we have to have in (9.31a) $\beta(m_0^2) = 1$. We cannot go as far as that: our perturbative $\beta \propto g^2/m^2$ is small. So in the perturbation $g\phi^3$ theory the scalar particle ϕ remains elementary: the partial-wave amplitude $\ell = 0$ corresponding to the t-channel scalar particle exchange *does not reggeize*,

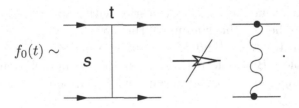

What we did get moving instead is another fixed pole, that at $\ell = -1$,

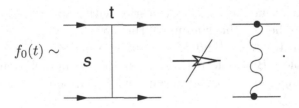

Indeed, let us calculate the partial wave corresponding to the s-channel exchange amplitude. Substituting

$$A_1(s,t) = \text{Im} \quad \frac{g^2}{m^2-s-\iota\varepsilon} \quad = g^2 \cdot \pi\delta(m^2 - s)$$

in the general expression (7.27b),

$$
\begin{aligned}
f_\ell(t) &= \frac{2}{\pi} \int_{z_0}^\infty dz_s\, A_1(t,s)Q_\ell(z_s) \\
&= \frac{4}{\pi(t-4m^2)} \int_{s_0}^\infty ds\, A_1(t,s)Q_\ell\left(1 + \frac{2s}{t-4m^2}\right),
\end{aligned}
$$

we obtain

$$f_\ell(t) = \frac{4g^2}{t-4m^2}Q_\ell\left(1 + \frac{2m^2}{t-4m^2}\right) \underset{\sim}{\overset{\ell\to-1}{}} \frac{4g^2}{t-4m^2}\cdot\frac{1}{\ell+1}. \tag{9.32a}$$

Comparing with the partial wave of the amplitude (9.30a),

$$f_\ell(t) \sim \frac{1}{\ell+1-\beta(t)}, \tag{9.32b}$$

we conclude that since $\beta(t) \ll 1$, generically our reggeon is connected to the fixed pole (9.32a) at $\ell = -1$.

One has to remember that the results of this section are valid only for $g^2/m^2 \ll 1$. The Regge trajectory which we got under this condition,

$$\alpha = -1 + \beta(t),$$

has no relation to the observed trajectories with hadron resonances placed on them. Nevertheless, the example of the $g\varphi^3$ theory is rather instructive, as it demonstrates what kind of s-channel processes may correspond to real Regge poles. But before turning to the s-channel structure of the reggeon exchange, let us briefly discuss what happens in other theories.

9.2.6 Reggeization in other theories

Thus, the scalar meson representing the ϕ field of the $g\phi^3$ quantum field theory remained an elementary particle. At the same time, a Regge trajectory appeared that corresponds in fact to a two-particle bound state (sort of 'positronium').

As for other theories, there are not many since we can operate only with renormalizable ones.

Fermion + scalar. If we take a renormalizable field theory based on spin $\frac{1}{2}$ fermions interacting with a scalar field, $\mathcal{L}_{\text{int}} \sim \bar{\psi}\psi\phi$, the answer is similar: the input objects stay elementary, while their bound states do reggeize.

Fermion + vector. A field theory of the type of quantum electrodynamics provides a more telling example. Once again, we take spin-$\frac{1}{2}$ 'electrons' and couple them to a vector field,

In this theory a curious thing happens. Look at the Compton process in the region of fixed u (backward scattering). Summing up perturbative radiative corrections in the approximation $e^2 \ll 1$, $e^2 \ln s \sim 1$ (LLA),

$$+ \cdots = r\xi_\alpha s^\alpha \qquad (9.33)$$

a reggeon emerges. What is even better, its trajectory satisfies the relation $\alpha(m_e^2) = \frac{1}{2}$. A remarkable phenomenon: in QED with a massive photon, the fermion *reggeizes*! How did it occur? In the Born approximation we have a fixed pole in the partial wave,

$$f_j^{\text{Born}} \propto \frac{e^2}{m^2 - u}\delta_{j,\frac{1}{2}}. \qquad (9.34)$$

Analogously to the scalar case, see (9.31b), the reggeon residue r in (9.33) is of the second order in the squared coupling, $r = \mathcal{O}(e^4)$. How can the first-order expression (9.34) be a part of it? What happens in higher orders is that (9.34) becomes a *limiting value* of the function

$$f_j = \text{const}\frac{e^4}{j - \frac{1}{2} - \beta(u)}, \qquad \beta \propto e^2, \ \beta(m^2) = 0. \qquad (9.35)$$

For angular momenta $j \neq \frac{1}{2}$, the specific contribution (9.35) to the partial wave is $\mathcal{O}(e^4)$ and can be neglected. At the same time, if we take the angular momentum (very close to) $j = \frac{1}{2}$,

$$f_{j \to \frac{1}{2}} = \text{const} \frac{e^4}{-\beta(u)} = \frac{e^2}{m^2 - u},$$

the magnitude of the partial wave becomes normal, and reproduces the Born amplitude (9.34).

In the usual electrodynamics where the photon is massless, $\mu = 0$, a complication emerges due to the standard infrared divergence.

In the $\mu \to 0$ limit, the one-particle pole collides with thresholds due to additional photons: $m + n \cdot \mu \to m$. As a result, the electron Green function at $p^2 = m^2$ does not have a pole anymore but develops a more tricky singularity. Nevertheless, in QED one can also claim that the electron lies on the Regge trajectory in the following sense. Whatever the nature of the singularity, the *position* of this whole thing *reggeizes*, that is moves with j, $m_e^2 \to m^2(j)$.

Formally speaking, $\alpha(t)$ describing the elastic $e\gamma$ scattering amplitude (9.33) diverges in the infrared region and becomes undefined. This is natural: purely elastic processes do not exist in QED; taking into account the infrared radiative corrections, elastic amplitudes vanish. However, if one considers observables that are insensitive to the emission of undetectable, infinitely soft photons, the physical cross sections become finite, and the electron trajectory can be properly defined,

$$\frac{d\sigma}{d\Omega} \propto s^{2(\alpha(u)-1)}, \quad \alpha(m_e^2) = \tfrac{1}{2}. \tag{9.36}$$

A QED photon does not reggeize.

Yang–Mills fields. Recently* a new class of very interesting renormalizable field theories was found, the Yang–Mills theories. The scheme includes self-interacting massless vector fields and fermions, as well as some additional scalars ('ghost'). In this theory both fermions and vector particles reggeize.

9.2.7 Non-pole singularities in perturbation theory

We have a couple of questions more to ask to the perturbation theory:

(1) are there singularities other than poles?

(2) any hints about the pomeron that we need for $\sigma_{\text{tot}} \to \text{const}$?

---

* The lecture was in 1975 (ed.)

The $g\phi^3$ field theory contains nothing but Regge poles. In all other theories, however, there are non-pole singularities too. Take QED as an example and consider once again the Compton scattering, but this time in the *forward* kinematics, $|t| \sim m^2 \ll s \simeq -u$. Interesting contributions will be the following:

$$e^2 \qquad\qquad e^4 \ln^2 s \qquad\qquad e^6 \ln^4 s$$

The analysis of the ladder diagrams of this type follows the steps of the scalar case, and in the leading approximation one arrives at the exponential of the reduced two-dimensional diagram describing the trajectory,

$$= \frac{e^2}{4\pi} \int \frac{d^2 \mathbf{k}_\perp}{(2\pi)^2} \frac{1}{\left[m - \hat{k} \right]\left[m - (\hat{q} - \hat{k}) \right]}. \qquad (9.37)$$

Contrary to the scalar loop, this integral *diverges* in the large momentum region! What does this mean? Our initial diagrams were convergent in the ultraviolet. So this must be not a real divergence but an artefact of the approximations made.

Indeed, we carried out the analysis of the ladder kinematics, estimated the virtualities etc. having *supposed* that transverse momenta are limited, $\mathbf{k}_\perp^2 \ll s$. In the scalar theory this working hypothesis found its confirmation in the end of the calculation: the integral for $\beta(q_\perp)$ converged at $\mathbf{k}_\perp^2 \sim m^2$. In QED, in a contrast to the super-convergent $g\phi^3$ theory, the coupling constant is dimensionless, and the $k_\perp$ integration produces another logarithm,

$$\sim e^2 \int_{m^2}^s \frac{d\mathbf{k}_\perp^2}{\mathbf{k}_\perp^2} = e^2 \ln\frac{s}{m^2}. \qquad (9.38)$$

So, in the second order, e.g. the 'box' diagram acquires two logs, $e^4 \ln^2 s$, instead of one, $e^4 \ln s$. Such a series corresponds in the angular momentum plane not to a Regge pole but to a different singularity.

From the t-channel point of view this problem resembles a non-relativistic system with a singular potential. You may remember that we have discussed how in non-relativistic quantum mechanics an interaction singular at small distances gave rise to fixed poles. We considered a non-relativistic potential $V \propto -c/r^2$ which corresponds to the 'falling on the centre' phenomenon. We noted that if the parameter c is not too large, $c < \frac{1}{4}$, then physically there is no catastrophic 'falling on the centre', but a fixed pole in the partial wave appears near zero, $j_0 = ce^2$. This is what happens in the *forward* Compton scattering problem.

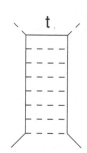

Another interesting question: is there not a singularity at $j = 1$ in QED? Look at the Feynman graphs with many electron loops inserted in the two photon exchange diagram.

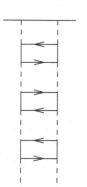

Cutting through such diagrams one gets the total cross section *increasing* with energy,

$$\operatorname{Im} A \sim e^4 s^{1+\gamma}, \quad \gamma \sim e^4; \qquad \sigma \propto s^{\gamma}. \qquad (9.39)$$

Such a behaviour corresponds to a fixed singularity at $j_0 > 1$. This high-energy behaviour is valid as long as $e^4 \ln s \lesssim 1$. At yet higher energies the power increase (9.39) must stop, according to the Froissart theorem, and the true position of the singularity should move to the left. A theoretical analysis of how this happens is lacking.

Theoretical studies of the high-energy behaviour of various QED processes left a number of unanswered questions. At the same time, they provided a certain experience which allows one to make conclusions about what can happen *beyond* the perturbation theory.

9.3 Inelastic processes at high energies

We embark on a discussion of a very important issue, namely what inelastic processes a Regge pole corresponds to.

We have seen that in the $g\varphi^3$ theory the Regge pole appears as a sum of multi-particle ladders:

$$A(s,t) = \sum \quad \boxed{}^{t} \quad = \quad \} \quad = \frac{g^2}{m^2}\left[\left(\frac{-s}{m^2}\right)^{-1+\beta(t)} + \left(\frac{s}{m^2}\right)^{-1+\beta(t)}\right].$$

Evaluating the imaginary part of the forward elastic amplitude we obtain, due to the optical theorem, the total cross section:

$$\sigma_{\text{tot}} \simeq s^{-1}\,\text{Im}\,A(s,0) \simeq \frac{g^2}{s^2}\left[\pi\beta(0)\left(\frac{s}{m^2}\right)^{\beta(0)}\right], \tag{9.40a}$$

where we invoked (9.19) to fix the phase of the complex factor $(-s)^\beta$ and expanded perturbatively $\sin\pi(1-\beta) \simeq \pi\beta$. Since $\beta(0) > 0$, the total cross section (9.40a) decreases slower with s and is therefore *much larger* at high energies than the Born elastic cross section,

$$\sigma^{\text{el}}_{\text{Born}} \propto \int \frac{d\Omega}{4\pi}\left|\frac{A(s,t)}{s}\right|^2 \sim \frac{g^4}{s^2}. \tag{9.40b}$$

This means that inelastic channels dominate.

9.3.1 Topological cross sections $2 \to 2+n$

Having our ladder pictures, it is straightforward to find not only total but all partial inelastic cross sections as well. Indeed, since σ_{tot} is determined by the sum

$$\text{Im}\,A(s,0) = \sum_{k=2} \text{Im}\,J_k(s,0),$$

where J_k is one of the ladder graphs, we have

$$\sigma_{n+2} = s^{-1}\,\text{Im}\,J_{n+2}(s,0) = -\frac{g^2}{s}\,\text{Im}\,\frac{[\beta(0)(\ln s - i\pi)]^{n+1}}{(n+1)!}$$

$$\simeq \frac{\pi g^2}{s^2}\frac{[\beta(0)]^{n+1}}{n!}\ln^n s \sim \sigma^{\text{el}}_{\text{Born}}\cdot\frac{[\beta(0)\ln s]^n}{n!}; \tag{9.41}$$

here σ_{n+2} (the so-called 'topological cross sections') is the cross section of the process $2 \to 2+n$ with the production of n additional particles. What is the characteristic number of produced particles in the subprocesses that have made σ_{tot} increase with respect to the elastic one? The expression (9.41) is nothing but a *Poisson distribution* in multiplicity

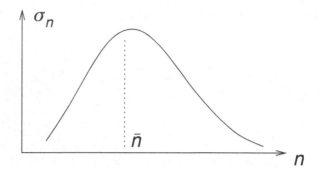

Fig. 9.5   Poisson multiplicity distribution of topological cross sections.

which we can represent invoking (9.40) as

$$\frac{\sigma_{n+2}}{\sigma_{\text{tot}}} = \frac{\bar{n}^n}{n!}e^{-\bar{n}}, \quad \sigma_{\text{tot}} = \sum_{n=0}^{\infty}\sigma_{n+2}. \tag{9.42a}$$

Here $\bar{n}(s)$ is the logarithmically increasing average particle multiplicity,

$$\bar{n} = \bar{n}(s) \simeq \beta(0)\ln\frac{s}{m^2}. \tag{9.42b}$$

So it is inelastic sub-processes with the number of particles increasing with energy, $n \sim \bar{n}$, that dominate the total cross section as shown in Fig. 9.5.

9.3.2 Multiperipheral kinematics

Another interesting question is how the produced particles are distributed. To answer it, we need to look into the internal structure of the ladders. Recall how we calculated above the ladder diagram J_n.
We have introduced $n-1$ momenta of the vertical lines, q_i, and took the residues in α_i putting on the mass shell all the rungs but the very bottom one. This last particle also becomes real when we take the imaginary part of J_n, see (9.25d),

$$\int d\beta_{n-1} \,\text{Im}\frac{1}{m_{n\perp}^2 - i\varepsilon - (1-\alpha_{n-1})(\beta_{n-1} + \gamma)s} = \frac{\pi}{s},$$

fixing the value of β_{n-1}.

   Since we are interested in the distribution of the produced particles, it is natural to express the ladder in terms of the final-state momenta, k_i.

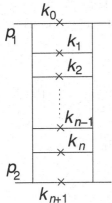

Let us take J_{n+2}. The top and bottom particles are 'leaders': the top one has a large β-component,

$$\beta_0^{(k)} = 1 - \beta_1 \simeq 1, \quad k_0 \simeq p_1 + \frac{m_{0\perp}^2}{s}p_2 + k_{0\perp},$$

and the bottom particle, correspondingly, carries away practically all the momentum of p_2,

$$\beta_{n+1}^{(k)} \simeq \frac{m_{n+1,\perp}^2}{s}, \quad k_{n+1} \simeq p_2 + \frac{m_{n+1,\perp}^2}{s}p_1 + k_{n+1,\perp}.$$

At the same time, the longitudinal momenta of all 'new' particles, $i = 1, 2, \ldots, n$, change freely, and it is these variations that enhance the cross section. Translating q_i into the rung momenta $k_i = q_i - q_{i+1}$, see (9.27),

$$\beta_i^{(k)} = \beta_i - \beta_{i+1} \simeq \beta_i, \tag{9.43a}$$

$$\alpha_i^{(k)} = \alpha_{i+1} - \alpha_i \simeq \alpha_{i+1}; \tag{9.43b}$$

for J_{n+2} we have

$$\operatorname{Im} J_{n+2} = \frac{\pi g^2}{s} \prod_{i=1}^{n+1} \left\{ \int \frac{d^2 k_{i\perp}}{2(2\pi)^3} \frac{g^2}{(i_\perp)^2} \right\} \int_\gamma^1 \frac{d\beta_1}{\beta_1} \cdots \int_\gamma^{\beta_{n-1}} \frac{d\beta_n}{\beta_n}. \tag{9.44}$$

Evaluating the integrals results in the multiplicity distribution (9.41). The integrand itself gives us the momentum spectra of final-state particles in the process $2 \to 2 + n$. We observe two important properties.

(1) Dependences on transverse and longitudinal variables of final-state particles, $k_{i\perp}$ and $k_{iz} \propto \beta_i$, factorized.

(2) The distribution is *uniform* in

$$\frac{d\beta i}{\beta_i} = d(\ln \beta_i). \tag{9.45}$$

Such a pattern of multi-particle production (not to forget the limited transverse momenta) is often referred to as 'multiperipheral kinematics'.

In the laboratory frame, with the target at rest, $\mathbf{p}_2 = 0$, and the projectile very fast, $p_{10} \simeq s/2m$, $k_i \simeq \beta_i p_1$, so that one may speak of the uniformity in $\ln k_{i0}$. This means that, descending the ladder, we will typically meet particles with energies

$$k_{10} \sim \lambda \cdot p_{10}, \; k_{20} \sim \lambda^2 \cdot p_{10}, \; \ldots, \; k_{n0} \sim \lambda^n \cdot p_{10}, \tag{9.46}$$

all the way down to the target, $k_{n+1,0} \sim \lambda^{n+1} \cdot p_{10} = m$. The energy fraction λ is a measure of the invariant mass squared of the neighbouring particles:

$$s_{12} = (k_1 + k_2)^2 \sim 2(k_1 k_2) = (\beta_1 \alpha_2 + \beta_2 \alpha_1)s$$

$$= \beta_1 \frac{m_{2\perp}^2}{\beta_2} + \beta_2 \frac{m_{1\perp}^2}{\beta_1} \sim (\lambda^{-1} + \lambda) m_\perp^2 \simeq \frac{m_\perp^2}{\lambda}, \quad (9.47)$$

where we substituted α_1, α_2 from the on-mass-shell conditions,

$$\alpha_i \beta_i s = m^2 + \mathbf{k}_{i\perp}^2 \equiv m_{i\perp}^2. \quad (9.48)$$

We have

$$(n+1) \cdot \ln \lambda^{-1} \simeq \ln \frac{P}{m} \simeq \ln \frac{s}{m^2};$$

substituting the average multiplicity (9.42b), we get an estimate

$$\ln \lambda^{-1} \simeq \ln \frac{\langle s_{i,i+1} \rangle}{\langle m_{i\perp}^2 \rangle} \simeq \frac{1}{\beta(0)} \propto \frac{1}{\bar{g}^2} \qquad \left(\bar{g}^2 \equiv \frac{g^2}{m^2} \right), \quad (9.49)$$

with $\bar{g}$ the dimensionless coupling constant. We see that the pair masses of the neighbours are very large in the perturbation theory where the coupling constant is small. It is worthwhile to mention here a kinematical relation linked to this observation. Let us construct the product of all pair invariants along the multiperipheral chain:

$$s_{01} s_{12} s_{23} \cdots s_{n-1,n} s_{n,n+1}.$$

Using $s_{i,i+1} \simeq \beta_i \alpha_{i+1} s$ (cf. (9.47)),

$$(\beta_0 \alpha_1 s)(\beta_1 \alpha_2 s) \cdots (\beta_{n-1} \alpha_n s)(\beta_n \alpha_{n+1} s),$$

and assembling the products (9.48), we get

$$\prod_{i=0}^{n} s_{i,i+1} = \beta_0 \cdot \prod_{i=1}^{n} (\alpha_i \beta_i s) \cdot \alpha_{n+1} s = s \cdot \prod_{i=1}^{n} m_{i\perp}^2, \quad (9.50)$$

where we have used $\beta_0 \simeq \alpha_{n+1} \simeq 1$ for the leading particles. The substitution of the average characteristics gives

$$\langle s_{i,i+1} \rangle^{\bar{n}} = \langle s \rangle^{\beta(0) \ln s} = s^{\beta(0) \ln \langle s \rangle} = s, \implies \beta(0) \ln \langle s \rangle = 1,$$

reproducing (9.49).

9.3.3 Inclusive particle spectrum

What sort of observables should we measure in the case of a large multiplicity? It is interesting to study the average characteristics of particle production, for example, to plot a histogram for the longitudinal momentum distribution. Let us introduce an important observable which is well suited for characterizing multi-particle production. Consider the density of particles per unit cell of the momentum phase space,

$$d^3\sigma^{\text{incl}}(\mathbf{k}) = f(k)\frac{d^3\mathbf{k}}{k_0}.$$

To measure this quantity, one has to register *one particle* with a given momentum $\mathbf{k}$ and ignore the number, and momenta, of other particles produced in the collision. Take for an example a pp collision and trigger one meson.

$$d\Gamma_{n-1} = d\sigma(k).$$

We square the n-particle production amplitude A, integrate indiscriminately over the momenta of all final-state particles but one π^- with momentum $\mathbf{k}$ and sum over all 'topologies' n. This quantity is called the 'inclusive spectrum', in contrast to 'exclusive' processes (like elastic scattering) where characteristics of all final-state particles are measured.

One should be aware of the fact that f *is not* a differential cross section in a sense, because its integral does not yield σ_{tot}. Write down the general expression for the inclusive one-particle spectrum,

$$d^3\sigma = \frac{1}{j}\sum_n \frac{d^3k}{2(2\pi)^3 k_0}\cdot\frac{1}{(n-1)!}\int\cdots\int\frac{d^3k_1\cdots d^3k_{n-1}}{(2\pi)^{3(n-1)}2k_{10}\cdots 2k_{n-1,0}} \tag{9.51}$$

$$\times \left|A_n(k;k_1,\ldots,k_{n-1})\right|^2 (2\pi)^4\delta(p_1 + p_2 - k - \sum k_i).$$

Essential here is the combinatorial factor $1/(n-1)!$ which takes care of multiple counting in the integration over the full phase space of $n-1$ identical particles. If we wanted to calculate σ_{tot}, we would have to integrate over all particle momenta, including k, with the factor $1/n!$. Therefore, integrating (9.51) over the momentum of the selected particle we will obtain not the total cross section $\sigma_{\text{tot}} = \sum_n \sigma_n$, but $\sum_n n\sigma_n \equiv \bar{n}\sigma_{\text{tot}}$, with $\bar{n}$ the average multiplicity.

9.3.4 Rapidity variable

How to examine inclusive cross sections? It makes little sense to plot it in the bins in k_z since, because of the strong ordering (9.27), (9.46), the bulk of particles is *soft*, $|k_z| \ll p_z$, and will fall into the single bin around $k_z \simeq 0$, in any reference frame. For example, in the cms we will not see anything but a huge peak in the centre of the distribution,

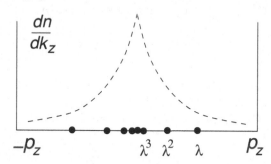

Uniformity of multiperipheral particle production in $d\beta/\beta$ means constant particle density per unit of phase space,

$$d\Gamma(k) = \frac{d^3\mathbf{k}}{k_0} = d^2\mathbf{k}_\perp \frac{dk_z}{k_0}, \quad \frac{dk_z}{k_0} = \frac{d\beta}{\beta}.$$

To characterize this key feature of multi-particle production in a Lorentz-covariant manner, one introduces a convenient variable called 'rapidity',

$$\eta = \frac{1}{2} \ln \frac{k_0 + k_z}{k_0 - k_z}. \tag{9.52a}$$

Using the on-mass-shell relation $(k_\mu)^2 = m^2$, we may rewrite (9.52a) as

$$\eta = \frac{1}{2} \ln \frac{(k_0 + k_z)^2}{k_0^2 - k_z^2} = \ln \frac{k_0 + k_z}{m_\perp}. \tag{9.52b}$$

In the frame where k is fast, $|\eta| \simeq \ln(2k_0/m_\perp)$ (the sign being that of k_z). This variable is special in the sense that it transforms *additively* under Lorentz boosts along the collision axis $\mathbf{z}$,

$$\eta \to \eta + \Delta\eta, \quad \Delta\eta = \frac{1}{2} \ln \frac{1+v}{1-v}.$$

The *relative* rapidity of two particles is therefore invariant under such a change of the reference frame. Observing that the energy and the longitudinal momentum can be expressed in terms of the rapidity (9.52) as

$$k_0 = m_\perp \cosh\eta, \quad k_z = m_\perp \sinh\eta,$$

we have for the invariant two-particle energy $s_{12} \equiv (k_1 + k_2)^2$

$$s_{12} - m_1^2 - m_2^2 = 2k_1 k_2 = 2(k_{10}k_{20} - k_{1z}k_{2z}) - 2(\mathbf{k}_1 \mathbf{k}_2)_\perp.$$

Transverse momenta do not change under the longitudinal Lorentz boost; the longitudinal part of the product of four-momenta depends on the difference of two rapidities (also an invariant)

$$(k_{10}k_{20} - k_{1z}k_{2z}) = m_{1\perp}m_{2\perp}\cosh(\eta_1 - \eta_2). \qquad (9.53)$$

In terms of the rapidity, the particle phase space reads

$$\frac{d^3\mathbf{k}}{k_0} = d^2\mathbf{k}_\perp \frac{dk_z}{k_0} = d^2\mathbf{k}_\perp \, d\eta. \qquad (9.54)$$

The property (9.45) translates then into *homogeneity* in rapidity. Let us introduce rapidities of the colliding particles,

$$\eta_+ = \ln\frac{p_{10} + p_{1z}}{m} \simeq \ln\frac{2p_{10}}{m},$$
$$\eta_- = \ln\frac{p_{20} + p_{2z}}{m} = -\ln\frac{p_{20} - p_{2z}}{m} \simeq -\ln\frac{2p_{20}}{m}. \qquad (9.55)$$

Each of them depends, obviously, on the frame. For example, in the laboratory frame $\eta_- = 0$, $\eta_+ = \ln(s/m^2)$; in the centre-of-mass frame of the collision $\eta_+ = -\eta_- = \frac{1}{2}\ln(s/m^2)$, etc. preserving the invariant difference

$$\eta_+ - \eta_- \simeq \ln\frac{4p_{10}p_{20}}{m^2} = \ln\frac{s}{m^2}.$$

The Sudakov variables are invariants too and are linked to the rapidity as follows

$$k = \beta \, p_1' + \alpha \, p_2' + k_\perp;$$

$$\beta = \sqrt{\frac{k^2 + \mathbf{k}_\perp^2}{s}} \, \exp\left\{ \eta - \tfrac{1}{2}(\eta_+ + \eta_-) \right\},$$
$$\alpha = \sqrt{\frac{k^2 + \mathbf{k}_\perp^2}{s}} \, \exp\left\{ -\eta + \tfrac{1}{2}(\eta_+ + \eta_-) \right\}. \qquad (9.56)$$

In terms of rapidities (9.44) simplifies. Introducing an additional $\mathbf{k}_\perp$ integration, we can represent the particle distribution in an exceptionally simple symmetric form:

$$\sigma_{n+2} = \frac{\sigma_2(s)}{\beta(0)} \prod_{i=0}^{n+1} \left(\int \frac{d^2\mathbf{k}_{i\perp}}{2(2\pi)^3} \int d\eta_i \right) \cdot \prod_{i=1}^{n+1} \frac{g^2}{[m^2 + \mathbf{q}_{i\perp}^2]^2}$$

$$\times 2(2\pi)^3 \delta^{(2)} \left(\sum_{i=0}^{n+1} \mathbf{k}_{i\perp} \right) \delta(\eta_0 - \eta_+) \delta(\eta_{n+1} - \eta_-), \qquad (9.57a)$$

where the ordering of rapidities is implied,

$$\eta_+ = \eta_0 > \eta_1 > \eta_2 > \cdots > \eta_n > \eta_{n+1} = \eta_-. \qquad (9.57b)$$

In (9.57a) $\mathbf{q}_\perp$ are transverse momenta of the vertical lines,

$$\mathbf{q}_{i\perp} = \sum_{m=0}^{i-1} \mathbf{k}_{i\perp}. \qquad (9.57c)$$

The only dependence on rapidities is contained in the ordering condition (9.57b); otherwise, the multi-particle distribution given by the integrand of (9.57a) depends only on the transverse momenta.

9.3.5 Rapidity plateau in the inclusive spectrum

The homogeneity of the particle distribution in rapidity is of extreme importance for the Regge-pole picture. How to verify this property experimentally? Let us look at the inclusive production cross section of a particle with momentum η, $\mathbf{k}_\perp$. We need to integrate the cumbersome expression (9.57a) over the variables of all particles but one,

$$\sigma_{n+2} \equiv \int d^2k_\perp \, d\eta \, f_n(k_\perp, \eta), \qquad (9.58a)$$

and construct the sum

$$\frac{d\sigma}{d^2\mathbf{k}_\perp \, d\eta} = f(k_\perp, \eta) \equiv \sum_n f_n(k_\perp, \eta). \qquad (9.58b)$$

As we have just discussed above, it is normalized as follows,

$$\sum_n \int d^2k_\perp \, d\eta \, f_n(k_\perp, \eta) = \sum_n n\sigma_{n+2} = \bar{n} \cdot \sigma_{\text{tot}}, \qquad (9.59)$$

where $\bar{n}$ is the average number of particles produced *in addition* to the two leading ones. If one includes the end points $\eta = \eta_\pm$ in the rapidity integration in (9.59), then $\bar{n}$ will be by two units bigger and count all the particles in the event.

We calculate now the inclusive cross section in perturbation theory.

We take the imaginary part of the forward scattering amplitude and choose one particle somewhere in the middle of the ladder, not too close in rapidity to the leaders. Above and below the selected rung there will be two shorter ladders again. Integrating over loop momenta inside these two ladders and summing over n_1 and n_2 results in the appearance of the product of the imaginary parts of two new forward amplitudes, and we get

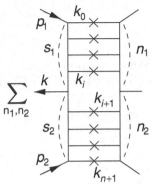

$$2(2\pi)^3 f(k) = \frac{1}{j} \int \frac{d^4 q}{(2\pi)^3} \frac{1}{[m^2 - q^2]^2} \frac{1}{[m^2 - (q-k)]^2} \qquad (9.60)$$
$$\cdot 2 \operatorname{Im} A((p_1 - q)^2, q^2) \cdot 2 \operatorname{Im} A((p_2 + q - k)^2, (q-k)^2).$$

Let us evaluate the invariants entering the two blocks. Casting the momenta in the Sudakov basis as

$$q = \beta_q p_1' - \alpha_q p_2' + q_\perp,$$
$$k = \beta p_1' + \alpha p_2' + k_\perp, \quad \alpha\beta s = m^2 + \mathbf{k}_\perp^2 \equiv m_\perp^2,$$

we have

$$s_1 = (p_1 - q)^2 = q^2 + \alpha_q s + \mathcal{O}(m^2),$$
$$s_2 = (p_2 + q - k)^2 = (q-k)^2 + (\beta_q - \beta)s + \mathcal{O}(m^2).$$

Let us express the longitudinal integration variables α_q and β_q in units of α, β by introducing the fractions $z_1, z_2 > 0$,

$$\alpha_q \equiv z_1 \cdot \alpha, \quad (\beta_q - \beta) \equiv z_2 \cdot \beta \quad (\beta_q = (1 + z_2)\beta);$$

$$d\alpha_q \, d\beta_q \frac{s}{2} = dz_1 \, dz_2 \frac{m_\perp^2}{2}. \qquad (9.61)$$

Then the momenta squared read

$$-q^2 \sim \alpha_q \beta_q s = z_1(1 + z_2)m_\perp^2,$$
$$-(q-k)^2 \sim (\alpha_q + \alpha)(\beta_q - \beta)s = (1 + z_1)z_2 m_\perp^2, \qquad (9.62)$$

where we have combined the product $\alpha\beta$ into the transverse mass.

The virtualities in (9.60) are limited. Therefore from (9.62) follows that z_i cannot run large. This tells us that the invariant energies of the two blocks are practically fixed by the parameters of the registered particle k,

$$s_1 \simeq \alpha_q s \equiv \alpha s \cdot z_1, \qquad z_1 \sim 1;$$
$$s_2 \simeq (\beta_q - \beta)s \equiv \beta s \cdot z_2, \qquad z_2 \sim 1. \tag{9.63}$$

We conclude that both blocks are in the asymptotic regime, $s_1, s_2 \gg m^2$, so that we can substitute the Regge pole amplitudes, see (9.40a),

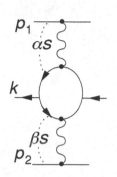

$$\operatorname{Im} A^{\mathrm{pole}}(p_1, q) = g^2 \pi \beta(0) \left(\frac{s_1}{m^2}\right)^{\alpha(0)},$$

$$\operatorname{Im} A^{\mathrm{pole}}(p_2, q - k) = g^2 \pi \beta(0) \left(\frac{s_2}{m^2}\right)^{\alpha(0)}.$$

In these blocks the s-dependence of the inclusive cross section is hidden; neither the propagators in (9.60) nor the phase space (9.61) contain s. The partial energies s_1 and s_2 in (9.63) obviously depend on the particle rapidity,

$$s_1 \sim \alpha s = m_\perp \sqrt{s}\, e^{-(\eta - \bar{\eta})}, \quad s_2 \sim \beta s = m_\perp \sqrt{s}\, e^{\eta - \bar{\eta}}; \quad \bar{\eta} \equiv \tfrac{1}{2}(\eta_+ + \eta_-),$$

but in their product this dependence disappears,

$$\left(\frac{s_1}{m^2}\frac{s_2}{m^2}\right)^{\alpha(0)} = \left(\frac{s}{m^2}\right)^{\alpha(0)} \cdot \left(z_1 z_2 \frac{m_\perp^2}{m^2}\right)^{\alpha(0)}.$$

Taken together with the flux factor, $j^{-1} \simeq s^{-1}$, we reproduce the s-behaviour of $\sigma_{\mathrm{tot}}(s)$ (9.40a). This being the only s-dependent ingredient of (9.60), for the inclusive cross section we finally obtain

$$f(k_\perp, \eta) = \sigma_{\mathrm{tot}}(s) \cdot \phi(\mathbf{k}_\perp^2), \tag{9.64}$$

where the normalized inclusive spectrum ϕ depends neither on s nor on the rapidity η of the triggered particle. The spectrum density ϕ has the following structure

$$\phi(\mathbf{k}_\perp^2) \propto (m_\perp^2)^{\alpha(0)+1} \int d^2\mathbf{q}_\perp \int_0^\infty dz_1\, z_1^{\alpha(0)} \int_0^\infty dz_2\, z_2^{\alpha(0)}$$
$$\cdot \frac{1}{[m^2 + \mathbf{q}_\perp^2 + z_1(1+z_2)m_\perp^2]^2 [m^2 + (\mathbf{q} - \mathbf{k})_\perp^2 + (1+z_1)z_2 m_\perp^2]^2}. \tag{9.65}$$

Its normalization is easy to fix using the general relation (9.59),

$$\int_{\eta_-}^{\eta_+} d\eta \int d^2\mathbf{k}_\perp\, f(k_\perp, \eta) = \bar{n}(s) \cdot \sigma_{\mathrm{tot}} = \beta(0) \ln \frac{s}{m^2} \cdot \sigma_{\mathrm{tot}},$$
$$\Longrightarrow \int d^2\mathbf{k}_\perp\, \phi(\mathbf{k}_\perp^2) = \beta(0). \tag{9.66}$$

In the spirit of the Regge pole picture, the inclusive spectrum $f(k_\perp, \eta)$ can be represented graphically as

$$f(k_\perp, \eta) = \qquad\qquad\qquad\qquad\qquad\qquad (9.67)$$

where ϕ plays the rôle of a new reggeon–reggeon vertex with the production of the particle with a given $\mathbf{k}$.

One can study multi-particle observables as well. Consider, for example, the double inclusive cross section, $f(k_1, k_2)$, which characterizes the production of two particles with fixed momenta in the same event. The consideration analogous to (9.51) which has led us to the normalization of the one-particle inclusive cross section,

$$\frac{1}{\sigma_{\text{tot}}} \int f(k_1)\, d\Gamma(k_1) = \bar{n}, \qquad\qquad (9.68a)$$

yields in the case of two registered particles

$$\frac{1}{\sigma_{\text{tot}}} \int d\Gamma(k_1) d\Gamma(k_2) f(k_1, k_2) = \langle n(n-1) \rangle = \langle n^2 \rangle - \bar{n}. \qquad (9.68b)$$

To see whether particles are correlated, one constructs the difference

$$C_2(k_1, k_2) = \frac{f(k_1, k_2)}{\sigma_{\text{tot}}} - \frac{f(k_1)}{\sigma_{\text{tot}}} \frac{f(k_2)}{\sigma_{\text{tot}}}.$$

The phase space integral of the correlation function,

$$\int d\Gamma(k_1)\, d\Gamma(k_2) C_2(k_1, k_2) = (\langle n^2 \rangle - \bar{n}^2) - \bar{n},$$

is zero for the Poisson distribution.

We take the ladders and assemble the intermediate lines into three Regge pole amplitudes, provided the pair energies are large. Invoking the kinematical relation (9.50),

$$(s_1 s_2 s_3)^{\alpha(0)} = s^{\alpha(0)} (m^2 + \mathbf{k}_{1\perp}^2)^{\alpha(0)} (m^2 + \mathbf{k}_{2\perp}^2)^{\alpha(0)},$$

we arrive at

$$f(k_1, k_2) = f(\mathbf{k}_{1\perp}, \eta_1; \mathbf{k}_{2\perp}, \eta_2) = s^{\alpha(0)-1} g_1^r \, \phi(\mathbf{k}_{1\perp}^2)\phi(\mathbf{k}_{2\perp}^2)g_2^r. \qquad (9.69)$$

In a full analogy with (9.67) the double inclusive cross section does not depend on particle rapidities. In perturbation theory particles (with large relative rapidity) are produced independently, $C_2(k_2, k_2) \equiv 0$.

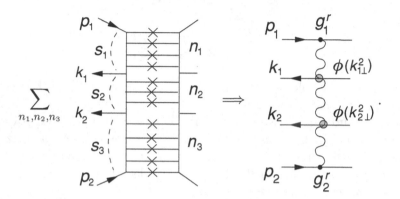

The fact that the inclusive cross sections do not depend on rapidity looks very natural; in the first place, there was no η_i dependence in the underlying multi-particle distribution (9.57) either. On the other hand, if we have indeed a Regge pole, it could not possibly be otherwise.

We have said before that the pole is *factorized*. This means that after a few steps away from the leading particle (say, descending the ladder) the system goes over into a definite state which no longer 'remembers' about the initial state, about the quantum numbers of the projectile and, in particular, about the initial momentum. But how can this be possible?

Let us change the reference frame by moving along the collision axis with some velocity v corresponding to $\Delta\eta$ in rapidity. Then the distribution $f(\eta)$ will shift as a whole, $f \to f(\eta + \Delta\eta)$. But if the particle deep inside the 'ladder' has forgotten about the energies of the colliding particles p_1 and p_2 (factorization!), then the probability to observe the particle should not change. (We cannot say much about the edges of the distribution, $\eta \simeq \eta_\pm$, at the moment. We will discuss this issue in what follows.)

Thus the uniformity in rapidity – the so-called *rapidity plateau* – is just the consequence of the factorization. We shall see later that homogeneity in rapidity follows from the factorization also beyond the perturbation theory, if the average transverse momentum of hadrons does not increase with the energy s.

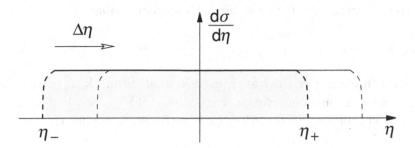

Fig. 9.6   Plateau in the inclusive cross section. Changing the reference frame does not affect particle density inside the ladder.

The mean particle multiplicity increases with the width of the distribution, $\eta_+ - \eta_- = \ln(s/m^2)$, in Fig. 9.6. The height equals $\beta(0) \propto \bar{g}^2$; in the perturbation theory it is small.

9.3.6 Particle distribution in the impact parameter space

The impact parameters of colliding particles do not change in the course of the high-energy scattering. And how are the *newly* produced particles distributed in the impact parameter space? Answering this question will allow us to understand the origin of the impact parameter structure of the pomeron amplitude (8.22):

$$\operatorname{Im} A^{\mathrm{pole}}(s, \boldsymbol{\rho}_{12}) = \frac{1}{4\pi\alpha'\xi} \exp\left\{-\frac{(\boldsymbol{\rho}_1 - \boldsymbol{\rho}_2)^2}{4\alpha'\xi}\right\}, \quad \xi = \ln\frac{s}{m^2}. \qquad (9.70)$$

In Lecture 5 we speculated about the possibility of having the interaction radius growing with energy. We saw that the random walk in the impact parameter, due to long-living multi-particle fluctuations, is capable of generating the growth of the radius characteristic for the Regge pole exchange (9.70), namely $\rho \propto \sqrt{\xi}$. Now we are in a position to verify this expectation within our perturbative model.

Take the ladder amplitude,

$$f(\mathbf{k}_0; \mathbf{k}_1, \dots, \mathbf{k}_n; \mathbf{k}_{n+1}) = g^{n+2} \frac{1}{m^2 + \mathbf{k}_{0\perp}^2} \frac{1}{m^2 + (\mathbf{k}_0 + \mathbf{k}_1)_\perp^2} \cdots$$

$$\cdots \frac{1}{m^2 + (\mathbf{k}_1 + \cdots + \mathbf{k}_{n-1})_\perp^2} \frac{1}{m^2 + (\mathbf{k}_1 + \cdots + \mathbf{k}_n)_\perp^2}$$

$$= g^n \frac{1}{m^2 + \mathbf{q}_{1\perp}^2} \frac{1}{m^2 + \mathbf{q}_{2\perp}^2} \cdots \frac{1}{m^2 + \mathbf{q}_{n,\perp}^2} \frac{1}{m^2 + \mathbf{q}_{n+1,\perp}^2}, \qquad (9.71)$$

and Fourier-transform it to the impact parameter space,

$$f(\rho_0, \rho_1, \ldots, \rho_{n+1}) = \int \frac{d^2 k_{1\perp}}{(2\pi)^2} \cdots \frac{d^2 k_{n\perp}}{(2\pi)^2} \, e^{i \sum_{i=0}^{n+1} (k_i \cdot \rho_i)} f(k_0, \ldots, k_{n+1}),$$

where the integral over the last transverse momentum, $k_{n+1,\perp}$ was traded for the momentum-conservation condition, $\delta^2(\sum_{i=0}^{n+1} k_{i\perp})$. The Fourier exponent can be cast in terms of the transferred momenta q_i,

$$\sum_{i=0}^{n+1} k_i \cdot \rho_i = q_1 \cdot (\rho_0 - \rho_1) + q_2 \cdot (\rho_1 - \rho_2) + \cdots + q_{n+1} \cdot (\rho_n - \rho_{n+1}).$$

Then, given the factorized structure of (9.71) in q_i, the transformed amplitude comes out very simple,

$$f(\rho_0, \ldots \rho_{n+1}) = g^{n+2} \varphi(\rho_0 - \rho_1) \varphi(\rho_1 - \rho_2) \cdots \varphi(\rho_n - \rho_{n+1}), \quad (9.72a)$$

$$\varphi(\rho) = \int \frac{d^2 q}{(2\pi)^2} \frac{e^{iq \cdot \rho}}{m^2 + q^2}. \quad (9.72b)$$

The probability of a given configuration will be given by $|f(\rho_1, \ldots, \rho_n)|^2$. Integration of the squared amplitude over all impact parameters ρ_i will obviously produce the ladder cross section σ_n.

Consider the probability $w_{\tau, \tau'}(\rho)$ of finding two particles, labelled τ and τ' at some distance $|\rho|$. This is a typical *inclusive* characteristic: we don't restrict other particles and integrate over their position in the transverse plane. The chain structure of the amplitude (9.72a) makes it clear that the positions of the ladder rungs with numbers i *outside* the interval between the triggered particles, $i \notin [\tau, \tau']$, will be integrated out freely without affecting the answer.

We can pick, for example, the last rung, $\tau' = n + 1$, and since the answer depends only on the difference of the coordinates, set $\rho_{n+1} = 0$. This way we will be studying the distance of the particle τ from the target p_2.

The normalized probability takes the form

$$w_\Delta(\rho) = \frac{1}{N^\Delta} \int d^2 \rho_{\tau+1} \cdots d^2 \rho_n \, \varphi^2(\rho - \rho_{\tau+1}) \, \varphi^2(\rho_{\tau+1} - \rho_{\tau+2}) \cdots \varphi^2(\rho_n),$$

$$\Delta = n + 1 - \tau; \quad \int d^2 \rho \, w_\Delta(\rho) = 1, \quad N \equiv \int d^2 \rho \, \varphi^2(\rho). \quad (9.73)$$

The integrand of w_Δ contains Δ factors φ; the number of integrations is $\Delta - 1$. In one step from the target, $\tau = n$, we have the probability $w_1(\rho) = \varphi^2(\rho)$ given by (9.72b). At distances larger than the characteristic radius $r_0 = m^{-1}$ it falls exponentially, $\varphi(\rho) \sim \exp(-|\rho|/r_0)$. It is natural to expect that with the number of steps Δ increasing, the probability of finding the particle will spread.

Indeed, (9.73) is a typical diffusion problem with $\varphi^2(\rho)$ the probability distribution of 'jumping' at a distance $\sim r_0$; the convolution (9.73) describes the result after Δ independent jumps. Whatever the form of the initial distribution w_1, after a few jumps $w_\Delta(\rho)$ becomes a *Gaussian*. Let us find its *dispersion* σ_Δ,

$$w_\Delta(\rho) \simeq \frac{1}{\pi \sigma_\Delta^2} \exp\left\{ -\frac{\rho^2}{\sigma_\Delta^2} \right\}, \tag{9.74a}$$

which measures the average squared distance from the target,

$$\sigma_\Delta^2 = \langle \rho^2 \rangle_\Delta = \int d^2\rho_\tau \int dw_\Delta(\rho_\tau) \left(\sum_{k=1}^{\Delta} [\rho_{\tau+k-1} - \rho_{\tau+k}] \right)^2. \tag{9.74b}$$

Since the directions of individual jump are uncorrelated, averaging the squared sum yields simply

$$\langle \rho^2 \rangle_\Delta = \Delta \cdot \langle \rho^2 \rangle_1, \tag{9.75a}$$

$$\langle \rho^2 \rangle_1 = \int dw_1(\rho)\rho^2 = \frac{1}{N} \int d^2\rho \, \varphi^2(\rho) \cdot \rho^2. \tag{9.75b}$$

To calculate the concrete number is of little interest. We will, instead, relate (9.75) directly to the Regge pole trajectory.

To do that, we go back for a moment to the momentum representation. From (9.72b) it immediately follows that the *Fourier image* of the probability distribution is nothing but the Regge trajectory,

$$\int d^2\rho \, \varphi^2(\rho) \, e^{-i\mathbf{q}\cdot\rho} = \int \frac{d^2\mathbf{q}'}{(2\pi)^2} \frac{1}{[m^2 + \mathbf{q}'^2][m^2 + (\mathbf{q}' - \mathbf{q})^2]} = c\,\beta(\mathbf{q}^2),$$

so that

$$\int dw_1(\rho) \, e^{-i\mathbf{q}\cdot\rho} = \frac{\beta(\mathbf{q}^2)}{\beta(0)}. \tag{9.76a}$$

Now the calculation of (9.75b) is very simple:

$$\langle \rho^2 \rangle_1 = \left\{ (i\boldsymbol{\nabla}_\mathbf{q})^2 \int dw_1(\rho) \, e^{-i\mathbf{q}\cdot\rho} \right\}_{\mathbf{q}=0}$$

$$= -\frac{4}{\beta(0)} \frac{d}{d\mathbf{q}^2} \beta(\mathbf{q}^2) \bigg|_{\mathbf{q}=0} = 4\frac{\alpha'(0)}{\beta(0)}, \tag{9.76b}$$

where $\alpha' = \beta'$ denotes the derivative of the trajectory over $t = -\mathbf{q}^2$.

Finally, what is the meaning of Δ in (9.75a)? The number of particles we can evaluate as the size of the rapidity interval between the triggered particles, the particle τ, and the target, multiplied by the average density of the plateau:

$$\sigma_\Delta^2 = \Delta \cdot \langle \rho^2 \rangle_1 = [(\eta - \eta_-) \times \beta(0)] \cdot 4 \frac{\alpha'}{\beta(0)}.$$

Substituting this dispersion into (9.75) and setting $\eta = \eta_+$ we obtain exactly the impact parameter profile of the vacuum pole (9.70):

$$w_\Delta(\boldsymbol{\rho}_{12}) \simeq \frac{1}{4\pi\alpha'\xi} \exp\left\{-\frac{\rho_{12}^2}{4\alpha'\xi}\right\} = A^{\text{pole}}(s, \boldsymbol{\rho}_{12}). \qquad (9.77)$$

9.3.7 Perturbative conclusions

Here we summarize the characteristic features of the inelastic processes which determine the Regge pole in perturbation theory.

(1) Topological cross sections follow the Poisson distribution

$$\sigma_n(s) = \sigma_{\text{tot}}(s) \frac{[\bar{n}(s)]^n}{n!} e^{-\bar{n}(s)}. \qquad (9.78a)$$

(2) The mean particle multiplicity grows logarithmically with energy,

$$\bar{n}(s) = \beta(0)\xi, \quad \xi = \eta_+ - \eta_- = \ln\frac{s}{m^2}. \qquad (9.78b)$$

(3) The inclusive particle spectrum is flat in rapidity (plateau) and does not depend on the total energy,

$$\frac{d^3n}{d^2k_\perp d\eta} = \frac{f(k_\perp^2, \eta; s)}{\sigma_{\text{tot}}(s)} = \quad \Bigg/ \quad = \phi(k_\perp^2). \qquad (9.78c)$$

(4) The production of particles with large rapidity intervals between them is uncorrelated,

$$\frac{d^6n}{d^2k_{1\perp} d\eta_1 d^2k_{2\perp} d\eta_2} = \quad \Bigg/ \quad = \phi(k_{1\perp}^2)\phi(k_{2\perp}^2). \qquad (9.78d)$$

(5) The particle density in the plateau is small,

$$\frac{dn}{d\eta} = \int d^2\mathbf{k}_\perp \, \phi(\mathbf{k}_\perp^2) = \beta(0) \sim \frac{g^2}{m^2} \ll 1. \tag{9.78e}$$

(6) The energy increase of the interaction radius characteristic for the Regge-pole exchange is due to diffusion in the impact parameter space,

$$\rho_0(s) \sim \sqrt{\beta'(0)\,\xi}. \tag{9.78f}$$

10

Regge pole beyond perturbation theory

We address now the most interesting question, namely how much of what we have found in the perturbation theory will survive in the real world where the hadrons interact strongly.

The hadrons and hadron resonances lie on Regge trajectories $\alpha(t)$ which relate the hadron spin σ and the mass, $\sigma = \alpha(m_h^2)$. These trajectories determine the asymptotic behaviour of the scattering in the crossing channel, $t < 0$, where the corresponding quantum numbers can be exchanged.

At the first sight, we succeeded in reggeizing the amplitude and obtained the Regge pole as expected. Unfortunately, the trajectory that we found in the perturbative $g\varphi^3$ theory,

$$\alpha(q^2) = -1 + \bar{g}^2 \int \frac{d^2\mathbf{k}_\perp}{2(2\pi)^3} \frac{m^2}{[m^2 + \mathbf{k}_\perp^2][m^2 + (\mathbf{k} - \mathbf{q})_\perp^2]}; \quad (10.1)$$

$$t = -\mathbf{q}_\perp^2; \quad \bar{g}^2 = \frac{g^2}{m^2},$$

does not possess a single particle; as long as the coupling is small, $\bar{g} \ll 1$, the trajectory stays below $j = 0$ at any t (at large $|t|$ it falls like $\ln t/t$). In principle we could get a bound state on the trajectory, or even enforce $\alpha(0) = 1$, if we took a large coupling, $\bar{g}^2 = \mathcal{O}(1)$.

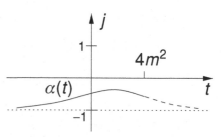

But this means moving outside the boundaries of perturbation theory where we are helpless. The selection of diagrams based on the leading log approximation is here not valid, since all sorts of corrections to the ladder become essential.

If we can hope at all that the realistic hadron Regge-pole picture will be established, then only if the interaction is non-perturbative. In spite of the impossibility to *calculate* the hadron trajectories, we will nevertheless be able to describe the structure of *inelastic processes* that constitute the base of the real Regge poles. And our experience with the investigation of the reggeons in perturbation theory will help us to do that.

10.1 Basic features of multiparticle production

10.1.1 Particle density

The underlying inelastic processes in perturbation theory were dominated by multiperipheral ladders. The simple reason for the ladder dominance was *large invariant pair energies* of the neighbouring particles (see Section 9.2):

$$\langle s_{i,i+1} \rangle = (k_i + k_{i+1})^2 \simeq \beta_i \alpha_{i+1} s \sim \frac{\beta_i}{\beta_{i+1}} m_{i+1,\perp}^2 \sim \frac{\beta_i}{\beta_{i+1}} m \gg m^2 \ . \quad (10.2)$$

Particles were distributed scarcely in rapidity (see Section 9.3):

$$\Delta\eta = \langle \eta_i - \eta_{i+1} \rangle \simeq \ln\frac{\langle s_{i,i+1} \rangle}{m^2} \simeq \frac{1}{\beta(0)} \gg 1, \quad (10.3)$$

and the permutation of two particles with large $\Delta\eta$ in (9.29) produced a significant recoil causing high virtuality of the propagator and the suppression of the amplitudes with crossed lines.

With the growth of $\bar{g}^2$ the decays become more frequent, the particle density increases and the pair energies $\langle s_{i,i+1} \rangle$ decrease. The time ordering of successive decays starts to disappear and processes with particle permutations cease to play the rôle of small corrections when we reach $\langle s_{i,i+1} \rangle \sim m^2$ at $\bar{g}^2 \sim 1$.

The key question is whether the particle density will continue to grow with the increase of the interaction strength, or will it stop and freeze at a certain value? Will the rapidity distribution remain homogeneous and independent of the total energy?

The problem we are dealing with reminds that of the one-dimensional gas. We plant a few points into an interval. If there is no repulsion between them, one can stuff in as many points as one wants. If there is, some mean density will be established, depending on the dynamics.

Two statements can be made.

(1) *If the asymptotic behaviour is determined by the Regge pole*, then in multi-particle production processes a certain constant rapidity density of final particles is reached.

(2) *Irrespectively of the Regge pole hypothesis*, a serious reason for arriving at a constant density lies in the astonishing, well established experimental fact that transverse momenta do not grow with energy (at least for energies up to 10^{13} eV).

The first statement can be rigourously proven. The second one is more intuitive but rather general too. Let us start from the latter.

10.1.2 Limited transverse momenta and rapidity plateau

In inelastic high energy hadron interactions particles are produced in two bunches following the directions of the colliding hadrons. If you measure the average transverse momentum of produced particles at different collision energies, you will get

$$\langle \mathbf{k}_\perp^2 \rangle = \text{const}(s). \tag{10.4}$$

This is just what happens in the perturbative model that we have used above. There the inclusive cross section decreased rapidly at large $k_\perp$,

$$\phi(\mathbf{k}_\perp^2) = \quad\sim \int \frac{d^2\mathbf{q}_\perp}{(\mathbf{q}_\perp^2)^4} \sim \frac{1}{\mathbf{k}_\perp^6}, \qquad \mathbf{k}_\perp^2 \gg m^2, \tag{10.5a}$$

so that the average transverse momentum was of the order of the mass,

$$\frac{\int d^2\mathbf{k}_\perp\, \phi(\mathbf{k}_\perp^2) \cdot \mathbf{k}_\perp^2}{\int d^2\mathbf{k}_\perp\, \phi(\mathbf{k}_\perp^2)} = \text{const} \cdot m^2. \tag{10.5b}$$

This is a unique feature of the $g\varphi^3$ theory which is too simplistic and specific to have any relation to the real world.

However, as we have already stated, if e.g. fermions are included in the scheme, the integral (10.5b) defining $\langle \mathbf{k}_\perp^2 \rangle$ becomes *logarithmically divergent*, and the transverse momenta increase with s. In fact, this happens in *any* renormalizable quantum field theory (with the only exception of the *superconvergent* $g\varphi^3$)! Thus it is the experimental situation alone that forces us to look for a theoretical description that would respect (10.4).

Let us show now that if we want $\langle \mathbf{k}_\perp^2 \rangle$ to be restricted, a finite particle density must be set up in the $s \to \infty$ limit.

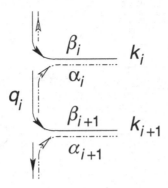

Fig. 10.1   Flow of longitudinal momenta in multiperipheral kinematics.

Recall that in perturbation theory transverse momenta were restricted by the virtual exchange propagators q_i in Fig. 10.1,

$$m^2 - q_i^2 = m^2 + \mathbf{q}_{i\perp}^2 + \alpha_q \beta_q s = m^2 + \mathbf{q}_{i\perp}^2 + \left(\sum_{j \leq i} \alpha_j\right)\left(\sum_{k \geq i+1} \beta_k\right)s.$$
(10.6)

In the *perturbative* multiperipheral kinematics (9.27) all αs and βs were strongly ordered. The longitudinal part of the virtuality,

$$\alpha_q \beta_q s \simeq \left(\alpha_i + \cdots\right)\left(\beta_{i+1} + \cdots\right)s \simeq m_{i+1,\perp}^2 \cdot \frac{\beta_{i+1}}{\beta_i} \sim m^2 \cdot \frac{\beta_{i+1}}{\beta_i}, \quad (10.7)$$

was then negligible, and it was for m^2 to set the upper bound for the variation of $\mathbf{q}_{i\perp}^2$ in (10.6). If we increase $\bar{g}^2$, the 'comb' gets denser and denser, and at $\bar{g}^2 \sim 1$ we eventually reach the situation when the neighbouring βs become comparable, $\beta_i \geq \beta_{i+1}$. They are still ordered, but not *strongly ordered* anymore, which makes the longitudinal virtuality (10.7) comparable with m^2. This situation corresponds to $\langle s_{i,i+1} \rangle \gtrsim m^2$, that is to a unit density in rapidity.

Imagine now that with the increase of s the particle density keeps growing as $dn/d\eta = D(s)$. Having $D(s)$ particles with comparable Sudakov components inside a unit rapidity interval, (10.6) will be modified as follows:

$$m^2 - q_i^2 \simeq \mathbf{q}_{i\perp}^2 + m^2 + \left(D(s)\tilde{\alpha}\right)\left(D(s)\tilde{\beta}\right)s, \quad \tilde{\alpha}\tilde{\beta}s \sim m^2. \quad (10.8)$$

As a result, the transverse momentum integrals will spread much broader, up to $\mathbf{q}_{i\perp}^2 \lesssim D^2(s)m^2$, and produce the average $\mathbf{k}_\perp^2$ increasing with energy together with the density $D(s)$.

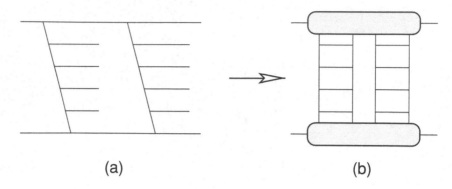

(a)   (b)

Fig. 10.2   Amplitude (a) and cross section (b) of the double-ladder process.

10.1.3 Large multiplicities and overlapping ladders

This does not mean, however, that in a collision of two hadrons there will be no events with multiplicities significantly larger than the average $\bar{n} \propto \ln s$. Even in the perturbative framework there is a simple way to obtain a large particle density, not breaking the restrictedness of $\langle \mathbf{k}_\perp^2 \rangle$. Let us draw a picture with two multiperipheral combs exchanged between the target and projectile. Squaring the diagram of Fig. 10.2(a) we will apply the previous analysis to the two ladders in Fig. 10.2(b) and will obtain the final-state multiplicity $\sim 2\bar{n}$, and therefore the double density in rapidity, while preserving limited transverse momenta inside each ladder.

When the interaction is strong, there is no reason for such a diagram to be any smaller than one ladder. There is, however, something bizarre about this picture.

Momentum distributions of particles in the two combs overlap perfectly. Why will they not interact, especially since the interaction is, once again, strong? On the other hand, if they *do* re-interact and become inseparable from the t-channel point of view, then our previous arguments will work linking the particle density to the average $\mathbf{k}_\perp^2$.

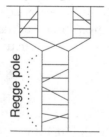

Actually, it is the t-channel we have to appeal to for the explanation. As we know, the Regge pole is a *bound state* in the t-channel.

But this implies that all the particles that 'propagate' in the t-channel must have *bounds* to each other; there cannot be two non-interacting groups of objects as in Fig. 10.2. In other words, such pictures do not belong to the *pole*.

The diagrams like that of Fig. 10.2 have a full right to exist but, if we believe in the pole approximation, they would better be small corrections describing density fluctuations on top of the underlying uniform plateau due to the Regge pole exchange.

10.1.4 Mueller–Kancheli diagram for inclusive spectrum

Let us calculate the inclusive spectrum corresponding to the Regge pole without appealing to the perturbation theory.

But first we make a qualitative remark to appreciate the key rôle played by the *factorization* feature of the Regge pole.

No matter how complicated the underlying diagrams are, due to the unitarity condition we have to take the imaginary part of the forward amplitude and extract one particle with a given momentum k in the intermediate multi-particle state,

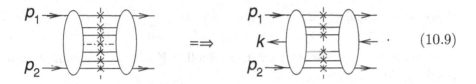

$$\tag{10.9}$$

What distinguishes the upper part of the full block from the scattering amplitude of particles p_1 and k is that it is connected with the lower part by some particle lines. If the number of these lines increased with the total energy s, the average transverse momentum

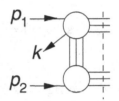

would also grow. Besides, if somewhere inside the process the number of exchanges depends on the initial energy, how can there be factorization on the l.h.s. of (10.9) which, as we have supposed, *is* described by the Regge pole?

Repeating the same argument we isolate the particle k as shown in the l.h.s. of (10.10),

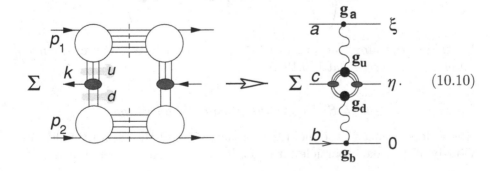

$$\tag{10.10}$$

It is connected to the top and bottom parts of the graph in a non-trivial manner via some particle 'bunches' u and d. If we additionally suppose that the interaction between particles is *local* in the rapidity space, then the central block $g^2_{c;u,d}$ that links the triggered particle (of type c) to the states u and d will span a finite rapidity interval of the order of unity.

Under these circumstances the invariant energies of the top and bottom blocks are large and we can substitute (imaginary parts of) the Regge pole amplitudes as shown on the r.h.s. of (10.10),

$$\sim \sum_{u,d} \cdot \left[g^r_a \, e^{\alpha(0)(\xi-\eta)} g^r_u \right] \cdot g^2_{c;u,d} \cdot \left[g^r_d \, e^{\alpha(0)\eta} g^r_b \right].$$

Since, due to factorization, the central part of the diagram does not depend on the total energy, we get the energy-independent rapidity plateau in the inclusive spectrum,

$$f(\mathbf{k}_\perp, \eta; s) = g^r_a g^r_b s^{\alpha(0)-1} \cdot \phi(\mathbf{k}_\perp); \quad \phi_c(\mathbf{k}_\perp) = \sum_{u,d} g^r_u \cdot g^2_{c;u,d} \cdot g^r_d. \quad (10.11)$$

The answer is represented by the Mueller–Kancheli reggeon diagram (Gribov, 2003),

$$f(\mathbf{k}_\perp, \eta; s) = \frac{1}{s} \cdot \quad \text{} \quad = \sigma^{ab}_{\text{tot}} \cdot \phi_c(\mathbf{k}_\perp). \quad (10.12)$$

To derive rigorously this important non-perturbative result, one considers the $3 \to 3$ scattering amplitude and continues it to complex angular momenta. If one supposes that there are Regge poles in $2 \to 2$ scattering at large s, then the asymptotic behaviour of $3 \to 3$ amplitude in the $s_1, s_2 \to \infty$, $s_1 s_2/s = \text{const}$ limit is determined by the exchange of two Regge poles i, j,

By taking the amplitude in which both i and j have vacuum quantum numbers, one arrives at (10.12).

10.1.5 Scaling in the fragmentation regions

The independence of the inclusive particle yield of the rapidity holds, obviously, if we take a particle far enough from the ends of the full rapidity interval, $\xi_\pm$. Only then the invariant energies between the triggered particle and the incoming ones are large, and we can substitute the reggeons for the corresponding scattering blocks.

What happens at the ends?

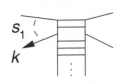

Let us take a particle with large rapidity, from the first ladder rungs on the side of the projectile. In this case we will be able to replace only the bottom part of the graph by a reggeon that covers a large rapidity interval $\eta - \xi_- = \eta$ (in the rest frame of the target p_2):

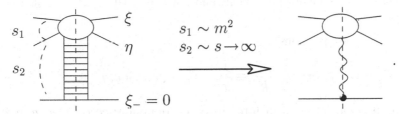

At the same time, there remains a serious dependence of the inclusive particle yield on the 'distance' $\xi - \eta$,

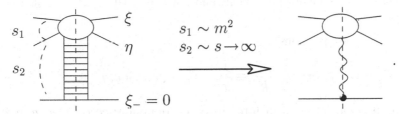

$$\eta \;=\; e^{\alpha(0)\eta}\psi(\mathbf{k}_\perp, s_1) \;=\; s^{\alpha(0)}g_b^r \cdot \phi_{ac}(\mathbf{k}_\perp, \xi - \eta). \quad (10.13)$$

Only after stepping away from the incident particle by about 2 units in rapidity, the system 'forgets' about the quantum numbers of the 'initiator' of the cascade, and the universal plateau starts developing.

Thus the inclusive spectrum consists of three regions: in addition to the plateau, two so-called *fragmentation regions* appear as shown in Fig. 10.3. They are called 'target fragmentation' and 'projectile fragmentation'.

The name *fragmentation* carries a deep meaning. According to (10.13), the structure of the fragmentation region of the *projectile*, in the interval between η_p and ξ in Fig. 10.3, depends on the type of the projectile a and that of the triggered particle c. Moving towards the kinematical boundary, the inclusive particle distribution may either increase as shown by the solid line (as in the reaction $\pi^- p \to \pi^- + X$), or drop ($\pi^- p \to \pi^+ + X$; dashed). At the same time it stays independent of the total energy s and of the type of the target b (the lower reggeon vertex g_b^r factors out into σ_{tot}^{ab}). The same is true for target fragmentation, $\eta < \eta_t$.

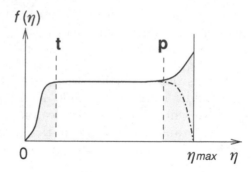

Fig. 10.3 Fragmentation regions and plateau in the inclusive spectum.

Since for fast particles

$$e^{\xi-\eta} \simeq \frac{p_{1z}}{k_z} \equiv \frac{1}{x}, \quad \phi_{ac} = \phi_{ac}(\mathbf{k}_\perp, x),$$

the dependence on s_1 translates into the dependence on the *momentum fraction* x of the incident momentum p_1 that is carried by the triggered particle k.

This feature is called the *Feynman scaling* or the 'limiting fragmentation hypothesis'. In our picture it is a direct consequence of the reggeon factorization. With the increase of s the fragmentation regions in Fig. 10.3 just separate further while preserving their specific shapes.

It is clear that the scaling must manifest itself earlier in energy than the plateau since for the latter one needs both s_1 and s_2 to be sufficiently large. Indeed, the limiting fragmentation sets in already for s of the order of a few GeV and is well established experimentally. At the same time, a flat plateau appears only for $\xi \gtrsim 5$ corresponding to $s \simeq 150\,\mathrm{GeV}^2$.

What should one expect when comparing, e.g. the inclusive reactions

$$pp \to \pi^+ + X \quad \text{and} \quad pp \to \pi^- + X \,? \tag{10.14}$$

When we register a particle in the fragmentation region, we take it from the *residue* of the cut pomeron, and not much can be said about it. In particular, $\pi^\pm$ production is different not only in the fragmentation of the incident *pion* as we have just mentioned above, but also in the *proton* fragmentation region, since the proton feels very well the difference between π^+ and π^-.

However, in the plateau region quantum numbers of produced particles must be 'well equilibrated'. Here the particle is taken from inside the *vacuum pole* itself which is 'blind' to I_3 or to (the sign of) the strangeness or the baryon charge. In this case the yields, e.g. of π^+ and π^- mesons must be identical. The symmetry between the reactions (10.14) holds

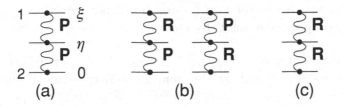

Fig. 10.4 Subleading corrections to the inclusive plateau density.

within a few percent. (One needs much higher energies to see protons and antiprotons 'equalize'.)*

Taking into consideration subleading reggeons **R** that also contribute to the forward scattering, such as ρ and $\mathbf{P}'$, one can study corrections to the asymptotics, as well as the transition between the plateau and fragmentation regions. Summing the diagrams of Fig. 10.4 gives for the particle density

$$\frac{f(\mathbf{k}_\perp, \eta; \xi)}{\sigma_{\text{tot}}} = \phi(\mathbf{k}_\perp) + c_1^R \phi'(\mathbf{k}_\perp) e^{-\kappa(\xi-\eta)}$$
$$+ c_2^R \phi'(\mathbf{k}_\perp) e^{-\kappa\eta} + c_1^R c_2^R \phi''(\mathbf{k}_\perp) e^{-\kappa\xi}, \tag{10.15}$$

where $\kappa = \alpha^P(0) - \alpha^R(0)$ is the shift between the pomeron intercept and that of the subleading reggeon R, and c_i are the reggeon residues normalized by the pomeron one, $c_i = g_i^R / g_i^P$. In reality, $\kappa \simeq \frac{1}{2}$ for the $\mathbf{f}(\mathbf{P}')$ and ρ trajectories, see Lecture 8. This shows that the graph Fig. 10.4(c) corresponding to the last term in (10.15) provides a 'flat', η-independent, pre-asymptotic correction $\propto 1/\sqrt{s}$ to the plateau height, while the magnitude of the corrections due to mixed graphs of Fig. 10.4(b) depends on rapidity. It increases towards the fragmentation region, introducing a *curvature* to the plateau–fragmentation transition.

Introducing an R pole into the two-particle inclusive cross section,

$$\propto e^{-\kappa|\eta-\eta'|}, \tag{10.16}$$

---

* At nucleon–nucleon energies $s = 10^4$ GeV^2, the yield of antiprotons became practically equal to that of protons, as we learnt from experiments at the heavy ion collider RHIC, Brookhaven, NY, USA (ed.).

results in a *positive* correlation ('attraction') between the particles since the non-leading reggeon tends to 'collapse', to reduce the difference of the rapidities.

In two different ways – by the extension of the perturbative analysis to the region $\bar{g}^2 \lesssim 1$ and by the analytic continuation of the six-point amplitude – under the assumption of the existence of the pomeron pole **P** in elastic scattering we arrive at the conclusion that multi-particle production processes at high energies have the following characteristic features.

(1) Final state hadrons are distributed homogeneously in η, away from the ends of the rapidity interval – the fragmentation regions,

$$f(k_\perp, \eta; s) = \quad \raisebox{0pt}{\includegraphics{}} \quad = g_a g_b \, \phi_c(\mathbf{k}_\perp). \qquad (10.17a)$$

(2) In the fragmentation of the incident particle i, the spectrum depends only on the relative rapidity $f = f(\eta_i - \eta)$ – Feynman scaling,

$$f(k_\perp, \eta; s) = \quad \raisebox{0pt}{\includegraphics{}} \quad = \phi_{ac}(\mathbf{k}_\perp, \xi - \eta) \, g_b. \qquad (10.17b)$$

(3) As a consequence, the average multiplicity increases logarithmically with collision energy,

$$\langle n \rangle \simeq \beta \ln \frac{s}{m^2} + \text{const}. \qquad (10.17c)$$

Essentially, the key hypothesis that ensures the existence of the asymptotically constant rapidity plateau is that the transverse momenta in hadron interactions are limited.

10.2 Inconsistency of the Regge pole approximation

Up to now we considered the Regge pole as the leading singularity in the complex angular momentum plane, assuming that all the other singularities of the partial amplitude are subleading poles, s^{α_R}, that give power-suppressed corrections and are irrelevant for the asymptotic behaviour.

We will show now that this assumption is contradictory.

10.2.1 Small-multiplicity events

Investigating perturbation theory, we have seen that the sum of ladder graphs had a Regge behaviour and, consequently, could provide a basis for the real Regge pole. Going beyond the perturbation theory, we have found that the qualitative features of typical inelastic processes with the production of $n \sim \bar{n}$ particles were similar to those of the multiperipheral ladder diagrams of the perturbative theory, although at $\bar{g}^2 \sim 1$.

What about small multiplicity events? As we know, in perturbation theory the topological cross sections σ_n follow the Poisson distribution

$$\sigma_{n+2} = \sigma_{\text{tot}} \cdot e^{-\bar{n}} \frac{\bar{n}^n}{n!}, \tag{10.18}$$

and fall fast when one takes n away from the maximum, in particular on the left wing, $n \ll \bar{n}$.

Elastic-scattering contribution to σ_{tot}. Let us look at the elastic scattering – the first among small multiplicity processes – and calculate its contribution to the total cross section:

$$\sigma_2 = \frac{1}{16\pi} \int dq^2 \left| \frac{A_{\text{el}}(s, q^2)}{s} \right|^2 = \frac{1}{2s} \quad \begin{array}{c} p_1 \longrightarrow p_1 \\ q \quad q \\ p_2 \longrightarrow p_2 \end{array} . \tag{10.19}$$

Substituting the asymptotic elastic amplitude determined by the Pomeranchuk pole **P**,

$$A_{\text{el}}(s, q^2) = \quad \begin{array}{c} q^2 \\ s \end{array} \quad = g^2(t)\xi_{\alpha(q^2)} \left(\frac{s}{m^2} \right)^{\alpha(q^2)}, \quad \alpha(q^2) \simeq 1 + \alpha' q^2,$$

we derive

$$\sigma_2 \simeq \left(\frac{g^2(0)}{m^2} \right)^2 \cdot \frac{1}{32\pi\alpha'\xi}, \quad \xi = \ln \frac{s}{m^2}. \tag{10.20}$$

Here we have used the fact that the momentum transfer in the integral is small since it is cut by the Regge radius, $|q^2| \simeq \mathbf{q}_\perp^2 \sim [\alpha' \ln(s/m^2)]^{-1}$, which allowed us to expand the pomeron trajectory, $\alpha(q^2) \simeq 1 + \alpha' q^2$, and to put $q^2 = 0$ in the reggeon vertex g and in the signature factor, $\xi \simeq i$.

We immediately see that the result we have just obtained is in a marked disagreement with the expectation based on the Poisson distribution (10.18) according to which the fraction of small multiplicity events is suppressed as a *power* of energy, $\sigma_2/\sigma_{\rm tot} \propto s^{-\beta_0}$, due to the logarithmic increase of the average $\bar{n} \simeq \beta_0 \ln s$. At the same time, (10.20) is suppressed at $s \to \infty$ only logarithmically.

10.2.2 Multiregge kinematics

This contradiction shows that the perturbative consideration fails when the number of final-state particles is small. It is clear what happened.

When the pair energy is small, two particles interact via the Born amplitude, $\succ\!\!-\!\!\prec = g^2/(m^2 - s_{12})$. (If we take the coupling $\bar{g}^2 \sim 1$, the exact amplitude is more complicated but not significantly different.) With the energy increasing, however, a new parameter appears, $\ln s$-enhanced terms become essential and the Born amplitude is replaced by the *reggeized amplitude* which corresponds to the 'floating spin' exchange in the t-channel. The standard $1/s$ amplitude gets enhanced:

$$s^{-1} \quad \Longrightarrow \quad s^{-1} \cdot s^{\beta(q^2)}.$$

When we treated final states with a number of particles of the order of the average multiplicity, typical pair energies,

$$\langle \ln s_{i,i+1} \rangle \simeq \frac{\ln s}{n} \sim \frac{\ln s}{\bar{n}} \sim (\bar{g}^2)^{-1},$$

were such that in the interaction between neighbours we could neglect the reggeization effects. In the elastic channel, on the contrary, we have a huge energy $s_{12} = s$ applied to two particles. In this situation we must modify the interaction amplitude by substituting the reggeon for the scalar particle exchange,

Obviously, the same substitution must be done also when encountering a large pair energy *inside the multiperipheral ladder*. This happens when one has a wide *gap* in the rapidity distribution of the produced particles. By making the ladder more and

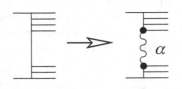

more sparse, one can form many rapidity gaps, and ultimately arrive at the picture with all final particles (or compact groups of particles) widely separated in rapidity and connected by reggeons,

$$s_{i,i+1} \gg m^2, \quad s_{01}s_{12}\cdots s_{n,n+1} \sim sm^{2n}. \qquad (10.21)$$

Such a situation is referred to as *multiregge kinematics* and corresponds to specific fluctuations in multi-particle production.

10.2.3 Multiregge amplitudes

We have to learn how various multiplicity fluctuations contribute to σ_{tot}.

Derivation of the $2 \to 3$ multiregge amplitude. We start with three final-state particles. If an additional hadron is produced in the fragmentation region of one of the colliding particles, the pomeron exchange dominates (see Lecture 8) and we get a contribution of the order of σ_2,

$$\delta\sigma_3 \sim \sigma_2.$$

Now we take the multiregge kinematics,

$$s_{12}, s_{23} \gg m^2, \quad s_{12} \cdot s_{23} \sim sm^2.$$

Omitting the complex phase, we can guess the answer straight away:

$$\sim g_1(\mathbf{k}_{1\perp})s_{12}^{\alpha(t_1)}\gamma(\mathbf{k}_{1\perp}, \mathbf{k}_{3\perp})s_{23}^{\alpha(t_2)}g_2(\mathbf{k}_{3\perp}),$$

$$(10.22)$$

$$t_1 = (k_1 - p_1)^2, \quad t_2 = (k_3 - p_2)^2.$$

It is not an easy task to derive rigorously the multiregge amplitudes by analytic continuation to complex j. The amplitude has specific analyticity in each of many sub-channels, and the *signature structure* of multi-point amplitudes becomes rather involved.

For our purpose of evaluating contributions to σ_{tot} in the s-channel, this subtlety is, however, irrelevant. So we can rely on the perturbative analogy and derive the $2 \to 3$ amplitude in the multiregge kinematics from the ladder picture. For the final particle momenta we write

$$k_1 = \beta_1 p_+ + \frac{m_{1\perp}^2}{\beta_1}p_- + k_{1\perp},$$

$$k_2 = \beta_2 p_+ + \alpha_2 p_- + k_{2\perp},$$

$$k_3 = \frac{m_{3\perp}^2}{\alpha_3}p_+ + \alpha_3 p_- + k_{3\perp}.$$

For momentum transfers q_i this gives

$$q_1 \simeq (1-\beta_1)p_+ - \frac{\mathbf{k}_{1\perp}^2 + (1-\beta_1)m^2}{\beta_1 s}p_- - k_{1\perp},$$

$$q_2 \simeq (1-\alpha_3)p_- - \frac{\mathbf{k}_{3\perp}^2 + (1-\alpha_3)m^2}{\alpha_3 s}p_+ - k_{3\perp}.$$

Then, due to the rapidity ordering and the momentum conservation,

$$\beta_3 \ll \beta_2 \ll \beta_1, \qquad \beta_3 + \beta_2 + \beta_1 = 1;$$
$$\alpha_1 \ll \alpha_2 \ll \alpha_3, \qquad \alpha_3 + \alpha_2 + \alpha_1 = 1,$$

we have $(1-\beta_1) \simeq \beta_2$, $(1-\alpha_3) \simeq \alpha_2$, and the longitudinal components of the transferred momenta q_i become expressed in the multiregge kinematics via the observed particle momentum k_2:

$$q_1 \simeq \beta_2 p_+ - \frac{\mathbf{k}_{1\perp}^2}{s}p_- - k_{1\perp}, \qquad q_1^2 \simeq -\mathbf{k}_{1\perp}^2;$$

$$q_2 \simeq \alpha_2 p_- - \frac{\mathbf{k}_{3\perp}^2}{s}p_+ - k_{3\perp}, \qquad q_2^2 \simeq -\mathbf{k}_{3\perp}^2.$$

We have to integrate over the loop momentum k,

$$k = \beta p_+ - \alpha p_- + k_\perp, \qquad d^4k = \frac{s}{2}\,d\alpha\,d\beta\,d^2\mathbf{k}_\perp,$$

whose Sudakov components determine the *invariant energies* of the top and bottom 'ladders',

$$s' \equiv (p_1 - k)^2 = (\alpha + \gamma)(1 - \beta)s + m^2 + k^2 \simeq \alpha s \equiv x \cdot s_{12},$$
$$s'' \equiv (p_2 + k)^2 = (1 - \alpha)(\beta + \gamma)s + m^2 + k^2 \simeq \beta s \equiv y \cdot s_{23}. \qquad (10.23)$$

In (10.23) we have introduced momentum fractions

$$x = \alpha/\alpha_2, \qquad y = \beta/\beta_2, \qquad (10.24)$$

and used $s_{12} = (k_1 + k_2)^2 \simeq \alpha_2 s$ and $s_{23} = (k_2 + k_3)^2 \simeq \beta_2 s$. Finally, let us have a look at the propagators:

$$\begin{aligned} m^2 - (k - q_1)^2 &= m^2 + (\mathbf{k} + \mathbf{k}_1)_\perp^2 + \alpha(\beta - \beta_2)s, \\ m^2 - k^2 &= m^2 + \mathbf{k}_\perp^2 + \alpha\beta s, \qquad (10.25) \\ m^2 - (k + q_2)^2 &= m^2 + (\mathbf{k} - \mathbf{k}_3)_\perp^2 + (\alpha - \alpha_2)\beta s. \end{aligned}$$

The integrals over x and y converge at $x \sim y \sim 1$; therefore the invariant energies (10.23) are large and we can substitute Regge poles for the 'ladder' amplitudes,

$$A(s', q_1^2; k^2, (k - q_1)^2) \sim g_a(q_1^2) \cdot \xi_{\alpha_1}(s')^{\alpha_1} \cdot \tilde{g}_1(q_1^2; k^2, (k - q_1)^2),$$
$$A(s'', q_2^2; k^2, (k + q_3)^2) \sim g_b(q_2^2) \cdot \xi_{\alpha_2}(s'')^{\alpha_2} \cdot \tilde{g}_2(q_2^2; k^2, (k + q_2)^2).$$

Here $\alpha_1 = \alpha_1(q_1^2)$ and $\alpha_2 = \alpha_2(q_2^2)$ are trajectories of the two reggeons, g_a, g_b are the standard reggeon–particle vertices, and the vertices $\tilde{g}$ contain the dependence on the virtualities of the participating particles. The answer has the form

$$A_{2 \to 3}(s, \mathbf{k}_{1\perp}, \mathbf{k}_{2\perp}, \eta_2) = g_1(q_1^2)g_2(q_2^2)\, \xi_{\alpha_1}\xi_{\alpha_2}\, m_{2\perp}^2 \cdot s_{12}^{\alpha_1} s_{23}^{\alpha_2} \times \gamma,$$
$$\gamma = \gamma(\mathbf{k}_{1\perp}, \mathbf{k}_{2\perp}) = \int \frac{d^2\mathbf{k}_\perp}{2(2\pi)^2} \int \frac{dx\,dy}{(2\pi)^2 i} \frac{x^{\alpha_1} y^{\alpha_2}}{(1)(2)(3)} \tilde{g}_1 \tilde{g}_2; \qquad (10.26a)$$

the propagators (10.25) in terms of the rescaled variables (10.24) read

$$\begin{aligned} (1) &= m^2 + (\mathbf{k} + \mathbf{k}_1)_\perp^2 + x(y - 1) \cdot m_{2\perp}^2 - i\varepsilon, \\ (2) &= m^2 + \mathbf{k}_\perp^2 + xy \cdot m_{2\perp}^2 - i\varepsilon, \qquad (10.26b) \\ (3) &= m^2 + (\mathbf{k} - \mathbf{k}_3)_\perp^2 + (x - 1)y \cdot m_{2\perp}^2 - i\varepsilon. \end{aligned}$$

The concrete form of the function γ depends on details of the interaction. Importantly, it is a function of the transverse momenta and is independent of the energy invariants. Therefore we can look upon γ as a new reggeon–reggeon–particle vertex.

Multiregge amplitudes $2 \to 2 + n$. The generalization of (10.26) to the case of many particles separated by large rapidity gaps is straightforward. For the amplitude $A_{2 \to 2 + n}$ in the multi-regge kinematics (10.21) we can write (modulo the complex phase factor)

$$
\begin{array}{c}
a \underset{q_1}{\overline{}} k_0 \\
\underset{q_2}{\vdots} \underset{}{\overline{}} k_1 \\
\vdots \\
\underset{q_{n+1}}{\overline{}} k_n \\
b \underset{}{\overline{}} k_{n+1}
\end{array}
\sim g_a(q_1^2) s_{01}^{\alpha_1} \gamma_{12} s_{12}^{\alpha_2} \cdots \gamma_{n,n+1} s_{n,n+1}^{\alpha_{n+1}} g_b(q_{n+1}^2), \qquad (10.27)
$$

where each subscript in the trajectory α_i and the vertex $\gamma_{i,i+1}$ marks the dependence on the corresponding transferred transverse momentum, $\alpha_i = \alpha(-q_{i\perp}^2)$, and $\gamma_{i,i+1} = \gamma(\mathbf{q}_{i\perp}, \mathbf{q}_{i+1\perp})$.

Contribution to σ_3 from the multiregge kinematics. Let us estimate the $2 \to 3$ cross section in the kinematical region (10.23). To begin with, the three-particle phase space volume is

$$
d\Gamma_3 = \frac{d^4 k_1}{(2\pi)^4} \frac{d^4 k_3}{(2\pi)^4} \prod_{i=1}^{3} [2\pi \delta_+ (m^2 - k_i^2)]
$$

$$
= \frac{d\beta_1}{\beta_1} \frac{d\alpha_3}{\alpha_3} \frac{d^2 \mathbf{k}_{1\perp}}{2(2\pi)^3} \frac{d^2 \mathbf{k}_{3\perp}}{2(2\pi)^3} 2\pi \delta_+ (m^2 + \mathbf{k}_{2\perp}^2 - \alpha_2 \beta_2 s) \qquad (10.28)
$$

$$
\simeq \frac{d\eta}{s} \frac{d^2 \mathbf{k}_{1\perp} d^2 \mathbf{k}_{3\perp}}{4(2\pi)^5}; \qquad d\eta = \frac{d\beta_2}{\beta_2}.
$$

Integrating the multiregge amplitude squared (10.26),

$$
\sigma_3 = \frac{1}{J} \int d\Gamma_3 |A_{2 \to 3}|^2,
$$

and omitting the constant normalization factor, we have

$$
\sigma_3 \sim s^{2(\alpha(0)-1)} \int d\eta \int d\mathbf{k}_{1\perp}^2 \, e^{-2\alpha'(\xi - \eta)\mathbf{k}_{1\perp}^2} \int d\mathbf{k}_{3\perp}^2 \, e^{-2\alpha' \eta \mathbf{k}_{3\perp}^2}
$$

$$
\sim \int_{\eta_0}^{\xi - \eta_0} \frac{d\eta}{\eta (\xi - \eta)} \sim \frac{\ln \xi}{\xi} \simeq \frac{\ln \ln s}{\ln s}; \qquad 1 \lesssim \eta_0 \ll \xi. \qquad (10.29)
$$

Although this estimate is valid only in an academic limit $\xi \gg 1$, it demonstrates that, at least formally, σ_3 is enhanced as compared to the elastic contribution σ_2. The origin of this slight enhancement is a broad integration over rapidity of the particle k_2 from the plateau region.

10.2.4 Multiplicity fluctuations and s-channel unitarity

The study of small multiplicities, $n \ll \bar{n}$, is a test of the Pomeranchuk pole hypothesis. If multiplicity fluctuations are weak and the pomeron is 'resistant' to the s-channel unitarity, we have a self-consistence theory at our disposal. If, however, fluctuations are strong and contribute 'too much' to σ_{tot}, then we have either to abandon altogether the initial idea that the pole is the rightmost singularity in the j plane, or to review the concept of **P** being an *isolated* singularity.

So, having started from the vacuum pole which from the s-channel point of view corresponds to an uniform particle distribution in rapidity,

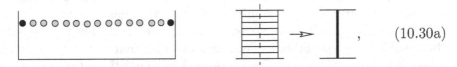

$$(10.30a)$$

we are driven by the unitarity relation in the s-channel to accept the existence of the processes that are difficult to accommodate into the pure Regge pole model. Thus, fluctuations in multi-particle production with large gaps in rapidity have led us to *reggeize* the amplitude by sending a reggeon between the two neighbouring particles with a large pair energy,

$$(10.30b)$$

This resulted in the cross section graphs containing the two poles which *coexist*, from the t-channel viewpoint. (Thick lines mark a cut-through **P**.)

By broadening the gap we eventually arrive at the elastic scattering

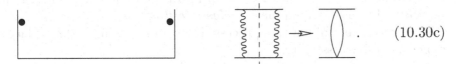

$$(10.30c)$$

As we have seen above, small multiplicity fluctuations contribute significantly to σ_{tot}; their contributions are suppressed only *logarithmically* in s, while we would expect a *power* smallness if there were only poles in the complex angular momentum plane.

The graphs with two 'parallel' reggeons, (10.30b) and (10.30c), make one think of *branch-cut singularities*, by an analogy with threshold branch cuts for usual particles. Another argument in favour of reggeon loops comes from the 'opposite side' – high multiplicity fluctuations. Indeed, we can imagine two multiperipheral cascades which will give rise to doubled

plateau density,

$$ \tag{10.30d} $$

In the cross section this corresponds again to a picture like (10.30b) but this time with all the pomerons being *cut*, $\mathbf{P} \to 2\,\mathrm{Im}\,\mathbf{P}$.

You may often hear people saying that the picture of multiple hadron production is very similar to statistics of a fluctuating gas. Literally, this analogy is wrong. The probability of finding in a gas just two molecules is exponentially small; in our case it *cannot* be small due to the optical theorem! It is the special rôle of elastic scattering that forces us to separate quasi-diffractive (fragmentation) processes from multi-particle production with $n \sim \bar{n}$. It is the latter for which the gas analogy works.

There are certain fluctuations that signal a catastrophic instability of our system. Let us discuss one particular very important fluctuation,

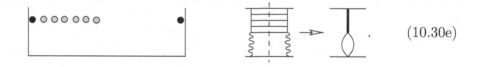

$$ \tag{10.30e} $$

10.2.5 Inelastic diffraction: triple-pomeron limit

This is the process in which one of the colliding particles scatters elastically, while the second one breaks up into a many-particle state. It may be referred to as the high-mass *inelastic diffraction*.

The reason for calling this process 'important' is, in the first place, its experimental accessibility.

It is not simple to investigate hadron collision events with rapidity gaps experimentally. One has to measure many particles (or rather their absence) to make sure that the event contains indeed rapidity gap(s). The process (10.30e) offers a much simpler option. Indeed, it is sufficient to register the leading scattered particle with sufficiently large energy, close to the initial one. Then the energy conservation law will ensure that there are no other energetic particles in the event, and we will get a gap in rapidity. Thus this process is a particular case of inclusive measurement with triggering of the particle from the very top, close to the kinematical boundary.

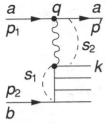

Let us look into kinematics. We choose for clarity the laboratory frame of the target, $\mathbf{p}_2 = 0$. Then

$$s = (p_1 + p_2)^2 \simeq 2mp_{10} \gg m^2,$$

$$s_1 = (q + p_2)^2 \simeq 2p_2q = 2mq_0 \gg m^2. \tag{10.31}$$

The invariant mass of the diffractive hadron system, $M^2 = s_1$, is determined by the energy transferred from the projectile to excite the target,

$$s_1 = \frac{q_0}{p_{10}} s \equiv (1 - x)s, \tag{10.32a}$$

where $x = p_0'/p_{10}$ is the energy fraction preserved by the scattered particle a. Practically the whole energy q_0 transferred to the target fragmentation block by the reggeon is carried by the fastest particle, the one on the top of the 'ladder', $k_0 \simeq q_0$. This allows us to evaluate the invariant energy s_2 corresponding to the size of the gap:

$$s_2 = (p' + k)^2 \simeq 2(p'k) \simeq p_0 \cdot \frac{m^2 + \mathbf{k}_\perp^2}{k_0} \simeq m_\perp^2 \frac{p_0}{q_0} = \frac{m_\perp^2}{1 - x}. \tag{10.32b}$$

By choosing x in the interval

$$\frac{m^2}{s} \ll 1 - x \ll 1 \tag{10.33}$$

we have the multiregge kinematics with s_1 and s_2 being both large, and

$$q^2 = -\frac{\mathbf{p}_\perp'^2}{x} - \frac{m^2(1 - x)^2}{x} \simeq -\mathbf{p}_\perp'^2.$$

We square the amplitude and, replacing the target fragmentation block summed over all possible hadron states by $2\,\mathrm{Im}\,\mathbf{P}$, arrive at the graph shown in Fig. 10.5. This is called the *triple-reggeon limit*, or $3\mathbf{P}$ since all three reggeons are pomerons in our case.

Substituting the pomeron trajectory, $\alpha_\mathbf{P}(q^2) = 1 + \alpha'q^2$, and integrating over q^2 and x we get the probability of these fluctuations *increasing* with the energy:

$$\frac{\sigma_{3\mathbf{P}}}{\sigma_{\mathrm{tot}}} \propto \ln\ln\frac{s}{m^2}. \tag{10.34}$$

We arrive at a contradiction: we supposed that σ_{tot} is asymptotically constant, and found a *part* of it that grows infinitely.

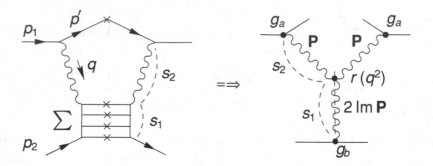

Fig. 10.5   Triple-regge limit for high-mass inelastic diffraction.

Let us calculate the contribution of the diagram of Fig. 10.5 to the total cross section. We have

$$\sigma_{3P} = \frac{1}{J} \int \frac{d^3\mathbf{p}'}{2p'_0(2\pi)^3} g_a^2(q^2) s_2^{2\alpha(q^2)} r(q^2) 2s_1^{\alpha(0)} g_b(0), \quad J = 2s, \quad (10.35a)$$

where g_a, g_b are couplings of $\mathbf{P}$ to the incoming particles 'a' and 'b', and r is a new three-reggeon vertex function. We did not write the signature factors $|\xi_\alpha|^2$, because at small q^2 which dominate the integral, $\xi_{\mathbf{P}} = i + \cot \frac{1}{2}\pi\alpha(q^2) \simeq i$.

Invoking (10.32) we absorb the factor $m_\perp^2$ from (10.32b) into redefining the vertex r to write

$$\begin{aligned}
\sigma_{3P} &= \frac{g_b(0)s^{\alpha(0)-1}}{16\pi^2} \int \frac{dx}{x} \int \frac{d^2\mathbf{q}_\perp}{\pi} g_a^2(q^2) \cdot r(q^2) \cdot (1-x)^{\alpha(0)-2\alpha(q^2)} \\
&\simeq \frac{\sigma_{\text{tot}}}{16\pi^2 \, g_a(0)} \int \frac{dx}{1-x} \int d\mathbf{q}_\perp^2 (1-x)^{2\alpha'\mathbf{q}_\perp^2} g_a^2(q^2) \cdot r(q^2) \\
&= \frac{\sigma_{\text{tot}}}{32\pi^2} \frac{g_a(0)}{\alpha'} \cdot r(0) \int \frac{dy}{y}, \qquad y = \ln\frac{1}{1-x}.
\end{aligned} \quad (10.35b)$$

Here we have extracted from under the integral the values of vertices g_a and r at $q^2 = -\mathbf{q}_\perp^2 = 0$, since the essential transverse momenta are small, due to the shrinkage of the diffractive cone, $\langle \mathbf{q}_\perp^2 \rangle \sim (\alpha'y)^{-1} \ll m^2$, at large y values. The integration over x in the multi-regge region (10.33) gives the catastrophic result announced above in (10.34):

$$\frac{\sigma_{3P}}{\sigma_{\text{tot}}} = r(0) \cdot \frac{g_a(0)}{32\pi^2\alpha'} \cdot \ln\ln\frac{s}{m^2}, \qquad \ln\frac{s}{m^2} \gg 1. \quad (10.36)$$

How can the apparent contradiction, $\sigma_{3P} > \sigma_{\text{tot}}$ in the $s \to \infty$ limit be handled? Let us examine the region of x in which the dangerous contribution has been accumulated. Actual integration limits are

$$s_1 = (1-x)s > \Lambda m^2 \gg m^2, \quad s_2 \sim \frac{m^2}{1-x} > \Lambda m^2 \gg m^2,$$

where Λ is a large number. First of all, in order to apply the reggeon amplitudes to both the top and the bottom parts of the diagram, we need to take at least $\Lambda \gtrsim 10$ which corresponds to stepping by about 2 units in rapidity from the projectile and the target (we know that only then the *plateau* starts emerging). So, we have to 'renormalize' the argument of the logarithm; conservatively,

$$\ln\frac{s}{m^2} \implies \ln\frac{s}{100\,m^2}.$$

Secondly, when we evaluated the integral over $\mathbf{q}_\perp$ we have ignored the size proper of the proton, embedded in $g_a(q^2)$. Hence, another modification:

$$\ln\ln s \implies \ln\left(\frac{R^2}{\alpha'} + \ln s\right), \quad \frac{R^2}{\alpha'_{\mathbf{P}}} \sim 4.$$

Moreover, the factor before $\ln\ln$ in (10.36) is numerically small. At present energies, the inelastic diffraction constitutes less than 10% of σ_{tot}. And $\ln\ln$ is hardly a function: it is indistinguishable from a constant. A real contradiction may appear only at fantastically high energies; in fact we have only $\ln\ln s = 5$ when s equals the mass squared of the Universe!

However, in order to accept this as an excuse we need to have a theory in which the mass of the Universe enters and solves this $\ln\ln$ phenomenon, which option is likely to belong to the domain of science fiction.

Therefore we must view this apparent contradiction as a serious fault of the theory we are constructing.

Fortunately, there is a more practical way to resolve the problem. In the three-reggeon diagram we have, in fact, introduced a new notion, that of the *reggeon interaction vertex* $r(q^2)$. We have supposed that the value $r(0)$ is finite. Further, we shall see that the vanishing of reggeon interaction vertices at $\mathbf{q}_\perp = 0$ in one of the possible solutions of the problem of taming multiplicity fluctuations.

In Lecture 15 we will return to the multiplicity fluctuation pattern and will discuss physical reasons for inelastic processes to *vanish* in the forward direction, $\mathbf{q}_\perp = 0$.

10.2.6 Multiparticle production with large rapidity gaps

High-mass inelastic diffraction is not the only multiplicity fluctuation going wild. Another example, due to K. Ter-Martirosyan, builds up on the multi-regge amplitudes (10.27) that describe the production of many particles with large rapidity gaps between them.

$$\propto g_a s_{01}^{\alpha_1} \gamma_{12} s_{12}^{\alpha_2} \cdots \gamma_{n,n+1} s_{n,n+1}^{\alpha_{n+1}} g_b.$$

The cross section $\sigma_{2 \to 2+n}$ can be calculated in the same way as for ordinary multiperipheral ladders. There is one essential difference: the Regge-dependence of the ladder cell amplitude on the pair energy has to be taken into account

$$M_i \sim \xi_{\alpha_i} s_{i-1,i}^{\alpha(q_i^2)} \sim s_{i-1,i}^{\alpha(0)} e^{-\alpha' \mathbf{q}_{i\perp}^2 y_i}; \quad y_i = \ln \frac{s_{i-1,i}}{m^2}, \ i = 1, \ldots, n+1. \tag{10.37}$$

Here y_i is the relative rapidity of two particles $i-1$ and i, which we will treat as large, $y_i \gg 1$. The cross section contains $n+1$ integrals over transverse momenta $\mathbf{q}_i$, $i = 1, \ldots, n+1$. These integrals are cut from above by Regge radii at sufficiently high energies $s_{i,i+1}$:

$$\mathbf{q}_{i\perp}^2 \lesssim \frac{1}{\alpha' \ln s_{i,i+1}} = \frac{1}{\alpha' y} \ll m^2, \tag{10.38}$$

therefore we put $\mathbf{q}_{i\perp} = 0$ in the vertices, g and $\gamma_{i,i+1}$, as we have done before in a number of occasions. Squaring the matrix element (10.37) and integrating over $\mathbf{q}_i$, we get the product of inverse relative rapidities. We are left with n integrations over ordered rapidities of final state particles k_i, $i = 1, \ldots, n$, which we represent as $n+1$ independent integrals over rapidity *differences*, y_i, satisfying the kinematical relation (10.21):

$$\frac{\delta\sigma_n}{\sigma_{\text{tot}}} \sim \frac{\gamma^n(0,0)}{(2\alpha')^{n+1}} \int_{\ln \Lambda}^{\xi} \frac{dy_1 \, dy_2 \cdots dy_{n+1}}{y_1 y_2 \cdots y_{n+1}} \, \delta\left(\sum_{i=1}^{n+1} y_i - \xi\right). \tag{10.39}$$

Here we have combined the product $\prod s_{i-1,i}^{\alpha(0)} = s^{\alpha(0)}$ and the vertices $g_{a,b}(0)$ into the total cross section and put $\ln \Lambda$ for the lower limits of rapidity integrals to justify the usage of the multi-reggeon approximation (see the discussion above). As far as the formal $s \to \infty$ asymptote is concerned, the leading contribution comes from the configurations when one of variables is much larger than the rest, $y_{(k)} \simeq \xi \gg y_i$, $i \neq k$, and we derive the answer for the fraction of the cross section due to the production of a sparse n-particle final state in the multi-regge kinematics:

$$\frac{\delta\sigma_n}{\sigma_{\text{tot}}} \simeq \text{const} \cdot (n+1)\frac{(\ln\xi)^n}{\xi}. \tag{10.40a}$$

(Importantly, there is no $n!$ in the denominator.) Our derivation implied

$$\ln\langle y_i \rangle \simeq \ln\frac{\xi}{n} \simeq \ln\frac{\bar{n}}{n} \gg \frac{R^2}{\alpha'} \simeq 4,$$

so we better 'soften' our result (10.40a) as follows,

$$\frac{\delta\sigma_n}{\sigma_{\text{tot}}} \simeq \text{const} \cdot \frac{n+1}{\xi}\left(\ln\frac{\xi}{n}\right)^n. \tag{10.40b}$$

Still, the formal contradiction is there. Let us demonstrate that at $s \to \infty$ we can always find such a 'sparse ladder' the cross section of which grows arbitrarily large. Expressing n as a fraction of the mean multiplicity, $n \equiv \xi/F$, $\ln F \gg 1$, and keeping F fixed, we obtain

$$\frac{\delta\sigma_{\beta\langle n\rangle}}{\sigma_{\text{tot}}} \sim \frac{\text{const}}{F}(\ln F)^{\xi/F} \propto s^{(\ln\ln F)/F},$$

that is, the fraction of events increasing as a (very small but finite) *power* of energy! Again, we arrive at a contradiction with the pomeron pole hypothesis: multiplicity fluctuations tend to ruin it (unless in (10.39) the pomeron–pomeron interaction vertex vanishes in the origin, $\gamma(0,0) = 0$).

10.3 Reggeon branch cuts and their rôle

What is the composition of the total cross section in our modified picture?

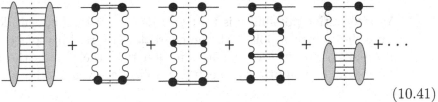

$$\tag{10.41}$$

In addition to the multiperipheral 'ladders' with $n \sim \bar{n}(s)$, we have the contribution of the (quasi)elastic scattering, as well as a series of terms describing multiparticle production with large rapidity gaps between (groups of) hadrons.

We have introduced the pomeron pole as the *rightmost singularity* in the vacuum channel. This implied that the total cross section was asymptotically given, with *power accuracy*, by the imaginary part of the pomeron amplitude,

$$\sigma_{\text{tot}} \to \operatorname{Im} A^{\mathbf{P}}(s,0) = \frac{1}{2} \;\; \times \left(1 + \mathcal{O}\!\left(\frac{1}{s^a}\right)\right), \quad s \to \infty. \quad (10.42a)$$

From the perturbation theory we learned that the 's-channel content' of the pomeron pole corresponds to the first term of (10.41),

$$\quad \simeq \quad . \qquad (10.42b)$$

Where in (10.42a) are then the *logarithmically behaving* 'fluctuation' terms that are present in (10.41)? To add an insult to injury, according to the logic that we have advocated before, the Regge pole must accommodate the structures that are *dynamically bound* in the *t*-channel; but the just-mentioned missing terms hardly satisfy this criterion, especially the first – elastic – one. There are two possible escapes:

(1) σ_{tot} in (10.41) differs from the multiperipheral model (10.42b) by a sum of positive terms. If we were to rescue the pure Regge pole approximation, we could imagine that the prescription (10.42b) was simply inaccurate, and in fact

$$A^{\mathbf{P}}_{\text{exact}} = A^{\mathbf{P}}_{\text{ladder}} - \Delta,$$

with Δ a *small correction* compensating the unwanted contributions in the optical theorem (10.41).

(2) An alternative possibility is that we made a mistake not in identifying the pomeron pole with the characteristic rapidity plateau as in (10.42b) but in the very assumption that the pole is the only significant contributor to the full amplitude, even in the $s \to \infty$ limit:

$$A_{\text{exact}} = A^{\mathbf{P}} - \Delta; \quad \frac{\Delta}{A^{\mathbf{P}}} \sim \frac{1}{\ln s}.$$

The second option – the only viable one, as it turns out – implies that the pomeron *is not* the leading singularity; more accurately, is not an *isolated* leading singularity. To legalize the existence of the corrections to $\operatorname{Im} \mathbf{P}$ that are suppressed only *logarithmically* in s in the unitarity condition, there must exist other singularities in the j-plane, weaker than the pole, positioned at the same point $j = 1$ at $t = 0$.

10.3.1 Enter reggeon branchings

Two-pomeron correction to the elastic amplitude. Let us examine the first correction to the elastic scattering amplitude due to s-channel iteration of the pomeron exchange:

$$
2\,\mathrm{Im}\quad\bigg\{ \quad \bigg\} \quad = \quad \begin{matrix} p_1 \rightarrow \\ q_1 - \tfrac{1}{2}q \\ p_2 \rightarrow \end{matrix} \bigg\{ \begin{matrix} \times \\ \\ \times \end{matrix} \bigg\} \begin{matrix} p_1 + q \\ q_1 + \tfrac{1}{2}q \\ \end{matrix}
$$

$$
= \int \frac{d^4 q_1}{(2\pi)^4} g^2(t_1) g^2(t_2) \xi_{\alpha(t_1)} s^{\alpha(t_1)} \cdot \xi^*_{\alpha(t_2)} s^{\alpha(t_2)}
$$

$$
\cdot\, 2\pi\delta((p_1 - q_1)^2 - m^2) 2\pi\delta((p_2 + q_1)^2 - m^2),
$$

$$\tag{10.43}$$

where

$$
t_1 \simeq -(\mathbf{q}_1 - \tfrac{1}{2}\mathbf{q})_\perp^2, \quad t_2 \simeq -(\mathbf{q}_1 + \tfrac{1}{2}\mathbf{q})_\perp^2.
$$

A simple calculation yields

$$
\bigg\{ \quad \bigg\} \simeq s^{2\alpha(t/4)-1} g^4 |\xi_\alpha|^2 \int \frac{d^2 \mathbf{q}_{1\perp}}{2(2\pi)^2} \exp\{-2\alpha' \mathbf{q}_{1\perp}^2 \cdot \ln s\}
$$

$$
\simeq g^4(t/4) \big|\xi_{\alpha(t/4)}\big|^2 \frac{s^{2\alpha(t/4)-1}}{16\pi\, \alpha' \ln s}.
$$

$$\tag{10.44}$$

Here we substituted $t_1 = t_2 = t/4$ in the pre-exponential factors since the transverse momentum integral at large $\xi = \ln s$ selects small $\mathbf{q}_{1\perp}$. Taking the forward scattering amplitude, $t \simeq -\mathbf{q}_\perp^2 = 0$, we recover the correction σ_2 that we have calculated above in (10.20). The energy exponent of (10.44) corresponds to a new singularity whose trajectory is

$$
j_2 = 2\alpha(t/4) - 1 \simeq 1 + \tfrac{1}{2}\alpha' t. \tag{10.45}
$$

It has a twice-smaller slope than the pomeron; at $t = 0$ the position of this new singularity coincides with that of the pomeron. The pre-exponential factor $(\ln s)^{-1}$ in (10.44) shows that this is not a pole but a *branch cut* in the angular momentum plane.

Multi-reggeon moving branch point singularities. In the next lecture we will demonstrate that the necessity of branch cuts in the complex angular momentum plane follows directly from the t-channel. They are driven by

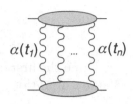

Regge poles. Once a Regge pole $\alpha(t)$ is introduced into the theory, it generates a series of moving branch-point singularities with trajectories

$$j_n(t) = n\,\alpha\left(\frac{t}{n^2}\right) - n + 1, \qquad (10.46)$$

the expression generalizing (10.45) to all n and originating from the exchange of n reggeons in the t-channel.

Historical remark, and a lesson. The question of branch-point singularities (branchings) has a curious history. First, people thought that such angular momentum singularities cannot exist, which is the case in the non-relativistic quantum mechanics. Then it was understood that they must be present. Searching for a model for branching, a diagram of Fig. 10.6(a) was suggested based on the perturbative picture of a pole as a ladder.

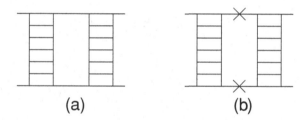

Fig. 10.6   The diagram (a) falls with s much faster than its imaginary part (b).

It was soon realized, however, that with s increasing this diagram is *falling fast*, as a power of s. This looks puzzling, since we have just seen above that its *imaginary part* shown in Fig. 10.6(b) is rather large and models the two-reggeon branching very well indeed!

The point is, the diagram (a) has *many cuts* in s, many 'imaginary parts', so to say, while we have selected in (b) one specific cut. In the full sum of all possible discontinuities of the diagram (a) different 'imaginary parts' cancel, making the picture with parallel ladders a wrong model for the branching. We will discuss this issue in detail in Lecture 12 where we will understand the physical reason behind this cancellation and construct the true s-channel image of the t-channel branch-point singularity.

But already at this stage there is an important message to take on board. When we have been discussing multiplicity fluctuations, we cut the diagrams for σ_{tot} in (10.41) 'in the middle', and avoided cutting through reggeons (where possible). At the same time, the reggeon amplitudes have a non-trivial complexity themselves. In the complete theory we must take these alternative discontinuities into full account. We had already a hint in this direction when we saw how different cuts of the same graph with a pomeron loop produced a rapidity-gap event, (10.30b), and a double multiplicity fluctuation, (10.30d).

10.3.2 Branchings in non-vacuum and vacuum channels

Let us look at the behaviour of (10.46) at large n and fixed t,

$$j_n(t) \simeq (n-1)(\alpha(0) - 1) + \alpha(0), \quad n \to \infty.$$

There are three qualitatively different patterns.

- $\alpha(0) < 1$. High order branchings move to the left and become insignificant.

- $\alpha(0) > 1$. No-go: $j_n \to +\infty$ violates analyticity/causality: the corresponding $A(s)$ would grow faster than any power of energy.[†]

- $\alpha(0) = 1$. In this exceptional case branch cuts accumulate at $j = 1$ and are all important. (From the consideration of s-channel phenomena, we already learnt that they have to be.)

The first case applies to all Regge trajectories but **P**.

A remarkable link between t-channel resonances and the asymptotic energy behaviour of corresponding scattering amplitudes in the s-channel stay intact after we take into account reggeon branchings.

For example, at large s the charge exchange reaction $\pi^- p \to \pi^0 n$ is dominated by the ρ Regge pole having trajectory $\beta(0) \simeq 0.5$. Repeating such a reggeon in the t channel produces a $1/\sqrt{s}$ suppressed correction:

$$j_2(0) - j_1(0) = (2\beta(0) - 1) - \beta(0) = -(1 - \beta(0)) \simeq 0.5.$$

Another possibility is to send a pomeron in parallel to the ρ pole. In this case we get

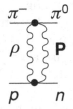

$$j_2(0) = \alpha(0) + \beta(0) - 1 = \beta(0).$$

The power falloff of the amplitude remains the same. However, the scattering angle dependence at small $t < 0$ will be affected by branching corrections.

[†] Dynamics of t-channel branch cuts *almost* contains the Froissart theorem!

Fig. 10.7   Relative position of the pomeron pole and branchings in j-plane.

Return to the most interesting vacuum channel case, $\alpha(0) = 1$. Let us draw what happens in the j-plane (Fig. 10.7). If $t > 0$, branchings are on the left from the pole and accumulate at $j = 1$. In the physical region of the s-channel, $t < 0$, the picture looks dramatic: the pole is no longer the rightmost singularity. This means that pomeron branchings are likely to seriously modify the t-dependence of the scattering amplitude.

As for the total cross section, here a difficult story starts. Prior to addressing the problem we must develop adequate means first. To connect colliding particles by parallel reggeons as we did before is not enough. We need to learn how reggeons *interact* among themselves.

This is exactly what happens, from the t-channel point of view, in high-mass inelastic diffraction and multi-gap events, in particular, those very fluctuations that we found most damaging for self-consistency of the pomeron picture:

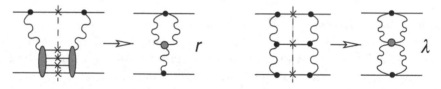

One possibility is that branchings are significant to such an extent that they turn out to play the *dominant* rôle, changing the energy behaviour of the cross section, of the plateau density, etc. (the so-called 'strong coupling' regime).

Another possible scenario preserves the asymptotic constancy of σ_{tot}. One can construct a self-consistent theory if all effective reggeon inter-action vertices (r, λ, etc.) vanish when the transverse momenta flowing through participating reggeons tend to zero. In this ('weak coupling') so-lution pomeron branchings, and multiplicity fluctuations along with them, are kept under control as corrections. A recently found unexpected con-sequence of this theory runs as follows: if total cross sections are asymp-totically constant, they must tend to *one and the same* constant for all scattering processes!

But first we have to return to the t-channel and to complex angular momenta.

11

Reggeon branchings

We considered strong interactions in the framework of the Regge poles; this picture applies, by the way, to weak and electromagnetic interactions as well.

It turns out, however, that the Regge poles are not enough to describe consistently the high-energy behaviour of scattering amplitudes. To see why and how more complicated singularities – branch cuts – emerge in the ℓ-plane, let us recall, what we learned about the possible energy dependence of two-particle scattering amplitudes at various t.

At $t > 0$ the elastic amplitude may grow as a power $s^{\alpha(t)}$ with $\alpha(t) > 1$. What do we know about the vacuum trajectory $\alpha(t)$? Below the t-channel threshold, $t = 4\mu^2$, the trajectory is real and decreases with t decreasing. Moreover, the Froissart theorem taught us that at $t = 0$ the ampitude cannot grow with energy faster than

$$A(s, t = 0) = \quad \text{———}\bigcirc\text{———} \quad < s \ln^2 s.$$

What happens at negative t? In non-relativistic quantum mechanics $\alpha(t)$ passes through integer points $-1, -2, \ldots$ and decreases indefinitely with t decreasing. Thus, in quantum mechanics we see no restrictions on the rate of the energy falloff of the amplitude: for large angles, $|t| \sim s$, it can fall arbitrarily fast with s increasing.

How about relativistic theory?

Recall that in fact we have already faced some difficulty with the Regge-pole picture. Indeed, we expected that the corrections to the leading vacuum pole (pomeron, $\mathbf{P}$) will be coming from other Regge poles and will be relatively suppressed as a *power* of s. However, having analysed contributions to the total cross section of various particle production topologies, we found in the r.h.s. of the equation for $\mathrm{Im}\,\mathbf{P}$ *logarithmically* behaving

287

contributions due to production processes of a few particles with large rapidity intervals between them:

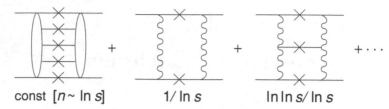

One cannot state a priori that these logarithmic terms will not sum up into a power. Still, this discrepancy is worrisome and worth bearing in mind.

Strangely enough, the apparent logarithmic nature of the corrections and the question of the possible falloff of the elastic amplitude turned out to be closely related.

11.1 $\ell = -1$ and restriction on the amplitude falloff with energy

Let us return to the partial wave amplitude with positive signature:

$$\varphi_\ell(t) = \frac{2}{\pi} \int_{z_0}^{\infty} \frac{dz}{(t - 4\mu^2)^\ell} Q_\ell(z) A_1(z, t).$$

We found simple analytic properties in t for $\varphi_\ell(t)$ at $t > 4\mu^2$; when $t < 0$, an additional complexity appeared:

$$\Delta\varphi_\ell = -\int_{z_0}^{-1} \frac{dz}{(t - 4\mu^2)^\ell} P_\ell(z) A_1(z, t); \qquad t_1 < t < 0.$$

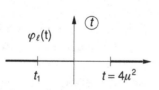

Moving further down in t we hit the third spectral function, where the 'imaginary part' A_1 became itself complex (see Fig. 7.3 on page 167). Then the expression for the discontinuity changed:

$$\Delta\varphi_\ell = -\int_{z_0}^{-1} \frac{dz}{(t-4\mu^2)^\ell} P_\ell(z) A_1^*(z, t) + \frac{2}{\pi} \int_{z_1}^{z_2} \frac{dz}{(t-4\mu^2)^\ell} Q_\ell(z) \rho_{su}.$$

$$(11.1)$$

11.1.1 Partial waves have poles at $\ell = -n$

At the first glance the additional term that appeared in the r.h.s. of (11.1) looks innocent. Indeed, although Q_ℓ has *poles* at negative integer $\ell = -n$,

$n > 0$, in the expression for the amplitude A,

$$A(s,t) = \frac{i}{4} \int \frac{d\ell}{\sin \pi\ell} \varphi_\ell(t) \cdot (t - 4\mu^2)^\ell [P_\ell(-z) + P_\ell(z)], \qquad (11.2)$$

the Legendre function P_ℓ has *zeros* in the same points,

$$P_\ell(z) \simeq \frac{\Gamma(\ell + \frac{1}{2})}{\sqrt{\pi}\Gamma(\ell + 1)}(2z)^\ell, \quad |z| \to \infty. \qquad (11.3)$$

In fact, if we chose the Mellin transformation instead of the Legendre one (with z^ℓ in place of P_ℓ), the poles would not appear at all. So these poles look to be artefacts of the passage from the amplitude $A(s,t)$ to the partial wave, $\varphi_\ell(t)$. It would have been the case if not for the t-channel unitarity: recall that the Legendre transformation was special in the sense that it diagonalized the *two-particle unitarity condition*.

11.1.2 Such poles contradict unitarity

In the interval between two- and four-pion thresholds, $4\mu^2 < t < 16\mu^2$, the partial wave is bounded from above by the unitarity condition:

$$\Delta\varphi_\ell = \rho_\ell \,\varphi_\ell\varphi_\ell^* \qquad \Longrightarrow \qquad |\varphi_\ell| < \frac{1}{2\rho_\ell}. \qquad (11.4)$$

How could its discontinuity then become infinite?

In order to appreciate that it is not easy for the amplitude to do so, let us turn to the dispersion relation

$$\varphi_\ell(t) = \frac{1}{\pi} \int_{4\mu^2}^{\infty} \frac{dt'}{t' - t} \operatorname{Im} \varphi_\ell(t') + \frac{1}{\pi} \int_{-\infty}^{t_0} \frac{dt'}{t' - t} \delta\varphi_\ell(t'). \qquad (11.5)$$

If $\delta\varphi_\ell \to \infty$ at some ℓ, the same is true for $\varphi_\ell(t)$ together with its discontinuity (unless there are special cancellations in the second integral in (11.5)).

Note that the contribution of the right cut (the first integral) is finite. Moreover, it could not possibly cancel the contribution of the left cut identically in t, since their analytic properties are different.

What could be a way out?

(1) In principle, the integral of $\rho(s, u)$ could turn to zero at $\ell = -n$. This is, however, not the case, since the double spectral function is positively definite, at least near the edge of the hyperbola.

(2) Taking into account multiparticle unitarity conditions would not help us either since, once again, the contribution coming from the

next threshold at $t = 16\mu^2$ has specific analytic features, and therefore it cannot cancel the pole identically, at arbitrary t, in the interval $4\mu^2 < t < 16\mu^2$.

Thus, the pole is there. Although the amplitude A itself is finite, the corresponding partial wave φ acquires the pole which would have been but an artefact of rewriting (P_ℓ in place of z^ℓ) if not for the *unitarity condition*.

How could we resolve this real contradiction with unitarity?

11.1.3 Condensing poles (quantum-mechanical analogy)

Let us recall the meaning of the singularities of the amplitude. As we have discussed in Lecture 7, singularities of t-channel particle exchange amplitudes determined the s-channel 'potential',

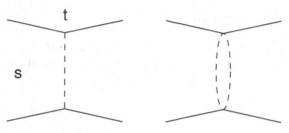

Correspondingly, singularities in s (and u)

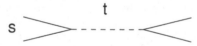

have to be understood as determining the interaction potential in the t-channel. The existence of the third spectral function ρ_{su} reflects the relativistic nature of the t-channel 'potential'. It is because of ρ_{su} that our 'potential' becomes infinitely large when $\ell \to -1$.

In NQM, potential scattering in the t-channel is described by the diagrams with successive particle exchange. The lines do not cross since the non-relativistic interaction is instantaneous; there is no retardation. There is only elastic scattering so at positive t we will have a single cut starting from $t = 4\mu^2$. The partial wave 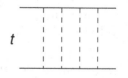 $f_\ell(t)$ will have a left cut too, with a rather complicated structure, reflecting multiple singularities of the amplitude in s, mimicking the 'potential'. However, the term with Q_ℓ in $\Delta\varphi_\ell$ will be absent since in non-relativistic quantum mechanics $\rho_{su} \equiv 0$, so that the pole in ℓ will be absent as well. In spite of this, we shall stick to NQM and try to imitate the catastrophic growth of the contribution of the left cut in the $\ell \to -1$ limit simply by increasing the magnitude of the non-relativistic potential. In so doing, the contradiction has to disappear, since a potential in quantum mechanics is arbitrary, and the answer must remain reasonable.

When V increases (attraction), sooner or later a level (bound state) appears from beneath the right cut. With the increase of the potential the number of levels and their binding energies are growing. In general, to the expression (11.5) for φ_ℓ a sum over bound states has to be added:

$$\varphi_\ell = \frac{1}{\pi} \int \operatorname{Im} \varphi_\ell + \frac{1}{\pi} \int \delta \varphi_\ell + \sum \frac{C_n}{t_n - t}.$$

It is this sum that compensates the growing contribution of the left cut ($\delta \varphi_\ell$) while the contribution of the right cut ($\operatorname{Im} \varphi_\ell$) and φ_ℓ itself stay finite. In the limit $V \to \infty$ the number of poles becomes infinite, and they eventually fill the t-axis below the threshold ($t < t_0$). Replacing the sum over poles by an integral,

$$\sum_n \frac{C_n}{t_n - t} \xrightarrow{V \to \infty} \int_{-\infty}^{t_0} \frac{\chi(t') \, dt'}{t' - t},$$

we will have, in the main part, $\chi(t') \approx -\delta \varphi(t')$ as $V \to \infty$.

This would have been the solution of the problem of infinitely large potential (which is our model for the singularity at $\ell \to -1$ of the left-cut contribution) if there were no *multiparticle thresholds* and related cuts.

11.1.4 Another possibility: moving branch points

In the relativistic theory there are inelastic sheets. Their existence provides another way out of the contradiction with unitarity. Instead of accumulating infinitely many poles, inelastic sheets may allow us to change the very analytic properties of the partial wave φ_ℓ instead. Indeed, we have arrived at the conclusion that the partial wave is bounded from above by equating in (11.4) its *discontinuity* $\Delta \varphi_\ell$ with the *imaginary part*, $\operatorname{Im} \varphi_\ell$:

$$\Delta_t \varphi_\ell(t) = \rho_\ell |\varphi_\ell(t)|^2; \tag{11.6}$$

$$\Delta_t \varphi_\ell(t) = \operatorname{Im} \varphi_\ell(t) \implies |\operatorname{Im} \varphi_\ell| \le |\varphi_\ell| < \text{const.}$$

Imagine now that from beneath a multiparticle sheet a *branch point* singularity emerged and moved to the left along the real axis. Then in the two-particle threshold region $4\mu^2 < t < 16\mu^2$ the discontinuity does not coincide with the imaginary part anymore, $\Delta \varphi \ne \operatorname{Im} \varphi$, and no restriction on the magnitude of φ_ℓ would follow from (11.6).

We shall explore this possibility later and will see that in the interesting cases it is moving branchings that resolve our contradiction. For the time being let us return to the first – pole – scenario: with $\ell \to -1$ we have more and more poles emerging on the physical sheet and passing through a given point t,

$$t_n(\ell) < t, \quad n = 1, 2, \ldots, N; \qquad N \to \infty \text{ with } \ell \to -1.$$

In the ℓ-plane this picture corresponds to the poles condensing towards $\ell = -1$ for any fixed t:

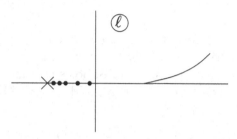

Thus, due to the unitarity condition, the *pole* in the partial wave φ_ℓ at $\ell = -1$ transforms into an *essential singularity*. I have demonstrated this, using quantum-mechanical analogy. Instead, one could just solve the integral equation for the unitarity condition (presuming that the inelastic cuts are not 'catastrophic') to come to the same conclusion rigourously.

From the fact that the point $\ell = -1$ is an essential singularity in the Sommerfeld–Watson integral,

$$A(s,t) \propto \int \frac{d\ell}{\sin \pi \ell} \varphi_\ell(t) \left[(-s)^\ell + (s)^\ell \right],$$

it immediately follows that for any t

$$|A(s,t)| > s^{-1-\varepsilon}, \quad \epsilon > 0. \tag{11.7}$$

We conclude that the amplitude $A(s,t)$ cannot decrease faster than $1/s$.

11.1.5 The origin of the $\ell = -1$ singularity

Let us have a closer look: how did it happen that $\ell = -1$ turned out to be a singular point and what was the rôle of the third spectral function ρ_{su} in this.

Consider the original amplitude $A(s,t)$ at negative t-values. Since I am interested in the possibility of A decreasing, I can write the dispersion relation without subtractions:

$$A(s,t) = \frac{1}{\pi} \int_{4\mu^2}^\infty \frac{ds'}{s'-s} A_1(s',t) + \frac{1}{\pi} \int_{4\mu^2}^\infty \frac{du'}{u'-u} A_2(u',t). \tag{11.8}$$

Let us consider a 'good' amplitude such that

$$A_1,\ A_2,\ A_3 < \frac{1}{s}.$$

For $t > t_1$ the imaginary part $\mathrm{Im}\,A$ was due to ($i\varepsilon$) in the denominators in (11.8): $\mathrm{Im}\,A = A_1$ for $s > 4\mu^2$ ($\mathrm{Im}\,A = A_2$ for $u > 4\mu^2$).

Decreasing t, we reach t_1 where A_1 and A_2 become complex and our amplitude seems to acquire an additional complexity

$$\delta\,\mathrm{Im}\,A(s,t) = \frac{1}{\pi}\int \frac{ds'}{s'-s}\rho_{su}(s',t) + \frac{1}{\pi}\int \frac{du'}{u'-u}\rho_{su}(u',t). \qquad (11.9)$$

How could this be? Isn't A_1, *by definition*, the full imaginary part of the amplitude in the physical region of the s-channel? There is no contradiction, of course, since $(s'-s) + (u'-u) = 0$.

At the same time we observe that, taken separately, neither of the contributions of the right and left cuts decreases faster than $1/s$:

$$\mathrm{Im}\,A_{\mathrm{right}} = \frac{1}{\pi}\int \frac{ds'}{s'-s}\rho_{su}(s') = -\frac{1}{\pi}\int \frac{ds'\,\rho_{su}(s')}{s} \sim \frac{1}{s}. \qquad (11.10)$$

Let us recall that we *were forced* to treat separately the contributions of the right and left cuts when we continued partial waves φ_ℓ onto the complex ℓ-plane. Hence, the nature of the pole in Q_ℓ is related to the $1/s$ falloff of the contribution of each cut.

The most important point remains to be understood. Namely, where did the essential singularity come from, when the pole itself seemed to be of "kinematical" nature? It is t-channel unitarity which is responsible.

Unitarity means that repetitions are needed. Let us take a scattering block $f = \ \diagdown\!\!\diagup\ $ (which does not include a two-particle intermediate state in the t channel) and see how the amplitude will behave when we start repeating it in the t-channel. We are going to demonstrate that the contributions of the right and left cuts of the block f enter separately, *one by one*, the asymptotics of the iterated amplitude.

Consider the high-energy limit of the amplitude

In terms of the Sudakov vectors,

$$k = \alpha p_1 + \beta p_2 + \mathbf{k}_\perp \qquad (p_+ \simeq p_1,\ p_- \simeq p_2),$$

we may write

$$A \sim \int \frac{d^4 k}{(2\pi)^4 i} \frac{f(s_1; k_\perp, q_\perp) f(s_2; k_\perp, q_\perp)}{[m^2 - \alpha\beta s + \mathbf{k}_\perp^2 - i\varepsilon][m^2 - \alpha\beta s + (\mathbf{q}-\mathbf{k})_\perp^2 - i\varepsilon]}. \tag{11.11}$$

For simplicity we omitted in the second propagator, $(\alpha - \alpha_q)(\beta - \beta_q)$, the small longitudinal Sudakov components of the total momentum transfer, $\alpha_q, \beta_q \sim m^2/s$; as always, we consider small momentum transfer $-q^2 \simeq \mathbf{q}_\perp^2 = \mathcal{O}(m^2)$. (In fact we have already analysed this integral when we calculated the box diagram in Section 9.2.3, see page 224.) The block energies are

$$s_1 \simeq 2p_1 k \simeq \beta s, \quad s_2 \simeq -2p_2 k \simeq -\alpha s.$$

Consider first the integral over α. In the α plane we have cuts of the lower block amplitude $f(s_2)$,

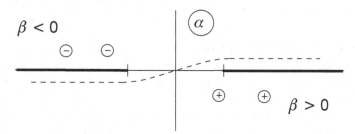

Depending on the sign of β, the poles of the propagators in α are either both below ($\beta > 0$) or above ($\beta < 0$) the real axis. If $\beta > 0$, the contour can be closed on the *left*, that is around the *right* cut in the invariant energy $s_2 = -\alpha s$ of the lower block:

$$A = \int \frac{d^2 \mathbf{k}_\perp}{(2\pi)^2} \int_0^\infty d\beta f(\beta s) \int_{s_0}^\infty \frac{ds_2}{2\pi} \frac{\text{Im} f_{\text{right}}(s_2)}{[\ ][\ ]} + \int_{\beta < 0} \{\text{Im} f_{\text{left}}(s_2)\}.$$

Each integral is complex due to a (right/left) cut of $f(\beta s)$. Evaluating the imaginary part we have

$$\text{Im} A = \int \frac{d^2 \mathbf{k}_\perp}{2\pi s} \int_{s_0}^\infty \frac{ds_1}{2\pi} \int_{s_0}^\infty \frac{ds_2}{2\pi} \frac{\text{Im} f(s_1)\,\text{Im} f(s_2)}{[\ ][\ ]} + \left\{ \begin{matrix} s_1 & \to -s_1 \\ s_2 & \to -s_2 \end{matrix} \right\}.$$

While the integrand of A in (11.11) contains the full block amplitudes, symbolically,

$$A \sim f \otimes f \sim (f_{\text{right}} + f_{\text{left}}) \otimes (f_{\text{right}} + f_{\text{left}}), \tag{11.12a}$$

in the imaginary part the cross-terms between right and left cuts cancel:

$$\text{Im } A \sim \frac{1}{s\,m^2} \left[\left(\int ds' \,\text{Im}\, f_{\text{right}}(s') \right)^2 + \left(\int ds' \,\text{Im}\, f_{\text{left}}(s') \right)^2 \right]. \quad (11.12\text{b})$$

We see that although each block amplitude in (11.12a) may be falling fast with energy due to cancellation between its two cuts, the imaginary part of the *iterated* amplitude (and therefore $A(s)$ itself) cannot decrease faster than $1/s$.

The expectation that an amplitude can fall arbitrarily fast owing to the cancellation between right- and left-cut contributions (each of which falls as $1/s$) turns out to be incorrect: it does not stand confrontation with the *t*-channel unitarity relation.

To understand better what is so special about the point $\ell = -1$ in the unitarity relation, let us return to the representation (11.2) for the amplitude and redefine, once more, the partial wave by embedding into it the ℓ-dependent normalization factor from the asymptotic expression for the Legendre function (11.3):

$$A(s,t) \simeq \frac{i}{4} \int \frac{d\ell}{\sin \pi \ell} \, \psi_\ell(t) \cdot \left[(-s)^\ell + s^\ell \right], \quad s \to \infty; \quad (11.13\text{a})$$

$$\psi_\ell(t) = \frac{\Gamma(\ell + \frac{1}{2})}{\sqrt{\pi}\,\Gamma(\ell+1)} \cdot \varphi_\ell(t). \quad (11.13\text{b})$$

Now that (11.13a) is 'clean', the normalization will reappear in the unitarity relation. Expressed in terms of new partial waves ψ_ℓ it now reads

$$\Delta\psi_\ell(t) = \rho_\ell(t) \cdot \psi_\ell(t)(\psi_\ell(t))^*, \quad 4\mu^2 < t < 16\mu^2,$$

$$\rho_\ell(t) = \frac{\sqrt{\pi}\,\Gamma(\ell+1)}{\Gamma(\ell + \frac{1}{2})} C_\ell(t), \quad C_\ell = \frac{1}{8\pi} \frac{k_t^{2\ell+1}}{\sqrt{t}}, \quad (11.14)$$

with C_ℓ given in (7.30). The factor ρ_ℓ can be looked upon as the phase space volume, continued to non-integer ℓ. We see that at negative integer values of ℓ it turns to infinity, making them singular points for the unitarity condition.

Conclusion: if there are only poles,

(1) the amplitude cannot fall faster than s^{-1}; and

(2) in the ℓ-plane the poles get closer and closer to each other as they approach $\ell = -1$.

This itself does not contradict the Regge-pole hypothesis, but in fact, as we will see now, the solution lies elsewhere.

11.2 Scattering of particles with non-zero spin

Up to now, we considered spinless particles; including spins does not modify essentially our previous considerations. The conclusion, however, turns out to be far more dramatic.

11.2.1 Energy behaviour of scattering amplitudes

Consider the scattering of a vector and a scalar particles:

$$= A_{\mu\nu}(p_1 + p_2, k_1, k_2).$$

The amplitude now carries vector indices and is built of Lorentz tensors,

$$A^{\mu\nu} = A_0 p^\mu p^\nu + A_1 g^{\mu\nu} + A_2(p^\mu k_1^\nu - k_2^\mu p_\nu) + \cdots \quad (p \equiv p_1 + p_2)$$

(we have taken into account the symmetry with respect to $k_1 \leftrightarrow -k_2$). Now we have to write dispersion relations for each invariant amplitude A_i (one has only to be careful not to include artificial singularities which may emerge while rewriting momenta in terms of each other). Repeating the above analysis, we would obtain again

$$A_0^{\text{right}} \sim A_0^{\text{left}} \sim s^{-1}.$$

Let us see, e.g. what gives the *longitudinal polarization* vector,

$$e_\mu^{(0)}(k) = (k_z; k_0, \mathbf{0}_\perp) \frac{1}{\sqrt{k^2}},$$

for the invariant matrix element

$$M^{\lambda_1 \lambda_2} = e_{\mu_1}^{\lambda_1} e_{\mu_2}^{\lambda_2} A^{\mu_1 \mu_2}.$$

For forward scattering, $k_1 = k_2$, in the laboratory frame ($p_1 = p_2 = (m_0; 0, \mathbf{0}_\perp)$) we obtain

$$M^{00} = (A_0 p^\mu p^\nu + \cdots) e_\nu^0 e_\mu^0 \sim A_0 \left(2m_0 \frac{k_{1z}}{m_1}\right)^2 \simeq A_0 \frac{s^2}{m_1^2}. \quad (11.15)$$

This means that considering, as before, contributions of each cut, M_{left}^{00} and M_{right}^{00}, separately, we will come to the conclusion that the scalar–vector scattering amplitude must *grow* in the large-s limit at least as

$M \gtrsim s!$ It is easy to arrive at the same conclusion analysing an interaction of spinless particles via vector particle exchange, $A = \mathcal{O}(s)$. That is, introducing particles with spin σ and following the same path, we can prove that the amplitude cannot be smaller than

$$A > s^{-1-\varepsilon+2\sigma}.$$

Already for $\sigma = 2$ this contradicts the Froissart bound. (In principle, some tricky cancellations cannot be excluded, but so far there are no indications for that.) One can interpret this conclusion as an observation of the contradictory character of the hypothesis of the existence of Regge poles *only*.

11.2.2 Azimov shift in terms of spiral amplitudes

This phenomenon known as 'the Azimov shift' can be explicitly seen in the t-channel as well. To parametrize the amplitude of the transition between two scalar and two vector particles, it is convenient to employ the formalism of spiral amplitudes. These are amplitudes with definite values of *helicities* of participating particles with $\sigma \neq 0$. Helicity is the projection of spin onto the direction of the particle momentum. In the cms of the t-channel, momenta of two vector particles are opposite, so that the *difference* of their helicities represents the projection of the total spin. The generalization of the t-channel partial-wave expansion reads

$$= A_{\lambda_1 \lambda_2} = \sum_j (2j+1) Y_{j\lambda}(\theta, \varphi) f_{j\lambda}(t); \quad \lambda = \lambda_1 - \lambda_2. \quad (11.16)$$

The total angular momentum vector, $\mathbf{j} = \boldsymbol{\ell} + \boldsymbol{\sigma}$, is a sum of the orbital momentum $\boldsymbol{\ell}$ and the total spin $\boldsymbol{\sigma}$. Since $\boldsymbol{\ell} = [\mathbf{r} \times \mathbf{p}]$ is orthogonal to the direction of the cms momentum $\mathbf{z}$, the helicity parameter λ in (11.16) equals $\lambda = \sigma_z = j_z$. Physically, the angular momentum projection is restricted, $|j_z| \leq j$. Indeed, the boundary $|\lambda| \leq j$ is contained in the normalization of the *spherical harmonics*:

$$Y_{j\lambda}(\theta, \varphi) \equiv \sqrt{\frac{\Gamma(j-\lambda+1)}{\Gamma(j+\lambda+1)}} P_{j\lambda}(z) e^{i\lambda\varphi}, \quad z = \cos\theta, \quad (11.17a)$$

where $P_{j\lambda}$ are the associated Legendre functions. At large $|z|$ we have

$$Y_{j\lambda} \simeq \frac{\Gamma(j+\frac{1}{2})}{\sqrt{\pi}\Gamma(j+\lambda+1)\Gamma(j-\lambda+1)}(2z)^j, \quad (11.17b)$$

where we have dropped the trivial dependence on the azimuth angle by setting $\varphi = 0$. We see that indeed, owing to the Γ factors, $Y = 0$ for $|\lambda| > j$. Mark that for $\lambda = 0$ the normalization factor in (11.17b) coincides with that of the spinless case, (11.13b).

The continuation of the series (11.16) to complex j with the help of the Sommerfeld–Watson integral does not pose difficulties:

$$A_\lambda(s,t) \simeq \frac{i}{4} \int \frac{dj}{\sin \pi j} \psi_{j\lambda}(t) \cdot [(-s)^j + s^j], \quad s \to \infty; \quad (11.18a)$$

$$\psi_{j\lambda}(t) = \frac{\Gamma(j+\frac{1}{2})}{\sqrt{\pi}\Gamma(j+\lambda+1)\Gamma(j-\lambda+1)} \cdot k_t^{-2j} \cdot f_{j\lambda}(t). \quad (11.18b)$$

The changes will affect the unitarity condition.

The first term in the r.h.s. with exchange of scalar particles will stay as before, so we concentrate on the new contribution describing particles with spins σ_1 and σ_2 in the intermediate state:

$$\Delta f_j(t) = \cdots + \sum_\lambda \tau(t) f_{j\lambda}(t+i0) f_{j\lambda}(t-i0).$$

The sum over λ has emerged since for a given j we can still have different helicities in the intermediate state. The sum runs up to $\lambda_{\max} = \sigma_1 + \sigma_2$. Due to conservation of parity, it is sufficient to sum over $\lambda \geq 0$, doubling the contributions of $\lambda \geq 1$.

Let us express the unitarity relation in terms of the partial waves ψ_j and $\psi_{j\lambda}$ defined by (11.13b) and (11.18b), correspondingly. Collecting the normalization factors, we get

$$\Delta \psi_j = \sum_\lambda^{\sigma_1+\sigma_2} C_j(t) \frac{\sqrt{\pi}\,\Gamma(j-\lambda+1)\Gamma(j+\lambda+1)}{\Gamma(j+1)\Gamma(j+\frac{1}{2})} \psi_{j\lambda}\psi_{j\lambda}^*. \quad (11.19)$$

In the spinless case, $\sigma_1 = \sigma_2 = 0$, we recover (11.14).

Decreasing j, we hit the first pole at $j = \sigma_1 + \sigma_2 - 1$,

$$\Delta \psi_j \propto \psi_{j\lambda} \frac{1}{j+1-\sigma_1-\sigma_2} \psi_{j\lambda}^*, \quad \lambda = \sigma_1 + \sigma_2. \quad (11.20)$$

In the first unphysical point in j (when the angular momentum is taken *smaller* than its projection), $j = \lambda - 1$, the phase space volume of the intermediate state becomes infinite. As we have discussed, this results in the lower limit of the high-energy behaviour of the scattering amplitude,

$A(s) \propto s^{\sigma_1+\sigma_2-1}$. Taking vector particles in the intermediate state, $\sigma_1 = \sigma_2 = 1$, we obtain $A \propto s$, the result that we have explicitly derived before. In general, (11.20) permits *elementary particles* to have spins 0, $\frac{1}{2}$ and 1, and no more.

What to do with particles with higher spins which exist, can be produced in a t-channel reaction and therefore participate in the unitarity relation? The problem was solved by Mandelstam. It turns out that when a *reggeized* particle is present in the intermediate state, a new singularity – a moving branch cut – appears from an unphysical sheet related to a multi-particle threshold, which *exactly compensates* the pole (11.20).

Imagine that we have a particle with spin σ which is a *bound state* of two scalar particles (with mass μ) and lies on the Regge trajectory $\alpha(t)$. Let m denote its mass, $\alpha(m^2) = \sigma$.

As we will demonstrate shortly, the position of the Mandelstam branch cut derives from the trajectory of the pole and reads

$$j = j_2(t) = 2\alpha\left(\frac{t}{4}\right) - 1. \tag{11.21}$$

The unitarity condition (11.20) holds above the two-particle threshold, $t \geq 4m^2$. Let us show that the contribution of this branching to the unitarity relation,

$$\Delta\psi_j(t) = \psi_{j,2\sigma}\frac{c}{j+1-2\sigma}\psi^*_{j,2\sigma} + \delta_{\text{branch}}\psi_j(t),$$

fully screens the unphysical pole due to exchange of two particles.

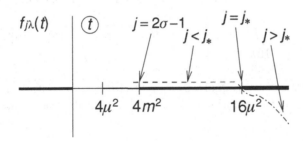

The trajectory $\alpha(t)$ is complex above the two-particle threshold, $t > 4\mu^2$. Therefore $\text{Im}\, j_2(t) > 0$ for $t > 16\mu^2$. This means that for real j larger than $j_* \equiv 2\alpha(4\mu^2) - 1$, the position $t_2(j)$ of the branch singularity (11.21) in the t-plane is complex. The branching then is on the unphysical sheet, since complex singularities on the physical one are forbidden by causality. It emerges on the physical sheet through the tip of the four-particle threshold, $t = (4\mu)^2$, and moves to the left with j decreasing. When we approach the troubling point $j = 2\sigma - 1$, the branching arrives precisely to $t = 4m^2$,

$$j = 2\alpha\left(\frac{t}{4}\right) - 1 \;\rightarrow\; 2\sigma - 1 \qquad \Longrightarrow \qquad \alpha\left(\frac{t}{4}\right) \rightarrow \sigma = \alpha(m^2),$$

covers fully the two-particle cut due to the $\sigma\sigma$ exchange and cancels the unphysical pole at $j < |j_z|$.

At the same time, in the *physical* integer values of j the branch cut contribution *vanishes*, leaving the two-particle exchange unperturbed. So, the Mandelstam branching is rather sophisticated. It is interesting that all this occurs automatically as soon as we suppose that the particle is reggeized.

11.2.3 A model for a moving branch-point singularity

Before we turn to the derivation of the reggeon branchings, let us try to guess the answer. Near the pole due to the exchange of particles with spins σ_1 and σ_2, the unitarity relation has the form

$$2\,\mathrm{Im}\,f_j = \quad + \quad \tau(t)\frac{c}{j + 1 - \sigma_1 - \sigma_2}\tilde{f}_{j\sigma}\tilde{f}_{j\sigma}^*, \qquad (11.22)$$

where we have extracted the singularity from the partial wave amplitudes $f_{j\sigma}$ describing the $0 + 0 \rightarrow \sigma_1 + \sigma_2$ transition, $f_{j\sigma} \rightarrow \tilde{f}_{j\sigma}$. It contains the phase space volume $\tau(t)$ which for particles with masses m_1 and m_2 reads

$$\tau(t, m_1^2, m_2^2) = \frac{k_c}{16\pi\omega_c} = \frac{\sqrt{t^2 - 2t(m_1^2 + m_2^2) + (m_1^2 - m_2^2)^2}}{16\pi\sqrt{t}}. \qquad (11.23)$$

When the particles are reggeized, $\sigma_i = \sigma_i(m_i^2)$, their masses become 'variable', suggesting to include into (11.22) integrals over masses, $t_1 = m_1^2$ and $t_2 = m_2^2$:

$$\int \frac{dt_1\,dt_2\,\tau(t, t_1, t_2)}{j + 1 - \sigma_1(t_1) - \sigma_2(t_2)}\tilde{f}_{j\sigma}\tilde{f}_{j\sigma}^*. \qquad (11.24)$$

Given the additional integrations, (11.24) would no longer be a pole but a branch cut in j. Adding the propagators of reggeized particles, either $[\sin\frac{\pi}{2}\sigma_i(t_i)]^{-1}$, or $[\cos\frac{\pi}{2}\sigma_i(t_i)]^{-1}$, depending on the signature, we would get a natural model of a moving branching. Included into the unitarity relation (11.22), this expression could compensate the 'elementary' pole.

11.3 Multiparticle unitarity and Mandelstam singularities

To understand how the reggeon branchings appear, we have to study for the first time multi-particle unitarity conditions.

11.3.1 Three-particle unitarity condition for partial waves

Let us consider the simplest example of $2 \to 2$ scattering of scalar particles above the three-particle threshold, $9\mu^2 < t < 16\mu^2$.

$$2\,\mathrm{Im}\,f_j = \quad \text{⬡⬡} \quad + \quad \text{⬡⬡}^{\sigma} \quad + \quad \text{⬡⬡} \qquad (11.25)$$

We suppose that a particle with spin σ shown by the dashed line is a bound state of two scalar ones and can be produced in the intermediate state together with a scalar. First of all, let us write the three-particle term in the unitarity condition. We can describe a three-particle system in two steps. First we group two particles and treat them as a composite object with an invariant mass $t_{12} = (k_1 + k_2)^2$ and an internal orbital momentum $\ell = \ell_{12}$ and its projection, $\lambda = \lambda_{12}$. Then we combine the pair, which looks as a particle with an arbitrary 'mass' and 'spin', with the remaining scalar particle k_3 into the system with the total angular momentum j and energy t. Such a representation can be rigorously derived by parameterizing by Euler angles the internal geometry and the orientation of the plane formed by three particles in the t-channel cms.

This makes five independent variables characterizing a five-point amplitude, The structure of the three-particle term is rather similar to that of the case of the unitarity condition with spin,

$$\Delta f_j^{(3)} = \sum_{\ell,\lambda} \int dt_{12}\, K(t, t_{12}) f_{j\ell\lambda}(t, t_{12}) f_{j\ell\lambda}^*(t, t_{12}), \qquad (11.26a)$$

with the only difference that now the 'mass' t_{12} of the composite particle (12) varies in the interval

$$4\mu^2 < t_{12} < (\sqrt{t} - \mu)^2. \qquad (11.26b)$$

The function K in (11.26a) is given by the product of the phase space volume functions (11.23),

$$K(t, t_{12}) = \tau(t, t_{12}, \mu^2)\tau(t_{12}, \mu^2, \mu^2). \qquad (11.26c)$$

Introducing the partial waves ψ_j and $\psi_{j\lambda}$ in order to extract the angular momentum singularities as before, we get

$$\Delta\psi_j^{(3)} = \sum_{\ell, \lambda} \int dt_{12}\, K(t, t_{12})\, \frac{c^2}{j+1-\lambda}\psi_{j\ell\lambda}\psi_{j\ell\lambda}^*, \qquad (11.27)$$

where we have explicitly extracted the main singularity from the Γ factors, cf. (11.19). Unlike (11.19), the sum over λ in (11.27) is not limited from above since the 'spin' ℓ of the composite object (12) can be arbitrary. We must replace the sum by an integral; otherwise we have poles at arbitrarily large j and the continuation to complex j is impossible. Let us write

$$\Delta\psi_j^{(3)} = \int dt_{12}\frac{K(t, t_{12})}{(2i)^2} \int_{C_\lambda} \frac{d\lambda}{\tan \pi\lambda} \int_{C_\ell} \frac{d\ell}{\tan \pi(\ell-\lambda)}\frac{c^2\psi_{j\ell\lambda}\psi_{j\ell\lambda}^*}{j+1-\lambda}. \qquad (11.28)$$

To correspond to the physical sum, $\sum_{\lambda=0}^{\infty}\sum_{\ell=\lambda}^{\infty}$, the contours have to be drawn as follows,

We will discuss the question of convergence of this representation later. In any case, as soon as the integration contours are deformed as in the standard Sommerfeld–Watson case, there are no problems with the poles at large j values anymore and the continuation can be carried out. Unfortunately, it is not unique. We could have added a function vanishing in integer points (for example, we could put sin instead of tan in the denominator). Let us suppose, nevertheless, that this continuation is reasonable and see what will be the structure of singularities in j. The pole is there in (11.28); it could not just disappear. However, some other singularities, not so apparent at the first glance, may emerge.

Let us examine the three-particle system in the intermediate state. Particles 1 and 2 will interact in the final state and can give a resonance which will manifest itself as a pole in $\ell = \ell_{12}$ of the partial wave amplitude $f_{j\ell\lambda}^{(2\to3)}$.

How can this possibility be extracted in a model-independent way?

The $2 \to 3$ amplitude has to satisfy the unitarity condition not only in t but also in t_{12}:

$$\Delta_{12} f^{(2\to3)}_{j\ell\lambda} \equiv \frac{1}{2i} [f^{(2\to3)}(t+i\varepsilon, t_{12}+i\varepsilon) - f^{(2\to3)}(t+i\varepsilon, t_{12}-i\varepsilon)]$$

$$= \underset{\text{}}{} = f^{(2\to3)}_{j\ell\lambda}(t^+, t_{12}) \tau(t_{12}, \mu^2, \mu^2) f^{(2)*}_{\ell}(t_{12}). \quad (11.29)$$

In the region $4\mu^2 < t_{12} < 9\mu^2$ the two-particle unitarity condition in t_{12} is valid. This makes (11.29) a linear equation for $f^{(2\to3)}$. Its solution is simple:

$$f^{(2\to3)}_{j\ell\lambda} = G_{j\ell\lambda}(t, t_{12}) f^{(2)}_{\ell}(t_{12}),$$

where G does not have a two-particle cut in t_{12} (two-particle irreducible amplitude). Indeed, evaluating the discontinuity in t_{12} and using the two-particle unitarity condition for the elastic amplitude $f^{(2)}_{\ell}$ we have

$$\Delta f^{(2\to3)} = G \cdot \Delta f^{(2)}_{\ell}(t_{12}) = G \cdot (f^{(2)}_{\ell} \tau f^{(2)*}_{\ell}) = f^{(2\to3)} \cdot \tau f^{(2)*}_{\ell},$$

which coincides with (11.29).

We know that $f^{(2)}_{\ell}$ can have a Regge pole,

$$f^{(2)}_{\ell}(t_{12}) \simeq \frac{g^2(t_{12})}{\ell - \alpha(t_{12})};$$

$$(11.30)$$

$$f^{(2\to3)}_{j\ell\lambda}(t, t_{12}) \simeq \left[G_{j\ell\lambda}(t, t_{12}) g(t_{12}) \right] \cdot \frac{1}{\ell - \alpha(t_{12})} \cdot g(t_{12}).$$

This expression has a clear diagrammatical meaning:

with the combination $N = Gg$ playing the rôle of the reggeon production amplitude, and g – the amplitude of its decay.

Before we substitute the Regge pole (11.30) into the unitarity relation (11.28) let us make the following simplifying observation. When we single out the final state interaction,

$$\psi_{j\ell\lambda} \implies G_{j\ell\lambda} \cdot f^{(2)}_{\ell}(t_{12}),$$

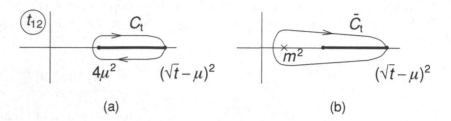

Fig. 11.1

the r.h.s. of (11.28) acquires the following structure,

$$K\psi_{j\ell\lambda}\psi^*_{j\ell\lambda} \to \tau(t,t_{12},\mu^2)GG^* \cdot \left(f_\ell^{(2)}\,\tau(t_{12},\mu^2,\mu^2)\,f_\ell^{(2)*}\right).$$

Here we have extracted from (11.26c) the phase space volume factor $\tau(t_{12})$ to form the *discontinuity* of the elastic scattering amplitude,

$$
\begin{aligned}
K\psi\psi^* &= \tau(t,t_{12},\mu^2)G \cdot \left\{\Delta_{12}f_\ell^{(2)}(t_{12})\right\}\cdot G^* \\
&= \Delta_{12}\left\{\tau(t,t_{12},\mu^2)Gf_\ell^{(2)}(t_{12})G^*\right\}.
\end{aligned}
\tag{11.31}
$$

This allows us to replace the integral over t_{12} along the real interval (11.26b) by a contour integration as shown in Fig. 11.1(a) and write

$$\int_{4\mu^2}^{(\sqrt{t}-\mu)^2} dt_{12}\,\Delta_{12}\{\tau Gf_\ell^{(2)}(t_{12})G^*\} \implies \int_{C_t}\frac{dt_{12}}{2i}\{\tau Gf_\ell^{(2)}(t_{12})G^*\}.$$

Now we can substitute the Regge pole (11.30) in the elastic final state scattering amplitude,

$$\Delta\psi_j^{(3)} = \int_{C_t}\frac{dt_{12}}{2i}\frac{\tau(t,t_{12},\mu^2)}{(2i)^2}\int_{C_\lambda}\frac{d\lambda}{\tan\pi\lambda}\frac{1}{j+1-\lambda}\int_{C_\ell}\frac{d\ell}{\tan\pi(\ell-\lambda)}\frac{N^2}{\ell-\alpha(t_{12})}.$$

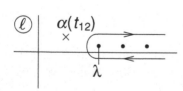

First we look at the integral over ℓ. Since $t_{12} > 4\mu^2$, the trajectory $\alpha(t_{12})$ has an imaginary part. When we change λ, the integral becomes singular when λ hits the point $\alpha(t_{12})$ and the two poles pinch the contour:

$$\Delta\psi_j^{(3)} = \int_{C_t}\frac{dt_{12}}{2i}\frac{\tau(t,t_{12},\mu^2)}{\tan\pi\alpha(t_{12})}\frac{N(t+)N(t-)}{j-\alpha(t_{12})+1}.\tag{11.32a}$$

This is just our model for a branching singularity (11.24), only for one spinless particle, and one with spin,

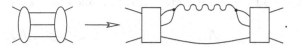

This three-particle contribution has to be compared with the second term in the unitarity condition (11.25), due to the exchange of two particles, one of which, σ, we suppose to be a bound state belonging to the trajectory,

$$\Delta \psi_j^{(\sigma)} = \tau(t, m^2, \mu^2) \frac{\psi(t+)\,\psi(t-)}{j - \sigma + 1}. \tag{11.32b}$$

In fact, the latter is nothing but the *residue* of the former integral (11.32a) at the pole $\tan \pi \alpha = 0$, $\alpha(t_{12}) = \sigma$. Therefore, we can take into account both contributions (11.32) by simply modifying the integration contour to include the particle pole, as shown in Fig. 11.1(b). By derivation, the

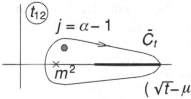

pole at $\alpha(t_{12}) = j + 1$ lies *inside* the contour $\bar{C}_t$. As a result, the point $j = \sigma - 1$ where the two poles collide, turns out to be a *regular* point in the j-plane: the three-particle singularity compensates the two-particle one.

The integral (11.32a) develops a singularity in j at $j = j_0$ such that the pole of the integrand hits the immobile endpoint of the integration contour:

$$t_{12} = (\sqrt{t}-\mu)^2, \quad \alpha(t_{12}) = j + 1 \quad \Longrightarrow \quad j_0(t) = \alpha((\sqrt{t}-\mu)^2) - 1. \tag{11.33}$$

This is the new branching singularity we are looking for. Actually, for t above the three-particle threshold, $t > (3\mu)^2$, the argument of α in (11.33) exceeds $(2\mu)^2$, so that the trajectory is complex, and the new singularity is hidden on the unphysical sheet beneath the three-particle unitary cut. For $t < 9\mu^2$ it emerges on the physical sheet. Its position in the t-plane, $t_0(j)$, moves on the left with j decreasing. If we take $j = \sigma - 1$, then from (11.33) follows $\sigma = \alpha((\sqrt{t} - \mu)^2)$, $t = (m + \mu)^2$, showing that the branching arrives at the tip of the two-particle threshold, to rescue the unitarity relation for the partial wave in spite of the '$\ell = -1$' singularity phenomenon.

11.3.2 Four-particle unitarity

The analysis of the four-particle unitarity condition proceeds along the same lines. We group intermediate state particles into two pairs,

and get the singularity $1/((j + 1 - \lambda_1 - \lambda_2)$ at the first unphysical value of their orbital angular momentum $L = -1$. Introducing Regge poles into the elastic final state rescattering amplitudes, we arrive at the integration over the pair masses,

$$\iint \frac{dt_1\, dt_2\, \tau(t, t_1, t_2)}{j - \alpha(t_1) - \alpha(t_2) + 1} \frac{\psi(t_1, t_2)\, \psi^*(t_1, t_2)}{\tan \pi\alpha(t_1) \tan \pi\alpha(t_2)}. \tag{11.34}$$

To extract singularities from the double integral is more difficult. In fact, there can be many singularities. Among them there is, however, one that does not depend on specific features of the Regge pole trajectory and does not contain masses explicitly. Let us outline the main steps of the derivation of this singularity which is the one we are looking for.

The integral over t_2 is singular in t_1 when the pole collides with the endpoint $t_2 \le (t_2)_{\max} = (\sqrt{t} - \sqrt{t_1})^2$, as before:

$$\implies \quad j = \alpha(t_1) + \alpha((\sqrt{t} - \sqrt{t_1})^2) - 1. \tag{11.35}$$

This is a non-linear equation for t_1. A singularity in j will appear when two solutions of (11.35) coincide, pinching the integration contour in the t_1-plane. The condition for having a multiple zero,

$$0 = \frac{d}{dt_1}[\alpha(t_1) + \alpha((\sqrt{t} - \sqrt{t_1})^2)] = \alpha'(t_1) - \alpha'((\sqrt{t} - \sqrt{t_1})^2) \cdot \frac{\sqrt{t} - \sqrt{t_1}}{\sqrt{t_1}},$$

has an obvious solution $\sqrt{t} - \sqrt{t_1} = \sqrt{t_1} = \frac{1}{2}\sqrt{t}$. Substituting it into (11.35) we get the branch singularity in the j-plane appearing at

$$j_2(t) = 2\alpha\left(\frac{t}{4}\right) - 1. \tag{11.36}$$

Restoring the signature factors that we have ignored in the preceding discussion, the discontinuity across the branch cut of the partial wave amplitude can be derived,

$$\delta_j f_j^{(2)}(t) = \frac{\pi}{2} \int \frac{dt_1\, dt_2}{2i} \tau(t, t_1, t_2) N_+^{(2)} \delta(j + 1 - \alpha_1 - \alpha_2) N_-^{(2)} \bar{\xi}_j,$$

$$\bar{\xi}_j = \xi_j\, \xi_{\alpha_1(t_1)} \xi_{\alpha_2(t_2)}. \tag{11.37}$$

We supposed that there was only one reggeon in the two-particle scattering amplitude. A generalization is straightforward. An inclusion of different Regge poles does not make the analysis much more complicated. If we insert two Regge poles α and β in the four-particle unitarity condition (11.34), the position of the two-reggeon branch-cut singularity,

$$j + 1 - \alpha(t_1) - \beta(t_2) = 0, \quad \sqrt{t_1} + \sqrt{t_2} = \sqrt{t},$$

will be determined by the extremum of the function

$$\alpha(t_1) + \beta(t_2) + \kappa[\sqrt{t_1} + \sqrt{t_2}]$$

(with κ the Lagrange multiplier), resulting in

$$j_2 = \alpha\left(\left[\frac{\beta'}{\alpha' + \beta'}\right]^2 t\right) + \beta\left(\left[\frac{\alpha'}{\alpha' + \beta'}\right]^2 t\right) - 1.$$

Taking α to be the pomeron, $\alpha(0) = 1$, in the linear approximation we have

$$j_{R+\mathbf{P}}(t) \simeq \beta(0) + \frac{\alpha'\beta'}{\alpha' + \beta'} \cdot t \simeq \beta(t) - \frac{\beta'^2}{\alpha' + \beta'} \cdot t, \tag{11.38}$$

showing that the branch point lies *between* $\beta(t)$ and $\beta(0)$.

Even if the trajectory is unique, there appear many singularities in the j-plane. Indeed, having obtained the branching singularity j_2 of (11.36), we can iterate it anew,

$\implies j_4(t) = 2j_2\left(\frac{t}{4}\right) - 1 = 4\alpha\left(\frac{t}{4}\right) - 3.$

This way an infinite series of *Mandelstam branchings* is generated by a single Regge pole,

$$j_n(t) = n\alpha\left(\frac{t}{n^2}\right) - n + 1. \tag{11.39}$$

The corresponding discontinuity across the n-reggeon branch is given by

$$\delta_j f_j^{(n)}(t) = \frac{\pi}{n!} \int d\Gamma_\tau \, N_+^{(n)} \delta\left(j - 1 - \sum_{i=1}^{n}[\alpha(t_i) - 1]\right) N_-^{(n)} \gamma_j^{(n)}. \tag{11.40}$$

As we have discussed in the previous lecture, the reggeon branchings are essential for the high-energy asymptotics in one case only, namely, that of the Pomeranchuk poles $\mathbf{P}$ with $\alpha_\mathbf{P}(0) = 1$, when the branchings condense to the point $j_n(t) = 1$ in the $n \to \infty$ limit. In the physical region of the s-channel, $t < 0$, this puts pomeron branchings in the *dominant* position, on the right of the pole $\alpha_\mathbf{P}(t)$, changing the asymptotic character of the elastic diffraction,

$$s^{\alpha(t)} \;\to\; s \cdot f_\alpha(\ln s, t).$$

Analogously, high pomeron branching corrections to *non-vacuum* poles,

$$j_{R+n\mathbf{P}}(t) \;\simeq\; \beta(0) + \frac{\alpha'\beta'}{\alpha' + n\beta'} \cdot t,$$

accumulate towards $j_n(t) \to \beta(0)$, and slow down the energy falloff of the 'charge exchange' cross sections,

$$s^{\beta(t)} \;\to\; s^{\beta(0)} \cdot f_\beta(\ln s, t).$$

We see that in order to describe these phenomena we need to know how to calculate and take into account multi-pomeron branchings. The problem is complicated by the fact that, according to (11.39), all the branch-point singularities are sitting at small t in one place, near $j = 1$. In this situation we cannot approximate the reggeon production block N by a constant since near $j = 1$ all the amplitudes are changing rapidly. We must localize and iterate all singular contributions which will lead us to the picture of *interacting reggeons*,

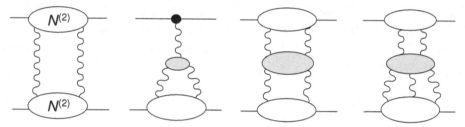

For ordinary hadrons all singularities were separated,

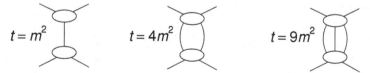

This permitted us to use the unitarity as a tool for *calculating* the inter-action amplitudes in the case of a small coupling constant. Even if the coupling is large, iterating the unitarity conditions allowed us to extract some valuable information. Now the situation looks much more difficult. From the unitarity viewpoint, our reggeons behave rather as *massless* objects (like photons).

In fact such an analogy can be drawn explicitly. The emerging reggeon picture is similar to a non-relativistic multi-body problem of statistical physics. Imagine a system of particles with the 'dispersion law' $\omega_i = \epsilon(\mathbf{k}_i)$ describing the dependence of the particle energy on the momentum. The amplitude of scattering via, e.g. a two-particle intermediate state n will have the structure

$$f_{a,b}(\omega, \mathbf{k}^2) \propto \sum_n \frac{V_{a,n} V_{n,b}}{\epsilon_1 + \epsilon_2 - \omega}.$$

It has a cut in energy running from the point

$$\omega_2(\mathbf{k}^2) = \min_{\mathbf{k}_1 + \mathbf{k}_2 = \mathbf{k}} \{\epsilon(\mathbf{k}_1) + \epsilon(\mathbf{k}_2)\}.$$

The discontinuity in energy reads

$$\delta_\omega f_{a,b}(\omega, \mathbf{k}^2) = \int d\Gamma(\mathbf{k}_1, \mathbf{k}_2) f_{a,n} \delta\big(\epsilon(\mathbf{k}_1) + \epsilon(\mathbf{k}_2) - \omega\big) f_{b,n}^*.$$

It resembles the two-reggeon unitarity condition (11.37) if we identify

$$j - 1 = \omega, \quad \alpha(t) - 1 = \epsilon(\mathbf{k}).$$

The pomeron case corresponds to *massless* excitations (like phonons in a solid state) with the dispersion law without a 'mass-gap': $\epsilon(\mathbf{0}) = 0$. Such multi-phonon thresholds accumulate to $\omega = 0$ as the pomeron branchings do.

There is, however, one but essential difference between the two problems, namely the signature factor $\bar{\xi}_j$ in (11.37). To have a complete analogy with a quantum-mechanical system, we would like the amplitude to be real below its singularities in energy. To try to get rid of $\bar{\xi}_j$ by simply absorbing $\sqrt{\bar{\xi}_j}$ into the reggeon production factors N is dangerous: $\bar{\xi}_j$ may be negative and this would introduce an unwanted singularity into the vertex function.

In non-relativistic quantum mechanics the discontinuity of the forward amplitude ($a = b$) is given by the product $f \times f^*$ and is positive. What about our problem? Near the most interesting point $\alpha_1 \approx \alpha_2 \approx j \approx 1$ we have $\bar{\xi}_j = \xi_j \xi_{\alpha_1} \xi_{\alpha_2} \simeq (i)^3$, producing in the unitarity condition (11.37)

$$\delta_j f_j^{(2)} \propto \frac{1}{i} \bar{\xi}_j \simeq -1.$$

This means that, contrary to common particles, the contribution of a two-reggeon branching is *negative*. (We shall see shortly that in fact every additional reggeon introduces (-1), and thus the signs are alternating.)

Still, except for the signs, the reggeon branching is similar to usual branching in a system of particles, only in an unusual space, with t-channel angular momentum j in the rôle of 'energy' ω.

We already remarked more than once that when $t < 0$ the branching of two pomerons is positioned *on the right* from the pole. Now that we have established an analogy with the non-relativistic theory, this observation becomes quite dramatic, since in NQM a pole cannot sit on top of a cut.

A multi-reggeon state in the t-channel, looks from s-channel as a *repetition* of the one-reggeon exchange, and repetitions is the domain of the unitarity. In the next lecture we start to construct the field theory of interacting reggeons using unitarity in the s-channel. This will allow us, in particular, to bypass the problem of non-uniqueness of the analytic continuation (11.28). We will also fix the signature factors γ_j in the reggeon unitarity conditions (11.40) which, as we have mentioned before, must vanish for physical integer values of j of the proper signature.

12

Branchings in the *s*-channel and shadowing

We have a serious task ahead: to analyse and sum up the branchings, taking into account how they influence each other, and to learn how to write the amplitudes respecting unitarity.

12.1 Reggeon branchings from the *s*-channel point of view

12.1.1 The s-channel approach

In Lecture 2 we discussed how, given a particle spectrum, we can construct interaction amplitudes with the help of unitarity conditions and analyticity (dispersion relations). The dispersive method is well formulated and straightforward but not very convenient in practice. So instead of iterating the amplitudes through the non-linear unitarity relations we draw series of Feynman diagrams which satisfies (at least term by term) the unitarity condition.

Now we face an opposite situation for the first time. We discovered the branchings and found the specific unitarity conditions they satisfy. Can we construct an effective field theory that would generate the reggeon unitarity conditions?

There are two ways of attacking the problem.

(1) One can explicitly construct an effective reggeon QFT directly in the *t*-channel, identifying reggeons with particles of a non-relativistic (anti-hermitian) field theory in $2 + 1$ dimensions.

(2) Alternatively, we can solve the problem of constructing the reggeon diagrams which solve the unitarity conditions, working directly in the *s*-channel.

The two routes yield the same results. We will follow the second one, for a number of reasons.

First of all, the t-channel reggeon unitarity conditions we studied in Lecture 11 were derived for positive t; in particular, the two-reggeon unitarity condition holds above $t = 16\mu^2$. The information so obtained has to be continued to the region $t < 0$ that really interests us. Moreover, the analysis of the discontinuity δf_j includes drawing complicated contours and is rather involved. In turns out that in the physical s-channel region, $t < 0$, everything looks much simpler, and the contours are trivial.

Secondly, if we work in the s-channel and exploit the s-channel unitarity, the problem of negative signature factors essentially disappears as well.

What did the Regge poles look like in the s-channel? We have seen that the simplest perturbative model for a pole was the *ladder*. From the point of view of the t-channel the ladder solves the two-particle unitarity condition. There are, of course, also corrections due to many-particle states in the t-channel.

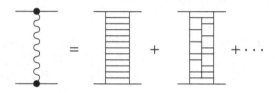

If we bear in mind the *exact* pole, rather than its naive perturbative image, it will contain everything.

Then, we considered a four-particle state and have accurately taken into account the interaction between particles (1) and (2), and (3) and (4), while neglecting cross-interactions between the pairs. This gave us a two-reggeon branch-cut singularity:

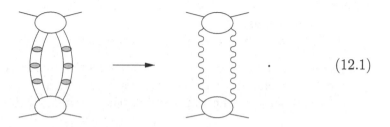

$$(12.1)$$

The question is, what does this picture correspond to in the s-channel?

So, we have two 'parallel ladders'. What we did here looks rather strange. Having suppressed cross-interactions in a four-particle state (which state in fact contributed to the *exact* pole) we obtained a *non-pole* singularity. To get it we had to pick some specific and, it seems, unimportant configurations when the usual strong interaction is not acting upon *all* the particles involved.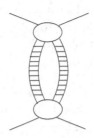

Recall that in perturbation theory the ladder diagram has emerged as the probability of a cascade production of many particles:

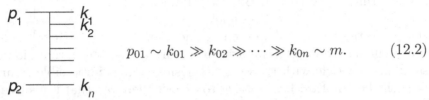

$$p_{01} \sim k_{01} \gg k_{02} \gg \cdots \gg k_{0n} \sim m. \qquad (12.2)$$

If I draw parallel cascades, momenta of particles from different chains will overlap, so that the two 'combs' may interact with each other, mixing everything in many ways. Squaring such an amplitude will produce a rather complicated picture.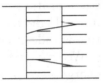

Certainly, by mixing up everything we would get a *large* contribution, even in perturbation theory. I hope, however, that this will be a contribution to the *pole*. Meanwhile, I want to find a particular piece, maybe not necessarily numerically large, but with *specific analyticity*.

Since in (12.2) particle momenta are gradually decreasing down the ladder, it is reasonable to expect that in our two-reggeon diagram (12.1) there are only particles with large momenta $k \sim p_1 \sim s/m$ in the upper

block ⧓ and small momenta $k \sim p_2 \sim m$ in the lower one ⧓

(in the laboratory frame where the target particle p_2 is at rest).

12.1.2 Coexisting ladders – s-channel image of branching

Once we decided to suppress the interaction between 'combs', I would draw the probability of such a process as follows:

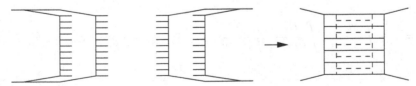

In the s-channel we have *coexisting* ladders; from the *t*-channel point of view we have two separate (non-interacting) pairs of interacting particles as shown in Fig. 12.1(a). The graph (a) has a particular *non-planar* topology. What about a simpler picture, also with two ladders, displayed in Fig. 12.1(b)? Will it also participate in the two-reggeon branching?

Let us ask ourselves, how does this process proceed in time? In the laboratory frame the slow target particle at the bottom of the graph gets the energy of the order of m so that the scattering process has to be over in a finite time $t = \mathcal{O}(m^{-1})$. But this implies that in a unit time the first fluctuation of the projectile has to collapse back, and the

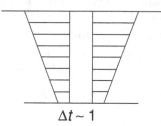

$\Delta t \sim 1$

other one has to develop in order to assure the second interaction of the same fast particle with the target. Physically, it is impossible to arrange at such short notice. It is natural to expect therefore that the space–time configurations of the type of Fig. 12.1(b) will not contribute significantly in the high-energy limit. In what follows we will discuss in detail whether this is indeed the case.

We will devote the rest of this lecture to the investigation of the high-energy asymptotics of the diagram of Fig. 12.1(a), which we consider the main candidate for the s-channel image of the two-reggeon branching.

12.2 Calculation of the reggeon–reggeon branching

12.2.1 Kinematics and factorization

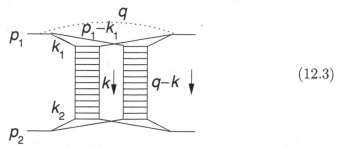

$$(12.3)$$

The diagram (12.3) contains three independent momentum integrations:

$$
f_2 = \frac{1}{2!} \int \frac{d^4k}{(2\pi)^4 i} \int \frac{d^4k_1}{(2\pi)^4 i} \int \frac{d^4k_2}{(2\pi)^4 i} f_1(k_1, k_2, k) f_1(p_1-k_1, p_2-k_2, q-k)
$$

$$
\times \frac{1}{m^2 - k_1^2} \frac{1}{m^2 - (p_1-k_1)^2} \frac{1}{m^2 - (k_1-k)^2} \frac{1}{m^2 - (p_1-k_1-q+k)^2}
$$

$$
\times \left\{ \text{four lower-part propagators} \right\}.
$$

$$(12.4)$$

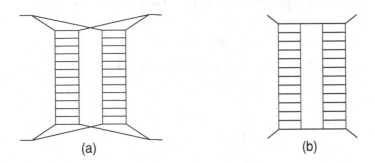

Fig. 12.1   Non-planar (a) and planar (b) two-ladder graphs.

We want to extract the contribution corresponding to the Regge poles in the ladder blocks f. The poles will appear, if the energy invariants of the ladders, $(k_1 + k_2)^2$ and $(p_1 - k_1 + p_2 - k_2)^2$, will tend to infinity in the $s \to \infty$ limit. In this asymptotics we hope to extract the contribution of the branching. We shall assume that the transverse momenta (and virtualities) of all participating particles stay finite,

$$\mathbf{k}_{i\perp}^2 \sim |k_i^2| = \mathcal{O}(m^2).$$
(12.5)

Introducing the usual Sudakov decomposition,

$$p_1^\mu = p_+^\mu + \gamma p_-^\mu, \quad p_2^\mu = \gamma p_+^\mu + p_-^\mu; \qquad \gamma = \frac{m^2}{s},$$

we have

$$k_i^\mu = \alpha_i p_+^\mu + \beta_i p_-^\mu + k_{i\perp}^\mu; \quad (k_\perp^\mu)^2 = -\mathbf{k}_\perp^2.$$

Let us look at the propagators:

$$k_1^2 = \alpha_1 \beta_1 s - \mathbf{k}_{1\perp}^2,$$
$$(p_1 - k_1)^2 = (1 - \alpha_1)(\gamma - \beta_1)s - \mathbf{k}_{1\perp}^2.$$

Applying the restriction (12.5) gives for the Sudakov components of k_1

$$\left| (p_1 - k_1)^2 \right| \sim 2p_1 k_1 \sim \beta_1 s \sim m^2,$$

so that

$$\beta_1 = \mathcal{O}(m^2/s), \quad \alpha_1 = \mathcal{O}(1).$$
(12.6a)

This means that the offspring move almost parallel to the direction of the incident particle p_1, sharing its large momentum in a finite proportion: $k_1 \simeq \alpha_1 p_1$, $(p_1 - k_1) \simeq (1 - \alpha_1) p_1$.

Considering analogously the bottom part of the graph,

$$k_2^2 = \alpha_2\beta_2 s - \mathbf{k}_{2\perp}^2 = \mathcal{O}(m^2),$$
$$(p_2 - k_2)^2 = (\gamma - \alpha_2)(1 - \beta_2)s - \mathbf{k}_{2\perp}^2 = \mathcal{O}(m^2),$$

we arrive at

$$\beta_2 = \mathcal{O}(1), \qquad \alpha_2 = \mathcal{O}(m^2/s). \tag{12.6b}$$

Now we analyse the momentum transferred along the first ladder,

$$k^\mu = \alpha_k p_+^\mu + \beta_k p_-^\mu + k_\perp^\mu.$$

From the finiteness of the propagators exiting the ladder we get

$$(k_1 - k)^2 \sim (\alpha_1 - \alpha_k)(\beta_1 - \beta_k)s = \mathcal{O}(m^2) \Rightarrow |\beta_k| \sim m^2/s,$$
$$(k_2 + k)^2 \sim (\alpha_2 + \alpha_k)(\beta_2 + \beta_k)s = \mathcal{O}(m^2) \Rightarrow |\alpha_k| \sim m^2/s. \tag{12.7}$$

This shows that the momentum transfer is practically *transversal* to the scattering plane:

$$k^2 = \alpha_k \beta_k s - \mathbf{k}_\perp^2 \simeq -\mathbf{k}_\perp^2, \quad (q - k)^2 \simeq -(\mathbf{q} - \mathbf{k})_\perp^2,$$

where we have used $|\alpha_q| \simeq |\beta_q| \simeq |q^2|/s = \mathcal{O}(m^2/s)$.

Since $|\alpha| \ll \alpha_1$, $|\beta| \ll \beta_2$, we can omit in all the propagators at the top part of the graph the α component of the momentum transfer and, correspondingly, β in the bottom. Then we observe that, thanks to the kinematical conditions (12.6), the integration variables have *factorized*:

$$\text{top propagators: } \alpha_1, \beta_1, \beta_k;$$
$$\text{bottom propagators: } \alpha_2, \beta_2, \alpha_k. \tag{12.8}$$

12.2.2 High-energy behaviour

The ladder amplitude f depends on the invariant energy of the pair of particles that enter the ladder, on the momentum transfer and four virtual particle 'masses', e.g.

$$f_1 = f\big((k_1 + k_2)^2, k^2; k_1^2, (k_1 - k)^2, k_2^2, (k_2 - k)^2\big). \tag{12.9}$$

We observe that in our kinematics the invariant energies are of the order of the total energy indeed:

$$
\begin{aligned}
s_{\mathrm{I}} &\equiv (k_1 + k_2)^2 = (\alpha_1 + \alpha_2)(\beta_1 + \beta_2)s - (\mathbf{k}_1 + \mathbf{k}_2)_\perp^2 \\
&\simeq \alpha_1 \beta_2\, s = \mathcal{O}(s)\,, \tag{12.10a}
\end{aligned}
$$

$$
\begin{aligned}
s_{\mathrm{II}} &\equiv (p_1 - k_1 + p_2 - k_2)^2 \\
&\simeq (1 - \alpha_1)(1 - \beta_2)s = \mathcal{O}(s)\,. \tag{12.10b}
\end{aligned}
$$

This allows us to substitute the asymptotic Regge pole expression f_1 for the ladder amplitudes (12.9),

$$
f_{1,\mathrm{I}} \simeq g_{\mathrm{I},1} \cdot \xi_{\alpha(k^2)} s_{\mathrm{I}}^{\alpha(k^2)} \cdot g_{\mathrm{I},2}\,, \tag{12.11a}
$$

$$
f_{1,\mathrm{II}} \simeq g_{\mathrm{II},1} \cdot \xi_{\alpha((q-k)^2)} s_{\mathrm{II}}^{\alpha((q-k)^2)} \cdot g_{\mathrm{II},2}\,, \tag{12.11b}
$$

where

$$
g_{\mathrm{I},1} = g(\mathbf{k}_\perp^2;\, k_1^2, (k_1-k)^2)\,, \quad g_{\mathrm{I},2} = g(\mathbf{k}_\perp^2;\, k_2^2, (k_2+k)^2)\,, \tag{12.11c}
$$

and the residues of the second pole, $g_{\mathrm{II},i}$ in (12.11b), differ from (12.11c) by the substitution $k \to q - k$, $k_1 \to p_1 - k_1$ and $k_2 \to p_2 - k_2$.

As far as the dependence on the longitudinal components of the reggeon-loop momentum k is concerned, the factorization property (12.8) holds also for the Regge residues: the top residues $g_{\mathrm{I},1}$ and $g_{\mathrm{II},1}$ depend only on β_k, while the bottom ones, $g_{\mathrm{I},2}$ and $g_{\mathrm{II},2}$, only on α_k.

Given such a splitting of the dependence of the integrand on α_k and β_k variables, we can represent the original integral (12.4) in the following compact form,

$$
f_2(s, q^2) = \frac{i}{2!} \int \frac{d^2 \mathbf{k}_\perp}{(2\pi)^2} \xi_{\alpha(k^2)} \xi_{\alpha((q-k)^2)} s^{\alpha(k^2)+\alpha((q-k)^2)-1} N_1 N_2\,. \tag{12.12a}
$$

Here

$$
\begin{aligned}
N_1 = N(q^2;\, k^2, (q-k)^2) &= \frac{1}{\sqrt{2}} \int \frac{d^4 k_1}{(2\pi)^4 i}\, \alpha_1^{\alpha(k^2)} (1-\alpha_1)^{\alpha((q-k)^2)} \\
&\times \int \frac{s\, d\beta_k}{2\pi i}\, \frac{g(k_1, k) g(p_1 - k_1, q - k)}{(\ \ )(\ \ )(\ \ )(\ \ )}\,, \tag{12.12b}
\end{aligned}
$$

with ( ) marking the four propagators written explicitly in (12.4). We have introduced the factor s in the β_k integral since from (12.7) we know that $\beta_k \propto 1/s$; so defined, N has a finite $s \to \infty$ limit. The expression for N_2 is similar to (12.12b) but with the internal integral running over α_k, instead of β_k. In fact, $N_1 = N_2$.

The factor $1/\sqrt{2}$ has emerged from the reggeon loop integral:

$$d^4k = \frac{s}{2}\,d\alpha_k\,d\beta_k\,d^2\mathbf{k}_\perp = d^2\mathbf{k}_\perp \cdot \left(\frac{s\,d\beta_k}{\sqrt{2}}\right) \cdot \left(\frac{s\,d\alpha_k}{\sqrt{2}}\right) \times \frac{1}{s}$$

(the last factor $1/s$ went into the shift of the energy exponent in (12.12a)).

12.3 Analytic structure of the particle–reggeon vertex

What is the diagrammatic meaning of our exercise? We have replaced the ladder amplitudes by reggeons:

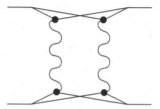

The function N contains an integral over the loop momentum k_1 of the product of four particle propagators and two reggeon vertices.

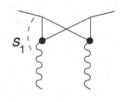

I would say that N describes the conversion of two particles into two reggeons, from the t-channel perspective, or the particle–reggeon scattering (in s channel). If I had to invent the corresponding amplitude A, it would depend on the energy of the 'colliding objects',

$$s_1 = (p_1 - k_1)^2 = (1 - \alpha_k)(\gamma - \beta_k)s \approx (\gamma - \beta_k)s.$$

This relation tells us that β_k determines the particle–reggeon pair energy so that $sd\beta_k = -ds_1$. Taking this into account, we can write N as

$$N = \int_{-\infty}^{\infty} \frac{ds_1}{2\pi i} A(s_1, q^2; k^2, (q-k)^2), \qquad (12.13)$$

where the particle–reggeon scattering amplitude A depends on the 'virtual masses' of the reggeons k^2, $(q-k)^2$, apart from the standard s and $t = q^2$ variables.

It is easy to extract the structure of A from (12.12b). Its only unusual feature is the presence of the powers of α_1 and $(1 - \alpha_1)$ which factors have emerged from the reggeon energies (12.10). The origin of these factors can be understood as follows. If in the t-channel exchange we had a *particle* k with spin σ, the vertex would have had tensor structure, $k_{\mu_1}\ldots k_{\mu_\sigma}$. Our expression contains, in fact, $(\cos)^{\alpha(k_\perp^2)}$; here $\alpha(k_\perp^2)$ plays the rôle of σ and 'counts the number of tensor indices'.

Let us have a deeper look at the function N. In the complex plane of the energy variable s_1, the amplitude A in (12.13) has cuts both at positive and negative s_1:

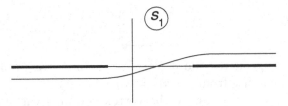

These cuts are described by two diagrams

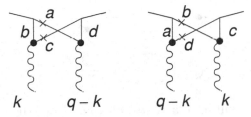

which differ by the exchange of the reggeon momenta $(s \leftrightarrow u)$. By putting the particles a and c in the first graph on-mass-shell we get the right cut; cutting through b and d in the second graph corresponds to the left one.

We have here a curious expression: an integral of the amplitude over the energy. If the behaviour at $s_1 \to \infty$ is suitable (which is so for this concrete diagram which decreases well), the contour can be closed, say, around the right cut and we get a finite, real answer for N:

$$N = \int_{s_0}^{\infty} \frac{ds_1}{\pi} A_1(s_1). \tag{12.14}$$

12.3.1 Amati–Fubini–Stanghellini puzzle

Let us return to the diagram of Fig. 12.1(b) which I did not like from the point of view of the space–time consideration. Repeating the above procedure literally, we arrive at a picture with a single particle separating two reggeons:

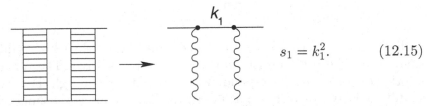

$$s_1 = k_1^2. \tag{12.15}$$

Amati, Fubini and Stanghellini (AFS) have considered elastic scattering and stated that there were branchings generated by the following

two-reggeon diagram

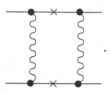

Indeed, if the reggeon residue g were a constant, we would get a finite answer $N \propto g^2$ coming from $A_1(s_1) = \pi g^2 \delta(s_1 - \mu^2)$.

However, $g(s_1) \neq$ const, since the vertex contains various singularities reflecting the dependence on the virtuality $k_1^2 \equiv s_1$,

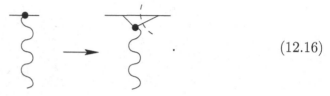

$$(12.16)$$

Because of this, N will have a *right cut* apart from the pole at $s_1 = \mu^2$:

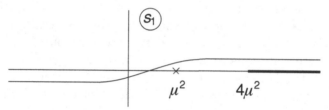

The main question is, how g behaves – does it decrease with k_1^2? If yes, we can close the contour *to the left* and get $N=0$. This is so in all reasonable theories (in $\lambda\varphi^3$, for example; in any case, we are actually interested in those theories in which the transverse momenta are limited).

It is interesting to see how N disappears. When we calculate the imaginary part of the AFS diagram,

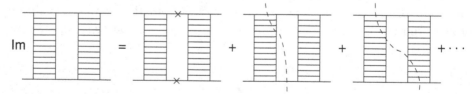

each discontinuity, corresponding to a definite cut on the r.h.s., is different from zero. The first contribution is obviously positive. The other have alternating signs and correspond, in fact, to cutting through the vertex (or simultaneously two vertices) as in (12.16). We see an astonishing picture: each specific cut of the amplitude decreases slowly with s, while the whole amplitude falls fast and does not contribute in the high-energy limit due to $N=0$. According to Mandelstam, there are branchings, indeed, but they are different from (12.15).

To understand the mechanism of the vanishing of the AFS graph is very important. It teaches us how analytic properties of Feynman diagrams are related to the space–time picture of the s-channel processes.

Let us return to the general case and represent the particle–reggeon scattering amplitude as a sum of right- and left-cut contributions:

$$A(s_1) = \frac{1}{\pi} \int_{\text{right}} \frac{ds'_1 \, A_1(s'_1)}{s'_1 - s_1} + \frac{1}{\pi} \int_{\text{left}} \frac{ds'_1 \, A_2(s'_1)}{s'_1 - s_1} . \tag{12.17}$$

Now substitute into the integral (12.13) for N the right- and left-cut contribution, separately. We could, it seems, close the contour to the *left* in the s_1-integral of the first (right cut) term, to the *right* of the second (left cut) one, and get $N = 0 + 0$. When would such a trick not work? We will have $N \neq 0$ in the only case when the *separate* contributions of two cuts do not decrease faster than $1/s_1$ on the large circle, $|s_1| \to \infty$. But these are just those diagrams (see section 1 of this lecture) which possess a *third spectral function* ρ_{su}.

Let us recall that it was actually the third spectral function which led us to a contradiction with the Regge pole picture: a problem of the partial wave having the pole at $\ell = -1$ emerged.

In a way, here the circle closes. Both from the t-channel and the s-channel we came to the conclusion that the reggeon branching emerges only in the *relativistic theory*, where we have $\rho_{su} \neq 0$ which guarantees a non-vanishing contribution, $N \neq 0$. This corresponds to diagrams in which the singularities in s and u cannot be separated.

12.3.2 Reggeon branching contribution to cross section

Finally, let us calculate f_2 given in (12.12) to see whether it corresponds indeed to a two-reggeon branching as we expect. For positive signature, at small transverse momenta where $\alpha \simeq 1$ we have

$$\xi_\alpha = -\frac{1 + e^{-i\pi\alpha}}{\sin \pi\alpha} = \frac{-e^{-i\frac{\pi\alpha}{2}}}{\sin \frac{\pi\alpha}{2}} \simeq i.$$

Since the integrand in (12.12a) contains an exponential factor

$$\exp\left\{ \left(\alpha(k^2) + \alpha((k-q)^2) \right) \ln s \right\},$$

at large $\ln s$ the answer will be dominated by the value of the transverse momentum at which the exponent is maximal, and can be evaluated by the steepest-descent method. In the linear approximation,

$$\alpha(k^2) + \alpha((k-q)^2) - 2\alpha(0) \simeq -\alpha'(\mathbf{k}_\perp^2 + (\mathbf{k}-\mathbf{q})_\perp^2)$$
$$= -\tfrac{1}{2}\alpha'\mathbf{q}_\perp^2 - 2\alpha'\left(\mathbf{k}-\tfrac{1}{2}\mathbf{q}\right)_\perp^2,$$

the $\mathbf{k}_\perp$ integration yields

$$\int \frac{d^2\mathbf{k}_\perp}{(2\pi)^2}\, e^{-\alpha'(\mathbf{k}_\perp^2 + (\mathbf{k}-\mathbf{q})_\perp^2)\ln s} = \frac{1}{8\pi \ln s}\, e^{-\frac{1}{2}\alpha'\mathbf{q}_\perp^2 \ln s}.$$

Observing that

$$2\alpha(0) - 1 + \tfrac{1}{2}\alpha't \simeq 2\alpha\,(t/4) - 1, \qquad t \equiv q^2 = -\mathbf{q}_\perp^2,$$

we finally arrive at

$$f_2(s,t) \simeq i\,\xi_{\alpha(t/4)}^2 N^2 \frac{s^{2\alpha(t/4)-1}}{16\pi\alpha'\ln s}; \qquad N = N(q/2, q/2). \tag{12.18}$$

The result of our s-channel calculation is perfectly satisfactory: the $(\ln s)^{-1}$ suppression tells us that we have indeed found a branch-cut singularity. Its position in the j-plane follows the Mandelstam rule (11.36); the sign of this expression is also correct: $\mathrm{Im}\, f_2 \propto \xi_\alpha^2 = -1$ (N is real).

12.3.3 Branching in the impact parameter space

At small t the two-reggeon contribution is suppressed as $1/\ln s$ as compared to the pomeron pole. It is easy to understand the origin of this suppression, if we turn to the impact parameter space. The image of the pole amplitude is

$$f_1(s,q^2) \simeq isg^2\, e^{-\alpha'k^2\xi} \equiv isg^2 \int d^2\boldsymbol{\rho}\, e^{i\mathbf{k}\cdot\boldsymbol{\rho}}\, G_1(\boldsymbol{\rho},\xi); \tag{12.19a}$$

$$G_1(\boldsymbol{\rho},\xi) = \frac{e^{-\rho^2/4\alpha'\xi}}{4\pi\alpha'\xi}, \qquad \xi \equiv \ln s. \tag{12.19b}$$

Let us evaluate the Fourier transform of the two-pomeron branching:

$$s \cdot G_2(\boldsymbol{\rho},\xi) \equiv \int \frac{d^2\mathbf{q}}{(2\pi)^2}\, e^{-i\mathbf{q}\cdot\boldsymbol{\rho}} f_2(s,q^2). \tag{12.20}$$

Substituting (12.19) in the amplitude (12.12),

$$\int \frac{d^2\mathbf{q}}{(2\pi)^2}\frac{d^2\mathbf{k}}{(2\pi)^2}\, e^{-i\mathbf{q}\cdot\boldsymbol{\rho}}\, e^{i\mathbf{k}\cdot\boldsymbol{\rho}_1 + i(\mathbf{q}-\mathbf{k})\cdot\boldsymbol{\rho}_2} G_1(\boldsymbol{\rho}_1,\xi) G_1(\boldsymbol{\rho}_2,\xi) N^2\, d^2\boldsymbol{\rho}_1\, d^2\boldsymbol{\rho}_2,$$

the integrals over momenta produce $\boldsymbol{\rho}_1 = \boldsymbol{\rho}_2 = \boldsymbol{\rho}$, and we derive

$$G_2(\boldsymbol{\rho},\xi) = -iN^2 G_1^2(\boldsymbol{\rho},\xi). \tag{12.21}$$

The pole amplitude (12.19b) describes *two-dimensional diffusion* in the impact parameter space. It gives the probability to find a particle (placed at $\rho=0$ at zero time) in a given point ρ after the time ξ (α' determines the diffusion rate). In this language, the expression (12.21) corresponds to finding *two particles* in the same point. It is clear that such a probability is inverse proportional to the area and decreases with time, $1/S \propto 1/\alpha'\xi$. This can be seen directly by integrating (12.21) over ρ:

$$f_2(t=0) \propto \int d^2\rho\, G_2(\rho,\xi) \propto \frac{1}{\xi}, \quad \text{while} \quad f_1(t=0) \propto \int d^2\rho\, G_1(\rho,\xi) = 1.$$

12.4 Branchings in quantum mechanics: screening

We got the reggeon branchings from the Mandelstam diagrams with 'crosses' in both vertices (Fig. 12.2(a)). In the previous section we have analysed the rôle of the third spectral function and understood, why a simpler picture, that of Fig. 12.2(b), would not produce a branch cut in the j-plane.

What is the difference between the two pictures? In the second case, the particle moves as an elementary object just repeating the interaction. As we have discussed above, the probability 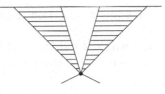 of such a process cannot be significant since the multi-particle fluctuation ('ladder') does not have enough time to collapse back into a single particle before it experiences the second scattering.

What is shown in the diagram of Fig. 12.2(a)?

Let us look at the lower part of the graph. Here the scattering occurs not on an elementary object, as in the diagram (b), but on the *decay products* of the target particle. We can model this picture in the non-relativistic

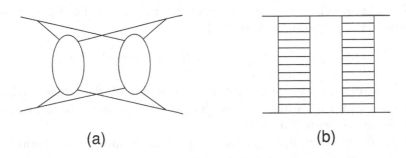

(a)   (b)

Fig. 12.2 Mandelstam two-reggeon branching (a) and the AFS graph (b).

scattering theory language if we take a deuteron D – a pn bound state – and study the scattering of the projectile off the proton and the neutron inside D.

There is a subtle point: in the relativistic theory both the upper and the lower parts of the diagram have to be complex – have to have 'crosses'. However, we will ignore this detail for the time being and consider the pion–deuteron scattering in non-relativistic quantum mechanics.

12.4.1 Deuteron scattering

Graphically, there are three possibilities for the πD scattering:

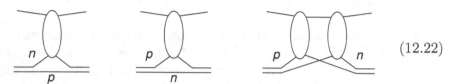

$$(12.22)$$

Since what happens in the upper part does not concern us for the time being, we can look upon this process as a scattering of a deuteron in the external potential (of the target pion).

Assume that a fast deuteron hits a potential:

$$\underline{} \, r_1$$
$$\underline{} \, r_2 \quad \bigcirc \quad \text{target}$$

How would I calculate this process quantum-mechanically?

In the initial state we have the wave function

$$\psi(\mathbf{r}_1, \mathbf{r}_2) = e^{i\mathbf{p}\cdot\frac{1}{2}(\mathbf{r}_1+\mathbf{r}_2)}\psi_D(\mathbf{r}_{12}), \qquad \mathbf{r}_{12} = \mathbf{r}_1 - \mathbf{r}_2, \qquad (12.23)$$

where $\mathbf{p}$ is the deuteron momentum, and $\psi_D(\mathbf{r}_{12})$ describes the relative motion of the proton p and neutron n inside it.

When D flies fast, it is clear that its nucleons have no time to interact with each other, so that the potential acts on p and n independently. Then in the final state the wave function acquires the scattering phase given by the sum of the p and n scattering phases:

$$\psi(\mathbf{r}_1, \mathbf{r}_2) \; \rightarrow \; \psi'(\mathbf{r}_1, \mathbf{r}_2) = \psi(\mathbf{r}_1, \mathbf{r}_2) \, e^{2i\delta(\boldsymbol{\rho}_1)+2i\delta(\boldsymbol{\rho}_2)}.$$

The phase depends on the impact parameters of the nucleons in the deuteron, $\boldsymbol{\rho}_1$ and $\boldsymbol{\rho}_2$, which do not change (stay 'frozen') in the course of the high-energy scattering.

To calculate the amplitude, I have to project ψ' onto the final state wave function ψ_f with a given momentum $\mathbf{p}_f = \mathbf{p} + \mathbf{q}$. If we are interested in D in the final state, we take $\psi_f = \psi_D$; if we investigate the decay, we look

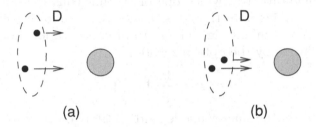

Fig. 12.3   Independent scattering (a) and shadowing configurations (b).

for $\psi_p \psi_n$. Since the S-matrix element does not depend on the longitudinal coordinates, z_i, the momentum transfer is purely transversal, $\mathbf{q} = \mathbf{q}_\perp$.

To derive the following expression is a very useful exercise (-1 subtracts the incoming beam):

$$f = \frac{p}{i} \int \frac{d^2 \boldsymbol{\rho}_c}{2\pi}\, e^{i\mathbf{q}_\perp \cdot \boldsymbol{\rho}_c} \int d^3 \mathbf{r}_{12}\, \psi_f^*(\mathbf{r}_{12}) \left[e^{2i\delta(\boldsymbol{\rho}_1)+2i\delta(\boldsymbol{\rho}_2)} - 1 \right] \psi_D(\mathbf{r}_{12}).$$

(12.24)

Let us take D in the final state and consider the forward scattering amplitude, $\mathbf{q}_\perp = 0$:

$$f(0) = ip \int \frac{d^2 \boldsymbol{\rho}_c}{2\pi} \int \psi_D^*(\mathbf{r}_{12}) \left[1 - e^{2i(\delta_1 + \delta_2)} \right] \psi_D(\mathbf{r}_{12})\, d^3 \mathbf{r}_{12}.$$

(12.25)

At the first glance nothing in the expression (12.25) resembles the diagrams of (12.22).

What is the difference between the two approaches?

The quantum-mechanical expression contains no information on which of the two particles has actually interacted and which has not, while in the language of Feynman graphs this is the main thing that enters.

To relate the two approaches, let us look at the simple algebraic identity

$$1 - S_1 S_2 = (1 - S_1) + (1 - S_2) - (1 - S_1)(1 - S_2), \quad S_i = e^{2i\delta(\boldsymbol{\rho}_i)}.$$

(12.26)

It can be interpreted as follows. The first two terms $(1 - S)$ describe independent interactions of p and n with the potential; the presence of the third one tells us that the sum of independent contributions apparently overestimates the answer, over-counts something. Indeed, when *one* of the nucleons interacts with the target as in Fig. 12.3(a), we get a sum of two contributions to the cross section.

However, the deuteron may also hit the potential in the configuration shown in Fig. 12.3(b). Here *both* nucleons interact though this contribution should be counted only once in the total cross section. It is the rôle of the quadratic term in (12.26) to correct for the double counting of the deuteron configuration in which one of the nucleons is *screened* by the other. It is clear therefore that this term – the 'shadowing correction' – must enter with a minus sign in the total cross section, related to the forward amplitude by the optical theorem,

$$\sigma_{\text{tot}} = \frac{4\pi}{p} \operatorname{Im} f(0). \tag{12.27}$$

The magnitude of the shadowing depends entirely on the geometry of the scattering process, on the size of the deuteron r_D as compared to the size of the potential, R. Let us analyse the two extreme cases.

$R \gg r_D$. This is the case of a very *broad* target. The total deuteron cross section, $\sigma_{\text{tot}} = 2\pi R^2$, will emerge from (12.26) as

$$2\pi R^2 = 2\pi R^2 + 2\pi R^2 - 2\pi R^2, \tag{12.28a}$$

meaning that the shadowing is 100% strong.

$R \ll r_D$. In this case the shadowing will occur only when the proton and neutron happen to have the same impact parameter, $|\boldsymbol{\rho}_1 - \boldsymbol{\rho}_2| \lesssim r_D$. The geometric weight of such rare configurations translates into a small shadowing correction

$$\frac{\Delta\sigma}{\sigma_1} \propto \frac{R^2}{r_D^2} \ll 1. \tag{12.28b}$$

12.4.2 Broad target

Consider first the case of a large target. One may have in mind, e.g. deuteron scattering off a heavy nucleus A, $A^{1/3} \propto R \gg r_D$. We obtain a big cross section by integrating the deuteron impact parameter $\boldsymbol{\rho}_c$ over the large area, $|\boldsymbol{\rho}_c| < R$. In this situation we can neglect the dependence on the relative coordinate, $\boldsymbol{\rho}_{12} = \boldsymbol{\rho}_1 - \boldsymbol{\rho}_2$, and approximate

$$\delta_1 = \delta(\boldsymbol{\rho}_c + \tfrac{1}{2}\boldsymbol{\rho}_{12}) \simeq \delta_2 = \delta(\boldsymbol{\rho}_c - \tfrac{1}{2}\boldsymbol{\rho}_{12}) \simeq \delta(\boldsymbol{\rho}_c).$$

Since in this approximation the S-matrix does not depend on $\mathbf{r}_{12}$, we get

$$f(0) = ip \int \frac{d^2\boldsymbol{\rho}_c}{2\pi} \left(1 - e^{4i\delta(\boldsymbol{\rho}_c)}\right) \times 1, \tag{12.29}$$

where the last '1' originates from the wave function normalization,

$$\int d^3\mathbf{r}_{12} |\psi_D(\mathbf{r}_{12})|^2 = 1. \tag{12.30}$$

If a deuteron hits a big nucleus head-on, it definitely interacts and disappears (is absorbed), feeding various inelastic channels. This corresponds to the elasticity coefficient $\eta(\boldsymbol{\rho})$ in the S-matrix element being vanishingly small at $|\boldsymbol{\rho}| < R$, so that

$$e^{2i\delta(\boldsymbol{\rho})} \equiv \eta(\boldsymbol{\rho})\, e^{2i\beta(\boldsymbol{\rho})} \simeq \begin{cases} 0, & \text{for } |\boldsymbol{\rho}| < R, \\ 1, & |\boldsymbol{\rho}| \gg R. \end{cases}$$

This is the 'black disc' picture,

$$\int_0^\infty \frac{d^2\boldsymbol{\rho}_c}{2\pi}\left(1 - e^{4i\delta(\boldsymbol{\rho}_c)}\right) \simeq \int_0^R \frac{d^2\boldsymbol{\rho}_c}{2\pi} \simeq \tfrac{1}{2}R^2,$$

yielding the total cross section

$$\sigma_{\text{tot}} = \frac{4\pi}{p}\,\text{Im}\, f(0) \simeq 2\pi R^2.$$

Half of this cross section is the diffractive elastic scattering,

$$\sigma_{\text{el}} = \int d\Omega\, |f(q)|^2 = \int d^2\mathbf{q}\frac{|f(q)|^2}{p^2} = \int d^2\boldsymbol{\rho}\left|1 - e^{4i\delta(\boldsymbol{\rho})}\right|^2 \simeq \pi R^2;$$

the other half is due to *inelastic* processes.

12.4.3 Diffractive dissociation of a deuteron

What if I am interested not in the elastic scattering, but in the deuteron break-up $D \to pn$? In principle, this is one of the inelastic channels. We are, however, interested in the specific inelastic process in which the target nucleus remains intact and scatters as a whole, $D + A \to p + n + A$, instead of being 'heated up'.

Such a *diffractive dissociation* of the deuteron is possible only if the momentum transfer is very small, $q \sim R^{-1}$, corresponding to scattering angles $\theta_s \sim 1/pR$; otherwise the nucleus A would break up too.

What will be the scale of such a cross section? Obviously, its amplitude cannot be as large as the elastic one. If it were, integrating over the same narrow angular cone as for the elastic scattering, we would get a diffractive dissociation cross section $\sigma_{\text{tot}}^{\text{dd}}$ as large as the elastic one:

$$\sigma_{\text{tot}}^{\text{dd}} = \int |f_{D \to pn}|^2\, d\Omega \sim \int d\theta_s^2 \left|pR^2\right|^2 \sim R^2, \quad d\theta_s^2 \sim \frac{1}{(pR)^2}.$$

But there is no place left in σ_{tot} to accommodate another contribution of the order of R^2. What is going on in the formula?

The point is that in the rough approximation of the scattering phases independent of $\boldsymbol{\rho}_{12}$, the main term in the forward scattering amplitude, $\mathbf{q} = 0$, cancels because of the *orthogonality of the initial- and final-state wave functions:*

$$f_{D \to pn}(0) = \frac{ip}{2\pi} \int d^2 \boldsymbol{\rho}_c [1 - S_1 S_2] \cdot \langle \psi_{pn} | \psi_D \rangle ;$$

$$\langle \psi_{pn} | \psi_D \rangle = \int d^3 \mathbf{r}_{12} \, \psi_{pn}^*(\mathbf{r}_{12}) \psi_D(\mathbf{r}_{12}) = 0. \tag{12.31}$$

To prevent this cancellation, the expansion of the phase in $\boldsymbol{\rho}_{12}$ has to be carried out,

$$\delta[1 - S_1 S_2] \simeq \tfrac{1}{4} [(\boldsymbol{\rho}_{12} \boldsymbol{\nabla}_{\boldsymbol{\rho}_c} S)^2 - S(\boldsymbol{\rho}_{12} \boldsymbol{\nabla}_{\boldsymbol{\rho}_c})^2 S], \quad S = S(\boldsymbol{\rho}_c). \tag{12.32}$$

If S changes smoothly, $|\boldsymbol{\nabla} S| \sim R^{-1}$, then

$$p^{-1} f_{D \to pn}(0) \sim \int^R \rho_c \, d\rho_c \frac{\langle \psi_{pn} | \boldsymbol{\rho}_{12}^2 | \psi_D \rangle}{R^2} \sim r_D^2. \tag{12.33}$$

If the target has a relatively *sharp edge*, in which region S only changes, the forward amplitude gets enhanced,

$$p^{-1} f_{D \to pn}(0) \sim \int \rho_c \, d\rho_c \frac{\delta(\rho_c - R)}{H} \cdot r_D^2 \sim \frac{R}{H} \cdot r_D^2, \tag{12.34}$$

where $H \sim 1/\mu$ is the width of the transition region, $R \gg H \gg r_D$. In this case the inelastic amplitude is still suppressed as $1/R \propto A^{-1/3}$ as compared to the elastic one.

The nature of the suppression of the forward dissociation amplitude (12.34) is consistent with the total contribution of inelastic diffraction to the interaction cross section, $\sigma_{\text{tot}}^{\text{dd}}$. When we take $\mathbf{q} \neq 0$ in the transition amplitude (12.24), the wave functions are no longer orthogonal; for small $\mathbf{q}$ one has

$$\langle \psi_{pn}(\mathbf{p} + \mathbf{q}) | \psi_D(\mathbf{p}) \rangle \sim (\mathbf{q} \cdot \mathbf{k}_{pn}) \, r_D^2,$$

where $\mathbf{k}_{pn}$ is the relative momentum of the nucleons. Integrating the amplitude squared over the scattering angle, one obtains

$$\int \frac{d^2 \mathbf{q}}{p^2} |f_{D \to pn}|^2 \sim r_D^4 \int d^2 \boldsymbol{\rho}_c |(\mathbf{k}_{pn} \cdot \boldsymbol{\nabla}_{\boldsymbol{\rho}_c} S^2)|^2 \sim \pi R \cdot \frac{r_D^2}{H} \cdot \mathbf{k}_{pn}^2 r_D^2 .$$

An integral over $\mathbf{k}_{pn}$ produces $\langle \mathbf{k}_{pn}^2 \rangle r_D^2 \sim 1$, and we get the estimate

$$\sigma_{\text{tot}}^{\text{dd}} \sim \pi R \cdot \frac{r_D^2}{H},$$

consistent with the magnitude of the forward amplitude (12.34).

Diffractive dissociation occurs only on the *periphery* of the target nucleus. It is very important to understand that the diffraction does not change the internal state of the system, unless the system is scattered as a whole, $\mathbf{q}_\perp \neq 0$.

12.4.4 Shadowing

We return to the discussion of the shadowing correction. Combining (12.24) and (12.26), we have

$$f(\mathbf{q}) = \frac{ip}{2\pi} \int d^2\boldsymbol{\rho}_c \, e^{i\mathbf{q}\boldsymbol{\rho}_c} \Big\langle (1-S_1) + (1-S_2) - (1-S_1)(1-S_2) \Big\rangle, \quad (12.35)$$

where $\langle\ \rangle$ stands for the average over the deuteron state. Let us consider this expression term by term, $f = f_1 + f_2 + f_{12}$. Recall that $S_1 = S_1(\boldsymbol{\rho}_1) = S_1(\boldsymbol{\rho}_c + \frac{1}{2}\boldsymbol{\rho}_{12})$. Introducing the Fourier transform of the profile of the nucleon scattering matrix element,

$$1 - S_1(\boldsymbol{\rho}) = \int \frac{d^2\mathbf{k}}{(2\pi)^2} e^{-i\mathbf{k}\cdot\boldsymbol{\rho}} \varphi_1(\mathbf{k}),$$

and performing the impact parameter integrals we have

$$f_1(\mathbf{q}) = \frac{ip}{2\pi} \varphi_1(\mathbf{q}) \cdot F_D(\mathbf{q}^2), \quad (12.36)$$

where F is the *electromagnetic charge form factor* of the deuteron, given by the Fourier transform of the probability density to find the electric charge (proton) inside the deuteron:

$$F_D(\mathbf{q}^2) = \int d^3\mathbf{r}_{12} |\psi_D(\mathbf{r}_{12})|^2 \, e^{-i\frac{\mathbf{q}}{2}\cdot\mathbf{r}_{12}}; \quad F(0) = 1.$$

The optical theorem (12.27) tells us that $\varphi(0)$ is simply related to the total interaction cross section of a single nucleon with the target,

$$\varphi(0) = \tfrac{1}{2}\sigma_N. \quad (12.37)$$

For the last term in (12.35) we have

$$(1-S_1)(1-S_2) = \int \frac{d^2\mathbf{k}_1}{(2\pi)^2} \frac{d^2\mathbf{k}_2}{(2\pi)^2} \varphi_1(\mathbf{k}_1)\varphi_2(\mathbf{k}_2) \, e^{-i(\mathbf{k}_1+\mathbf{k}_2)\boldsymbol{\rho}_c} \, e^{-i\frac{\mathbf{k}_1-\mathbf{k}_2}{2}\boldsymbol{\rho}_{12}},$$

and derive analogously

$$f_{12}(\mathbf{q}) = -\frac{ip}{2\pi} \int \frac{d^2\mathbf{k}}{(2\pi)^2} \varphi_1(\tfrac{1}{2}\mathbf{q}+\mathbf{k}) \, \varphi_2(\tfrac{1}{2}\mathbf{q}-\mathbf{k}) \cdot F_D(4\mathbf{k}^2). \quad (12.38)$$

Assembling the three terms of (12.35), for the total cross section we obtain

$$\sigma_D = \sigma_p + \sigma_n - 2\,\mathrm{Re} \int \frac{d^2\mathbf{k}}{(2\pi)^2} \varphi_p(\mathbf{k})\varphi_n(-\mathbf{k}) \cdot F_D(4\mathbf{k}^2). \quad (12.39)$$

Now we are ready to check our expectations (12.28) about the magnitude of the shadowing correction that were based on the classical geometric considerations.

Broad potential. In this case p and n almost always hit the target together, and we expected in (12.28a) a 100% negative shadowing correction. Within the black-disc model, the nucleon amplitude f_1 is purely imaginary, so that φ is real and we can replace $\operatorname{Re}\varphi^2$ by $|\varphi|^2$. Using

$$\frac{|\varphi(\mathbf{k})|^2}{(2\pi)^2} = \frac{1}{p^2}|f_1(\mathbf{k})|^2 = \frac{1}{p^2}\frac{d\sigma_{\mathrm{el}}^N}{d\Omega} \simeq \frac{d\sigma_{\mathrm{el}}^N}{d^2\mathbf{k}},$$

this allows us to represent (12.39) in terms of the differential elastic nucleon scattering cross section as

$$\sigma_{\mathrm{tot}}^D = 2\sigma_{\mathrm{tot}}^N - 2\int d\mathbf{k}^2 \frac{d\sigma_{\mathrm{el}}^N}{d\mathbf{k}^2} F_D(4\mathbf{k}^2). \tag{12.40}$$

The characteristic momenta in the integral (12.40), $\mathbf{k}^2 \sim R^{-2}$ are much smaller than the internal scale of the form factor, r_D^{-2}, therefore we can put $F = F(0) = 1$ to obtain

$$\sigma_{\mathrm{tot}}^D = 2\sigma_{\mathrm{tot}}^N - 2\sigma_{\mathrm{el}}^N = 2\sigma_{\mathrm{tot}}^N - \sigma_{\mathrm{tot}}^N = \sigma_{\mathrm{tot}}^N.$$

Compact potential, large-size projectile. The product $\varphi_p(\boldsymbol{\rho}_1)\varphi_n(\boldsymbol{\rho}_2)$ requires p and n to be at the same impact parameter in order to simultaneously aim at the small-size target, $R \ll r_D$, and screen one another. The deuteron form factor falling sharply above $\mathbf{k}^2 \sim r_D^{-2} \ll R^{-2}$, the factors φ in (12.39) can be taken out,

$$\sigma_D \simeq \sigma_p + \sigma_n - 2\,\varphi_p(0)\varphi_n(0)\int \frac{d\mathbf{k}^2}{4\pi} F_D(4\mathbf{k}^2).$$

Invoking (12.37), this immediately gives the Glauber formula

$$\sigma_D \simeq \sigma_p + \sigma_n - \frac{\sigma_p\sigma_n}{4\pi}\cdot\overline{r_D^{-2}}, \tag{12.41a}$$

where we have used

$$\overline{r_D^{-2}} \equiv \int d^3\mathbf{r}_{12}\frac{|\psi_D(\mathbf{r}_{12})|^2}{\mathbf{r}_{12}^2} = \tfrac{1}{2}\int d\mathbf{k}^2\, F_D(4\mathbf{k}^2). \tag{12.41b}$$

12.5 Back to relativistic theory

The analogy between the quantum-mechanical shadowing phenomenon and the two-reggeon branching can be made explicit if we substitute the relativistic amplitude in the expression (12.38) for the double-scattering graph:

$$2i\varphi_1(\mathbf{q}) = \frac{4\pi f_1(\mathbf{q})}{p} \implies \frac{A(s,t)}{s}.$$

We have used the deuteron as a model for a composite target, in order to understand the origin of the negative correction due to the exchange of two reggeons. Let us say a few words about the real deuteron, about high-energy πD scattering.

12.5.1 Glauber scattering

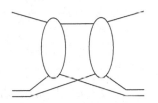

We have analysed the double scattering in quantum mechanics and found a large correction. We could carry out this calculation directly from relativistic diagrams, in the non-relativistic approximation (treating the masses of the particles as large parameters).

But, in the framework of our former logic, this correction has to be *zero*, since there is *no cross* in the upper part of the corresponding diagram! On the other hand, the non-zero answer was legitimately obtained in the non-relativistic theory. What is happening here?

Normally there are no small parameters in the diagrams describing relativistic hadron interactions. The deuteron problem is, however, special: calculating the diagrams for πD scattering we encounter a *small parameter*, namely the ratio of the size of the pion, $R \sim r_0 \sim \mu^{-1}$, to that of the deuteron, $R/r_D \ll 1$. It is the presence of this new parameter that is responsible for the survival of the *semi*-AFS diagram.

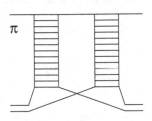

What is the reason for that, in classical terms? Consider the process in a reference frame in which the π meson is fast. Assuming that the situation is non-relativistic, we have obtained three contributions: π collided with either of p or n, or there was screening as a result of a double interaction. Why were there no virtual particles involved; why was the answer expressed via the on-mass-shell amplitudes, i.e. via real particles?

We assumed r_D to be large, and this was the reason why, in the non-relativistic theory, π propagated between the two successive collisions as a real particle. The fact that the π meson had to pass a large distance, limited the energy uncertainty, $\Delta E \sim 1/\Delta t \sim 1/r_D \ll 1/R \sim \mu$, and forced π to be real. But this argument disregards the relativistic retardation effect (as if virtual objects could not propagate at large distances!). In the relativistic situation, $E_\pi \gg \mu$, the lifetime of the virtual pion fluctuation, $\Delta t \sim E_\pi / (\Delta m)^2$, may become comparable with r_D even for large invariant masses of the virtual state, Δm. Depending on the pion energy, we have three scenarios.

(1) $E/\mu^2 \ll r_D$. There is just one π, and the non-relativistic quantum mechanics gives a correct answer.

(2) $E/\mu^2 \sim r_D$. A transitional regime: the pion can be accompanied by a small number of additional particles.

(3) $E/\mu^2 \gg r_D$. In the space between the collisions with p and n, multi-particle showers with large invariant masses $(\Delta m)^2 \gg \mu^2$ can propagate, and the probability of just one π becomes vanishingly small.

12.5.2 Relativistic inelastic corrections to Glauber scattering

The deuteron example shows that the vertex blocks N that we have treated as constant, may contain, in specific cases, small internal parameters (like the deuteron binding energy) and therefore may still be changing fast at small momenta, well below a typical hadronic scale $\mathbf{k}^2 \lesssim \mu^2$.

There is another very important lesson. In the non-relativistic theory the screening means that one particle is an obstacle for another one just geometrically, i.e. the screening is the result of the geometry of subsequent collisions. It looks natural to discuss the scattering of a particle off a *nucleus* in terms of successive interactions with nucleons. In so doing, it is usually implied that the projectile preserves its identity when it propagates between successive collisions. This picture takes into account the Glauber corrections due to *elastic screening*. In the reggeon language, these diagrams correspond to *non-enhanced branchings*.

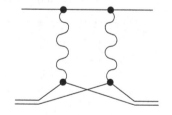

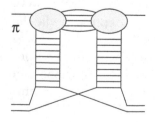

In the relativistic case there may be a whole shower propagating between the interaction points, since when the energy of the projectile $E \to \infty$, it itself fluctuates at distances exceeding the size of the target. We get another contribution to the screening phenomenon – an *inelastic* screening.

At very high energies the processes with large cascades become dominant. But these are again 'ladders'. Summing up high-mass intermediate inelastic states we arrive at the *enhanced branchings*,

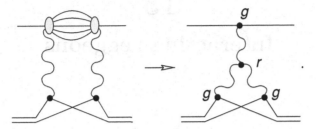

13

Interacting reggeons

13.1 Constructing effective field theory of interacting reggeons

Now we are in the position to address an important question, namely: what can be said about the vertex functions N for other, more complicated, processes.

13.1.1 Stuffing up the vertex: $R \to RR$ transition

Let us draw a more complicated diagram for N. Again, I will carry out the factorization and obtain the same structure expressing N as the energy integral (12.14) of some new amplitude (which now has a three-particle threshold in s_1), etc.

Continuing the process I will face the following problem. As the diagram becomes more complicated, it may start to decrease slower and slower with the growth of energy. For a concrete diagram, N is a number. Summing up a set of diagrams may rend, however, a *diverging answer* for N (if faraway multi-particle thresholds happen to dominate the internal integrals in A_1). Stuffing the particles into N, we effectively increase s_1 and may arrive, e.g. at the graph shown in Fig. 13.1(a). This one has all the reasons to contradict my initial expectation that it was profitable to ascribe a *finite energy* $\langle s_1 \rangle$ to the particle–reggeon scattering blocks, and the bulk of the total energy – to the parallel reggeons, $s_{\mathrm{I}} \sim s_{\mathrm{II}} \sim s$.

This observation exposes another flaw in our logic. In the previous lecture we have analysed a branch cut singularity in the angular momentum plane due to two reggeons. But who ever told us that the blocks

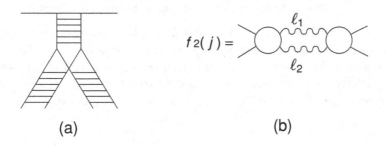

(a)   (b)

Fig. 13.1   (a) A ladder in the two-particle–two-reggeon vertex block; (b) partial
wave for the two-reggeon branching.

in Fig. 13.1(b) did not contain singularities themselves? In particular, if
 contains, e.g. *a pole* in the total angular momentum j (as
the ladder in Fig. 13.1(a) suggests), the asymptotics of $N(s_1)$ would be
bad, and the integral (12.14) would diverge. Therefore, it was a mistake
to suppose that the reggeon production amplitude was a smooth function
in the j-plane.

Let us write, once again, the integral for the vertex function N:

$$N = \int_{s_0}^{\infty} \frac{ds_1}{2\pi} A_1(s_1, q^2; \ldots) . \qquad (13.1a)$$

Does this not look familiar?

Recall, how in Lecture 7 we have related the t-channel partial wave to
the imaginary part of the amplitude, cf. (7.27),

$$f_j(t) \sim \int Q_j(z) A_1(z, t) \, dz \simeq \int_{s_0}^{\infty} ds \frac{A_1(s, t)}{s^{j+1}} . \qquad (13.1b)$$

Comparing the two expressions (13.1) we conclude that, if the wavy lines
in Fig. 13.1(b) were ordinary particles rather then reggeons, N would be
just the value of the partial wave f_j for the $2 \to 2$ scattering amplitude
at $j = -1$. Thus, our N is the analogue of the partial wave f_{-1} for a
two particles $\to$ two reggeons transition. In this language, the problem
of convergence reduces to the question where the singularities in j are:
if they all lie on the left of $j = -1$ then N is finite; if the rightmost
singularity is above -1 then $N = \infty$. So, the divergence of the vertex N
is connected to the structure of angular-momentum singularities of the
reggeon production amplitude.

13.1.2 Reggeon field theory: construction logic

So, what is the basic idea? We have found the branchings, and realized that all of them are relevant for the asymptotic behaviour. To construct an effective field theory describing high-energy scattering phenomena, I have to have, I would say, an initial (bare) reggeon, reggeon emission amplitudes, amplitudes of inter-reggeon interactions, etc.

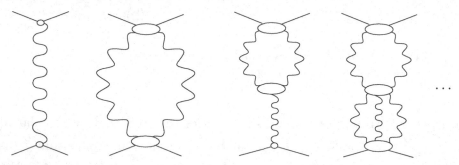

Dealing with a field theory we assume that once we plug in bare quantities (propagators, vertices), the true scattering amplitudes can be calculated, taking into account all renormalization corrections, repetitions, etc. From the point of view of dispersion relations, *all* the branchings are important. Obviously, if such a situation persists at the level of 'true' (renormalized) objects and interactions, we would have failed the task. In practice, we would rather have an answer that is self-consistent within a scheme that contains but a small number of bare (unknown) quantities.

Let us divide the diagrams into two classes.

(1) Diagrams with only a few lines, in configurations with not too large relative momenta and virtualities (so that the s_1 integral converges), yielding finite functions N. These we will call 'bare reggeon vertices', and will treat them as such.

(2) Diagrams with large pair energies $s_1 \gg \mu^2$. As to these diagrams, we will assume that they can be expressed, in turn, in terms of reggeons, since s_1 is large and the amplitude is in the asymptotic regime.

By so doing we make the second class of diagrams, again, subject to *calculation*. This fits well the idea of constructing a self-consistent field theory. There is, of course, a problem: how to actually separate diagrams into the two classes (when few becomes a good few?).

In the diagrammatic language, the main hypothesis can be formulated as follows. I *suppose* that after I reformulate the theory in terms of *exact* (renormalized) reggeons, the diagrams will naturally fall into two classes.

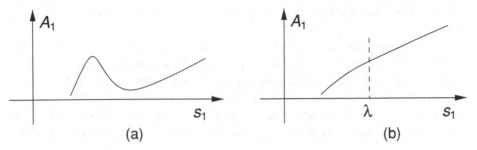

Fig. 13.2   Dividing vertex diagrams into two classes: (a) a clear-cut case; (b) setting an arbitrary separation parameter $s_1 = \lambda$.

The ideal case is when the two energy regions separate naturally as in Fig. 13.2(a). But even if the discontinuity A_1 does not have a prominent structure as shown in Fig. 13.2(b), we can introduce an arbitrary finite cut-off λ to formally divide the two regions.

We hope that in the *high energy* region, $s_1 \gg \lambda = \mathcal{O}(1)$, the picture will get simpler in the asymptotics, and we will be able to construct a self-consistent calculation scheme. At the same time, we will treat particle–reggeon and reggeon–reggeon interaction blocks at *low energies*, $s_1 \lesssim \lambda$, as input (bare) vertices. The introduction of bare vertices is a way to separate the finite energy domain about which nothing definite can be said from first principles.

Now that we have virtually 'calculated' *all* the diagrams of the first class ($s_1 \leq \lambda$, $s_2 \leq \lambda$, $s_{\text{internal}} \sim s \to \infty$), we can make an important statement: $N \neq 0$.

Since my 'calculation' is valid only for $s_1 \leq \lambda$ anyway, I can deform the contour around the right cut in spite of the divergence of the integral, and define a 'bare' vertex

$$N_{\text{bare}} \equiv \int_{4\mu^2}^{\lambda} \frac{ds_1}{\pi} A_1(s_1). \tag{13.2}$$

About the integrand we can say that $A_1 > 0$, at least in a certain region: for (near to) forward scattering it is given by a sum of positive contributions

$$\text{Im}\, A = A_1 = \;\vcenter{\hbox{\includegraphics{fig1}}}\; = \sum_n a_n\, a_n* > 0,$$

$$\vcenter{\hbox{\includegraphics{fig2}}} = \vcenter{\hbox{\includegraphics{fig3}}} + \vcenter{\hbox{\includegraphics{fig4}}} + \vcenter{\hbox{\includegraphics{fig5}}} + \cdots \tag{13.3}$$

This allows us to state that the particle–reggeon scattering does exist: $N_{\text{bare}} \neq 0$.

At this point you may wonder: how did we manage to get $N = 0$ for the AFS amplitude? The total contribution of each multi-particle state is positive since the same (full) amplitude stands there on the left and on the right from the discontinuity in (13.3). To observe the AFS cancellation, we have picked *specific pieces* from the multi-particle cuts through (corrections to) the *reggeon vertex*, which cuts are not positively definite:

$$A_1^{\text{AFS}} = \quad\rightarrowtail\!\!\cdot\!\!\cdot\!\!\leftarrowtail\quad + \quad 2\text{Re}\,\rightarrowtail\!\!\cdot\!\!\cdot\!\!\cdot\!\!\bullet\!\!\!\!\!\!\!< \quad + \cdots \qquad (13.4)$$

The series of selected terms cancelled, upon integration over s_1, the positive contribution $a_1 a_1^*$ of the one-particle state. Such a selection procedure eliminates not only the pole but some other positive terms on the r.h.s. of the unitarity relation (13.3) as well. For example,

$$A_1^{(2)} = \quad\rightarrowtail\!\!\bullet\!\!\cdot\!\!\cdot\!\!\bullet\!\!\leftarrowtail\quad + \cdots$$

which diagram, being *planar*, has no third spectral function either. At the same time, the diagrams with $\rho_{su} \neq 0$ do survive and contribute to N. For them an artificial procedure of extracting negative pieces from the farther terms of the unitarity relation makes no sense: the series of potentially compensating contributions diverges.

13.2 Feynman diagrams for reggeon branchings

Now we have to examine different diagrams and learn to calculate them.

How can one describe the contribution of branchings in a transparent way? We start from a two-reggeon branching,

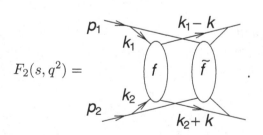

At high energies this diagram can be expressed in terms of the asymptotics of the blocks f and $\tilde{f}$. For one of the blocks, introducing the complex angular momentum variable ℓ_1 and the reggeon Green function G_{ℓ_1}, we

can write

$$f = g(k_1, k)g(k_2, k) \cdot (-1) \underbrace{\int \frac{d\ell_1}{2\pi i} \xi_{\ell_1} [2k_1 k_2]^{\ell_1} G_{\ell_1}(\mathbf{k})}_{\text{block}} .$$

If G has a simple pole, $G = 1/(\ell_1 - \alpha(\mathbf{k}^2))$, the integration can be carried out producing $(2k_1 k_2)^\alpha \xi_\alpha$. Similarly, for the second block we have

$$\begin{aligned}
\tilde{f} &= -g(p_1 - k_1, q - k)g(p_2 - k_2, q - k) \\
&\times \int \frac{d\ell_2}{2\pi i} \xi_{\ell_2} [2(p_1 - k_1)(p_2 - k_2)]^{\ell_2} G_{\ell_2}(\mathbf{q} - \mathbf{k}).
\end{aligned}$$

Taking into account that

$$\alpha_1 \sim 1, \ \beta_1 \sim s^{-1}; \qquad \alpha_2 \sim s^{-1}, \ \beta_2 \sim 1;$$
$$\alpha_k \sim \beta_k \sim s^{-1}, \qquad k^2 \simeq -\mathbf{k}_\perp^2,$$

$$2k_1 k_2 = \alpha_1 \beta_2 s, \quad 2(p_1 - k_1)(p_2 - k_2) = (1 - \alpha_1)(1 - \beta_2)s,$$

everything becomes factorized, and we obtain

$$\begin{aligned}
F_2(s, q^2) &= \frac{1}{2!} \frac{i\pi}{2} \int \frac{d\ell_1}{2\pi i} \int \frac{d\ell_2}{2\pi i} \int \frac{d^2\mathbf{k}}{(2\pi)^2} \xi_{\ell_1} G_{\ell_1}(\mathbf{k}) \xi_{\ell_2} G_{\ell_2}(\mathbf{q} - \mathbf{k}) \\
&\quad \cdot N_{\ell_1 \ell_2}(\mathbf{k}, \mathbf{q}) N^*_{\ell_1 \ell_2}(\mathbf{k}, \mathbf{q}) s^{\ell_1 + \ell_2 - 1},
\end{aligned} \tag{13.5}$$

where

$$N_{\ell_1 \ell_2} = \frac{1}{\sqrt{2}} \int \frac{d^4 k_1}{(2\pi)^4 i} \int \frac{s \, d\beta_k}{2\pi i} \frac{g(k_1, k)g(p_1 - k_1, q - k)\alpha_1^{\ell_1}(1 - \alpha_1)^{\ell_2}}{(\ \ )(\ \ )(\ \ )(\ \ )}.$$

We have discussed that this expression is analogous to a Feynman diagram for particles with spins; $\alpha_1^{\ell_1}(1 - \alpha_1)^{\ell_2}$ are 'cosines of angles'. Since α_1 changes in the interval $0 < \alpha_1 < 1$ (cf. calculation of the box diagram in Section 9.2.3 of Lecture 9), this factor does not pose any problem.

13.2.1 Two-reggeon branching as a Feynman integral

How to find the t-channel partial wave corresponding to the amplitude (13.5)? This is done using the 'relativistic projection' (7.19):

$$f_j^{(2)}(q^2) = \frac{2}{\pi} \int_{s_0}^\infty \frac{ds}{s^{j+1}} \operatorname{Im} F_2(s, q^2); \tag{13.6a}$$

$$F_2(s, q^2) = -\frac{1}{4i} \int dj \, \xi_j s^j f_j^{(2)}(q^2). \tag{13.6b}$$

The integral is very simple:

$$f_j(q^2) = \frac{1}{2!} \int \frac{d^2\mathbf{k}}{(2\pi)^2} \int \frac{d\ell_1 d\ell_2}{(2\pi i)^2} \gamma_{\ell_1 \ell_2} N_{\ell_1 \ell_2}^2 G_{\ell_1}(\mathbf{k}) G_{\ell_2}(\mathbf{q}-\mathbf{k}) \frac{s_0^{\ell_1+\ell_2-j-1}}{j+1-\ell_1-\ell_2}.$$

Here $\gamma_{\ell_1 \ell_2} \equiv \mathrm{Im}(i\xi_{\ell_1}\xi_{\ell_2})$.

One may always choose units such that $s_0 = 1$. More importantly, I will always be interested in the region j close to the singularity so that $j + 1 - \ell_1 - \ell_2 \approx 0$ and the factor $s_0^{\ell_1+\ell_2-j-1}$ can be dropped, independently of the value of s_0.

The expression we have obtained reminds us very much of the old perturbation theory. Indeed, we took two 'particles' with propagators

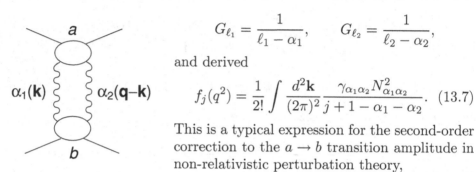

$$G_{\ell_1} = \frac{1}{\ell_1 - \alpha_1}, \qquad G_{\ell_2} = \frac{1}{\ell_2 - \alpha_2},$$

and derived

$$f_j(q^2) = \frac{1}{2!} \int \frac{d^2\mathbf{k}}{(2\pi)^2} \frac{\gamma_{\alpha_1\alpha_2} N_{\alpha_1\alpha_2}^2}{j+1-\alpha_1-\alpha_2}. \quad (13.7)$$

This is a typical expression for the second-order correction to the $a \to b$ transition amplitude in non-relativistic perturbation theory,

$$\delta f_{a\to b}(E) = \sum_n \frac{V_{an}V_{nb}}{E_n - E}. \quad (13.8)$$

We get a direct correspondence if we treat

$$\alpha_1(\mathbf{k}) - 1 \equiv \varepsilon_1, \qquad \alpha_2(\mathbf{q}-\mathbf{k}) - 1 \equiv \varepsilon_2$$

as energies of the two particles, $E_n = \varepsilon_1 + \varepsilon_2$, and $j - 1 \equiv \omega$ as the total energy (E):

$$E - E_n \equiv \omega - (\varepsilon_1 + \varepsilon_2) = (j-1)-(\alpha_1-1)-(\alpha_2-1) = j+1-\alpha_1-\alpha_2.$$

The two-dimensional momentum $\mathbf{k}$ in (13.7) plays the rôle of the index n of the intermediate state in (13.8).

This result can be rewritten using Feynman's techniques (covariant perturbation theory). To this end we return to the original representation containing integration over ℓ_1 and ℓ_2. The contours in ℓ_i run parallel to the imaginary axis, *on the right* of the singularities of the Green functions

G_{ℓ_i}. Consider, e.g., the ℓ_2-plane:

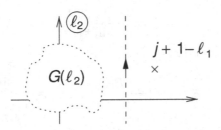

For the energy integral in (13.6a) to converge, j has to be sufficiently large: $j + 1 - \ell_1 - \ell_2 > 0$. Then, in the ℓ_2-plane the contour lies *on the left* of the pole: $\ell_2 < \ell_2^{\text{pole}} = j + 1 - \ell_1$. Therefore we may evaluate the integral by closing the contour on the right half-plane, around the pole. Introducing $\omega_1 = \ell_1 - 1$ and $\omega_2 = \ell_2 - 1$ as new variables, $G_{\ell_i} \to G_{\omega_i}$, the relation $j + 1 - \ell_1 - \ell_2 = 0$ translates into $\omega_2 = \omega - \omega_1$ and we have

$$f_\omega(q^2) = \frac{1}{2!} \int \frac{d^2\mathbf{k}}{(2\pi)^2} \int \frac{d\omega_1}{2\pi i} \gamma_{\omega_1\omega_2} N_{\omega_1\omega_2}^2 G_{\omega_1}(\mathbf{k}) G_{\omega - \omega_1}(\mathbf{q} - \mathbf{k}). \quad (13.9\text{a})$$

This expression can be cast in a more symmetric form by introducing two sets of integrations,

$$f_\omega(q^2) = \frac{1}{2!} \int \frac{d\omega_1 \, d^2\mathbf{k}_1}{(2\pi)^3 i} \int \frac{d\omega_2 \, d^2\mathbf{k}_2}{(2\pi)^3 i} \gamma_{\omega_1\omega_2} N_{\omega_1\omega_2}^2 G_{\omega_1}(\mathbf{k}_1) G_{\omega_2}(\mathbf{k}_2)$$
$$\cdot (2\pi i) \, \delta(\omega - \omega_1 - \omega_2) \cdot (2\pi)^2 \delta(\mathbf{q} - \mathbf{k}_1 - \mathbf{k}_2). \quad (13.9\text{b})$$

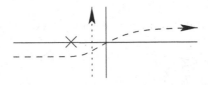

This is a Feynman diagram in the exact sense of the word; a 'particle' with 'energy–momentum' $(\omega_1, \mathbf{k}_1)$ is described by the propagator $G(\omega_1, \mathbf{k}_1)$. The only unfamiliar feature is the form of the integration in 'energy' ω: we have now the integral running along the *imaginary axis* instead of the usual Feynman integration contour,

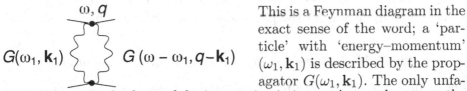

This does not make much of a difference, it is just more convenient. It is important that the singularities of the propagator $G(\omega, \mathbf{k})$ lie on the one side of the contour, i.e. our 'particles' are *non-relativistic*.

Recall the reggeon signature factor (8.8c),

$$\xi_\ell = -\frac{e^{-i\pi\ell} \pm 1}{\sin \pi\ell},$$

which can be represented as

$$\xi_\ell = -\zeta_\ell^{-1} \cdot \exp\left\{-i\frac{\pi}{2}\left(\ell + \frac{1-P_\ell}{2}\right)\right\}, \qquad \zeta_\ell = \begin{cases} \sin\frac{\pi\ell}{2}, & P_\ell = +1, \\ \cos\frac{\pi\ell}{2}, & P_\ell = -1, \end{cases}$$

where $P_\ell = \pm 1$ corresponds to positive (negative) signature. The factors ζ_ℓ may be included into reggeon propagators, G_ℓ; this way we would have poles in ℓ both for reggeons and *particles*: $\ell = 2n$ for positive, and $\ell = 2n+1$ for negative signature trajectories.

Let us look at the numerator of the γ factor in (13.9),

$$\mathrm{Im}\left(i\xi_1\xi_2\right) \propto \mathrm{Re}\ \exp\left\{-i\frac{\pi}{2}\left(\ell_1 + \ell_2 + \tfrac{1}{2}(1 - P_1) + \tfrac{1}{2}(1 - P_2)\right)\right\},$$

giving

$$\gamma_{\ell_1\ell_2} = \zeta_{\ell_1}^{-1}\zeta_{\ell_2}^{-1} \cos\frac{\pi}{2}\left(j + 1 + \tfrac{1}{2}(1 - P_1) + \tfrac{1}{2}(1 - P_2)\right), \qquad (13.10)$$

where we have taken into account that $j + 1 = \ell_1 + \ell_2$. The cos factor can either be absorbed into vertices, $\sqrt{\cos} \cdot \sqrt{\cos}$, or ascribed to the intermediate state of the two reggeons.

The signature of the reggeon branching is given by the product of signatures of participating reggeons, $P^{(2)} = P_1 P_2$. Indeed, the factor $s^{\ell_1 + \ell_2 - 1}$ in the expression (13.5) contains two reggeon amplitudes, $\xi_\ell s^\ell$, each of which transforms under the $s \to -s$ reflection according to its proper signature, while s^{-1} originates from the phase space volume and is in fact $|s|^{-1}$.

13.2.2 Multi-reggeon exchange: conservation of signature

Let us take into account more complicated processes. We have considered a diagram with two blocks with Regge pole asymptotics, and arrived at the two-reggeon branching contribution. What if the asymptotic behaviour of the block amplitude itself corresponds to the *branching* rather than to the pole?

Let us insert $f^{(2)}$ that we have just calculated in one of the blocks,

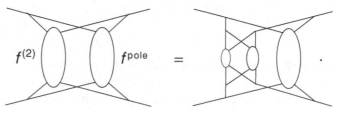

The vertex will now contain $N \cdot g$ instead of $g_1 \cdot g_2$. However, since everything remains factorized in the integrand, denoting as $N_{\ell_1 \ell_2 \ell_3}$, for the amplitude we obtain

$$F_3(s, q^2) = \frac{i^2}{3!} \int \frac{d^2 \mathbf{k}' d^2 \mathbf{k}}{(2\pi)^4} \int \frac{d\ell_1 \, d\ell_2 \, d\ell_3}{(2\pi i)^3} \xi_{\ell_1} \xi_{\ell_2} \xi_{\ell_3}$$
$$\cdot G_{\ell_1} G_{\ell_2} G_{\ell_3} s^{\ell_1 + \ell_2 + \ell_3 - 2} (N_{\ell_1 \ell_2 \ell_3})^2. \tag{13.11}$$

Here 3! accounts for different equivalent insertions, and s^{-2} originates from the phase space volume.

It is clear now how to repeat the procedure an arbitrary number of times. For the n-reggeon branching we have

$$F_n(s, q^2) = \frac{\pi}{2} \frac{(-1)^n i^{n-1}}{n!} \int \prod_1^{n-1} \frac{d^2 k_i}{(2\pi)^2} \prod_1^n \frac{d\ell_1}{2\pi i} \xi_{\ell_i} G_{\ell_i} \cdot s^{\sum_i \ell_i - n + 1} N_{\ell_1 \ldots \ell_n}^2.$$
$$\tag{13.12}$$

Let us look into the origin of the phase in (13.12).

In the case of two reggeons we had $i\xi_1 \xi_2$; a three-reggeon amplitude contains the factor $i^2 \xi_1 \xi_2 \xi_3$. Each additional transverse momentum integration brings in a factor i:

$$\frac{d^4 k_i}{(2\pi)^4 i} = \frac{i}{2|s|} \cdot \frac{d\alpha_i s}{2\pi i} \frac{d\beta_i s}{2\pi i} \frac{d^2 \mathbf{k}}{(2\pi)^2},$$

since integrals over α_i and β_i, as we have learned before, reduce to integrations of discontinuities of reggeon creation amplitudes and produce real-valued vertex functions. Evaluating the corresponding partial wave using (13.6a) and introducing an integral over k_n in order to symmetrize the expression, we get

$$f_j^{(n)}(q^2) = \frac{1}{n!} \int \prod_{i=1}^n \frac{d^2 \mathbf{k}_i d\ell_i}{(2\pi)^3 i} G_{\ell_i}(k_i) \cdot (2\pi)^2 \delta \left(\sum \mathbf{k}_i - \mathbf{q} \right)$$
$$\cdot \operatorname{Im} \left(i^{n-1} \xi_{\ell_1} \ldots \xi_{\ell_n} \right) \frac{N^2}{j + n - 1 - \sum \ell_i}. \tag{13.13}$$

Finally, we may introduce ω_i and one more integration,

$$f_j^{(n)}(q^2) = \frac{1}{n!} \int \prod_{i=1}^n \frac{d^2 \mathbf{k}_i \, d\omega_i}{(2\pi)^3 i} \cdot (2\pi)^3 \, i\delta \left(\omega - \sum \omega_i \right) \delta \left(\sum \mathbf{k}_i - \mathbf{q} \right)$$
$$\cdot \gamma_{\omega_1 \ldots \omega_n} \cdot N_{\omega_1 \ldots \omega_n}^2 \prod_{i=1}^n G_{\omega_i}(\mathbf{k}_i), \tag{13.14}$$

to obtain, again, a standard expression for the Feynman diagram. The only difference is in the signature factor γ:

$$\gamma \equiv \text{Im}\left[(-1)^n i^{n-1} e^{-\frac{i\pi}{2}\left(\sum \ell_i + \sum_i \frac{1-P_i}{2}\right)} \right] \cdot \prod_{i=1}^{n} \zeta_{\ell_i}^{-1}. \qquad (13.15)$$

Recalling that we have taken the residue $j + n - 1 - \sum_i \ell_i = 0$, for (13.15) it is straightforward to derive

$$\gamma = (-1)^{n-1} \sin\frac{\pi}{2}\left[j + \sum_{i=1}^{n} \frac{1-P_i}{2} \right] \cdot \prod_{i=1}^{n} \zeta_i^{-1}. \qquad (13.16)$$

What is the meaning of this factor? Depending on the sign of the product $P^{(n)} = \prod_{i=1}^{n} P_i$, the amplitude is proportional to $\sin\frac{\pi}{2}j$ ($P^{(n)} = +1$) or $\cos\frac{\pi}{2}j$ ($P^{(n)} = -1$). This is the manifestation of the *conservation of signature*: the symmetry of the n-reggeon amplitude is determined by the 'product' of symmetries of all the poles.

Because of this factor, the contribution of the n-reggeon branching *vanishes* in the points of its proper signature, $\gamma_+ \propto \sin\frac{\pi}{2}j = 0$ for $j = 2k$, and $\gamma_- \propto \cos\frac{\pi}{2}j = 0$ for $j = 2k + 1$.

The signature of a branching of n_+ positive and n_- negative signature reggeons is

$$P^{(n)} \equiv P^{(n_+ + n_-)} = (-1)^{n_-}.$$

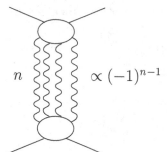 $\propto (-1)^{n-1}$

Let us examine the most important case when all the poles are pomerons **P**. Then, as we know, $j_n(t) = n\alpha(t/n^2) - n + 1$. At small t-values the branching $j_n(t) \simeq 1 + (\alpha' t/n)$ is positioned near 1, as well as all the poles ℓ_i, and (13.16) gives

$$\gamma_{\ell_1 \dots \ell_n} \equiv \gamma_{n\mathbf{P}} \simeq (-1)^{n-1}. \qquad (13.17)$$

In another interesting case when we have n pomerons and one non-vacuum pole with some trajectory $\beta(t)$,

$$j_{n+1}(t) \simeq \beta(0) + \frac{\alpha'}{\alpha' + n\beta'}t \simeq \beta(0),$$

the branching signature factor reads

$$\gamma_{n\mathbf{P}+\beta} = (-1)^n \zeta_{\beta(0)}^{-1} \sin\frac{\pi}{2}\left[\beta(0) + \frac{1-P_\beta}{2} \right] = (-1)^n P_\beta. \qquad (13.18)$$

Thus, in both cases when all (or all but one) poles are **P**, adding one pomeron changes the sign of the partial wave amplitude of the branching.

We conclude that the contributions of the (simplest) branchings are very similar to Feynman diagrams, except for the alternating signs. In terms of the field theory, the fact that contributions to the unitarity condition have alternating signs means that the Hamiltonian corresponding to our theory is *anti-Hermitian*.

Where does this oscillation come from? Recall that the characteristic feature of the vacuum pole was that the corresponding scattering amplitude A was *purely imaginary* at small t values, $A \propto i$. When we iterate n vacuum amplitudes, $n-1$ loops each produce the factor i,

$$F^{(n)} \sim A^n \frac{d^4 k_1 \ldots d^4 k_{n-1}}{[(2\pi)^4 i]^{n-1}} \sim A^n \left(\frac{i}{i^2}\right)^{n-1}$$

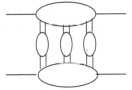

(i^2 in the denominator participates in forming the real multi-reggeon production vertices, as we have just discussed), and we get

$$F^{(n)} \sim (iA)^{n-1} A \sim (-1)^{n-1} A.$$

What if I considered a photon?

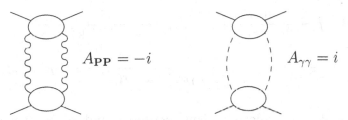

$$A_{\mathbf{PP}} = -i \qquad A_{\gamma\gamma} = i$$

Both processes are diffraction scatterings, but in the case of the photon the basic amplitude is real. Thus the alternating sign takes its origin from the complexity of the vacuum pole amplitude.

As we have discussed in the previous lecture, from the s-channel point of view the opposite sign of the two-pomeron branching is nothing but the *screening* phenomenon.

13.3 Enhanced branchings

Till now we supposed that the reggeon creation vertices N contained no singularities in ℓ and treated them as constants. What we have obtained this way is known as 'non-enhanced' reggeon branchings.

Now it is time to move further and try to combine poles with branch cut singularities.

For example, what will be the contribution to the asymptotics of a diagram like this one? This is an example of the so-called *enhanced branchings*.

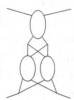

To learn how to deal with enhanced branchings, it is sufficient to consider a general graph (13.19), without specifying the details of the blocks:

$$F(q^2, s) = \qquad (13.19)$$

For the time being we assume the simplest case, with two particles in the intermediate t-channel state; more complicated configurations will be considered later.

$$F = \int \frac{d^2\mathbf{k}_\perp \, d\alpha \, d\beta \, s}{2(2\pi)^4 i} f(p_1, q, k) \frac{1}{m^2 - \alpha\beta s + \mathbf{k}_\perp^2 - i\varepsilon}$$

$$\frac{1}{m^2 - (\alpha - \alpha_q)(\beta - \beta_q)s + (\mathbf{k} - \mathbf{q})_\perp^2 - i\varepsilon} f'(p_2, q, k). \quad (13.20)$$

The key question is, which invariant energies s_1 and s_2 are relevant? If one of them is small, we arrive at the situation we have already considered. So, we will suppose that both energies are in the asymptotic regime, $s_1, s_2 \gg \mu^2$.

13.3.1 Renormalization of the Regge pole

First, we write for the blocks f and f' in (13.20) just the pole expressions:

$$f = g(q, p_1) \int \frac{d\ell_1}{2\pi i} \xi_{\ell_1} G_{\ell_1}(q) (2p_1 k)^{\ell_1} g(q, k) = \qquad (13.21)$$

This leads to the picture of two reggeons connected by a particle loop,

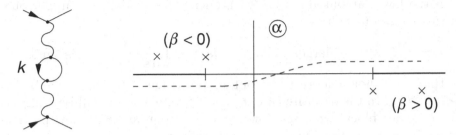

Let us investigate the singularities in α. The poles are

$$\alpha_1 = \frac{m_\perp^2 - i\varepsilon}{\beta s}, \qquad \alpha_2 = \frac{m_\perp'^2 - i\varepsilon}{(\beta - \beta_q)s} + \alpha_q.$$

The cuts in α come from the lower block amplitude $f'(s_2)$:

$$\begin{aligned}
s_1 &= (p_1 - k)^2 = (1 - \alpha)(\gamma - \beta)s - \mathbf{k}_\perp^2 \simeq -\beta s, \\
s_2 &= (p_2 + k)^2 = (\gamma + \alpha)(1 + \beta)s - \mathbf{k}_\perp^2 \simeq \alpha s.
\end{aligned}$$

Depending on the sign of β, I close the contour around the left $(\beta > 0)$ or the right $(\beta < 0)$ cut. As always, to determine the asymptotics, we have to consider the s- and u-cuts independently (see Lecture 11):

$$F(q^2, s) = \int_{\beta<0} \frac{d\beta s\, d\alpha\, d^2\mathbf{k}}{2(2\pi)^3\pi} \frac{f(p_1, q, k)}{(\ )(\ )} \operatorname{Im}_s f'(p_2, q, k) + \int_{\beta>0} \operatorname{Im}_u f'(p_2, q, k). \tag{13.22}$$

Consider the first integral. It includes

$$\xi_{\ell_1}(-\beta s)^{\ell_1}(\alpha s)^{\ell_2};$$

the second signature factor, ξ_{ℓ_2}, is absent since we took the s-channel imaginary part of the lower amplitude. Everywhere in the propagators enters $\alpha\beta s = \mathcal{O}(1)$, thus it is reasonable to introduce $x = -\alpha\beta s$ as an integration variable, instead of β:

$$\begin{aligned}
F(q^2, s) =\ & g(q, p_1)g(q, p_2) \int \frac{d\ell_1}{2\pi i} \xi_{\ell_1} G_{\ell_1}(q) \int \frac{d\ell_2}{2\pi i} G_{\ell_2}(q) \\
& \cdot \int \frac{d^2\mathbf{k}\, dx}{(2\pi)^4} \frac{g(q, k)g(q, k)}{(\ )(\ )} \int \frac{d\alpha}{\alpha} \left(\frac{x}{\alpha}\right)^{\ell_1} (\alpha s)^{\ell_2}. \tag{13.23}
\end{aligned}$$

Since the propagators depend only on x, the integral in α can be easily taken:

$$\int \frac{d\alpha}{\alpha} \alpha^{\ell_2 - \ell_1}, \qquad \frac{1}{s} < \alpha < 1.$$

The result depends on the magnitude of the difference $\ell_2 - \ell_1$.

If the angular momenta are significantly different, for example, $\ell_2 > \ell_1$, then the lower amplitude grows with energy faster than the upper one. In this situation we have

$$\alpha \sim 1 \; ; (\alpha s)^{\ell_2} \quad \text{is large, while} \quad (\beta s)^{\ell_1} \sim \left(\frac{x}{\alpha s} s\right)^{\ell_1} \sim 1 \quad \text{is small,}$$

so that the whole energy turns out to be assigned to the lower block, $s \sim s_2 \gg s_1$. In the opposite case, $\ell_2 < \ell_1$, a large energy will be assigned to the upper block, while the lower one will contain only a few particle interactions, away from the asymptotic regime.

An interesting case is $\ell_2 \approx \ell_1$. This gives a possibility to get an enhanced contribution by playing on the redistribution of energy between the two blocks:

$$\int_{1/s}^{1} \frac{d\alpha}{\alpha} = \frac{1}{\ell_2 - \ell_1}(1 - s^{(\ell_1 - \ell_2)}) \sim \ln s. \tag{13.24}$$

This is just the phenomenon which prompted me to carry out the calculation. According to (13.24), the singularity in j of our simplest enhanced diagram is *not identical* to that of the initial reggeon amplitude. Let us rewrite the expression so that this can be clearly seen:

$$F(q^2, s) = g^2 \int \frac{d\ell_1}{2\pi i} \xi_{\ell_1} G_{\ell_1}(q) \int \frac{d\ell_2}{2\pi i} r_{\ell_1 \ell_2} G_{\ell_2}(q) \frac{s^{\ell_1} - s^{\ell_2}}{\ell_1 - \ell_2}. \tag{13.25a}$$

Here

$$r_{\ell_1 \ell_2} = 2 \int \frac{d^2 \mathbf{k} \, dx}{(2\pi)^4} \frac{g^2(q, k)}{(\;)(\;)} x^{\ell_1}. \tag{13.25b}$$

The partial wave is given by the integral

$$f_j(q^2) = g^2 \int \frac{d\ell_1}{2\pi i} G_{\ell_1}(q) \int \frac{d\ell_2}{2\pi i} G_{\ell_2}(q) \, r_{\ell_1 \ell_2} \int_{s_0}^{\infty} \frac{ds}{s^{j+1}} \frac{s^{\ell_1} - s^{\ell_2}}{\ell_1 - \ell_2}. \tag{13.26}$$

Let us calculate it at $\ell_1 \approx \ell_2$:

$$\int_{s_0}^{\infty} ds = \frac{1}{\ell_1 - \ell_2} \left[\frac{s_0^{j - \ell_1}}{j - \ell_1} - \frac{s_0^{j - \ell_2}}{j - \ell_2} \right] \simeq \frac{1}{(j - \ell_1)(j - \ell_2)}.$$

Finally, closing the ℓ-contours around the poles, we obtain

$$f_j(q^2) = g(q^2) G_j(q) \, r_{jj} G_j(q) \, g(q^2). \tag{13.27}$$

Depending on the signatures of the two reggeons, the second term of (13.22) due to the u-channel cut either cancels the s-channel contribution (opposite signatures, $P_1 = -P_2$) or doubles it ($P_1 = P_2$; the corresponding factor 2 is already included in the formula (13.25b) for $r_{\ell_1 \ell_2}$).

The expression (13.27) is rather transparent from the point of view of Feynman diagrams. The two reggeons belonging to the same scattering amplitude, having the same quantum numbers and the same signature, can transform one into another; they 'mix'.

If we insert the calculated expression in the upper block of our general graph (13.19) and repeat the analysis, we obtain a graph with two successive reggeon–reggeon transitions.

If initially the reggeon Green function has the form

$$G_0 = \frac{1}{j - \alpha_0},$$

then after taking into account all iterations of this sort we arrive at

$$G = \frac{1}{j - \alpha_0} + \frac{r}{(j - \alpha_0)^2} + \frac{r^2}{(j - \alpha_0)^3} + \cdots = \frac{1}{j - \alpha_0 - r}.$$

The trajectory has changed.

By the way, this means that if the pole is a genuine one, there cannot be such diagrams ($r = 0$). (This is just an aside.)

13.3.2 Basic reggeon interactions

What truly interests us is diagrams like

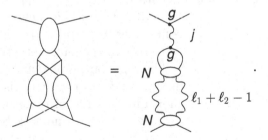

Using the results of our previous calculations, we obtain

$$f_j(q) = g\, G_j(q) \int \frac{d\ell_1\, d^2\mathbf{k}}{(2\pi)^3 i} r_{j\ell_1\ell_2} G_{\ell_1}(k) G_{\ell_2}(q - k) \gamma_{\ell_1\ell_2} N_{\ell_1\ell_2}(q, k), \quad (13.28)$$

where $\ell_2 = j + 1 - \ell_1$. The new vertex $r_{j\ell_1\ell_2}$ is analogous to $r_{\ell_1\ell_2}$, only with the two-reggeon creation function N replacing one of the $g(q, k)$ factors in (13.25b).

We come to the conclusion that one reggeon can transform into two: there is a mixing. Such a transition can be important only when the upper and the lower blocks have the same degree of the s behaviour, i.e. $\alpha_{\text{pole}} \approx \alpha_{\text{branching}}$. But as we already know this is just the situation that we have, at small t-values, for the vacuum pole $\mathbf{P}$ with $\alpha_{\mathbf{P}}(0) = 1$.

Similarly, one arrives at more complicated diagrams. For example, taking a three-reggeon branching for one of the blocks in (13.19), we will obtain diagrams

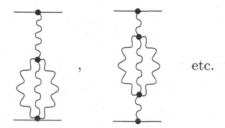

, etc.

Thus, having started from a pole, we obtain all diagrams with different groups of reggeons following one another in the t-channel. These diagrams modify particle–reggeon vertex functions and the reggeon propagator.

Will there be graphs of another sort, which look like corrections to reggeon–reggeon interaction vertices?

Should a reggeon diagram like this one, (13.29), exist this would generate,

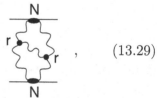

(13.29)

via (13.19), a whole new family of graphs containing 'reggeon corrections' to r as well as to other reggeon interaction vertices.

The reggeon graphs of the type of (13.29) do appear from the analysis of the diagram shown in the l.h.s. of (13.30). To show this is a rather cumbersome task because of many regions in the α_i, β_i variables in an amplitude with five scattering blocks.

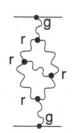

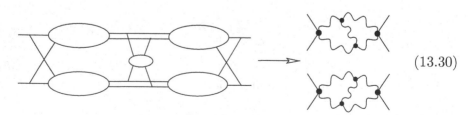

(13.30)

The answer, however, turns out to be quite simple: just the sum of two terms shown on the r.h.s. of (13.30). If the outer reggeons forming the basic loop are different, these contributions are not identical, exactly as it would have been the case in a non-relativistic quantum mechanical problem.

Essentially, the answer reduces to a simple statement that using reggeons we can draw all diagrams that can be imagined in a field theory. There is one important additional rule: every reggeon vertex r has its own signature factor γ, which is characterized by the set of reggeon complex angular momenta ℓ_i in each intermediate state,

We have investigated only the structures of the diagrams. What are the coefficients these diagrams should be added up with? Is there any correspondence with a (non-relativistic) field theory? The answer is simple: the coefficients should be such that the *unitarity conditions* are satisfied, since the series of Feynman diagrams is a formal solution of the unitarity.

The asymptotic behaviour of the amplitudes at $s \to \infty$ can be expressed in terms of the asymptotics of the internal blocks. If we assume that the basic scattering block is just a Regge pole, then, applying the diagrammatical rules, everything can be constructed from this reggeon.

13.4 Feynman diagrams and reggeon unitarity conditions

In Lecture 11 we derived the reggeon unitarity condition in the t-channel, at $t > 0$, which contained unknown reggeon creation amplitudes. Now we know that the branchings can be obtained directly from the physical region of the s-channel ($t < 0$) where all the ingredients are well defined, diagrammatically. It is the series of these diagrams that allows us to simplify the unitarity conditions.

Recall the two-reggeon unitarity condition that we have written in the t-channel,

$$\delta f_j = -\pi \int dt_1 \, dt_2 \, \tau(t, t_1, t_2) N_j^+ \cdot \delta(j + 1 - \alpha_1 - \alpha_2) \cdot N_j^-, \qquad (13.31)$$

where τ is the two particle phase space volume, and $N_j^\pm = N_j^\pm(t, t_1, t_2)$ are the values of the full two-particle–two-reggeon transition amplitude above and below the cut in j, cf. (11.37).

We have calculated above a similar expression in terms of Feynman diagrams, using the simplest explicit vertex functions N:

$$f_j = \quad\raisebox{-1em}{\text{[diagram with }N\text{ vertices]}}\quad = -\int \frac{d^2\mathbf{k}\, d\ell_1}{(2\pi)^3 i} G_{\ell_1}(k) G_{j+1-\ell_1}(q-k) N\, N\,. \qquad (13.32)$$

Substituting $G = 1/\ell - \alpha$,

$$f_j = -\int \frac{d^2\mathbf{k}}{(2\pi)^2} \frac{N N}{j+1-\alpha_1-\alpha_2}\,.$$

The imaginary part of the partial wave,

$$\delta f_j = -\pi \int \frac{d^2\mathbf{k}}{(2\pi)^2} N\delta(j+1-\alpha_1-\alpha_2)N, \qquad (13.33)$$

has a structure resembling the t-channel unitarity condition. The only difference is that now we have $d^2\mathbf{k}$ instead of $dt_1\, dt_2\, \tau$ in (13.31).

So, is there a correspondence between the two formulae? Let us trade the two-dimensional transverse momentum integration,

$$d^2 k = k\, dk\, d\varphi = \tfrac{1}{2} d(k^2)\, d\varphi,$$

for the arguments of the reggeon Green functions,

$$-t_1 = \mathbf{k}^2, \quad -t_2 = \mathbf{k}'^2 \equiv (\mathbf{q}-\mathbf{k})^2; \qquad (-t = \mathbf{q}^2)$$

$$k'^2 = q^2 + k^2 - 2kq\cos\varphi\,, \quad dk'^2 = 2kq\sin\varphi\, d\varphi\,.$$

We obtain

$$\sin^2\varphi = 1 - \left(\frac{k'^2 - q^2 - k^2}{2kq}\right)^2, \quad d^2\mathbf{k} = \frac{dt_1\, dt_2}{2kq|\sin\varphi|} = \frac{dt_1\, dt_2}{2\sqrt{-t}\cdot p_c},$$

where

$$p_c = p_c(t, t_1, t_2) = \left[\frac{t^2 - 2t(t_1 + t_2) + (t_1 - t_2)^2}{4t}\right]^{\frac{1}{2}}\,.$$

The factor p_c appearing in the *denominator* of (13.33) is nothing but the relative momentum of the two reggeons in the centre-of-mass of the t-channel. Curiously, the same factor p_c is present in the unitarity condition (13.31) but in the *numerator*, $\tau \propto p_c$.

The two expressions would match if we could extract $1/p_c$ from the full vertex $N^\pm$ by introducing $N^\pm = \tilde{N}/p_c(t, t_1, t_2)$, and claim that $\tilde{N}$ behaves similarly to N from (13.33).

Such an operation, however, looks intrinsically dangerous, since the singularity we are studying emerges just near $p_c = 0$.

Let us recall the origin of the two-reggeon creation amplitude N:

The production amplitude of *usual particles* with the orbital momentum L behaves at small p_c as $N \sim p_c^L$. Our branch point corresponds to $j = \sigma_1 + \sigma_2 - 1$. Comparing with

$$\mathbf{j} = \boldsymbol{\sigma}_1 + \boldsymbol{\sigma}_2 + \mathbf{L} \;, \quad |\mathbf{j}|_{\max} = \sigma_1 + \sigma_2 + L,$$

we observe that the *two-reggeon* singularity appears when the orbital momentum assumes the first unphysical value $L = -1$. Consequently, the vertex $N^{\pm}$ is *obliged* to behave like $1/p_c$, and this is just what we need: $\tilde{N}$ turns to a constant in the $p_c \to 0$ limit.

As for multi-reggeon diagrams, using in the partial wave (13.14) the Regge-pole expression for $G_{\ell_i}(k_i)$ and evaluating the imaginary part gives

$$\delta f_j^{(n)} = \frac{(-1)^{n-1}\pi}{n!} \int \prod_{i=1}^{n-1} \frac{d^2 k_i}{(2\pi)^2} N \delta\left(j + n - 1 - \sum_{m=1}^{n} \alpha(k_m)\right) N. \quad (13.34)$$

This expression describes the simplest contribution to the multi-reggeon unitarity condition, corresponding to $N = $ const. To satisfy the genuine unitarity condition (11.40) one has to draw and analyse the full set of diagrams for the creation of n reggeons in a two-particle interaction, which determines the exact vertex $N^{\pm}$.

14

Reggeon field theory

Reggeons turn out to be similar to particles with varying spins not only in the sense of the 'pole' contribution to the asymptotics, but also in the interaction picture.

We have started with a Regge pole and generated series of *non-enhanced* reggeon diagrams characterized by non-singular particle–reggeon vertices N,

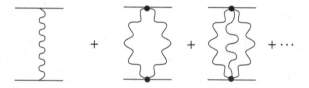

Angular momentum singularities in reggeon creation amplitudes gave rise to various *enhanced* reggeon diagrams,

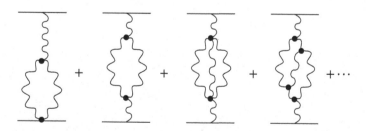

We are looking for an effective field theory that would solve the reggeon unitarity. The usual field theory contains a hypothesis about the form of the interaction. If I chose three-reggeon, or only four-reggeon, interactions

to build up the theory,

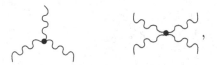

or even employed them together, I would be able to fulfil the corresponding reggeon unitarity conditions. But I have no reason to restrict myself to such vertices. In other words, my 'theory' contains in principle an infinite number of unknown constants.

Everything would have been fine if we had $\alpha(0) < 1$, that is, σ_{tot} decreasing with energy. In this case the branchings are separated from the pole, produce but small controllable corrections to the asymptotics, and the interaction can be looked upon as being weak.

In the interesting case of $\alpha(0) = 1$, however, the multi-reggeon interactions are absolutely essential: at $t = 0$ branchings accumulate to $j = 1$, and for $t < 0$ even move *to the right* from the pomeron pole. This means that from a practical point of view we have lost. From the point of view of the theory, however, the problem, although a complicated one, remains sensible: the iteration of poles and branchings produced but new reggeon branchings, and the picture remained self-consistent.

It is somewhat distressing that the interactions of reggeons cannot be considered as weak ones, so the beauty of reggeons, as objects embodying strong interaction, apparently disappears. Nevertheless, one can hope that in certain cases just a finite number of vertices will be relevant. This being the case, it will allow one to relate different observables and use the reggeon field theory in order to make quantitative as well as qualitative predictions.

14.1 Prescriptions for reggeon diagram technique

To construct a field theory we have to start with the bare propagator and interaction vertices. To describe interacting pomerons it is convenient to introduce the bare **P** Green function as

$$G_0(k) = \frac{-1}{\omega + \varepsilon(\mathbf{k})}; \qquad \omega = j - 1, \quad \varepsilon(\mathbf{k}) = -\alpha(k^2) + 1 \simeq \alpha' \mathbf{k}^2. \quad (14.1a)$$

Changing the overall sign of the propagator eliminates the oscillating factor $(-1)^{n-1}$ in the n-pomeron contribution to the reggeon unitarity (all terms in the unitarity condition for $-f_j$ enter with positive sign).

For the vertices we choose

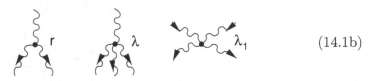

$$\text{(14.1b)}$$

Let us stress that the two four-particle vertices would have been identical in a relativistic theory; here they are not, $\lambda \neq \lambda_1$.

In the expression for the scattering amplitude

$$A(s, q^2) = s \int \frac{d\omega}{2\pi i} \xi_j \, e^{\omega \xi} f_j(q^2), \quad \xi = \ln s,$$

expanding the signature factor ξ_j at small ω values,

$$\xi_j = -\frac{e^{-i\frac{\pi}{2}j}}{\sin \frac{\pi}{2}j} = i \frac{e^{-i\frac{\pi}{2}\omega}}{\cos \frac{\pi}{2}\omega} \simeq i + \frac{\pi}{2}\omega,$$

the real part of the amplitude is conveniently represented by the derivative of the imaginary part,

$$A(s, q^2) \simeq \left[i + \frac{\pi}{2} \frac{\partial}{\partial \xi} \right] \text{Im} \, A(s, q^2). \quad \text{(14.2a)}$$

Therefore it suffices to know the imaginary part. We will calculate the function

$$F(\xi, q^2) \equiv \frac{1}{s} \, \text{Im} \, A(s, q^2),$$

whose Fourier transform, $f(\omega, q^2)$, is given by the sum of diagrams with reggeon propagators and vertices (14.1)

$$F(\xi, \mathbf{k}^2) = -\int \frac{d\omega}{2\pi i} \, e^{\omega \xi} f(\omega, \mathbf{k}^2). \quad \text{(14.2b)}$$

14.1.1 Reggeons and branchings in the impact parameter space

We start from the contribution of the non-enhanced branchings,

$$f(\omega, \mathbf{k}^2) = -\sum_{n=1}^{\infty} \int \frac{d\omega_1 \dots d\omega_n \, d^2\mathbf{k}_1 \dots d^2\mathbf{k}_n}{n! \, [(2\pi)^3 i]^n} G(\omega_1, \mathbf{k}_1) \cdots G(\omega_n, \mathbf{k}_n)$$

$$\cdot (2\pi)^3 \, i\delta(\omega - \sum \omega_i)\delta\left(\mathbf{k} - \sum \mathbf{k}_i\right) N_n^2(\omega_i, \mathbf{k}_i). \qquad (14.3)$$

The fact that this diagram is *not enhanced* means that the vertex blocks N_n are not singular and at small ω and k values, they can be replaced by some numbers, $N_n \approx$ const.

By using the Fourier transformation to the impact parameter space,

$$(2\pi)^2 \delta(\mathbf{k} - \sum \mathbf{k}_i) = \int d^2\rho \, e^{i\mathbf{k}\cdot\rho - i\sum \mathbf{k}_i\cdot\rho},$$

we factorize the transverse momentum integrations to get

$$F(\xi, \mathbf{k}^2) = -\sum_{n=1}^{\infty} \frac{1}{n!} \int d^2\rho \, e^{i\mathbf{k}\cdot\rho} \left(\int \frac{d\omega_1}{2\pi i} e^{\omega_1 \xi} \int \frac{d^2\mathbf{k}_1}{(2\pi)^2} G(\omega_1, \mathbf{k}_1) e^{-i\mathbf{k}_1\cdot\rho} \right)^n.$$

The integrals over ω_i run along the imaginary axis,

$$\int \frac{d\omega_1}{2\pi i} e^{\omega_1 \xi} G(\omega_1, \mathbf{k}_1) = -\int \frac{d\omega_1}{2\pi i} \frac{e^{\omega_1 \xi}}{\omega_1 + \alpha'\mathbf{k}_1^2} = -e^{-\alpha'\mathbf{k}_1^2 \xi}.$$

Integrating over $\mathbf{k}_1$ we have

$$-\int \frac{d^2\mathbf{k}_1}{(2\pi)^2} e^{-\alpha'\mathbf{k}_1^2 \xi - i\mathbf{k}_1\cdot\rho} = -\int \frac{d^2\mathbf{k}_1}{(2\pi)^2} e^{-\alpha'\xi\left(\mathbf{k}_1 + \frac{i\rho}{2\alpha'\xi}\right)^2 - \frac{\rho^2}{4\alpha'\xi}}$$

$$= -\frac{1}{4\pi\alpha'\xi} \exp\left\{-\frac{\rho^2}{4\alpha'\xi}\right\} \equiv \tilde{G}(\xi, \rho). \qquad (14.4)$$

The function $\tilde{G}(\xi, \rho)$ is an important distribution whose physical meaning we will discuss later in this lecture.

Substituting (14.4) into the n-reggeon branching amplitude,

$$F^{(n)}(\xi, \mathbf{k}^2) = -(-1)^n \frac{N_n^2}{n!} \int d^2\rho \, e^{i\mathbf{k}\cdot\rho} \frac{1}{(4\pi\alpha'\xi)^n} e^{-\frac{n\rho^2}{4\alpha'\xi}}$$

$$= \frac{N_n^2}{nn!} \frac{(-1)^{n-1}}{(4\pi\alpha'\xi)^{n-1}} \int d^2\rho \, e^{i\mathbf{k}\cdot\rho} \cdot \frac{-n}{4\pi\alpha'\xi} e^{\frac{-n\rho^2}{4\alpha'\xi}},$$

we observe that the integrand here is the one we have just calculated above, i.e. the Fourier transform of the Green function $G(\xi, \mathbf{k})$, with the only difference that α' is substituted by α'/n. Hence, we obtain

$$F^{(n)}(\xi, \mathbf{k}^2) = \frac{(-1)^{n-1} N_n^2}{nn!(4\pi\alpha'\xi)^{n-1}} e^{-\frac{\alpha'}{n} \mathbf{k}^2 \xi}. \qquad (14.5)$$

In the particular case of $n = 1$ we recover the pole expression, $F^{(1)} = e^{-\alpha' \mathbf{k}^2 \xi}$. What is the magnitude of the nth branching contribution like? Is it large or small?

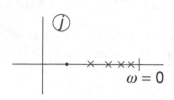

The (modulus of the) exponent in (14.5) is at its maximum for $n = 1$, i.e. the contributions of the branchings, $n \geq 2$, are *larger* than that of the pole, in the sense of the position of the singularity in the j-plane. On the other hand, the first term is larger owing to the suppression of the higher terms by the pre-exponential factor $F^{(n)} \propto 1/\xi^{n-1}$. Thus one cannot state a priori that $n \to \infty$, $j_n \to 1$ are the most important contributions.

14.1.2 Qualitative estimate of the series

Let us try to estimate the sum

$$F = \sum_n F^{(n)}(\xi, \mathbf{k}^2). \tag{14.6}$$

Our estimate is going to be rough since the we don't know the dependence of the vertex functions N_n on n, and other – enhanced – diagrams are important too. Nevertheless, just for curiosity's sake, let us look at which n are relevant in the series (14.6).

If $\mathbf{k}^2 = 0$, everything is very simple: all singularities are at the same point $j = 1$ and the pole dominates:

$$F(\xi, \mathbf{k}^2) = g^2 - \frac{N_2^2}{4\pi\alpha'\xi} + \mathcal{O}(\xi^{-2}).$$

This is true not only for $\mathbf{k}^2 = 0$, but in an interval of momenta where the condition $\alpha' \mathbf{k}^2 \xi \ll 1$ is satisfied.

In the opposite case, when $\alpha' \mathbf{k}^2 \xi \gg 1$, the far terms of the series become important. How can we estimate the sum? It is clear that one has to find $n = n_{\max}$ which marks the maximal contribution, $\max_n\{F_n\} = F_{n_{\max}}$.

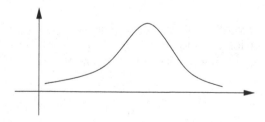

Since the number of relevant terms is large, we can write

$$F \simeq \sum_n \mathrm{e}^{\varphi(n)} \simeq \int dn\, \mathrm{e}^{\varphi(n)} \,,$$

and attempt to apply the steepest-descent method to the exponent

$$\varphi(n) = -\frac{\alpha' \mathbf{k}^2 \xi}{n} - (n-1)\ln(4\pi\alpha'\xi) - n\ln n + i\pi n, \qquad (14.7)$$

where $-n\ln n$ originated from the combinatorial $1/n!$ factor. There is one delicate point here; the series has alternating signs, therefore the term $i\pi n$ in (14.7). This is not a very sensible way to estimate oscillating series but is good enough to illustrate the key feature of the answer.

The saddle-point equation,

$$\frac{d\varphi(n)}{dn} = \frac{\alpha' \mathbf{k}^2 \xi}{n^2} - \ln(4\pi\alpha'\xi) - \ln n - 1 + i\pi = 0,$$

determines the scale of n at which the terms of the series are large:

$$\frac{\alpha' \mathbf{k}^2 \xi}{n_{\max}^2} = \ln(\alpha'\xi n_{\max}) + \mathcal{O}(1) \,, \qquad n_{\max}^2 \approx \frac{\alpha' \mathbf{k}^2 \xi}{\ln(\alpha'\xi n_{\max})} \sim \frac{2\alpha' \mathbf{k}^2 \xi}{\ln(\xi^3 \mathbf{k}^2)},$$

where we omitted the constant in the argument of the logarithm. Now we approximate $\varphi(n_{\max})$,

$$\varphi(n_{\max}) \sim \sqrt{2\alpha' \mathbf{k}^2 \xi \ln(\xi^3 \mathbf{k}^2)},$$

and obtain the scale of the answer in the region $\alpha' \mathbf{k}^2 \xi \gg 1$:

$$F(\xi, \mathbf{k}^2) \sim F^{(n_{\max})} \sim \mathrm{e}^{-c\sqrt{\alpha' \mathbf{k}^2 \xi}}. \qquad (14.8)$$

Here we have dropped the ln factor in the exponent, since our estimation is rough anyway. The decrease of (14.8) with ξ is fast, but *slower* than any exponent, $\exp(-\gamma\xi)$, $\gamma > 0$.

What is the correct way of calculating the sign-alternating series? One has to write down the representation of the Sommerfeld–Watson type,

$$\sum_n (-1)^{n-1} f_n = \frac{1}{2i} \int \frac{dn}{\sin \pi n} f_n \,,$$

and include $\ln \sin \pi n$ into the exponent $\varphi(n)$. Its presence produces a pair of complex conjugate points as the saddle-point solution. As a result, in addition to the exponent of $\sqrt{\alpha'\xi \mathbf{k}^2}$, the answer also acquires a factor $\cos(\alpha'\xi \mathbf{k}^2)$ which leads to *oscillations* in the scattering amplitude.

Let us see, what we obtain by fixing ξ and increasing k:

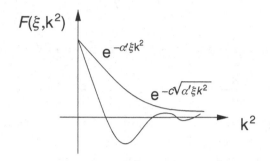

We have here a curious result, just like in classical diffraction: the pole term leads to a diffraction cone; taking into account the branchings, the slope of the amplitude falloff decreases, and oscillations appear.

We can conclude that in the region of small momentum transfers non-enhanced branchings do not alter the pole picture. With the growth of k^2, however, the angular distribution changes drastically; due to the alternating signs of the multi-reggeon branching contributions, maxima and minima appear naturally in the differential scattering cross section.

This would have been the answer, and a rather simple one, if not for enhanced reggeon diagrams.

14.2 Enhanced diagrams for reggeon propagator

To write down everything is impossible, so we restrict ourselves to the simplest interaction ⌇ . The exact reggeon Green function is given by the sum of diagrams with various reggeon loops the bare reggeon can mix with, and their repetitions,

$$G = \;\; + \;\; + \;\; + \cdots$$

14.2.1 Reggeon loop in the reggeon propagator

Consider the first correction to the propagator:

$$G^{(1)} = \frac{1}{\omega + \alpha' \mathbf{k}^2} \Sigma(\omega, \mathbf{k}^2) \frac{1}{\omega + \alpha' \mathbf{k}^2}, \tag{14.9}$$

$$\Sigma(\omega, \mathbf{k}^2) = r^2 \int \frac{d\omega' \, d^2\mathbf{k}'}{(2\pi)^3 \, i} \frac{1}{\omega' + \alpha' \mathbf{k}'^2} \frac{1}{\omega - \omega' + \alpha'(\mathbf{k} - \mathbf{k}')^2}. \tag{14.10}$$

To calculate the 'self-energy' insertion (14.10) it is natural to take the ω' integral first. The integration contour lies between the two poles; closing the contour around one of them we obtain

$$\Sigma(\omega, \mathbf{k}^2) = r^2 \int \frac{d^2\mathbf{k}'}{(2\pi)^2} \frac{1}{\omega + \alpha' \mathbf{k}'^2 + \alpha'(\mathbf{k} - \mathbf{k}')^2}. \tag{14.11}$$

This is a typical expression for the two-reggeon branching. The first thing we observe is that this integral diverges at $\mathbf{k}^2 \to \infty$. There is nothing strange nor terrible in this, since we treated all the vertices as constants and expanded $\varepsilon(\mathbf{k}^2)$, being interested in region of small transverse momenta. The integration has to be carried out up to a certain value $\mathbf{k}_{\text{max}}^2$.

In any case, not this is the source of our problems. More important is the strong singularity in ω: taking $\omega = 0$, the integral tends to infinity in the limit of *small* momentum transfer, $\mathbf{k}^2 \to 0$. This is a logarithmic divergence corresponding to a branch-cut singularity.

Let us calculate the integral (14.11). Introducing a symmetric integration variable $\mathbf{q}$ such that $\mathbf{k}' = \mathbf{q} + \frac{1}{2}\mathbf{k}$ ($\mathbf{k} - \mathbf{k}' = \mathbf{q} - \frac{1}{2}\mathbf{k}$) we derive

$$\Sigma(\omega, \mathbf{k}^2) = r^2 \int \frac{d^2\mathbf{q}}{(2\pi)^2} \frac{1}{\omega + \frac{1}{2}\alpha' \mathbf{k}^2 + 2\alpha' \mathbf{q}^2} = \frac{r^2}{8\pi\alpha'} \ln \frac{\Lambda}{\omega + \frac{1}{2}\alpha' \mathbf{k}^2}, \tag{14.12}$$

where parameter Λ limits from above the q-integration, $\mathbf{q}^2 \le \Lambda/2\alpha'$. The position of the singularity, $\omega = -\frac{1}{2}\alpha' \mathbf{k}^2 \equiv \omega_2$ is just that of the two-reggeon branching, cf. (11.36).

Let us imagine that we sum up the series of 'self-energy' corrections to the propagator:

$$\begin{aligned} G(\omega, \mathbf{k}^2) &= G_0 + G_0 \Sigma G_0 + G_0 \Sigma G_0 \Sigma G_0 + \cdots = G_0 + G_0 \Sigma \cdot G \\ &= \frac{1}{G_0^{-1} - \Sigma(\omega, \mathbf{k}^2)} = -\frac{1}{\omega + \alpha' \mathbf{k}^2 + \Sigma(\omega, \mathbf{k}^2)}. \end{aligned} \tag{14.13}$$

We obtained an expression not having any pole at $\omega \simeq 0$ at small $\mathbf{k}^2$ values; the correction (14.12) is huge (*infinite* at $\omega \propto \mathbf{k}^2 \to 0$). What happened could have been foreseen: when the pole is on the cut, in the same point as the branchings, one cannot expect to get anything nice.

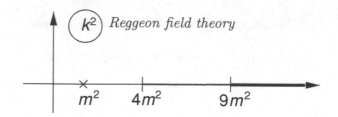

Fig. 14.1   The particle pole and multi-particle branch cuts in QFT with $m \neq 0$.

14.2.2 Analogy with $m = 0$ infrared singularity in QFT

What phenomenon in the field theory could correspond to this catastrophe? Recall the usual ϕ^3 field theory. What sort of singularities did we have there? The bare Green function contained the mass parameter m_0; the corrections lead to the appearance of the renormalized mass m entering observable phenomena. In the momentum transfer plane the particle pole is separated from threshold branchings as shown in Fig. 14.1.

In order to feel the analogy, let us imagine that the intercept of our pomeron is not exactly unity: $\alpha_{\mathbf{P}}(0) \neq 1$.

$$j \simeq \alpha(0) + \alpha'\mathbf{k}^2 \,, \qquad \omega = \alpha(0) - 1 - \alpha'\mathbf{k}^2 = \Delta - \alpha'\mathbf{k}^2.$$

In this case the multi-pomeron branchings would be in the points

$$\omega_n = n\Delta - \frac{\alpha'\mathbf{k}^2}{n}.$$

If $\Delta < 0$, higher branchings move away from the pole, and the structure of singularities displayed in Fig. 14.2 is analogous to that in particle theory (Fig. 14.1). In terms of particles $\Delta \to 0$ means that the *mass* of the particle is approaching zero. What would happen in field theory then? The same trouble as in our pomeron problem.

Take $m_0 = 0$ in the particle propagator,

$$G_0 = \frac{\quad}{\quad} = \frac{1}{-k^2} \,;$$

the first self-energy correction —⟨◯⟩— gives $G = -(\mathbf{k}^2 + \Sigma(\mathbf{k}^2))^{-1}$. Generally speaking, the pole at $k^2 = 0$ would disappear – the particle

Fig. 14.2   Regge pole $\alpha(0) - 1 = \Delta < 0$ and corresponding branchings.

acquires a mass, if only one does not take special measures to prevent it from doing so. We have met such a situation in electrodynamics:

$$D^0_{\mu\nu}(k) = \frac{g_{\mu\nu}}{k^2} \implies D_{\mu\nu}(k) = \frac{g_{\mu\nu}}{k^2 + \Pi(k^2)} .$$

Making use of the conservation of electromagnetic current we have shown that the polarization operator vanishes in the origin, $\Pi(k^2) \propto k^2$ at $k^2 \to 0$. In this sense we can say that the photon did not acquire a mass owing to the symmetry – to gauge invariance in this case.

We see that our situation with $\alpha_{\mathbf{P}}(0) = 1$ is just the same; taking the bare Green function with zero mass, the corrections diverge. This is actually the main problem of the theory of interacting pomerons. We do not know, have not formulated any reason *why* the total cross sections are asymptotically constant. Having not understood this, having not imposed the additional condition, we will always face the problem that the pole does not 'hold' at $\alpha(0) = 1$.

Can it go to the right? No, since we took into consideration the branchings and respected the Froissart theorem which forbade the power growth of the total cross section. However, there is no way to prevent it from *decreasing* with energy.

Indeed, it is easy to see that the pole bounces to the left, since, owing to the anti-hermiticity of the theory, the correction to the position of the singularity comes with a negative sign,

$$\delta\omega_0 = -\frac{r^2}{8\pi\alpha'} \ln\frac{\Lambda}{\omega_0} < 0.$$

Consequently, without understanding *why* the cross section is constant, we cannot ensure the self-consistence of the theory.

14.3 $\sigma_{\text{tot}} \simeq$ **const as an infrared singular point**

We have come to two conclusions, namely that

(1) the pomeron pole does not stay at $\alpha_{\mathbf{P}} = 1$; and

(2) taking into account the simplest interactions, it shifts *to the left*.

Nevertheless the question arises what Green function and interaction vertices do we have to use in order to have the *renormalized* pole at 1. Evidently, the initial position of the pole has to be chosen to the right from unity:

$$G_0 = \frac{-1}{\omega - \Delta + \alpha' \mathbf{k}^2}, \quad \Delta = \alpha(0) - 1 > 0.$$

This, however, does not solve the problem. Indeed, inside the two-reggeon loop Σ we have to use the *renormalized* poles, in order to avoid the self-energy corrections to the loop propagators,

$$\Sigma = r^2 \int \frac{d^2\omega \, d^2\mathbf{k}'}{(2\pi)^3 i} G(\omega', \mathbf{k}') G(\omega - \omega', \mathbf{k} - \mathbf{k}'). \qquad (14.14)$$

The exact Green functions, however, are supposed to have poles in the origin, $\omega \propto \mathbf{k}^2 \to 0$, and the loop integral acquires singularity at the point where the two poles pinch,

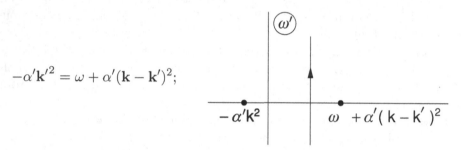

$$-\alpha' \mathbf{k}'^2 = \omega + \alpha' (\mathbf{k} - \mathbf{k}')^2;$$

The condition for the pinch in $\mathbf{k}'$ gives $\omega + \frac{1}{2}\alpha' \mathbf{k}^2 = 0$, and we get

$$G(\omega, \mathbf{k}^2) = -\left(\omega - \Delta + \alpha' \mathbf{k}^2 + \frac{r^2}{8\pi\alpha'} \ln \frac{\Lambda}{\omega + \frac{1}{2}\alpha' \mathbf{k}^2}\right)^{-1}. \qquad (14.15)$$

We see that $\Delta > 0$ did not help: the infrared singularity is *too strong* to be taken care of by mere introduction of a constant shift in the position of the bare pole. One has to look deeper, to take into account more complicated diagrams, to address the question of the possible behaviour of the renormalized vertices, etc.

We face the following problem: *what the Green function has to be in order to ensure that the theory contains a massless excitation?*

This is a general question which appears also in the condensed matter physics context. In our case it is about $\mathbf{P}$ in 1.

14.3.1 Corrections to the vertex part

Before we start summing the diagrams, let us see what is the scale of the correction to the *vertex* (as we will discover shortly, it is actually the vertex that plays a determining rôle).

$$\Gamma \equiv \quad = \quad + \quad + \cdots$$

In terms of renormalized reggeon Green functions, the first correction takes the form

$$\Gamma_2 \quad = \quad \simeq \quad r^3 \int \frac{d\omega' \, d^2\mathbf{k}'}{(2\pi)^3 \, i} \frac{1}{\omega' + \alpha' \mathbf{k}'^2}$$

$$\cdot \frac{1}{\omega - \omega' + \alpha'(\mathbf{k} - \mathbf{k}')^2} \cdot \frac{1}{\omega' - \omega_1 + \alpha'(\mathbf{k}' - \mathbf{k}_1)^2} \cdot \quad (14.16\text{a})$$

Now at large $\mathbf{k}'^2$ values everything is all right. We will estimate this expression by simply counting the powers. Let all the external variables be of the same order of magnitude, $\omega_1 \sim \omega_2 \sim \alpha' \mathbf{k}_{1,2}^2 \sim \alpha' \mathbf{k}^2 \sim \omega$. Then

$$\Gamma_2 \sim \frac{r^3}{\omega}, \qquad \frac{\Gamma_2}{r} \sim \frac{\delta\Gamma}{\Gamma} \sim \frac{r^2}{\omega}. \qquad (14.16\text{b})$$

For $\omega > r^2$ the correction is small, and we can use perturbation theory all right. However, at small values of ω, $\omega < r^2$, we face a rather catastrophic situation: the correction becomes large, and diverges in the $\omega \to 0$ limit.

What happens here is almost identical to the problem of the second-order phase transitions ('almost' because our specific problem is marked by the anti-Hermiticity of the effective Hamiltonian). Normally, correlation functions in a thermodynamical system fall exponentially with the distance,

$$\int d^3\mathbf{k} \frac{e^{i\mathbf{k}\cdot\mathbf{r}}}{\mathbf{k}^2 + \Delta^2} \sim \frac{e^{-\Delta r}}{r}. \qquad (14.17)$$

If at a certain temperature T the value of Δ turns to zero, the correlation radius goes to infinity, and the system undergoes the phase transition. At temperatures close to the critical, $T \approx T_c$, the excitations interact strongly and the perturbation theory becomes divergent.

Historically, the understanding of the problems of interacting pomerons and of the second order phase transitions in condensed matter physics was being gained practically simultaneously. The two problems turn out to be very similar; the only difference is that, let us stress it again, the pomeron dynamics corresponds effectively to an anti-Hermitian Hamiltonian.

14.3.2 Equations for the renormalized vertex and the reggeon propagator

What can be said about the exact vertex? It can be shown that all the diagrams for the corrections to the vertex part can be combined into *skeleton diagrams* built of *exact* Green functions and vertices. (Skeleton diagrams are those which contain no block that would represent a correction to an internal vertex.)

$$(14.18)$$

This is one equation for two quantities, the propagator and the vertex. A second equation seems to be easy to write – the Dyson equation,

$$G(\omega, \mathbf{k}) = G_0(\omega, \mathbf{k}) + \quad \text{(14.19)}$$

The impression that this equation may be more informative than that for the vertex function (14.18), containing an infinite number of terms, is deceiving. In fact, (14.19) is *not an equation*: the loop integral diverges, and the expression contains $G(\mathbf{q})$ and $\Gamma(\mathbf{q}, \mathbf{q}, \mathbf{k})$ in the region of large momenta $\mathbf{q}$ about which we do not know anything.

One usually performs the *renormalization*. This, however, does not solve the problem but rather hides it. At the same time, there is a simple way to derive the second equation we need. Indeed, the divergence is a consequence of the fact that G is a 'dimensional' quantity: $[G] \sim 1/\omega$. It is sufficient therefore to take a derivative and write down an equation for the 'dimensionless' quantity $\partial G^{-1}(\omega, \mathbf{k})/\partial\omega$.

Imagine one of the diagrams participating in Σ, . The external 'energy' ω flows through some of the internal lines. Differentiation over ω leads to *doubling* one of these internal lines,

$$\frac{\partial}{\partial\omega}\frac{1}{(\omega - \sum_k \omega'_k + \cdots)} = -\frac{1}{(\omega - \sum_k \omega'_k + \cdots)} \cdot \frac{1}{(\omega - \sum_k \omega'_k + \cdots)},$$

and resembles the vertex with zero 'energy–momentum' transfer,

$$\frac{\partial}{\partial \omega}\ (\text{———})\ =\ -\ \text{⊗}\ .$$

Indeed, it can be checked that the equation for the derivative of the inverse reggeon Green function looks diagrammatically exactly the same as the equation for the exact vertex:

$$\frac{\partial G^{-1}}{\partial \omega} \equiv \otimes = -1 + \ \ + \ \ +\cdots \qquad (14.20)$$

Unfortunately, equations (14.18) and (14.20) are represented in the form of infinite series. Therefore, there is always a danger that, owing to the possible divergence of the series, conclusions that one would derive from these equations may turn out to be wrong.

14.4 Weak and strong coupling regimes

Leaving aside the problem of the convergence of the series, we can guess the structure of the solution. But first, having expressed the equations for the vertex (14.18) and the propagator (14.20) in terms of *exact*, renormalized quantities, we have to re-examine the size of corrections. In other words, we have to estimate the magnitude of the effective expansion parameter.

Each subsequent term on the r.h.s. of the equation contains, compared to the previous one, three reggeon Green functions, two vertices and an additional integration over the reggeon loop:

$$\frac{\delta \Gamma}{\Gamma} \sim \frac{\delta \Sigma}{G^{-1}} \sim \int d^2\mathbf{k}'\, d\omega'\, G^3 \Gamma^2 \sim e^2. \qquad (14.21a)$$

We may represent this parameter as

$$e^2 \sim \frac{\langle \mathbf{k}^2 \rangle}{\omega} \cdot (\omega G)^3 \cdot \frac{\Gamma^2}{\omega}, \qquad (14.21b)$$

with $\langle \mathbf{k}^2 \rangle$ the characteristic transverse momentum in the integral. If renormalization effects were moderate and did not drastically change the behaviour at small ω, so that $\Gamma \sim r$, $\omega G \sim 1$, $\langle \mathbf{k}^2 \rangle \sim \omega$, this would lead us back to the original estimate (14.16b), $e^2 \sim r^2/\omega$.

The quantity e^2 depends on the external variables ω, $\mathbf{k}^2$ and can be looked upon as the 'invariant charge' – the true measure of the 'interaction strength' in the theory of interacting pomerons. Our theory may have a

self-consistent solution in two regimes, characterized by the magnitude of
the 'invariant charge' in the $\omega \to 0$ limit:

- Weak coupling, $e^2 \ll 1$. This regime is possible if the interaction
  modifies the interaction vertex so that Γ *vanishes* when $\mathbf{k}^2 \to 0$,
  $\omega \to 0$. Given a small effective interaction strength, one can use perturbative expansion to control the corrections.

- Strong coupling, $e^2 \sim 1$. In this case all terms in the equations are of
  the same order, and perturbation theory is not applicable.

14.4.1 Scaling solution. Strong coupling

We have seen that having taken the bare quantities and having started
the iterations of the equation, we have obtained the growing corrections.
How can we, nevertheless, make the l.h.s. and the r.h.s. of the equation
equal?

Since correction terms appear to be singular in the origin, let us suppose that the *bare terms* could be neglected in the equations. If so, we
would arrive at a homogeneous non-linear integral equation for the singular quantities (while the constant bare terms may be cancelled by non-
singular pieces of the skeleton diagrams on the r.h.s.).

When we search for a self-consistent solution, not only the matching of
numerical values of the l.h.s. and r.h.s. of the equation is required, but also
that of the *singularities*. If the singularity is weak, then the perturbation
theory can be used. If, on the contrary, the singularity is strong, then the
Born terms drop out and we get a homogeneous equation.

Bearing in mind the second case, let us look for a solution in the following form:

$$G(\omega, \mathbf{k}^2) \;=\; \omega^\mu g\left(\frac{\mathbf{k}^2}{\omega^\nu}\right), \tag{14.22a}$$

$$\Gamma(\omega, \mathbf{k}^2; \omega_1, \mathbf{k}_1^2; \omega_2, \mathbf{k}_2^2) \;=\; \omega^\rho \gamma\left(\frac{\mathbf{k}_1^2}{\omega_1^\nu}, \frac{\mathbf{k}_2^2}{\omega_2^\nu}, \frac{\mathbf{k}^2}{\omega^\nu}, \frac{\omega_1}{\omega}\right). \tag{14.22b}$$

This is a statement of the 'scaling' type: we extracted the overall powers
of ω and introduced the functions g and γ depending only on the *ratios*
of all other variables. We have to substitute this scaling solution in the
equation (omitting the finite bare terms) and see if it reproduces itself
under iterations.

It is clear that the 'homogeneous matching' can be achieved only if
$e^2 \sim 1$. Substituting (14.22) into, for example, the equation for Γ, the

first diagram in (14.18) gives

$$\Gamma \simeq \int \frac{d\omega'}{\omega'} \frac{d^2\mathbf{k}'}{\omega'^{\nu}} \cdot (\omega')^{\nu+1} (\omega'^{\mu} g)^3 (\omega'^{\rho} \gamma)^3. \qquad (14.23)$$

Due to the scale invariance of the functions g and γ in (14.22) under the transformation

$$\omega \to \lambda\omega, \qquad \mathbf{k}^2 \to \lambda^{\nu}\mathbf{k}^2,$$

the integral (14.23) will behave as $\lambda^{3(\rho+\mu)+\nu+1}$. This means that it can always be written in the form

$$\Gamma \simeq \omega^{3(\mu+\rho)+\nu+1} F(\text{ratios})$$

(provided the integral is convergent). Now, equating the exponents, our only requirement is that the relation $\rho = 3(\mu + \rho) + \nu + 1$ should be satisfied:

$$3\mu + 2\rho + \nu + 1 = 0.$$

Let us note that this condition actually means $e^2 = \mathcal{O}(1)$. Indeed,

$$e^2 \sim \int d^2\mathbf{k}\, d\omega\, G^3\Gamma^2 \sim \omega^{\nu+1}\omega^{3\mu}\omega^{2\rho}\, \Phi(\text{ratios}) = \text{const}.$$

This shows that in the scaling solution the 'effective charge' e^2 does not depend on the small quantity ω. Hence, $e^2 \sim 1$, corresponding to the *strong coupling* regime.

It can be easily verified that in *all* diagrams in the equations for the Green function and the vertex, the scaling solution (14.22) reproduces itself.

Our equations in the strong coupling regime do not contain parameters at all. The phase space volume $d^2\mathbf{k}\, d\omega \sim (d^3k)$ is the only quantity which reflects the specific features of the theory. K. Wilson suggested an interesting method for the investigation of such equations, namely, the continuation in the *number of dimensions*. By treating the deviation (ϵ) from the actual number of dimensions (three in our case) as a small parameter, one can approximate the solution by series in ϵ.

14.4.2 The cross section seems to change inevitably

Let us demonstrate that the scaling ('strong coupling') solution is incompatible with the $\sigma_{\text{tot}} \to \text{const}$ asymptotic behaviour. Evaluating the imaginary part of the reggeon Green function to get the total cross section we obtain

$$\sigma_{\text{tot}}(\xi) \propto \text{Im}\, A(\xi, \mathbf{k}^2 = 0) \sim \int d\omega\, e^{\omega\xi} \cdot \omega^\mu \propto \xi^{-(\mu+1)}. \qquad (14.24)$$

Consider first the case $\mu + 1 > 0$ corresponding to the total cross section *decreasing* with energy. In this case,

$$\frac{\partial G^{-1}}{\partial \omega} \to \infty \quad (\omega \to 0),$$

so that the bare term (-1) in (14.20) for the derivative of the inverse pomeron Green function can be neglected. But then my equations would not know that the Hamiltonian of my theory was actually anti-Hermitian: with the Born terms dropped, the equations become insensitive to the sign of G. A solution with $G > 0$ would not satisfy us, since in this case the contributions of the branchings would be of the same sign, and the unitarity condition would be not of the reggeonic type. The series must be alternating, corresponding to $G < 0$. As a more detailed analysis shows, there is no satisfactory solution for $\mu + 1 > 0$.

Now we take $\mu + 1 < 0$:

$$\frac{\partial G^{-1}}{\partial \omega} \to 0 \quad (\omega \to 0).$$

Now the unity can not be thrown away in the equation. In this case a solution exists corresponding to $G < 0$. This means that the total cross sections might *grow logarithmically* with energy.

Neither of the two cases is what we have been looking for. The scaling solution does not allow us to have a constant cross section, $\mu + 1 = 0$. In fact, all negative integer μ values are forbidden by the strong coupling equations.

How can this be seen? If $\mu = -1$, in the $\omega \to 0$ limit we have

$$\frac{\partial G^{-1}}{\partial \omega} \sim \omega^0 = \text{const.} \qquad (14.25)$$

Will this constant be reproduced by the equation (14.20)?

Let us consider the simplest diagram,

$$\Delta\frac{\partial G^{-1}}{\partial\omega} = \quad \sim \int d\omega'\, d^2\mathbf{k}'\frac{\partial G^{-1}}{\partial\omega}G^3\Gamma^2,$$

and rewrite the integrand as follows,

$$\int\frac{d\omega'}{\omega'}\frac{\partial G^{-1}}{\partial\omega}\Big[d^2\mathbf{k}'\cdot\omega'\cdot G^3\Gamma^2\Big] \sim \int\frac{d\omega'}{\omega'}\frac{\partial G^{-1}}{\partial\omega}\cdot e^2. \tag{14.26}$$

Since $e^2 = \mathcal{O}(1)$, substituting the constant for the derivative of G leads to a logarithmic divergence,

$$\Delta\frac{\partial G^{-1}}{\partial\omega} \simeq \mathrm{const}\int\frac{d\omega'}{\omega'} \propto \ln\omega,$$

incompatible with the assumption (14.25). This is a general feature of dimensionless integrals.

14.4.3 How to enforce $\sigma_{\mathrm{tot}} \simeq$ const? Weak coupling

The question arises, whether, in fact, the cross section can be still asymptotically constant? What would be necessary to achieve such a behaviour? We have to force the corrections to be small, despite the divergence of the perturbation corrections.

We have supposed that the three-reggeon vertex is finite, $\Gamma = \mathcal{O}(1)$, and estimated the contribution of the reggeon loop in (14.16) as $\delta\Gamma \propto \omega^{-1}$ – this is, obviously, not suitable. However, if we took for the renormalized vertex Γ an expression that *vanishes* with $\omega, \mathbf{k} \to 0$,

$$= a\omega + b\big(\mathbf{k}_1^2 + \mathbf{k}_2^2\big) + c\mathbf{k}^2, \tag{14.27}$$

we might achieve the matching.

If the increasing solution is analogous to that in phase transitions, the present – *weak coupling* – situation corresponds to quantum electrodynamics: the photon does not acquire a mass because its emission vertex contains momenta in the numerator:

$$\Pi_{\mu\nu} = (k_\mu k_\nu - g_{\mu\nu}k^2)\Pi(k^2) \quad\Longrightarrow\quad m_\gamma = 0.$$

Up to now we have considered only the three-pomeron vertex. For the self-consistence of the weak-coupling picture it is necessary and sufficient that the four-reggeon interaction vertices in (14.1b) also vanish in the origin, $\lambda, \lambda_1 \to 0$ with $\omega_i, \mathbf{k}_i \to 0$.

It can be shown that all possibilities for the solutions of the interacting pomeron problem are exhausted by the described above cases of 'strong' and 'weak' coupling.

A few words about the structure of the j-plane in the two regimes.

Weak coupling. The pomeron Green function,

$$G(\omega, \mathbf{k}^2) = -\frac{1}{\omega + \alpha' \mathbf{k}^2 - \Sigma(\omega, \mathbf{k}^2)}, \quad \Sigma(\omega, \mathbf{k}^2) \sim \;\; \Gamma \bigcirc \Gamma \;, \quad (14.28a)$$

acquires a small but *complex* self-energy term, $\Sigma \propto \mathbf{k}^4$. Owing to the complexity of the correction, the initial pole transforms into two conjugate

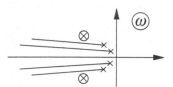

poles. (If the theory were Hermitian, these poles would have had to move onto an unphysical sheet.) Apart from these poles, we have a family of branch cuts accumulating to $\omega = 0$.

Strong coupling. Here the Green function contains branch point singularities embedded from the start, due to its complicated structure,

$$G(\omega, \mathbf{k}^2) = \omega^\mu g\left(\frac{\mathbf{k}^2}{\omega^\nu}\right). \quad (14.28b)$$

For $t > 0$ all singularities are on the right of, and condensing to, $\omega = 0$; at $t = 0$ they form a continuous cut starting at $\omega = 0$. In the physical region, $t < 0$, this cut is accompanied by a strong accumulation of branch cuts, whose presence strongly affects the angular dependence of the scattering amplitude. We made the hypothesis that 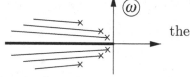 vacuum Pomeranchuk pole (pomeron $\mathbf{P}$) exists and studied the total cross section, and amplitudes of elastic and quasi-elastic processes of production of a small number of particles with large rapidity gaps between them. From this study a picture has emerged in which one has to include, apart from the pole $\mathbf{P}$, also branchings, both non-enhanced,

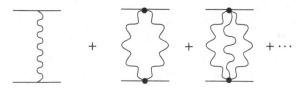

and enhanced

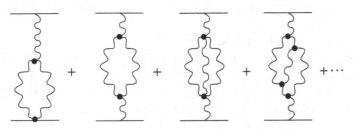

We have found two scenarios that may lead to a self-consistent solution of the problem of interacting pomerons.

(1) If we insist on $\sigma_{\text{tot}} \to$ const in the $s \to \infty$ asymptotics, then we need to consider the weak coupling regime (14.28a) characterized by vanishing three- and four-pomeron vertices, see (14.27).

(2) If we release the asymptotic *constancy* condition and allow the total cross section to increase logarithmically, $\sigma_{\text{tot}} \sim \ln^a s$ with $a \leq 2$, then we may turn to the strong coupling regime (14.28b).

One may think that in both scenarios it is the simplest graph that mainly determines the cross sections, while the multi-pomeron diagrams provide controllable corrections.

14.5 Weak and strong coupling: view from the s channel

Now we are going to formulate our results in terms of rapidities and impact parameters. Then we will turn to the main question: how the picture of interacting pomerons manifests itself in the s-channel. For example, what s-channel processes correspond to G in the strong coupling regime?

Let us make, first of all, a technical remark. Recall representation (14.2) for the scattering amplitude,

$$A(s, \mathbf{k}^2) = s \left(i + \frac{\pi}{2} \frac{\partial}{\partial \xi} \right) \int \frac{d\omega}{2\pi i} \, e^{\omega \xi} f(\omega, \mathbf{k}^2), \qquad (14.29)$$

where $f(\omega, \mathbf{k}^2)$ is the sum of reggeons diagrams. Recalling that ω lies on the imaginary axis, this integral can be considered as the transition from the frequency to the time representation. In this representation, the image of the pomeron propagator,

$$G(\xi, \mathbf{k}^2) = -e^{-\alpha' \mathbf{k}^2 \xi} \theta(\xi),$$

resembles the Green function $G(t) = \exp(-i(k^2/2m)t)\theta(t)$ describing a free non-relativistic particle, with $\xi = \ln s$ the analogue of (imaginary) time.

Since there is a full analogy with time, an arbitrary diagram can be written without any calculations, using the rules of non-relativistic field theory. For example, for a semi-enhanced diagram with the transition of one reggeon into two, we introduce integration over the transition 'time' ξ_2 to write

$$g \int_{\xi_3}^{\xi_1} d\xi_2\, G(\xi_1 - \xi_2, \mathbf{k}) \int \frac{d^2 k_1}{(2\pi)^2} r G(\xi_2 - \xi_3, \mathbf{k}_1) G(\xi_2 - \xi_3, \mathbf{k} - \mathbf{k}_1) N.$$

14.5.1 Diffusion in the impact parameter space

One can go even further, moving to the transverse coordinates, see (14.4):

$$\tilde{G}(\xi, \boldsymbol{\rho}) = \int \frac{d^2 k}{(2\pi)^2}\, \mathrm{e}^{-i\mathbf{k}\cdot\boldsymbol{\rho}} G(\xi, \mathbf{k}) = -\frac{\mathrm{e}^{-\rho^2/4\alpha'\xi}}{4\pi\alpha'\xi}.$$

Once again, this is the Green function of a non-relativistic particle propagating in imaginary time $t = i\xi$, or, better to say, the Green function of an equation describing *two-dimensional diffusion*.

Let us consider in these terms the simplest non-enhanced branching:

$$= \; g^2 \tilde{G}(\xi_1 - \xi_2, \boldsymbol{\rho}_1 - \boldsymbol{\rho}_2), \qquad\qquad (14.30\mathrm{a})$$

$$= \; N^2 \tilde{G}^2(\xi_1 - \xi_2, \boldsymbol{\rho}_1 - \boldsymbol{\rho}_2). \qquad\qquad (14.30\mathrm{b})$$

In this language, the expression for the two-reggeon branching (14.30a) is no more difficult to put down than that for one reggeon, (14.30a). This also makes it immediately clear, why the branching contributes less than the pole.

Consider the contribution to σ_{tot}, which corresponds to setting $\mathbf{k} = \mathbf{0}$. Then in the impact parameter space representation we need to evaluate the integral

$$\int \tilde{G}^2(\xi, \rho)\, d^2\rho = \int d^2\rho \frac{e^{-2\cdot\rho^2/4\alpha'\xi}}{(4\pi\alpha'\xi)^2}.$$

In the pomeron propagation, characteristic impact parameters grow with 'time', $\rho^2 \sim \alpha'\xi$, but the normalization in (14.30) is chosen such that $\int \tilde{G}(\xi, \rho)\, d^2\rho = -1$ (by the very nature of the Green function). So,

$$\int \tilde{G}^2(\xi, \rho)\, d^2\rho = \frac{1}{4\pi\alpha'\xi} \cdot \left(\frac{1}{2}\right) \int \frac{e^{-2\cdot\rho^2/4\alpha'\xi}}{4\pi\alpha'\xi} \cdot 2\, d^2\rho = \frac{1}{8\pi\alpha'\xi}. \quad (14.31)$$

The two-pomeron branching contribution falls with 'time' as $1/\xi$. What is the meaning of this smallness?

In (14.30a), a certain source produces a diffusive distribution (in our case, $\delta(\rho - \rho_1)$ at $\xi = 0$). With the increase of ξ, the integral over ρ of the distribution stays constant; the total probability to find a particle anywhere in space is determined solely by the power of the source and does not depend on time.

In (14.30a) we create two diffusion waves (with N their emission amplitude). If at time $\xi > 0$ I measured them *independently*, the probability conservation would have been intact, as in the one-pomeron case.

However, the amplitude we are interested in is given by the probability to find the two particles *in the same point* ρ_2, ξ_2. Such probability is inversely proportional to the typical area the distribution spreads over in time ξ, that is $1/\langle\rho^2\rangle \sim 1/\alpha'\xi$.

Investigating the flow coming from the centre, at large distances a single particle (**P**) gives the leading contribution $\propto \exp(-\rho^2/4\alpha'\xi)$. The corrections (**PP**) fall faster with the distance, $\propto \exp(-2\cdot\rho^2/4\alpha'\xi)$, although they may be significant at large times (and finite ρ^2). Such an interpretation looks quite satisfactory. Recall, however, how the same situation looked in the momentum representation:

$$G^{(\mathbf{P})}(\xi, \mathbf{k}^2) \sim \exp\left(-\alpha'\xi\mathbf{k}^2\right), \qquad G^{(\mathbf{PP})}(\xi, \mathbf{k}^2) \sim \exp\left(-\tfrac{1}{2}\cdot\alpha'\xi\mathbf{k}^2\right).$$

Now, when ξ is large, the second contribution is larger than the first one! The picture in the impact parameter space turned out to be intuitively more satisfactory than that in the momentum representation.

Similarly, we can 'spell out' any diagram in the language of diffusion. For example, the enhanced graph shown in (14.32) corresponds to the creation of one particle at the impact parameter point ρ_1 at time ξ_1, which splits into two particles at ρ_2, ξ_2. These particles then combine at a space–time point ρ_3, ξ_3 into one, which one is then registered at time ξ_4 in ρ_4.

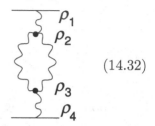

$$(14.32)$$

14.5.2 Energy dependence of σ_{tot}

We are ready to make an important general statement about the energy behaviour of σ_{tot}.

Consider the weak coupling regime and collect all relevant leading corrections:

$$f \simeq - \quad + \quad - \quad - \quad + \quad .$$

$$(14.33)$$

Let us examine the semi-enhanced diagram. Taking $\mathbf{k} = 0$ and using the expression (14.27) for the three-pomeron weak coupling vertex, for the third graph on the r.h.s. of (14.33) we have

$$g \cdot \frac{1}{\omega} \cdot \int \frac{d^2\mathbf{k}_1 \, d\omega_1}{(2\pi)^3 i} \frac{a\omega + 2 \cdot b\mathbf{k}_1^2}{(\omega_1 + \alpha'\mathbf{k}_1^2)(\omega - \omega_1 + \alpha'\mathbf{k}_1^2)} \cdot N \sim \ln \omega . \quad (14.34)$$

But if we take the Fourier transform of $\ln \omega$,

$$\int \frac{d\omega}{2\pi i} \, e^{\omega \xi} \ln \omega \;\to\; \int_{-\infty}^{0} e^{\omega \xi} \, d\omega \;\to\; \frac{1}{\xi},$$

we get the same $1/\xi$ behaviour as we had for the **PP** branching above. We see that the **P** pole $1/\omega$ in (14.34) has cancelled, and the contribution of the semi-enhanced diagram reduced to that of the non-enhanced one:

Analogous cancellation of both **P** poles occurs also in the enhanced diagram (14.32) **P** → **PP** → **P**, the last term in (14.33). There is, however, a subtle point, namely the signs. Assembling all contributions to the r.h.s. of (14.33), for the total cross section we derive

$$\sigma_{\text{tot}} = g^2 - \frac{(N - g \cdot c)^2}{8\alpha'\xi}, \tag{14.35}$$

where we denoted by c the constant that emerges from the **P** → **PP** vertex in (14.34). This shows that if the cross section tends to a constant, then it approaches this limit *from below.*

In the strong coupling case we do not need to calculate anything, since we know already that the cross section grows logarithmically with s.

We may thus conclude that at sufficiently high energies the total cross sections *should slowly grow* with s, at least temporarily (weak coupling), if not forever (strong coupling).

14.5.3 Experimental situation

What is the experimental situation?

Up to $s \sim 30\,\text{GeV}^2$ all the cross sections decrease (but for σ_{tot}^{pp} and $\sigma_{\text{tot}}^{K^+p}$ that stay nearly constant) (Fig. 14.3). Between 70 and 2000 GeV^2 a new phenomenon takes place: the cross sections flatten off and start to slowly increase.[*] The fact that σ_{tot}^{pp} has a minimum, means that there exists a definite energy at which the matter is maximally transparent.

For the first time the 'complication' of the theory reveals itself; without branchings nothing of this kind would happen.

14.5.4 Would one ever reach the true asymptotics?

An impressive body of predictions of the theory has been experimentally confirmed. Among them are the following statements.

(1) Factorization of scattering amplitudes and cross sections.

(2) Universal nature of particle production in the plateau region.

(3) High-mass inelastic diffraction in the three-pomeron limit, whose mass distribution is spectacularly different from expectations based on the statistical model of the hadron production.

(4) Shrinkage of the diffractive cone.

---

[*] From the RFT perspective, growing total cross sections (Fig. 14.3) may shift ones preference towards the strong coupling regime (ed.).

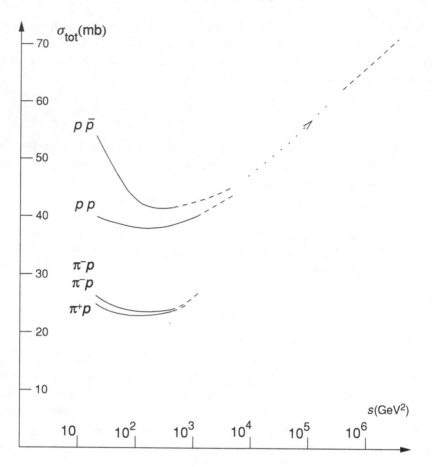

Fig. 14.3   Sketch of the total hadron cross sections $\sigma_{tot}(s)$. Hadron accelerator data that appeared after middle-1970s are shown by dashed lines.

(5) Increase of the total cross sections with $\ln s$, which signals either an approach to asymptotic constant values (weak coupling) or a steady growth characteristic of the strong coupling regime.

Many attempts were made to describe also the *angular distributions*. Qualitatively, it has become apparent that the branchings play an essential rôle here. At the same time, quantitative description of the t-dependence has not been achieved.

At first sight, the procedure seems well defined and simple; one has to substitute known expressions into given formulae, calculate the effect and compare the prediction with the data. Strangely enough (accidentally or, may be, for a deep reason beyond our understanding) this happens to be an impossible task.

We assume ξ to be large. The true parameter of the problem is, however, $\alpha'\xi$. The easiest way to see this is by recalling the two-pomeron branching,

$$= \tilde{G}^2(\xi, \boldsymbol{\rho}_1 - \boldsymbol{\rho}_2)N^2. \tag{14.36}$$

Dealing with this diagram, a number is substituted for the reggeon emission amplitude N. My motivation was that I am close to the singularity in ω. If so, ξ is large, and so is, consequently, the transverse distance $|\boldsymbol{\rho}_1 - \boldsymbol{\rho}_2|$ due to diffusive nature of the pomeron, $G \sim \exp(-|\Delta\rho|^2/4\alpha'\xi)$.

In fact, the hadron projectile has its own transverse size R, so that inside the vertex part N there are various diagrams in which impact parameters of participating particles vary within $|\Delta\rho| \lesssim R$. Therefore we should have introduced separate transverse coordinates for the reggeons and should have written a multiple integral over the impact parameters,

$$\sim \int N(\boldsymbol{\rho}_1; \boldsymbol{\rho}_1', \boldsymbol{\rho}_1'')\tilde{G}(\xi, \boldsymbol{\rho}_1' - \boldsymbol{\rho}_2')\tilde{G}(\xi, \boldsymbol{\rho}_1'' - \boldsymbol{\rho}_2'')N(\boldsymbol{\rho}_2; \boldsymbol{\rho}_2', \boldsymbol{\rho}_2'').$$

From the point of view of an asymptotic behaviour, we have acted correctly by treating N as a constant. This is, however, justified only if $R^2 \ll 4\alpha'\xi$ so that $|\boldsymbol{\rho}_1 - \boldsymbol{\rho}_2| \gg |\boldsymbol{\rho}_1 - \boldsymbol{\rho}_1'| \sim R$. And just here lies the catastrophe.

Strictly speaking, it is difficult to estimate the radius of a hadron; I would say that $R \sim 1/2m_\pi$. At the same time, the value of α' is measured from the shrinkage of the diffractive cone. It turns out to be rather small, $\alpha' \approx 1/4m_N^2$, about four times smaller than the slopes of non-vacuum Regge trajectories, $\alpha'_R \simeq 1\,\mathrm{GeV}^{-2}$. Hence, our inequality becomes

$$4\alpha'\xi \gg R^2 \quad \Longrightarrow \quad \xi \gg \xi_{\mathrm{crit}} \equiv \frac{R^2}{4\,\alpha'} \sim \frac{m_N^2}{4m_\pi^2} \simeq 10.$$

Obviously, the condition $\xi \gg \xi_{\mathrm{crit}}$ can by no means be satisfied, ever. At best we may reach $\xi \sim \xi_{\mathrm{crit}}$ ($s = 2.000\,\mathrm{GeV}^2$ corresponds to $\xi \approx 8$).

For reasons we do not understand, there is no way to reach the true asymptotic regime.

Examining the parameters in the strong coupling regime, the situation is even worse: the corresponding diffusion coefficient (effective α') turns out to be even smaller.

The appearance of a new small parameter, $1/\xi_{\mathrm{crit}} = 4\alpha'/R^2 \ll 1$, on the other hand, suggests certain simplifications for the region of moderately high energies, $\xi \lesssim \xi_{\mathrm{crit}}$. Since in the $\alpha' \to 0$ limit the impact parameter diffusion disappears and parallel showers do not separate but keep interacting with each other; employing α' as a small parameter leads to a picture of 'heavy pomeron', whose image is no longer a two-particle 'ladder' but a more complicated t-channel state of many re-interacting particles (Gribov, 1976).

To conclude, our object diffuses in unit time at a typical transverse distance $|\Delta\rho|^2 \sim 4\alpha'$ (the diffusion coefficient). Why did this distance turn out to be *much smaller* than the size proper of the hadron? Qualitatively, this phenomenon may be due to the fact that hadrons are *composite*. If the true fundamental objects – quarks – are more compact, their smaller size would introduce a relatively large mass scale parameter that might explain the smallness of the slope α' of the t-channel vacuum exchange.

It is just the smallness of α' which impairs quantitative description of the t-dependence of hadron–hadron scattering amplitudes. There is, however, a whole complex of problems not related to the angular distributions, like multi-particle production processes, and it is necessary to understand the physics which corresponds to them.

15

Particle density fluctuations and RFT

15.1 Reggeon branchings and AGK cutting rules

15.1.1 Inelastic processes corresponding to reggeon branchings

In this lecture we will investigate the correspondence between various inelastic processes and reggeon diagrams. We begin with the simplest object, the pole.

As before, we assume an essential property, namely: that the particle distribution emerging from cutting the pomeron pole is *homogeneous*, i.e. the inclusive spectrum $\varphi(k_\perp^2)$ does not depend on the rapidity η:

$$\frac{d^3\sigma}{d\eta \, d^2\mathbf{k}_\perp} = \quad = g_1 G(\xi - \eta)\varphi G(\eta)g_2 = \sigma_{\text{tot}} \cdot \varphi(k_\perp^2). \quad (15.1)$$

For $\xi - \eta \sim 1$ ($\eta \sim 1$) when the particle is close in rapidity to one of the fragmentation regions, the shape of the particle yield depends on the quantum numbers of the registered particle and of the projectile (target):

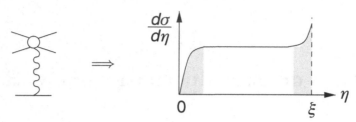

If each pole corresponds to the
uniform distribution, what sort
of inelastic processes are contained in the branching of two
such poles?

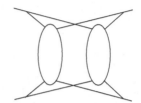

In Lecture 12 we have calculated the two-reggeon branching diagram
as a whole. Now I would like to find out, what sort of imaginary parts it
has; in other words, how can the diagram be cut?

(1) First of all, we can have a quasi-elastic process by making a cut
between the reggeons:

$$(15.2a)$$

(2) One can cut one of the two ladders:

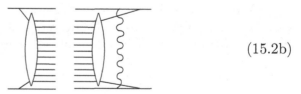

$$(15.2b)$$

which gives the correction to the probability of having the usual
final particle distribution.

(3) Finally, both ladders can be cut:

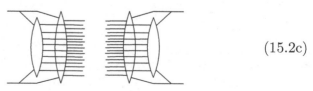

$$(15.2c)$$

I get two new processes, (1) and (3), in addition to the correction (2). In case (3) I observe two ladders at the same time, i.e. the particle density is twice as high as before. So, the branching describes fluctuations of the number of final state particles.

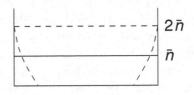

Curiously, the changes are rather sharp – there is either no particle in the plateau region, or a 100% enhancement. Obviously, we have chosen too simple a diagram.

By taking enhanced diagrams, we obtain various *local* fluctuations. For example, by cutting the two-pomeron loop in the Green function, we get:

(1) a central rapidity gap;

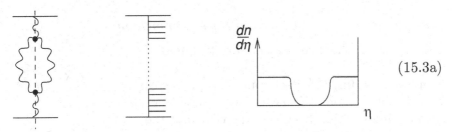

$$(15.3a)$$

(2) another correction to the uniform ladder;

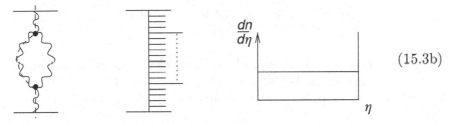

$$(15.3b)$$

(3) a local double density fluctuation.

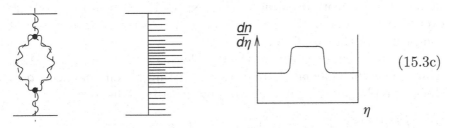

$$(15.3c)$$

Considering a diagram with three reggeons in the *t*-channel, we will have *triple particle density* as a new fluctuation (plus extra corrections to the distributions that we already had).

From the point of view of the s-channel, the self-consistency of our picture implies moderate fluctuations in the particle density. Recall an analogy with 'particle gas' that we have discussed in Lecture 10. A statistical system is stable when fluctuations are small. On the other hand, we are near the critical point where fluctuations of arbitrary size emerge. As we already know, the weak coupling corresponds to small fluctuations. In the strong coupling regime, on the contrary, fluctuations are large, so large that there is no average density at all.

Not having solved the interacting pomeron theory, we do not know how to cut the exact Green function; and this is the problem. In spite of this, it turns out to be possible to understand the pattern of fluctuations in multi-particle production that is induced by the presence of pomeron branchings.

15.1.2 Two-reggeon branching

Let us study the two-reggeon branching diagram, $F^{(2)}$. Now I will be interested not in the expression for $\mathrm{Im}_s \, F^{(2)}$ itself, but what processes it is assembled from. We have to learn to extract such an information from our knowledge of the expression for the diagram as a whole.

Recall that at high energies the amplitude became factorized,

$$\frac{d^4 k}{(2\pi)^4 \, i} \to i \cdot \frac{d^2 \mathbf{k}_\perp}{(2\pi)^2} \cdot \frac{d\alpha}{2\pi i} \frac{d\beta}{2\pi \, i},$$

integrals over α and β produced real particle–reggeon vertex functions, N, on the top and the bottom of the diagram, and left us with a factor i for the reggeon loop. So, the branching can be written as

$$F^{(2)} = N \cdot f_1 \, i f_2 \cdot N, \qquad (15.4)$$

with f the reggeon amplitude. This expression is symbolic; integration over $d^2\mathbf{k}$ is implied. However, it represents correctly the nature of the complexity of the amplitude.

It is convenient to calculate the double imaginary part (discontinuity), $2\,\mathrm{Im}$. Then the calculation reduces to putting all cut particles on mass shell by replacing their propagators by delta-fuctions, e.g.

$$2\,\mathrm{Im} \int \frac{d^4 \ell}{(2\pi)^4 \, i} \frac{1}{m^2 - \ell^2} \frac{1}{m^2 - (P-\ell)^2} = \int \frac{d^4 \ell}{(2\pi)^2} \delta(m^2 - \ell^2)\delta(m^2 - (P-\ell)^2).$$

Let us see how our diagram can be cut.

(1) The simplest imaginary part arises from cutting *between* the reggeons and corresponds to quasi-elastic scattering:

$$2\,\mathrm{Im}^{(1)} F^{(2)} = N^2 \big(f_1 f_2^* + f_1^* f_2\big) = 2N^2 |f|^2 . \qquad (15.5a)$$

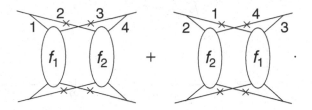

(2) We may cut through one reggeon, say, f_1. The second one may then stand either on the left from the cut (and describe rescattering of particle 2 in the amplitude) or on the right (rescattering in the conjugated amplitude),

$$2\,\mathrm{Im} F^{(2)} \Longrightarrow 2\,\mathrm{Im}\, f_1 \cdot \big(i f_2 + (i f_2)^*\big).$$

We have to add the cut through f_2 (and rescattering of particle 1):

$$2\,\mathrm{Im}^{(2)} F^{(2)} = N^2 \big[2\,\mathrm{Im} f_1\big(i f_2 + (i f_2)^*\big) + 2\,\mathrm{Im}\, f_2 \big(i f_1 + (i f_1)^*\big)\big]$$
$$= -8N^2 (\mathrm{Im}\, f)^2 . \qquad (15.5b)$$

(3) What remains to be done is simple: the last cut has to be made through both blocks, replacing all particle propagators inside the reggeon 'ladders' by delta functions:

$$2\,\mathrm{Im}^{(3)} F^{(2)} = N^2 \cdot 2\,\mathrm{Im}\, f_1 \cdot 2\,\mathrm{Im}\, f_2 = 4N^2 (\mathrm{Im}\, f)^2 . \qquad (15.5c)$$

Adding together (15.5), we obtain the imaginary part of the branching as consisting of three pieces,

$$2\,\mathrm{Im}_s F^{(2)} = N^2 \big[2 f f^* - 8(\mathrm{Im}\, f)^2 + 4(\mathrm{Im}\, f)^2\big]. \qquad (15.6)$$

We may verify our conclusion by directly evaluating the imaginary part of the symbolic expression (15.4):

$$2\,\mathrm{Im}\, F^{(2)} = 2\,\mathrm{Im}\, \big\{N f i f N\big\} = 2N^2\,\mathrm{Re}\, f^2 = 2N^2 \big\{(\mathrm{Re}\, f)^2 - (\mathrm{Im}\, f)^2\big\}$$
$$= 2N^2 \big[f f^* - 2(\mathrm{Im}\, f)^2\big] \equiv N^2 \big[2 f f^* - 8(\mathrm{Im}\, f)^2 + 4(\mathrm{Im}\, f)^2\big].$$

The first and third terms in the r.h.s. of (15.6) are positive because they represent cross sections of two processes: quasi-diffractive scattering and double density particle production. The second term is a correction to the

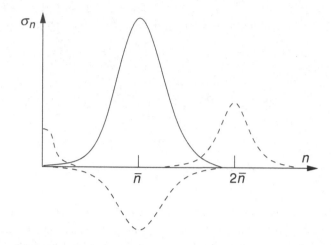

Fig. 15.1 Topological cross section distribution corresponding to single pomeron (solid line) and multiplicity fluctuation pattern induced by two-pomeron branching (dashed).

pole; it may be negative (though not very). The relation $2 : 8 : 4$ expresses the share of different final states contained in the imaginary part.

This simple example illustrates an important pattern of fluctuations in the multiplicity distribution induced by branchings. The cross section in the main region, $n \sim \bar{n}$, *decreases* (-8) to make room for the new particle production processes (characterized by the shares $+2$ and $+4$) as shown in Fig. 15.1. For the pomeron pole we have $|\mathrm{Re}\, f| \ll \mathrm{Im}\, f$, so that

$$2\,\mathrm{Im}_s F \simeq N^2(2 - 8 + 4)(\mathrm{Im}\, f)^2 = -2N^2(\mathrm{Im}\, f)^2. \tag{15.7}$$

The overall effect of the branching is *negative*; the total cross section decreases. This is screening.

This is an example of how we can sort out the content of $\mathrm{Im}_s F$ of arbitrary multi-reggeon diagrams. It is important that we did so according to the cuts of f, not touching N. This means that the we have carried the procedure in a universal way, and did not need to worry about the (potentially complicated) internal structure of the vertices.

15.1.3 Universality of the vertex function N

In the derivation we implied only one (but essential) thing, namely that the vertex block N remains the same in all cases (15.5). I would prove that the expression (15.6) is correct if N, indeed, does not depend on the way we cut the diagram.

To see that this is indeed the case, let me remind you what we did before. We expressed all particle momenta in terms of Sudakov variables using the momentum vectors p_1, p_2 of colliding particles as

$$k_i^\mu = \alpha_i p_1^\mu + \beta_i p_2^\mu + k_{i\perp}^\mu.$$

In the upper part of the graph, close to p_1, we have $\alpha_i \sim 1$, $\beta_i \sim 1/s$. The small components, β_i, do not affect the lower part of the diagram and hence, the β_i-integrations concern only the upper vertex block,

$$(15.8)$$

Recall

$$s_1 = (p_1 + k_1)^2 = (1 + \alpha_1)(\gamma + \beta_1)s \quad \Longrightarrow \quad s\,d\beta_1 = ds_1.$$

This shows that the block is integrated over the invariant energies (because of energy–momentum conservation, there are three independent integrations). The integration reduces to the closing of the contour around the physical cut and thus to the transition to real states:

In fact, after integration over energies the function N becomes symmetric with respect to the transmutation of particles 1, 2, 3, 4. I said that to cut the diagram means to make a certain set of internal particles real. But inside the block N all particles are already on the mass shell; it can be treated as a real function, in a deep sense: there is no way to cut it any further.

To clarify this important point, let us imagine that external particles are represented by operators of some field theory. Then the function N is given by the time-product

$$A(k_i) = \int \langle p_1' | T \varphi_1(x_1)\varphi_2(x_2)\varphi_3(x_3)\varphi_4(0) | p_1 \rangle \prod_{i=1}^{3} e^{ik_i x_i} \, d^4 x_i \,. \qquad (15.9)$$

As we have already seen, this expression does not itself enter our calculations, just its integral over all energies:

$$N = \int dk_{10} \int dk_{20} \int dk_{30} \, A.$$

It gives $\delta(t_1)\delta(t_2)\delta(t_3)$, so that the time ordering sign, T, can be omitted and we obtain an equal-time product of *commuting* operators,

$$N = \int \langle |\varphi_1(\mathbf{r}_1, 0)\varphi_2(\mathbf{r}_2, 0)\varphi_3(\mathbf{r}_3, 0)\varphi_4(0, 0)| \rangle \prod_{i=1}^{3} e^{-i\mathbf{k}_i \cdot \mathbf{r}_i} d^3\mathbf{r}_i.$$

This expression can be written in terms of real intermediate states, in various equivalent ways. In particular,

In the notation of (15.8), the first representation enters the quasi-diffractive imaginary part (15.5a), the second – (one of the four) one-reggeon cuts (15.5b), and the third – the cut through both pomerons (15.5c).

From our derivation it is clear that this property applies to particle–reggeon blocks with an arbitrary number of reggeons attached. It also holds for reggeon–reggeon interactions, like the three-reggeon vertices r in the enhanced correction graph (15.3).

Now we will generalize the result (15.4),

$$F^{(2)} \sim N \cdot f i f \cdot N,$$

to multi-reggeon branchings. Astonishing cancellations will allow us to calculate *inclusive particle spectra* in the most general manner, even for the case of strong coupling.

15.1.4 Cutting through many reggeons

What happens in multi-reggeon branchings? Let us write, in analogy with the two-reggeon case, (15.4),

$$= N_{(n)} \underbrace{f\,if\,i\ldots if}_{n} N_{(n)}. \qquad (15.10)$$

In the same way as before, the particle–reggeon vertices $N_{(n)}$ do not change when we cut the diagram. What is the contribution to different processes of a non-enhanced n-reggeon diagram with n_1 cut reggeons? This is a simple combinatorial problem. We put n_2 uncut reggeons on the left of the cut ones, and the remaining $n - n_1 - n_2$ on the right, as shown

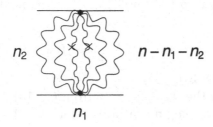

Fig. 15.2 n-reggeon diagram with n_1 cut reggeons.

in Fig. 15.2, and write

$$2\operatorname{Im} F^{(n)}_{n_1,n_2} = N^2_{(n)}(2\operatorname{Im} f)^{n_1}(if)^{n_2}(-if^*)^{n-n_1-n_2} C^{n_1}_n C^{n_2}_{n-n_1}. \quad (15.11)$$

The combinatorial factors C count the number of ways to choose n_1 cut reggeons from n, and to divide $n - n_1$ uncut ones into n_2 to be put into the amplitude (left) and $n - n_1 - n_2$ into the amplitude conjugate; $C^k_m = m!/k!(m-k)!$. We are interested in the sum over n_2:

$$\sum_{n_2=0}^{n-n_1} C^{n_2}_{n-n_1}(if)^{n_2}(-if^*)^{n-n_1-n_2} = (if - if^*)^{n-n_1} = (-1)^{n-n_1}(2\operatorname{Im} f)^{n-n_1}.$$

Substituting into (15.11) yields the contribution to the cross section with n_1 cut reggeons ($n_1 \geq 1$):

$$2\operatorname{Im} F^{(n)}_{n_1} = \sum_{n_2} 2\operatorname{Im} F^{(n)}_{n_1,n_2} = (-1)^{n-n_1} C^{n_1}_n (2\operatorname{Im} f)^n N^2_{(n)}. \quad (15.12)$$

We have to consider the case when no reggeon is cut ($n_1 = 0$) separately. To this end we calculate the number of ways to split n reggeons into $n_2 \geq 1$ to the left of the cut (with $n - n_2 \geq 1$ to the right), and take the imaginary part of the factor i separating the two groups in (15.10),

$$\sum_{n_2=1}^{n-1} (if)^{n_2}((if)^*)^{n-n_2} C^{n_2}_n = \underbrace{(if - if^*)^n}_{(-2\operatorname{Im} f)^n} - (if)^n - ((if)^*)^n,$$

producing

$$2\operatorname{Im} F^{(n)}_0 = (-1)^n(2\operatorname{Im} f)^n N_{(n)} - 2\operatorname{Re}[(if)^n]N_{(n)}. \quad (15.13)$$

Adding together (15.12) and (15.13) we have

$$\sum_{n_1=0}^{n} C^{n_1}_n(-1)^{n_1} \cdot (-2\operatorname{Im} f)^n - 2\operatorname{Re}[(if)^n] = -2\operatorname{Re}[(if)^n],$$

since the sum equals zero $((1+x)^n|_{x=-1})$. Thus, for the total imaginary part we obtain

$$2\operatorname{Im} F^{(n)} = -2\operatorname{Re}(if)^n N_{(n)}^2 = 2\operatorname{Im}\big[-i(if)^n\big]N_{(n)}^2,$$

in agreement with (15.10).

Let us note that in the case of $n = 2$, from (15.13) ($n_1 = 0$) and (15.12) ($n_1 = 1, 2$) we rederive the known proportion (15.7):

$$\operatorname{Im} F_0^{(2)} : \operatorname{Im} F_1^{(2)} : \operatorname{Im} F_2^{(2)} = 2 : -8 : 4.$$

Thus, a simple calculation gave us an important result for the composition of the n-reggeon branching: the contribution to the cross section with n_1 cut reggeons, $0 \leq n_1 \leq n$, is

$$2\operatorname{Im}_s F_{n_1}^{(n)} = \big((-1)^{n-n_1} C_n^{n_1}(2\operatorname{Im} f)^n - 2\delta_{n_1,0}\operatorname{Re}[(if)^n]\big)N_{(n)}^2. \quad (15.14)$$

15.2 Absence of branching corrections to inclusive spectrum

Let us study the single particle inclusive spectrum. In Lecture 10 we saw that in the pole approximation the inclusive spectrum is given by the Mueller–Kancheli diagram (15.1). Take now a two-reggeon branching. Since I need to register a particle in the central region, the quasi-elastic cut (15.2a) does not contribute. In the cross section, the screening correction diagram (15.2b) was twice as large as the graph with two cut reggeons, (15.2c). However, when both ladders are cut, I can take the necessary particle from either of them, and hence, this imaginary part enters with a factor 2 the inclusive spectrum:

$$0 \times 2 - 8 + 2 \times 4 = 0$$

The meaning of an inclusive spectrum is the multiplicity multiplied by the cross section:

$$\int f(\eta)\, d\eta = \sum_n n\sigma_n \equiv \bar{n}\sigma.$$

Recall how we arrived at this: the phase volume of k identical particles contains $1/k!$. Fixing the momentum of one of the particles, I have for the remaining ones $\frac{1}{(n-1)!} \prod_{i=1}^{n-1} dk_i$.

Let us show that there will never be any corrections to the inclusive spectrum in the central region. As we already know, n-reggeon branching with $n_1 > 0$ cut reggeons contributes to the cross section as

$$2\operatorname{Im} F_{n_1}^{(n)} = C_{n_1}^n(-1)^{n-n_1}(2\operatorname{Im} f)^n. \quad (15.15)$$

Calculating the inclusive spectrum, I have to multiply this expression by n_1, and to sum over n_1 in order to obtain the total contribution of the branching:

$$\delta f^{(n)} \propto \sum_{n_1=1}^{n} n_1 \cdot C_n^{n_1}(-1)^{n-n_1} = \sum_{n_1=1}^{n} \left[n - (n - n_1) \right] \cdot C_n^{n_1}(-1)^{n-n_1}$$

$$= \left\{ n(1+\kappa)^n - \kappa \frac{\partial}{\partial \kappa}(1+\kappa)^n \right\}_{\kappa=-1} = n \cdot (1+\kappa)^{n-1} \Big|_{\kappa=-1}.$$

Since $n \geq 2$ (we do not consider a pole), the total contribution of an arbitrary non-enhanced branching to the inclusive spectrum equals zero.

Let us consider a more complicated diagram. The diagram is integrated over rapidities η_1, η_2 at which the reggeons interact, while we are looking for a particle with a certain rapidity η. Depending on the order of the rapidities, we have three situations:

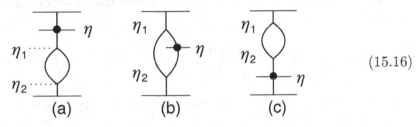

$$(15.16)$$

(1) $\eta_2 < \eta_1 < \eta$, (15.16) graph (a), or $\eta < \eta_2 < \eta_1$ (c). There are no cancellations; we have a correction to the inclusive spectrum.

(2) $\eta_1 > \eta > \eta_2$, (15.16) graph (b). In this region the contributions of one and two cut regions cancel; $\delta f(\eta) = 0$.

This consideration shows that I can draw any reggeon corrections from above or from below of the registered particle, but never embrace the point η. Summing up the total set of such diagrams leads, obviously, to replacing the Green functions, G, of the bare pomeron pole by the *exact* Green functions, $\mathbf{G}$:

$$\frac{d\sigma}{d\eta \, d^2\mathbf{k}_\perp} = g_1 \mathbf{G}(\xi - \eta)\varphi(k_\perp^2)\mathbf{G}(\eta)g_2. \qquad (15.17)$$

15.2.1 Shape of the inclusive spectrum

How do reggeon branchings affect the shape of the inclusive spectrum?

Weak coupling. In the pomeron pole approximation, the inclusive spectrum is flat in the central rapidity region, $f(\eta) = \text{const}$. Reggeon loops

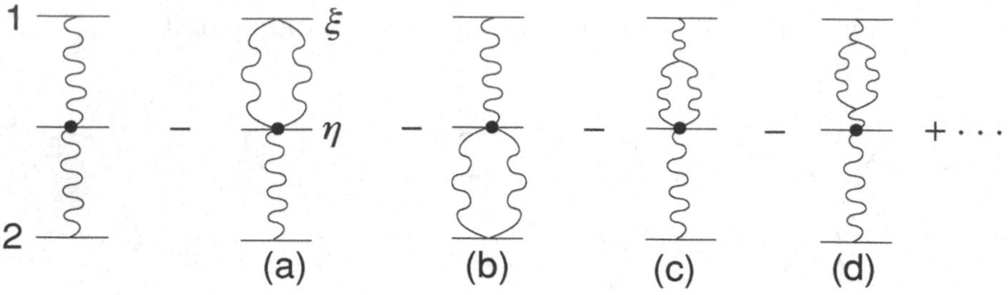

Fig. 15.3   One-loop corrections to inclusive spectrum (weak coupling).

provide sub-asymptotic corrections to the plateau. In addition to the corrections to the top and bottom pomeron propagators (Fig. 15.3(d)), we may take the triggered particle right from the three-reggeon vertex r as shown in Fig. 15.3(a–c). Let us evaluate the first correction term, Fig. 15.3(a):

$$f(\eta, \mathbf{k}^2) = g_1 \varphi(\mathbf{k}^2) g_2 - \int \frac{d^2 \mathbf{k}'}{(2\pi)^2} N_1 \cdot e^{-2\alpha' \mathbf{k}'^2 (\xi - \eta)} \tilde{\psi}(\mathbf{k}, \mathbf{k}') \cdot g_2 + \cdots,$$

where $\mathbf{k}'$ is the transverse momentum flowing through the reggeon loop. Since the rapidity interval $\xi - \eta$ is large, $\mathbf{k}'$ is small, and we may replace $\tilde{\psi}(\mathbf{k}, \mathbf{k}') \simeq \tilde{\psi}(\mathbf{k}, 0) \Rightarrow \psi(\mathbf{k}^2)$.

$$f(\eta, \mathbf{k}^2) \simeq g_1 g_2 \varphi(\mathbf{k}^2) - \frac{N_1 g_2}{4\pi} \frac{\psi(\mathbf{k}^2)}{2\alpha' \cdot (\xi - \eta)} + \cdots. \tag{15.18}$$

In the weak coupling regime the vertex vanishes, $r \propto \omega \sim \mathbf{k}^2$. Therefore in Fig. 15.3(c,d) the pomeron propagators neighbouring the loop cancel, so that these graphs reduce, in fact, to Fig. 15.3(a), modifying the function ψ in (15.18).

We obtain the distribution

$$f(\eta, \mathbf{k}^2) \simeq g_1 g_2 \varphi(k_\perp) - \frac{C_1(\mathbf{k}^2)}{\xi - \eta} - \frac{C_2(\mathbf{k}^2)}{\eta}, \tag{15.19}$$

where we added the contribution of the symmetric graph (b). We conclude that the plateau must have a *positive curvature*.

Strong coupling. In the strong coupling regime this property is expressed in a much more manifest way. Here essential multi-reggeon corrections are contained in the Green function itself, $G(\xi) \propto \xi^\mu$, so that the inclusive spectrum,

$$f(\xi; \eta, \mathbf{k}^2) \simeq g_1 G(\xi - \eta)\varphi(\mathbf{k}^2)G(\eta)g_2,$$

increases with energy, as does the total cross section. It make sense to measure the inclusive cross section in units of σ_{tot}:

$$\phi(\xi; \eta, \mathbf{k}^2) \equiv \sigma_{\text{tot}}^{-1}(\xi) \cdot f(\xi; \eta, \mathbf{k}^2), \quad \sigma_{\text{tot}}(\xi) \propto G(\xi) \propto \xi^\mu;$$

$$\phi(\xi; \eta, \mathbf{k}^2) \propto \frac{G(\xi - \eta)G(\eta)}{G(\xi)} \propto \xi^\mu \cdot z^\mu(1 - z)^\mu, \quad z \equiv \frac{\eta}{\xi}.$$

Unlike the parton model, where the system was 'forgetting' about its boundaries, here the situation is different. There is now neither a 'zeroth approximation', as in the weak-coupling case, nor an asymptotic plateau. The Feynman scaling is broken: the probability of finding a particle depends seriously not only on its place in the rapidity but also on the total energy.

Let us calculate the total multiplicity:

$$\int_0^\xi d\eta \int d^2\mathbf{k}\, \phi(\xi; \eta, \mathbf{k}^2) \equiv \bar{n}(\xi) \propto \xi^{\mu+1} \cdot \int_0^1 dx\, x^\mu(1 - x)^\mu.$$

The mean multiplicity has increased significantly – where did this come from? In the strong coupling case enhanced multi-reggeon diagrams are essential. This fact alone makes it evident that $\bar{n}$ must grow.

Indeed, recall the distribution for topological cross sections, σ_n. From Fig. 15.1 for multiplicity fluctuations due to the two-reggeon branching, it is not obvious, a priori, in what direction the average value moves, to the left or to the right. But, the more complicated branchings are included, the more the number of cuts grows compared with the number of uncut reggeons, resulting in the multiplicity increase.

From the space–time picture, in strong coupling our 'ladder' consists often of a few ladders of normal density tied together. Hence, in the course of the interaction a larger number of particles is 'shaken off'.

15.3 Two-particle correlations

To study multiplicity fluctuations, not only average quantities can be investigated but also the correlations of particles which allow us to find the dispersion of the multiplicity distribution. Let us construct the cross section for inclusive production of two particles with momenta k_1, k_2,

$$f(k_1, k_2) = \sum_n \frac{1}{(n-2)!} \int d\Gamma_{n-2} |F_n(k_1, k_2; q_1, \ldots, q_{n-2})|^2,$$

where

$$dΓ_{n-2} = \prod_{i=1}^{n-2} dΓ(q_i), \quad dΓ(q) = \frac{d^3q}{2q_0(2π)^3},$$

is the phase space volume for $n-2$ unregistered particles with momenta q_i.

If we now integrate the double differential distribution f over momenta k_1, k_2 of the triggered particles, we obtain the second multiplicity moment:

$$\int dΓ(k_1)\, dΓ(k_2) f(k_1, k_2) = \sum_n \frac{1}{(n-2)!} \int dΓ_n |F|^2$$

$$= \sum_n n(n-1) \int \frac{dΓ_n}{n!} |F|^2 = \sum_n n(n-1)σ_n ≡ \overline{n(n-1)}σ_{\text{tot}}. \quad (15.20)$$

Recall that the integral of a *single particle* inclusive cross section gives the average multiplicity,

$$\int dΓ(k_1) f(k_1) = \bar{n}σ_{\text{tot}}.$$

Constructing the *correlation function*

$$φ(k_1, k_2) ≡ \frac{f(k_1, k_2)}{σ_{\text{tot}}} - \frac{f(k_1)f(k_2)}{σ_{\text{tot}}^2}, \quad (15.21)$$

we have

$$\int dΓ_1\, dΓ_2\, φ(k_1, k_2) = \overline{n(n-1)} - \bar{n}^2 = \overline{n^2} - \bar{n}^2 - \bar{n}. \quad (15.22)$$

If particles are produced independently then, obviously,

$$φ(k_1, k_2) = 0 \quad \Longrightarrow \quad \overline{n^2} - \bar{n}^2 = \bar{n},$$

which is a property of the Poisson distribution. Hence, $φ \neq 0$ describes the deviation from the independent emission.

If rapidities $η_1$, $η_2$ are close, nothing definite can be said. If, however, the relative rapidity $η_1 - η_2$ is large, and both particles are far away from the fragmentation regions ($η_2 \gg 1$, $ξ - η_1 \gg 1$), in zeroth approximation we can draw a pole picture and see that $φ(η_1, η_2) = 0$. In an approximation without corrections, pomeron exchange gives rise to a homogeneous

Poisson distribution:

$$f(\eta_1, \mathbf{k}_1; \eta_2, \mathbf{k}_2) = \quad = g_1 \varphi(k_1^2) \varphi(k_2^2) g_2. \qquad (15.23)$$

We take now two-reggeon branching:

$$-8 \quad + 4 \left(\quad + \quad + \quad \right) \implies 4 \qquad (15.24)$$

What will we get from more complicated branchings? Again, we take an n-reggeon branching diagram, cut n_1 reggeons, and choose two particles from them:

$$2 \operatorname{Im} F_{n_1}^{(n)} = C_n^{n_1} \cdot (-1)^{n-n_1} (2 \operatorname{Im} f)^n \cdot \left[\frac{n_1(n_1 - 1)}{2} + n_1 \right], \qquad (15.25)$$

where $n_1(n_1 - 1)/2$ is the number of ways to select two particles from different reggeons, and n_1 from the same reggeon. We know, however, that $\sum_{n_1} n_1 \cdot 2 \operatorname{Im} F$ gives zero, cf. (15.24); hence,

$$n_1(n_1 - 1) \implies \left. \frac{\partial^2}{\partial x^2} (1 - x)^n \right|_{x=1},$$

and only the branching $n = 2$ contributes. A general statement can be verified: non-enhanced branchings up to the nth order contribute to the cross section of the production of n particles.

In order to generalize this result for enhanced branchings, one has to be somewhat careful. Indeed, for the standard cancellation to take place, we have to make sure that the reggeon interaction vertices do not depend on the way the diagrams are cut, e.g. the vertex λ in (15.26),

$$\lambda \quad \overset{?}{=} \quad \qquad (15.26)$$

As a result, a very important picture emerges: in the full set of reggeon diagrams for two-particle inclusive spectrum any interactions between the reggeons are possible except those in the interval between the rapidities η_2 and η_1 of the triggered particles.

In the weak coupling regime one can verify that the correction (15.24) *broadens* the multiplicity distribution of particles compared to the Poisson distribution.

To analyse the multiplicity fluctuation pattern in the case of the strong coupling is a much more difficult task. To see qualitatively, what is taking place let us consider the simplest diagram with the scaling Green functions, $G(\xi) \propto \xi^\mu$ substituted for the pomeron poles in (15.23). For the double inclusive distribution we then have

$$\phi(k_1, k_2) = \frac{G(\xi-\eta_1)G(\eta_1-\eta_2)G(\eta_2)}{G(\xi)} - \frac{G(\xi-\eta_1)G(\eta_1)}{G(\xi)} \cdot \frac{G(\xi-\eta_2)G(\eta_2)}{G(\xi)}.$$

Keeping $x_1 = \eta_1/\xi$ and $x_2 = \eta_2/\xi$ fixed, both terms behave as $\xi^{2\mu}$. Now we calculate the integral (15.22), extracting the overall ξ-dependence:

$$\xi^{-2\mu-2} \cdot \int d\Gamma_1 \, d\Gamma_2 \, \phi(k_1, k_2)$$

$$\propto \int dx_1 \, dx_2 (1-x_1)^\mu |x_1 - x_2|^\mu x_2^\mu - \left(\int dy (1-y)^\mu y^\mu \right)^2$$

$$= 2 \int_0^1 dx_1 (1-x_1)^\mu x_1^{2\mu+1} \int_0^1 dy (1-y)^\mu y^\mu - \left(\int_0^1 dy (1-y)^\mu y^\mu \right)^2$$

$$= \int_0^1 dy (1-y)^\mu y^\mu \cdot \left[2B(\mu+1, 2\mu+2) - B(\mu+1, \mu+1) \right] < 0, \quad (15.27)$$

where $B(a, b) = \Gamma(a)\Gamma(b)/\Gamma(a+b)$. The problem is that the combination of the B-functions in the square brackets is *negative* ($\mu > 0$). This means that the distribution became *narrower* than the poissonic one, which is rather strange. Moreover, for a sufficiently large ξ (15.27) would violate the mathematical fact that $\langle n^2 \rangle - \langle n \rangle^2 \geq 0$. Consequently, the branching corrections must be significant. In any case, there is a large positive correlation between particles in the strong coupling scenario.

15.4 How to tame fluctuations

We return to the discussion of fluctuations.

The simplest one is just the elastic scattering. There is an interesting class of fluctuations of a similar nature, which we considered in Lecture 10

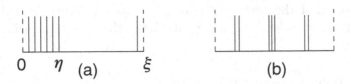

Fig. 15.4   Inelastic diffraction (a) and an event with large rapidity gaps (b).

when we discussed the inconsistency of the Regge-pole approximation. These fluctuations are characterized by large intervals in the rapidity distribution of produced hadrons, as shown in Fig. 15.4.

15.4.1 High-mass inelastic diffraction: triple-reggeon limit

The first process, Fig. 15.4(a), is simple to measure. Indeed, to make sure that there are no other fast particles, it is sufficient to observe a particle with an energy very close to the initial one. (There can be no uniform plateau when we register too energetic a particle in the final state.) This is a correlation forced by the energy conservation.

The invariant mass of particles produced on the side of the target is

$$M^2 = (k_1 + k_2 + \cdots + k_n)^2 \simeq m^2 \, e^{\eta}, \tag{15.28}$$

where η can be equated with the rapidity of the fastest particle k_1 in the bunch ($k_{10} \gg k_{20} \gg \cdots \gg k_{n0} \sim m$), cf. (10.32b). In the kinematics of Fig. 15.4(a), the mass is small compared to the total energy, $M^2 \ll s$. At the same time, it can be large in absolute terms, $M^2 \gg m^2$, so that the condition $\xi \gg \eta \gg 1$ can be satisfied. Then the cross section transforms into the three-reggeon picture as shown in Fig. 15.5. Hence, selecting a particle with a large momentum, I virtually measure the three-reggeon vertex directly. This is just the same vertex r that enters the reggeon diagrams, since, as we know, it does not change when the diagram is cut.

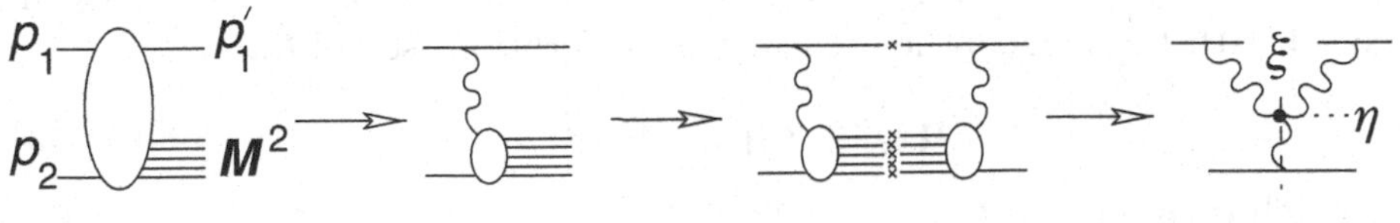

Fig. 15.5   Inclusive spectrum in the triple-reggeon limit.

Let us see that the vertex r, indeed, has to go to zero, as this follows from the reggeon field theory. We calculated this diagram in Lecture 10, see (10.35):

$$d\sigma_{3P} \propto \frac{1}{s} \frac{d^3 p'_1}{2E'_1} \cdot g_1^2(\mathbf{q}_\perp^2) r(\mathbf{q}_\perp^2) g_2(0)$$

$$\cdot \left| i \frac{s}{M^2} G(\xi - \eta, \mathbf{q}_\perp^2) \right|^2 \cdot \left[M^2 G(\eta, 0) \right] \tag{15.29}$$

$$\sim d^2 \mathbf{q}_\perp \, d\eta \, g_1^2 r g_2 \cdot e^{-2\alpha' \mathbf{q}_\perp^2 (\xi - \eta)}; \quad d\eta = \frac{dx}{1-x}.$$

Here x is the energy fraction carried by the fast registered particle which has to be chosen in the interval $m^2/s \ll 1 - x \ll 1$ in order to satisfy the conditions of applicability of the reggeon approximation, $\xi - \eta \gg 1$, $\eta \gg 1$ (see (10.32)).

Strictly speaking, the inclusive spectrum has the meaning of cross section *weighted* by particle multiplicity, $n \cdot \sigma$. However, when measuring $x > \frac{1}{2}$, I am sure there may be only one particle with such a large energy in the event. Consequently, by measuring the inclusive particle yield in the three-reggeon kinematics we measure not multiplicity, but directly the contribution of the total cross section.

How large is this contribution? We have to integrate (15.29) over a large rapidity interval, η up to ξ. If the cone did not shrink, $\alpha' = 0$, we would have $e^{-2\alpha' \mathbf{q}^2 (\xi - \eta)} = 1$, and

$$\int^\xi d\eta \int d^2 \mathbf{q}_\perp \frac{d\sigma_{3P}}{d\eta \, d^2 \mathbf{q}_\perp} \sim r \cdot \xi, \tag{15.30}$$

i.e. an infinitely increasing with energy partial contribution to the cross section! This is, essentially, the same contradiction as the one we faced when we discussed the black disc model, $A = s \cdot F(t)$, in Lecture 6. But even taking account of $\alpha' \neq 0$ the result is still unacceptable:

$$\sigma_{3P} = \int^\xi d\eta \int d^2 \mathbf{q}_\perp \frac{d\sigma_{3P}}{d\eta \, d^2 \mathbf{q}_\perp} \sim \int_0^{\xi - \text{const}} \frac{d\eta}{\alpha'(\xi - \eta)} \sim \ln \xi. \tag{15.31}$$

Although growing much slower with s that in (15.30), σ_{3P} eventually takes over $\sigma_{\text{tot}} = \text{const}$. To avoid the contradiction, we have to have, in accord with the reggeon field theory consideration, see (14.27),

$$r(\mathbf{q}_\perp) = c \mathbf{q}_\perp^2, \quad \text{for } \mathbf{q} \to 0, \tag{15.32}$$

in which case

$$\int d^2 \mathbf{q}_\perp \frac{d\sigma_{3P}}{d^2 \mathbf{q}_\perp} \sim \frac{d\eta}{|\xi - \eta|^2}, \quad \text{and} \quad \sigma_{3P} = \mathcal{O}(1).$$

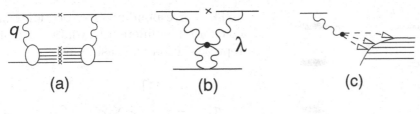

Fig. 15.6

In the *strong coupling* regime, the reggeon–reggeon interaction vertices also effectively vanish at small $\mathbf{q}_\perp$.

15.4.2 Vanishing reggeon–reggeon vertices

Let us turn to the discussion of an important question.

We see from (15.32) that the scattering cross section vanishes in the *forward direction*, $\mathbf{q}_\perp \to 0$. This is a strong conclusion which has to have serious consequences.

Indeed, is it not strange that the total cross section of the inelastic diffraction processes in Fig. 15.6(a) has to turn into zero for $\mathbf{q}_\perp = 0$? (Strictly speaking, there should be no *pomeron pole* in the bottom part of Fig. 15.6(a), while something like Fig. 15.6(b) could, in principle, be there.) If I draw a ladder, would not the expression be positively definite? One can imagine playing on the non-locality of the vertex in attempt to effectively *screen* it, by integrating over the place where the reggeon is attached, as shown in Fig. 15.6(c). As we will see shortly, this is not an easy thing to do.

Nevertheless, let us suppose that we managed somehow to force the three-pomeron vertex vanish for the forward scattering, $r(0) = 0$. Then a two-pomeron branching in Fig. 15.6(b) gives the leading contribution to the inclusive cross section which, obviously, has to be positive. But we know that the total imaginary part of the branching diagram is *negative*: $2 - 8 + 4 = -2$. To avoid the contradiction, the four-pomeron vertex λ also has to be put to zero at $\mathbf{q} = 0$.

15.4.3 Multi-particle production with large rapidity gaps

One arrives at the same conclusion, $\lambda \to 0$, by considering another curious fluctuation, when there is a large number of 'holes' in the distribution in rapidity, see Fig. 15.4(b) above. (Historically, this was the first example of a fluctuation that has led to the contradiction.)

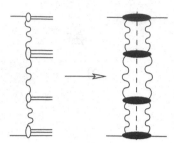

The corresponding cross section can be calculated in different ways. In the j-plane representation one easily gets

$$\sigma(\omega) = \left\{ \vphantom{\int} \right\} \propto \lambda^2 \ln \omega^{-1}. \qquad (15.33)$$

Summing up two-reggeon loops,

$$\sum_{n=1}^{\infty} \lambda^n \ln^n \omega^{-1} = \frac{\lambda \ln \omega^{-1}}{1 - \lambda \ln \omega^{-1}},$$

we get a pole *to the right* from unity. Although the loop diagram (15.33) itself amounts to a relatively small correction,

$$\sigma(\xi) \sim \int \frac{d\omega}{2\pi i} \ln \omega^{-1} e^{\omega \xi} \sim \xi^{-1};$$

its repetition in the t-channel gives rise to a very strong exponential enhancement, $\exp(\omega_0 \xi)$, $\omega_0 > 0$. This contradiction (the existence of a subprocess whose cross section grows faster than the total one) requires, once again, to have $\lambda(0) = 0$.

15.4.4 Fluctuations in the weak and strong coupling regimes

In the weak coupling case the s-channel unitarity imposes very serious restrictions on the theory. The conditions $r(0) = \lambda(0) = 0$ are needed to make sure that large fluctuations would not ruin the stationary uniform rapidity distribution that characterizes the pomeron. In the language of the reggeon field theory, these are actually the requirements that keep our system away from the critical point.

The case of strong coupling does not contain contradictions, either in the '3P limit', or for multiple large rapidity gaps.

Given the σ_{tot} increasing with energy, the large deviations from the scaling solution (14.22) turn out to be (relatively) suppressed. The screening of dangerous fluctuations is achieved here by summing diagrams with 'long' *reggeons* substituted for the dashed lines in Fig. 15.6(c). This tells us (as strange as it may sound) that it is *easier* to verify the strong-coupling scenario than the weak coupling one.

In the first case we have a reggeon field theory (RFT) at our disposal. We can plug in a bare pomeron pole and carry out the analysis* (although we do not know a priori whether there is a small parameter in the theory, we can at least try to calculate things within this hypothesis).

In the weak coupling, the vanishing of the vertex in Fig. 15.6(c) has to be demonstrated at the level of *particles* (hadrons) rather than reggeons. Therefore, in order to check the weak coupling regime, detailed knowledge of the true theory of hadrons is necessary. This makes a world of a difference: while the *reggeons* enable us to carry out calculations, at least in principle, *hadrons* do not: we do not have a theory of strongly interacting *hadrons*.

The weak coupling corresponds to a rather simple physical picture:

(1) total cross sections are asymptotically constant;

(2) there is a uniform rapidity distribution;

(3) the multiplicity grows logarithmically with energy;

(4) fluctuations in particle production are relatively rare; and

(5) branching corrections to the leading pole approximation are small.

In contrast with that, the strong coupling regime has a rather complicated structure. Here:

(1) the total cross sections increase, $\sigma_{\text{tot}}(s) \sim \ln^{\nu} s$ (with $0 < \nu \leq 2$);

(2) there is no asymptotic particle density;

(3) no factorization;

(4) not much is left of the initial pole; branchings are 100% important; and

(5) the leading approximation for the reggeon Green function G can be fixed only roughly, in a 'dimensional sense'.

These two versions are different also in the 'microscopic' language.

In the *weak coupling* there exists the asymptotic parton wave function of a fast hadron, which ensures *factorization* of high-energy interactions. This wave function is characterized by a finite rapidity density of particles and is boost invariant.

In the *strong coupling* case the systems never forgets about its initial energy. Long-range correlations are omnipresent, and the parton content

---

* Indeed, an example is known based on a definite relation between the bare vertex r and the pole trajectory $\alpha(t)$, in which the strong-coupling solution does emerge (ed.).

of colliding hadrons keeps growing with the energy of the interaction. One might say, there is a specific Lorentz frame characterizing each collision.

On the physics side, the first reggeon field theory scenario looks more elegant. However, from the point of view of a pure theory – summing up the pictures – the impression is just the opposite.

Indeed, the strong coupling can be obtained directly from the diagrams. Meanwhile, the weak coupling calls for additional requirements ($r(0) = \lambda(0) = 0$) which, in terms of diagrams, do not follow from anything: they are just necessary to impose in order to preserve the s-channel unitarity.

15.5 Weak coupling: vanishing pomeron–particle vertices

The study of the s-channel picture of particle production is very efficient in understanding the conditions for the self-consistency of the theory. We see that for the consistency of the weak coupling regime, rather strong additional conditions must be satisfied namely, that the reggeon interaction vertices r and λ vanish in the $\mathbf{q}_\perp \to 0$ limit. But if they do, what will remain? Generally speaking, it is not so easy to set something to zero by hand in a field theory; will this not result, e.g. in all cross sections becoming zero too?

The hypothesis $\sigma_{\text{tot}} \to$ const looks rather attractive. So, we will recall the main steps of the analysis of fluctuations we have carried above, to see what are the consequences of the weak coupling conditions $r(0) = \lambda(0) = 0$.

15.5.1 π-meson production vertex

Let us fix a large η value and keep increasing $\xi - \eta$. Taking into account all corrections to the pole on the bottom part of the three-reggeon graph,

$$\tag{15.34}$$

we arrive at the so-called reggeon–particle scattering amplitude:

$$\tag{15.35}$$

We are interested in the forward scattering ($t = 0$). This amplitude can be studied in terms of the partial wave, $f(\omega, \mathbf{k})$. The limit $\eta \to \infty$ corresponds to $\omega \to 0$, in which limit the vertex r goes to zero. The pomeron contribution to the scattering block disappears, and there remain only non-pole contributions (branching corrections). This means that the reggeon–particle amplitude (15.35) *decreases* with the growth of energy, η. Let us recall, however, the *s*-channel meaning of the branchings. The total cross section for the scattering of the pomeron **P** on a particle, $\sigma_{\text{tot}}^{(\mathbf{P})}$,

$$\sigma_{\text{tot}}^{(\mathbf{P})} = \sum_n \sigma_n,$$

is a sum of topological cross sections, σ_n, with the production of a given number of hadrons. We have discussed that the branchings lead to the appearance of processes with particle multiplicity smaller ($n < \bar{n}(\eta)$) and larger than the average ($n > \bar{n}(\eta)$), as well as to negative corrections to the pole term. Now that the leading term has disappeared, these corrections dominate the answer and would result in the *negative cross section* for the processes with the average multiplicity, $\sigma_{n \sim \bar{n}} < 0$.

We conclude that λ in (15.35) must also go to zero. The necessity of the vanishing of the four-reggeon vertex followed also from the consideration of multiplicity fluctuations with multiple large rapidity gaps. From that consideration it was not clear, however, under what specific conditions λ has to be zero.

The three-reggeon vertex r on the r.h.s. of (15.35) is zero when the momentum **k**, transferred along the pomeron, vanishes (and $\omega = 0$, corresponding to the $\eta \to \infty$ limit). This makes it evident that we have to have $\lambda = 0$ as soon as $\mathbf{k} = 0$, independently of the value of the momentum $\mathbf{k}'$ in the lower part of the branching graph.

Let us construct the four-reggeon vertex λ in the ω-representation:

$$\lambda(\omega; \mathbf{k}, \mathbf{k}') = \quad \text{} \quad = \int d\eta \, e^{-\omega \eta} \, \text{Im} \, A_{\mathbf{PP}}(\eta; \mathbf{k}, \mathbf{k}'), \qquad (15.36)$$

where $A_{\mathbf{PP}}$ is the **PP** scattering amplitude. The integral on the r.h.s. of (15.36) must vanish at $\omega = 0$, $\mathbf{k} = 0$. However, by virtue of the *s*-channel unitarity condition (optical theorem for **PP** scattering), the imaginary part of the forward amplitude can be represented as a sum of cross sections,

$$0 = \lambda(0; \mathbf{0}, \mathbf{k}') = \int d\eta \, \text{Im} \, A_{\mathbf{PP}}(\eta; \mathbf{0}, \mathbf{k}') = \int d\eta \sum_n \overbrace{}^{n}. \qquad (15.37a)$$

We integrate a sum of the amplitudes squared. Hence, we come to the conclusion that *each* of these amplitudes has to be zero at $\mathbf{k} = 0$:

$$\mathbf{k} \Big\} \mathsf{n} = 0 \quad \text{when } \mathbf{k}{=}0, \text{ for any } \eta \text{ and } \mathbf{k}', \qquad (15.37\text{b})$$

identically for all energies. This is actually what we saw in the case of the multi-gap events, with only one addition: to have $\lambda = 0$ it is in fact sufficient to put to zero the transverse momentum of *one* of the reggeons.

Take a particular case $n = 2$, and consider the production of two pions:

$$\mathbf{k}{=}0 \Big\} \equiv 0 = \sum_{\mathbf{R}} \mathbf{R} \Big\{ \begin{smallmatrix} \pi \\ \pi \end{smallmatrix}, \qquad (15.37\text{c})$$

where we have expanded the amplitude over Regge poles $\mathbf{R}$ (by choosing π we exclude from the sum the vacuum pole $\mathbf{P}$). But different reggeons have different energy behaviour, hence *each term* of the sum is zero.

Next, the Regge-pole amplitudes factorize; therefore from (15.37c) we conclude that *each* $\mathbf{PR}\pi$ *vertex* vanishes in this point,

$$\begin{matrix} \mathbf{P} \Big\langle \mathbf{k}{=}0 \\ \mathbf{R} \end{matrix} \pi = 0. \qquad (15.37\text{d})$$

Now, let us continue this 'zero' in the virtuality $t' = -\mathbf{k}'^2$ of the reggeon $\mathbf{R}$ to the point $t' = m_b^2$, to obtain a particle–particle–pomeron amplitude,

$$a \underline{\quad\quad} b \Big\} = 0. \qquad (15.37\text{e})$$
$$\mathbf{P} \Big\{ \mathbf{k}{=}0$$

15.5.2 Goodbye pomeron?

We arrived at a fantastic result: the transition amplitude $b \to a$ must vanish with the transferred transverse momentum, $\mathbf{k}_\perp \to 0$, independently of the type of the target! Moreover, if we take $a = b$, (15.37e) would mean that the forward elastic amplitude is zero, and so is its imaginary part, and the total cross section. The pomeron does not hook onto any hadron, and therefore does not exist!

We learned that the high mass inelastic diffraction of the target hadron does not contain the pomeron pole when the projectile particle scatters *forward*, $r(\mathbf{k}_\perp = 0) = 0$, and were hoping that as a result the amplitude (15.34) would slowly decrease with the increase of $\eta \simeq \ln M^2$. However, having carried the reggeon logic through the consecutive steps (15.37), we seem to have arrived at a totally bizarre conclusion.

15.6 How to rescue a pomeron

We must address the two cases contained in (15.37e) separately:

- $b \neq a$ – a surprising prediction;

- $b = a$ – a catastrophe; the end of the pomeron hypothesis.

In the first case we deal with an inelastic process, and no formal objection can be raised against vanishing of inelastic amplitudes on a pomeron.

In order to rescue the *elastic* scattering, $b = a$, we could have the following situation: the genuine vertex, strictly speaking, could be *singular* at $m_a = m_b$. The singularity which prevents the elastic vertex from becoming zero could be of the form

$$\gamma(\mathbf{k}, \mathbf{q}) \propto \frac{2(\mathbf{k}_\perp \mathbf{q}_\perp)}{m_a^2 + \mathbf{k}_{1\perp}^2}. \qquad (15.38)$$

(Such an object – the *transverse mass* of the produced particle, $m_a^2 + \mathbf{k}_{1\perp}^2$, – always enters the reggeon cross sections.) Let us introduce $t = -\mathbf{q}^2$,

$$\frac{2(\mathbf{k}_\perp \mathbf{q}_\perp)}{m_a^2 + (\mathbf{q} + \mathbf{k})_\perp^2} = \frac{2(\mathbf{k}_\perp \mathbf{q}_\perp)}{m_a^2 - t_2 + 2(\mathbf{q}_\perp \mathbf{k}_\perp) + \mathbf{k}_\perp^2},$$

and analytically continue the amplitude into the point $t = m_b^2 > 0$, where the dashed reggeon line in (15.38) represents a particle b:

$$\frac{2(\mathbf{k}_\perp \mathbf{q}_\perp)}{m_a^2 - m_b^2 + 2(\mathbf{k}_\perp \mathbf{q}_\perp) + \mathbf{k}_\perp^2} \overset{\mathbf{k}_\perp \to 0}{\longrightarrow} \left\{ \begin{array}{cc} 1, & a = b \\ 0, & a \neq b \end{array} \right\}. \qquad (15.39)$$

Hence, formally, our aim can be achieved. This is, however, not a real proof, although this may be more than just a mathematical trick. In fact, it is exactly what happens in quantum electrodynamics.

15.6.1 Vanishing of the vertex in quantum electrodynamics

Consider some inelastic QED process, for example photon conversion into an e^+e^- pair in the external electromagnetic field:

$$= (+e) + (-e) = 0.$$

The incoming particle creates a current which is scattered on the field. If $\mathbf{k}_\perp$ is small, this field is almost homogeneous and therefore interacts only with the *integral* of the current, i.e. with the total electric charge. But the charge is conserved; it does not change when the system fluctuates (e.g. $\gamma \to e^+ e^- \to \gamma$). As a result, the scattering on a *homogeneous* electromagnetic field ($\mathbf{k}_\perp \to 0$) does not alter the internal state of the projectile.

How can this be seen mathematically? Writing the Weizsäcker–Williams formula, we get just the singularity that we invented in (15.39).

Take an arbitrary process in which an incident particle a transforms into an object (a system of particles) with the total momentum p_b and invariant mass $p_b^2 = m_b^2 \geq m_a^2$. A stationary field does not transfer energy, $A_\mu(k) \propto \delta_{\mu,0} \delta(k_0)$, and we have

$$p_a \longrightarrow\!\!\bigcirc\!\!\Longrightarrow p_b$$
$$k \to 0$$
$$k_0 = 0;$$
$$p_b^2 = (p_a + k)^2 = m_a^2 - 2p_z k_z - \mathbf{k}_\perp^2 - k_z^2. \tag{15.40}$$

When the energy is large, $p_{a0} \simeq p_{az} \equiv p$, the longitudinal momentum absorbed by the field becomes small,

$$k_z = \sqrt{p^2 - (m_b^2 - m_a^2 + \mathbf{k}_\perp^2)} - p \simeq -\frac{m_b^2 - m_a^2 + \mathbf{k}_\perp^2}{2p}, \tag{15.41}$$

and the transferred momentum is transversal, $k^2 \simeq -\mathbf{k}_\perp^2$.

M_μ is the matrix element of the sub-process $a + \gamma(k) \to b$ shown by the grey blob in (15.40). It must satisfy the current conservation condition,

$$k^\mu M_\mu = k_0 M_0 - k_z M_z - (\mathbf{k}_\perp \mathbf{M}_\perp) = 0, \tag{15.42}$$

from which we derive

$$M_z = -\frac{(\mathbf{k}_\perp \mathbf{M}_\perp)}{k_z} \simeq 2p \cdot \frac{(\mathbf{k}_\perp \mathbf{M}_\perp)}{m_b^2 - m_a^2 + \mathbf{k}_\perp^2} \simeq M_0,$$

where we used (15.41). Finally, for the amplitude of the process we obtain

$$A^\mu M_\mu \propto M_0 \propto \frac{(\mathbf{k}_\perp \mathbf{M}_\perp)}{m_b^2 - m_a^2 + \mathbf{k}_\perp^2}, \tag{15.43}$$

were we have exploited the fact that in a relativistic situation, $M_0 \simeq M_z$ ($M_0 - M_z = \mathcal{O}(1/p) \to 0$).

This is precisely the structure of the vertex we are seeking to rescue the pomeron. With $\mathbf{k}_\perp \to 0$, it goes to zero for $m_a \neq m_b$ but stays finite for the elastic scattering, $a = b$ (in which case $\mathbf{M}_\perp \propto \mathbf{k}_\perp$, for lack of any other transverse vector in the problem).

We could discuss this phenomenon in the language of Lorentz-invariant form factors. If the object b has no internal structure (is characterized only by its total four-momentum p_b), we may write

$$M_\mu(p_a, p_b) = \Gamma_1(k^2)(p_a + p_b)_\mu + \Gamma_2(k^2)(p_a - p_b)_\mu.$$

The condition (15.42) gives $\Gamma_1(m_a^2 - m_b^2) + \Gamma_2 k^2 = 0$, and we obtain

$$a = b: \quad M_\mu = \Gamma_1(k^2)(p_1 + p_2)_\mu,$$

$$a \neq b: \quad M_\mu = \Gamma_2(k^2)\left[\frac{k^2}{m_b^2 - m_a^2}(p_1 + p_2)_\mu + k_\mu\right]; \quad \Gamma_i(k^2) \xrightarrow{k^2 \to 0} \text{const.}$$

In the latter case the high-energy inelastic transition vanishes in the forward direction, $M_0(a \to b) \propto \mathbf{k}_\perp^2 \cdot p + \mathcal{O}(1/p)$, while the elastic amplitude is finite, $M_0(a \to a) \simeq e \cdot 2p$.

Hence, although this example looks somewhat artificial, it demonstrates that the solution we are looking for may exist. Actually, this is not an argument in favour of the pomeron, since in quantum electrodynamics we have a conservation law in the game (that of electric charge).

Nevertheless, the elastic scattering in the reggeon problem is not zero either, and there is a reason for that.

15.6.2 Diffraction on a broad potential (deuteron)

Let us study another example.

In Lecture 12 we discussed the diffraction on a large size target. We considered the scattering of a deuteron off the external field of size R (Fig. 15.7) and wrote the transition amplitude from the initial state i to the final state k:

$$f_{ik} \sim \int d^3 r_{12}\, d^2\rho_c\, \psi_k^*(r_{12}) \left[e^{2i\delta(\boldsymbol{\rho}_1) + 2i\delta(\boldsymbol{\rho}_2)} - 1\right] \psi_i(r_{12}). \qquad (15.44a)$$

When the spread of the potential is much larger than the internal size of the projectile, $R \gg r_D$, in the scattering phases $\delta(\boldsymbol{\rho}_1)$, $\delta(\boldsymbol{\rho}_2)$ we can neglect the separation $\boldsymbol{\rho}_{12}$ between the proton and neutron inside the

Fig. 15.7

deuteron, $\rho_1 = \rho_c + \frac{1}{2}\rho_{12} \simeq \rho_2 = \rho_c - \frac{1}{2}\rho_{12} \simeq \rho_c$, to obtain

$$f_{ik} \sim \underbrace{\int d^3 r_{12}\, \psi_k^*(r_{12})\psi_i(r_{12})}_{\delta_{ik}} \cdot \int d^2\rho_c \left[e^{4i\delta(\rho_c)} - 1 \right]. \qquad (15.44\text{b})$$

Thus, while the *elastic* process $(i = k)$ goes, the amplitude becomes zero when $i \neq k$ that is for scattering with excitation – inelastic channel, Fig. 15.7(b).

We have established a correspondence of these formulae with 'pictures'. Recall the representation for the scattering matrix element that we have used in order to translate (15.44) in the language of the diagrams:

$$S_1 S_2 - 1 = (S_1 - 1) + (S_2 - 1) + (S_1 - 1)(S_2 - 1). \qquad (15.45)$$

Consider two concrete cases.

First we take $i = \psi_D$, $k = \psi_D$, corresponding to the elastic scattering:

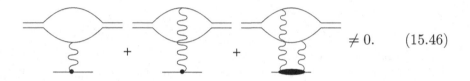

$$\neq 0. \qquad (15.46)$$

Now let us draw the *decay*, when ψ_k is the wave function of the continuous spectrum, $k = (n, p)$. What is the corresponding graphical representation? The wave function of the produced n, p can be written as

$$\psi_k = e^{ik \cdot r_{12}} + \psi_{n,p}', \qquad (15.47)$$

where $\psi_{n,p}'$ is an addition due to the interaction between n and p in the final state. Let us insert it in the integral (15.44b) for the transition amplitude f_{ik} represented in the form of (15.45).

The free propagation gives

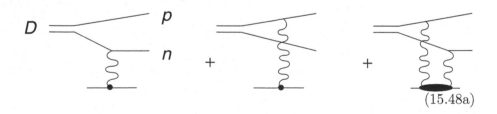

$$(15.48\text{a})$$

Owing to the additional term ψ'_{np} in (15.47), after the interaction with the target there is an interaction between the two outgoing nucleons:

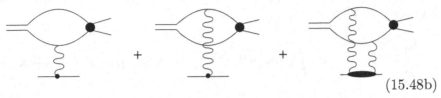

$$(15.48b)$$

Having drawn the diagrams, we may hope that this picture can be applied in a more general context (not only to a deuteron).

From the properties of the wave function, (15.44b), we know that the dissociation amplitude is zero at $\mathbf{k}_\perp = 0$. How does this happen in terms of diagrams? The mechanism of the cancellation between the graphs (15.48a) and (15.48b) turns out to be very similar to the case of a conserved current.

The final state interaction amplitude shown by black circles in (15.48b) may be rather complicated. However, I can expand it over intermediate states, and I know that at $\mathbf{k} = 0$ only one of them would survive namely, the *pole term* corresponding to D. Then the sum of the diagrams (15.48) reduces to two graphs displayed in Fig. 15.8, where the dashed block symbolizes the sum of the amplitudes for scattering of the two nucleons off the target.

The two graphs of Fig. 15.8 must cancel one another in the $\mathbf{k} \to 0$ limit. How is this possible?

The second graph (Fig. 15.8(b)) describes forward scattering of the deuteron (followed by its dissociation); therefore its amplitude is proportional to $\sigma_{\text{tot}}(D)$ – the cross section of deuteron interaction with the target.

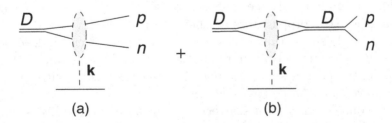

Fig. 15.8　(a) $D \to pn$ dissociation followed by interaction of nucleons with the target, (15.48a); (b) D pole in the final state pn interaction amplitude, (15.48b).

The first graph (Fig. 15.8a) can be also expressed via the total cross section, this time, of the $p + n$ pair. Indeed, if the potential is broad, we may neglect, once again, its dependence on ρ_{12} and approximate $V(\rho_1, \rho_2) \simeq V(\rho_c)$:

$$\text{(diagram)} = \int d^2\rho_{12}\, e^{i(\mathbf{k}_{12} - \mathbf{k}'_{12}) \cdot \boldsymbol{\rho}_{12}} \cdot V(\rho_1, \rho_2) \propto \delta(\mathbf{k}_{12} - \mathbf{k}'_{12}).$$

$$(15.49)$$

The internal state momenta $\mathbf{k}'_i$ coincide with the final momenta $\mathbf{k}_i$; the proton and neutron scatter *forward*, and the amplitude yields $\sigma_{\text{tot}}(np)$.

So it becomes clear that for the cancellation to take place, the equality of the cross sections is necessary,

$$\sigma_{\text{tot}}(D) = \sigma_{\text{tot}}(np). \qquad (15.50)$$

This relation obviously holds in our model: both D and pn interaction amplitudes are determined by the standard formula (15.44a), and in case of scattering on a large target it is irrelevant, whether p and n are flying separately, or together, inside D. If we were to carry out the real calculation, we would see that the normalization factors conspire in such a way that (15.50) turns out to be not only *necessary* but also *sufficient* for the cancellation of graphs (Fig. 15.8), which describe pn and D interacting with the target, correspondingly.

Let us demonstrate how it happens, by applying the Feynman rules to the forward $a \to b$ amplitude in the external field:

$$\text{(diagrams)} = \gamma \frac{1}{m_b^2 - p^2}\sigma_{\text{tot}}^{(b)} + \sigma_{\text{tot}}^{(a)} \frac{1}{m_a^2 - p^2}\gamma. \quad (15.51)$$

γ is the transition amplitude between the states ($a = D$, $b = p + n$). The virtuality p^2 of the intermediate state (shown by a thick line) is defined differently in the two diagrams (15.51). Namely, $p^2 = m_a^2$ in the first term (state a is on the mass shell), and $p^2 = m_b^2$ in the second (on-mass-shell state b). We immediately see that as soon as $\sigma_{\text{tot}}(a) = \sigma_{\text{tot}}(b)$, the amplitude vanishes (the propagators do their job properly).

This example shows that inelastic transitions $a \to b$ can indeed be zero at $\mathbf{k}_\perp \to 0$ if the interactions in the initial and final states cancel. To make it possible, the particles a and b must *interact identically* with the target, have to have *equal total cross sections*.

In the considered case of scattering on a large target this is true not only for the total cross sections, $\sigma_{\text{tot}}(a) = \sigma_{\text{tot}}(b)$, but even in a more subtle sense: scattering does not lead to a redistribution of momenta in the continuous spectrum, see (15.49).

15.7 Vanishing of forward inelastic diffraction in RFT

How could one discuss this situation without appealing to the quantum-mechanical deuteron analogy?

Let us take our hypothesis that the amplitudes of inelastic processes vanish when the transverse momentum **k** transferred along the pomeron goes to zero,

$$= 0, \qquad (15.52)$$

and ask ourselves how to derive consequences of this condition, not knowing the theory of hadron interactions?

15.7.1 'Sharp screening' in RFT

I put to zero not a constant but a function of the pair energy (invariant mass) s_{bc}. As any other amplitude, (15.52) has singularities. In particular, it may have a pole at $s_{bc} = m_a^2$. It is not difficult to write the pole term explicitly:

$$\propto \sigma_{\text{tot}}(a) \neq 0. \qquad (15.53)$$

This contribution to the amplitude is non-zero, since it is proportional to $\sigma_{\text{tot}}(a)$. How could a function having a pole be zero? There may be other intermediate states too,

$$\propto \sigma(a \to d) = 0. \qquad (15.54)$$

However, by our hypothesis, the $a \to d$ transition on the pomeron is forbidden, and therefore no other states contribute to the imaginary part (the discontinuity over s_{bc}). Have we not 'killed' the hypothesis (15.52)? The answer is 'no', and for a subtle reason.

The amplitude (15.52) is a four-point function, having many variables. Usually, after making use of the Lorentz invariance, we talk about two independent variables and, correspondingly, two independent imaginary parts (discontinuities). But a Regge amplitude is *not* Lorentz invariant; it depends on the direction of the collision. We are concerned with $\mathbf{k}_\perp = 0$.

If I set to zero the *total* four-vector $k_\mu = 0$, the amplitude (15.52) would turn into a three-point function. The latter is fully determined by the particle masses, $A(m_a^2, m_b^2, m_c^2)$, and has no free variables left.

If I take $(k_\mu)^2 = 0$, this is one condition, equivalent to fixing the mass of one external line of the four-point amplitude; two independent variables are at our disposal.

Setting $\mathbf{k}_\perp = 0$ means, however, *two* conditions, not one. Therefore, the amplitude (15.52) at this point possesses only *one* independent variable. As a result, when studying the s_{bc} imaginary part, we have to take into consideration simultaneously discontinuities over s_{ab} and s_{ac}! We can draw two additional simple graphs straight away:

$$
\begin{array}{c}
a \xrightarrow{\hspace{1cm}} b \\
 \underset{s_{bc}}{\diagup} c
\end{array}
\; - \;
\begin{array}{c}
a \xrightarrow{\hspace{1cm}} b \\
 s_{ab} \diagdown c
\end{array}
\; - \;
\begin{array}{c}
s_{ac} \xrightarrow{\hspace{1cm}} b \\
a \diagup c
\end{array}
\tag{15.55}
$$

All three amplitudes have poles at the same point. Suppose, this were the whole answer. Then the condition (15.52) would read

$$
\sigma_{\text{tot}}(a) \; = \; \sigma_{\text{tot}}(b) + \sigma_{\text{tot}}(c).
$$

But this is nonsense.

In the deuteron story it was not like that. For the system of two nucleons we had $\sigma(pn) = 2\pi R^2$, and not $\sigma(pn) = 2 \times 2\pi R^2$.

What is missing in (15.55)? The screening correction. It is this additional term that saved the day in the deuteron case:

$$
\begin{array}{c}
a \\ \\ b \\ c
\end{array}
\quad \Longrightarrow \quad \sigma(pn) = 4\pi R^2 - 2\pi R^2 = 2\pi R^2. \tag{15.56}
$$

The battle is not won yet. Indeed, each of three diagrams in (15.55) has a *pole*, while the screening graph (15.56) does not seem to have one. But it has to have a pole in order to participate in the cancellation we are looking for.

Actually, there *is* a pole which emerges in a very peculiar way. On a broad potential, the double interaction is 'sharp'. As I have stressed above, there emerges $\delta(\mathbf{k}' - \mathbf{k})$: momenta of particles in the intermediate state coincide with the final state ones, $\mathbf{k}'_i = \mathbf{k}_i$ ($i = 1, 2$). The energy denominator of the intermediate state peaks, producing the pole.

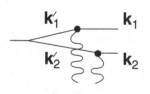

Before we turn to the question whether such a phenomenon appears in the reggeon problem, let me make a remark. As in the deuteron case, see

(15.50), the self-consistency condition (15.37e) will imply

$$
a \underbrace{}_{\mathbf{P} \; k=0} b = 0 \quad \Longrightarrow \quad \sigma_{\text{tot}}(a) = \sigma_{\text{tot}}(b). \tag{15.57}
$$

Cross sections of different objects (hadrons) that can turn into one another on the pomeron (having the same quantum numbers) must be asymptotically equal.

15.7.2 The essence of reggeon screening

We have to discuss whether we can obtain a necessary screening contribution in the reggeon framework. Let us draw the corresponding reggeon diagram,

$$
f(\boldsymbol{\rho}_1, \boldsymbol{\rho}_2) \;=\;
$$

and calculate it presuming $r(0) \neq 0$, for a start. Formally speaking, we would not expect a branching diagram to have a pole in ω; in other words, we would expect it to be small in the high-energy limit, $\xi \to \infty$.

In the impact parameter language,

$$
f(\boldsymbol{\rho}_1, \boldsymbol{\rho}_2) = r \int d^2\rho \, d\eta \, G(\boldsymbol{\rho}_1 - \boldsymbol{\rho}, \xi_1 - \eta) G(\boldsymbol{\rho}_2 - \boldsymbol{\rho}, \xi_2 - \eta) G(\boldsymbol{\rho}, \eta). \tag{15.58}
$$

This is the contribution to the amplitude from the three-pomeron diagram. Since the Green function G is Gaussian in ρ, see (12.19), the integral can be easily calculated. But even without calculation we can answer the question whether there will remain the dependence on the relative distance ρ_{12}.

What sort of a problem is this? The rapidities $\xi - \eta$ and η play the rôle of time in the diffusion process. We deal with a probability for two objects, emerging from points $\boldsymbol{\rho}_1$ and $\boldsymbol{\rho}_2$, to meet after time $\xi - \eta$ in the point $\boldsymbol{\rho}$, and keep propagating, stuck together. The integral over η and ρ means that I am looking for the total probability of such a meeting anywhere in the 'space–time'. The point is, in the two-dimensional space this *always happens*.

The brownian motion formula, $\langle \rho^2 \rangle \sim T$, means that the size of the region where the diffusing object can be found at time T, increases linearly with T. On the other hand, the path length is also proportional to T.

This tells us that our object, typically, visits *each point* inside the disc of increasing radius. Indeed, the probability to be found in a given point ρ any time *before T*,

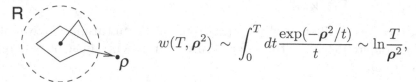

$$w(T, \rho^2) \sim \int_0^T dt \frac{\exp(-\rho^2/t)}{t} \sim \ln\frac{T}{\rho^2},$$

grows with the time elapsed. It makes it look plausible that the probability for the two objects to meet (anywhere, some time) will grow too: for sufficiently large $T \gg \rho_{12}^2$, the collision occurs inevitably, irrespectively of the size of the initial separation. Let us look at the formula to see if our expectation materializes.

Substituting in (15.58) $\boldsymbol{\rho}_{1(2)} = \boldsymbol{\rho}_c \pm \frac{1}{2}\boldsymbol{\rho}_{12}$,

$$f \sim \int \frac{d^2\boldsymbol{\rho}\, d\eta}{(\xi - \eta)^2} \exp\left(-\frac{(\boldsymbol{\rho}_c - \boldsymbol{\rho})^2}{2\alpha'(\xi - \eta)} - \frac{\rho_{12}^2}{8\alpha'(\xi - \eta)}\right) \times \frac{1}{\eta} \exp\left(-\frac{\rho^2}{4\alpha'\eta}\right).$$

We are interested in the forward amplitude, $\int d^2\boldsymbol{\rho}_c \exp(i\mathbf{k} \cdot \boldsymbol{\rho}_c)$ at $\mathbf{k} = 0$. Integrals over $\boldsymbol{\rho}_c$ and $\boldsymbol{\rho}$ compensate two nomalization factors, and we are left with

$$f(\rho_{12}^2, \mathbf{k} = 0) \sim \int_0^\xi \frac{d\eta}{\xi - \eta} \exp\left(-\frac{\rho_{12}^2}{8\alpha'(\xi - \eta)}\right) \simeq \ln\frac{8\alpha'\xi}{\rho_{12}^2}.$$

The dependence on $\boldsymbol{\rho}_{12}$ drops out in the dominant piece of the amplitude, $f \sim \ln \xi$, and therefore the integration over $\boldsymbol{\rho}_{12}$ produces the delta-function contribution $\delta(\mathbf{k} - \mathbf{k}')$ that we are looking for.

Well, I deceived you, of course. If I substitute in (15.58) the vanishing **PPP** vertex, $r(0) = 0$ (as I should have done) this graph is simply small.

15.7.3 Bare and renormalized three-pomeron vertices

We are discussing whether particles screen each other when they interact via pomeron exchange.

What is the difference between **P** and a simply large target? On one hand, the pomeron resembles a large target because its radius grows with energy as $\ln s$. On the other hand, the cross section on a large target is also very large $\sigma \sim R^2$, while in the pomeron case the disc becomes more and more transparent with the increase of energy.

The fact that screening corrections are large on a large target is essentially trivial; as soon as one particle hits the target, so does the second one inside the projectile. If I had $r \neq 0$, this situation could reproduce itself in the pomeron problem, owing to the blackening of the disc that

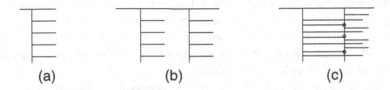

Fig. 15.9 (a) Image of a bare pole; (b) branching; (c) interacting ladders.

corresponds to the 'true pole'. However, if $r(0) = 0$, then the pomeron disc remains transparent and does not provide the necessary sharp screening.

But what is a 'true pole'?

We need to reflect upon the meaning of a 'pole' and 'branching'. If there is an interaction between 'ladders', a new mixed coherent state emerges, representing the 'true pomeron' (Fig. 15.9(c)). Cutting through such an object, I would not necessarily find large fluctuations.

And what would be a branching then?

It describes a situation when two such objects (the true poles) do not interact during a long period. The weak coupling condition, $r \to 0$, means that such long-living fluctuations are not frequent, i.e. the uniform rapidity distribution is quasi-stable – it does not fluctuate much. From this perspective, the vertex r is responsible for *deviations* from the stable asymptotic distribution. This vertex function r has to go to zero.

When discussing various reggeon branching issues, we drew diagrams and used to write the same letter r for the three-pomeron vertex. However, there were essentially different quantities!

In particular, in the case of a deuteron, the ladders surely do not overlap for a long time since p and n inside the deuteron are far apart. Many diffusion step are necessary in order for the ladders to meet. On average, $\alpha'(\xi - \eta) \simeq \langle \rho_{12}^2 \rangle \sim r_D^2$. Thus, during a *very long* 'time' $r_D^2/\alpha' \gg 1$ they typically do not mix together and develop independently.[†]

The proper size of the deuteron is an external factor. Thus, the properties of the deuteron wave function impose themselves on the reggeon theory.

The vertex r that governs the magnitude of fluctuations around the stationary distribution I would call *renormalized*, r_{ren}. It is this vertex that has to vanish in the forward kinematics, for the sake of self-consistency of the weak-coupling pomeron theory.

[†] There is a possibility for p, n to be initially close, but it has a small probability (ed.).

At the same time, the quantity r that enters the deuteron scattering problem I would treat as a *bare* reggeon interaction vertex, $r_{\text{bare}} \to \text{const}$, because in this concrete case a (seldom fluctuating) asymptotic coherent parton distribution does not have enough time to be formed (and hardly ever will, at any energies imaginable).

Constructing the theory of interacting reggeons, I would have to start from a bare vertex, $r_{\text{bare}} = \text{const}$, and then (provided I really knew how to calculate things) to *demonstrate* that the renormalized one indeed vanishes, $r_{\text{ren}}(\mathbf{k} \to 0) \to 0$. The analysis of the triple-reggeon limit in (15.58) was actually a step in this direction; by calculating such three-reggeon graph we were calculating in fact the *true* (renormalized) deuteron–pomeron vertex via the bare **PPP** constant r_{bare}.

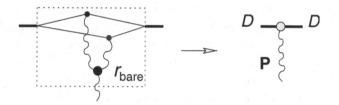

Whether this program can be realized in reality remains unclear; the full understanding is lacking of how, essentially, the renormalization of the pomeron constants takes place.

With somewhat less certainty, we can make the statement (15.57) even stronger; total interaction cross sections of all particles (and not only those having identical quantum numbers) must tend to one and the same constant in the limit of asymptotically high energies.

15.8 All σ_{tot} are asymptotically equal?

In order for the screening phenomenon to be independent of the specificity of the state, it is necessary to be able to calculate the screening correction graph (see, e.g. (15.56)) in a universal manner via reggeons (and not particles!).

Suppose that in the consistency condition (15.57),

$$a \underline{\qquad} b \quad \mathbf{P} \underset{k=0}{} \; = 0 \implies a\underline{\;\;}a \;\; b\underline{\;\;}b \;\; \mathbf{P} = \mathbf{P} \;,$$

we take the particle b to be a composite object, $b = c + d$. Then the interaction of this object with the target I will represent as a sum of three contributions:

$$\sigma_b = \sigma_{(cd)} =$$

where the interaction block V is universal, not depending on the types of the participating particles. In terms of pomeron–particle interaction constants, this is a non-linear relation,

$$g_b = g_c + g_d + g_c g_d \cdot V. \tag{15.59}$$

But c and d can be chosen differently; a hadron can be 'composed' in multiple different ways. A proton, for example: $p = n + \pi^+ = \Delta^{++} + \pi^-$, etc. Therefore, the only solution (15.59) admits is $g_c = g_d = g_b$.

Thus, having supposed that the high-energy screening is determined by the pure reggeon physics, we come to the conclusion

$$\sigma_{\text{tot}}^{(p)}(s \to \infty) = \sigma_{\text{tot}}^{(\pi)}(\infty) = \sigma_{\text{tot}}^{(D)}(\infty) = \sigma_{\text{tot}}^{(He^4)}(\infty) = \cdots$$

Is it really so weird to have the interaction cross sections of all hadrons with a given target to be asymptotically equal?

In order for this to happen in reality, one would need, of course, grandiose colliding energies, such that the reggeon interaction processes are fully developed: $\alpha \ln s \gg R_h^2$, the Regge radius of the pomeron *much larger* than the proper radii of the projectile hadrons.

Parton clouds belonging to the constituents inside the hadron must all mix together, resulting in the coherent universal parton distribution whose density no longer depends on the type of the incident hadron. When this limit is reached, the interaction cross section in the centre-of-mass frame of the initial particles A and B is expressed in a universal way, via the interaction cross section of slow ('wee') partons.

In the framework of the parton picture, the hypotheses which guarantees an asymptotic equality of *all* cross sections looks rather natural. It presumes that, due to the fact that the hadron interaction is *strong*, such a universal distribution is reached, eventually, with the unit probability, and does not depend on the nature of colliding objects.

16

Strong interactions and field theory

16.1 Overview

16.1.1 Phenomenological approach to hadron scattering

We have considered the phenomenological theory of strong interactions. In the last 20 years the theory was developing in two parallel directions. The questions were:

- how the interaction takes place in a relativistic situation, irrespectively to *who* is interacting;

- how the mass spectrum of hadrons is built up; what type of laws can be extracted not knowing the microscopic dynamics?

Qualitatively, high-energy processes are understood. Such general consequences of the complex angular momentum theory as the uniformity of particle production in rapidity, the shrinkage of the diffractive cone and the logarithmic multiplicity growth, are in good qualitative agreement with the experiment. Why is there no good *quantitative* agreement?

In the expression for the Regge radius,

$$R^2 = R_0^2 + \alpha' \ln\frac{s}{m^2} \,,$$

the pomeron slope – the impact parameter diffusion coefficient – turned out to be *numerically small*:

$$\alpha'/R_0^2 \sim 1/12 \,.$$

The hadron radius increases very slowly with the energy, and hence different simplifying properties that we have discussed in these lectures will never manifest themselves; the true high-energy asymptotics will never be reached. We do not understand the *reason* why this is so.

Nevertheless, the *qualitative* agreement exists, since, although the growth of the radius is not sufficient, a relatively homogeneous distribution in rapidity is established in nature.

In what concerns the possible structure of the mass spectrum of hadrons, exploring this problem has led to the notion of 'duality'.*

16.1.2 Duality

Take some field theory based on a certain number of interacting fields (bare particles). If the coupling constant is sufficiently large, bound states (and resonances) will appear and enter the theory on equal footing with the input particles. A complicated, but legitimate, structure of the physical particle spectrum will emerge, driven by unitarity and analyticity.

We know that each Regge pole in the j-plane feels unitarity thresholds which, generally speaking, essentially deform its trajectory. However, experimentally this is not true. In a large interval of t all known Regge trajectories are *linear* with quite a good accuracy.

This fact looks rather strange and calls for explanation. How is it that all along a distance of several GeV the linearity persists?

One possibility is that a large mass scale may be there, hidden at the quark level (quark masses?)[†] the fact that it is mesons and baryons – bound states of quarks – that propagate at macroscopic distances is a secondary thing, less important for the dynamics.

Another attempt to explain the linearity consists of introducing into the theory, from the very beginning, an infinitely large number of particles and connecting their spins (placing them on the Regge trajectories). This way we make a step beyond the QFT framework not by abandoning locality but by introducing an infinite number of fields.

Why does this lead to linear trajectories specifically?

We know that all the singularities (including trajectories) are determined by unitarity, i.e. by the interaction. Hence, to insert, e.g. a $\sqrt{t}$ singularity by hand, does not seem clever or harmless.

The simplest analytic choice is a straight line.

---

* The concept of duality gave birth to string theory.
† On the appearance of a large mass scale in the vacuum channel, see Shifman and Vainshtein, 2005 (ed.).

Let us approximate the scattering amplitude by the sum over s-channel particle poles:

$$A(s,t) = \bigcirc = \sum_n^{\infty} \underset{\sigma_n}{\overset{m_n}{\rightthreetimes}} + \text{corrections.} \qquad (16.1)$$

One could think that the hadron interaction that we observe is *strong* already because there are many particles to exchange in (16.1). Within this logic, one might hope the corrections to such 'Born' approximation to be small.

Did we draw in (16.1) everything we had to? Should we not include also the poles in the t- and u-channels?

This would have been necessary if the number of particles was finite, or, at least, if the series in (16.1) converged well. However, the situation here is not so simple: since we are planning to include particles with arbitrarily high spins σ_n, the behaviour of $P_\sigma(\cos\theta_s)$ at $\sigma \to \infty$ is a worry. At $t > 0$ ($\cos\theta_s > 1$) the Legendre polynomials P_σ *increase* as a power of σ. Hence, the series makes sense only in the physical region of the s-channel ($t < 0$) and diverges at $t > 0$.

What does the divergence of the series mean for the amplitude $A(s,t)$? It is just a singularity in t. We know, however, that all the singularities are either poles or thresholds. We want to construct the Born amplitude which does not take into account the interactions; consequently, there may be only poles. Thus, we are unable to write an amplitude that would have poles *only* in s.

Here lies the basic idea of *duality*: the requirement that the series has in t just the necessary poles, those corresponding to particles (resonances) that can be exchanged in the t-channel of the reaction. The scattering amplitude can be alternatively expanded in terms of the t-channel poles:

$$A(s,t) = \sum_n^{\infty} \rightthreetimes = \sum_n^{\infty} \mathord{\Yup} . \qquad (16.2)$$

The equality of the sums over s- and t-channel resonances (*duality* of the two representations) guarantees the self-consistency of the construction. Essentially, the duality idea is the only way to introduce an infinite number of particles in the theory.

Obviously, the duality relation (16.2) imposes severe restrictions on the possible structure of the particle spectrum. How can we write a suitable meromorphic function having only poles? As it turns out, the problem can be solved quite easily (Veneziano, 1968):

$$A(s,t) = \frac{\Gamma(-\alpha(s))\Gamma(-\alpha(t))}{\Gamma(-\alpha(s) - \alpha(t))}$$

$$= B(-\alpha(s), -\alpha(t)) \equiv \int_0^1 dx(1-x)^{-\alpha(s)-1}x^{-\alpha(t)-1}$$

(16.3)

where α is a linear trajectory: $\alpha(t) = t$ (in units $\alpha' = 1$). This function has poles in each integer point both in the s- and t-channels. It is clear that having written a function symmetric in s and t, I have satisfied the duality relation automatically.

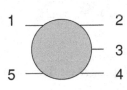

Actually, one can go even further and construct, in particular, a $2 \to 3$ scattering amplitude. It contains many pair invariants; continuing into each particular channel, $s_{ik} > 0$, we must obtain a series of the corresponding resonance poles.

It is easy to draw a diagram having poles, say, in s_{12} and s_{45}. Can one invent a contribution that would have, in addition, poles in s_{35}? It is clear that this is impossible: as soon as particles 4 and 5 combine into a pole, there is no

information left about the momentum p_5 that could correlate with p_3 (only information about spins can be carried up the graph). As a result, dual multi-particle amplitudes consist of a sum of all possible tree diagrams with poles in all sub-channels.

There is a technical difficulty: the amplitude must factorize; moreover, the residues must be *positive*. The task of constructing the meson amplitude close to the real hadron spectrum is almost completed. The case of baryons turns out to be more difficult.

How will the Born dual amplitude behave at $s \to \infty$? If the asymptotic behaviour were arbitrary, there would be no hope that the pole approximation is any good. In this case the correction terms to (16.1) which have the form of *loops* and are responsible for the *unitarization* of the amplitude would have to be significant.

From the point of view of the t-channel pole expansion in (16.2), the reggeon behaviour $A \propto s^{\alpha(t)}$ seems to be very natural. Indeed, the Veneziano formula matches this expectation, meaning that the theory embedded a priori not just many particles but the true Regge families. To see this we take the limit of large $(-s)$ in the integral representation

(16.3) and make use that $-\alpha(s) \gg 1$:

$$\int_0^1 dx(1-x)^{-\alpha(s)-1}x^{-\alpha(t)-1} = \int_0^1 dx\, x^{-\alpha(t)-1}\,e^{-\alpha(s)\ln(1-x)}$$

$$\approx \int_0^\infty \frac{dx}{x}\, x^{-\alpha(t)}\,e^{\alpha(s)x} = \int_0^\infty \frac{dy}{y}\left(\frac{y}{-\alpha(s)}\right)^{-\alpha(t)}\,e^{-y} \qquad (16.4)$$

$$\propto (-\alpha(s))^{\alpha(t)} = (-s)^{\alpha(t)}.$$

Actually, we could use the well-known Stirling formula for Γ in order to derive the asymptotics of the B function. The integral representation is, however, more suitable for generalizations.

Multi-particle diagrams also exhibit multi-regge behaviour:

$\sim (-s_{23})^{\alpha(t_{12})} \cdot (-s_{34})^{\alpha(t_{45})}$

As for the problem of unitarization, almost no progress has been achieved apart from a telling technical achievement.

It turns out that in the case of an infinite number of particles, in the same way as in the usual QFT, instead of inserting the imaginary part in the dispersion integral one can draw a Feynman diagram automatically satisfying the unitarity conditions. However, here the loop diagrams have more subtle properties. Due to (16.2), there is a variety of remarkable equalities between absolutely different graphs, for example,

The sharp turn from the attempts to do without the internal structure of hadrons (local features of the objects) was, I think, due to experiments, namely, those on deep inelastic scattering.

16.2 Parton picture

The concept of Feynman–Bjorken partons was invented to explain the deep inelastic scattering phenomenon but may have a deeper significance.

Using perturbative language, we have seen that at high energies there emerged significant simplifications in a wide class of hadron interaction processes. Is it possible to see these simplifications without appealing to the perturbation theory, not with the help of reggeons, but directly from the field theory?

16.2.1 Parton wave function of a high-energy hadron

How does field-theoretical description relate to the usual quantum mechanics? Why do we use Feynman graphs rather than the wave function as in the non-relativistic theory? Can one return to the wave function in QFT?

In a field theory, if we make a snapshot of a particle we will see many particles whose number changes with time. The NQM wave function depends on the coordinates of all the particles in the system; even if we manage to invent a QFT wave function, it would be a multi-component object.

What is a physical particle?

Even if I start with a bare object, it 'dresses up', and in the course of propagation develops into a multi-particle system:

$$\Psi = \begin{pmatrix} \psi_1(t, \mathbf{x}_1) \\ \psi_2(t, \mathbf{x}_1, \mathbf{x}_2) \\ \cdots \\ \psi_n(t, \mathbf{x}_1, \mathbf{x}_2, \ldots, \mathbf{x}_n) \\ \cdots \end{pmatrix}.$$

An ensemble of such Green functions determines probability amplitudes for finding 2, 3, etc. particles at a given time in definite points in space.

A wave function is a convenient object to use; it is normalized in a definite way, the integrals of the squared wave function determine probabilities of various physical processes.

So why don't we use a wave function in quantum field theory?

Recall that the QFT diagrams necessarily contain the space–time configurations in which some particles originate directly from the *vacuum* rather than from the original particle itself. Even in the simplest case of the self-energy correction diagram, the order of interaction times is arbitrary, and I would have to include into consideration all sort of virtual process going on in the vacuum.

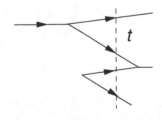

In a quantum field theory, a free particle is not a dynamically closed system. Rather, it is like an object in a medium – in the 'external' field of vacuum fluctuations.

Fig. 16.1   Configurations $t_{21} > 0$ and $t_{21} < 0$ in the self-energy graph.

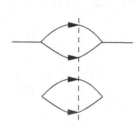

For certain graphs, it is straightforward to distinguish vacuum processes that do not affect the particle propagation. However, as soon as our particle and a vacuum fluctuation interact with one another (even in the future!), the picture becomes confusing; it is no longer possible to tell what belongs to the particle under consideration, and what to the vacuum.

If the longitudinal momenta of particles in the intermediate state of a virtual decay of their energetic parent, $E \gg m$, in Fig. 16.1(a) are of the same order, $x_1/x_2 = \mathcal{O}(1)$, the energy difference becomes small:

$$\Delta E = \sqrt{m^2 + \mathbf{p}^2} - \sqrt{m_1^2 + \mathbf{p}_1^2} - \sqrt{m_2^2 + \mathbf{p}_2^2} \simeq \frac{m^2}{2p} - \frac{m_{1\perp}^2}{2x_1 p} - \frac{m_{2\perp}^2}{2x_2 p}.$$

If you now integrate over t_1, t_2, an essential time interval between the interaction points will be very large – proportional to the initial energy, $t_2 - t_1 \sim 1/\Delta E \propto E$. As for the configuration of Fig. 16.1(b), the energy defect here is enormous:

$$\Delta E = 0 - \left(\sqrt{[1]} + \sqrt{[2]} + \sqrt{[3]} \right) \propto E,$$

and the lifetime of such a fluctuation is instead small: $|t_2 - t_1| \sim 1/E$. In order for our conclusion about the first graph (Fig. 16.1(a)) to hold, transverse momenta of intermediate state particles must be limited; otherwise, there would be no cancellation.

Let us sketch the calculation of contributions of these two time-ordered regions. The self-energy diagram contains two-particle Green functions:

$$\Sigma(p) = \int d^4 x_{12} \, e^{ipx_{21}} G(x_{21}) G(x_{21}).$$

If $x_0 = t > 0$, we close the integration contour in the lower half-plane of the energy component k_0 to obtain

$$G(x) = \int \frac{d^4 k}{(2\pi)^4 i} \frac{e^{-ikx}}{m^2 - k^2 - i\epsilon} = \int \frac{d^3 \mathbf{k}}{(2\pi)^3} \frac{e^{-i(t\sqrt{m^2+\mathbf{k}^2} - \mathbf{x}\cdot\mathbf{k})}}{2\sqrt{m^2 + \mathbf{k}^2}}. \qquad (16.5)$$

Thus, when we integrate over *positive* times, $t_{21} > 0$ (Fig. 16.1(a)),

$$\int_0^\infty dt_{21}\, e^{it_{21}\left(p_0 - \sqrt{[1]} - \sqrt{[2]}\right)} \sim \frac{1}{p_0 - \sqrt{m_1^2 + \mathbf{k}_1^2} - \sqrt{m_2^2 + \mathbf{k}_2^2}}.$$

When $t_{21} < 0$, the sign of the phase in the Green functions (16.5) flips:

$$\int_{-\infty}^0 dt_{21}\, e^{it_{21}\left(p_0 + \sqrt{[1]} + \sqrt{[2]}\right)} \sim \frac{1}{p_0 + \sqrt{m_1^2 + \mathbf{k}_1^2} + \sqrt{m_2^2 + \mathbf{k}_2^2}}.$$

(Clearly, the integral over $d^3\mathbf{k}$ must converge for our estimate to make sense.) With the growth of energy, the real time-ordered processes of the type (a) give larger and larger contributions as compared to vacuum fluctuations (b).

You may ask: does it make sense to separate two time-ordering regions in the diagram which is relativistic invariant *as a whole*? If I choose to sit in a reference frame where the energy is finite, $E = \mathcal{O}(m)$, the vacuum processes would again be inseparable from the particle, and the simplification would disappear.

The point is, in a high-energy scattering process there is always at least one fast particle, and our consideration is valid. Already at this point it becomes apparent that the wave function Ψ that we are about to introduce will be by no means a Lorentz invariant object.

16.2.2 Feynman scaling

Let us see what will happen in more complicated diagrams. I want to stress again that our first and basic assumption is that the transverse momenta of the particles involved are bounded from above: $\mathbf{k}_\perp^2 \lesssim m^2$.

In Lectures 5 and 10 we have discussed the 'ladder' (multiperipheral) kinematics and saw that it is most efficient to share the large momentum roughly equally at each step of the particle multiplication: $k_{1\|} \sim k_{2\|} \sim \frac{1}{2} p_\|$. Obviously, this is so only on average; there are always fluctuations, imbalanced configurations which give, however, a smaller contribution. Thus, when we consider more and more complicated diagrams with an increasing number of particles, we get an ever-slower particle in the intermediate state. After

$$\bar{n} \sim \ln\frac{p}{m} \Big/ \ln 2$$

steps we get a slow particle with $k_n \sim m$, and my logic of unimportance of the interaction with the vacuum breaks down.

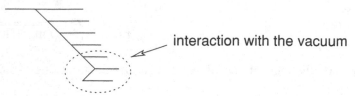

interaction with the vacuum

We come to the picture of a system of point-like particles – a 'comb' of partons inside an incident fast hadron – which 'scratches' the vacuum by its soft end (the so-called 'wee' partons).

Such a picture can be characterized by a probability amplitude ψ which is *almost* a wave function. In the coordinate space, it can be normalized:

$$\psi_n(t; \mathbf{r}_1, \mathbf{r}_2, \ldots, \mathbf{r}_n), \qquad \sum_n \int \prod_{i=1}^n d^3 \mathbf{r}_i \cdot |\psi_n(t, \{\mathbf{r}_i\})|^2 = 1.$$

In the momentum space, we have

$$\psi_n(\mathbf{k}_1, \mathbf{k}_2, \ldots, \mathbf{k}_n), \qquad \sum_i \mathbf{k}_i = \mathbf{p}.$$

The introduction of such an object does break the Lorentz invariance, but only slightly – at the level of the *slowest parton* in the ensemble. To see how this comes about, let us try to invent a *Hamiltonian* for our multi-component wave function. We take for simplicity the $\lambda \varphi^3$ interaction, represent the field as the sum of creation and annihilation operators, $\varphi = \varphi^\dagger + \varphi$, and drop the two terms corresponding to the vacuum processes:

$$(\varphi^\dagger + \varphi)^3 = (\varphi^\dagger)^3 + \varphi^3 + 3\left[(\varphi^\dagger)^2 \varphi + \varphi^\dagger \varphi^2\right] \implies (\varphi^\dagger)^2 \varphi + \varphi^\dagger \varphi^2.$$

The two remaining terms describe the splitting, $1 \to 2$, and the fusion processes, $2 \to 1$. Let us look at a stationary state,

$$E\psi = \hat{\mathsf{H}}\psi,$$

and write down the dynamical equation for the n-parton wave function component:

$$E\psi_n = \sum_{i=1}^n E_i \psi_n + g \sum_i \psi_{n-1}(\mathbf{k}_1, \ldots, \mathbf{k}_i + \mathbf{k}_{i+1}, \ldots, \mathbf{k}_n)$$

$$+ g \sum_i \int d\mathbf{p}' \, \psi_{n+1}(\mathbf{k}_1, \ldots, \mathbf{k}_{i-1}, \mathbf{p}', \mathbf{k}_i - \mathbf{p}', \ldots, \mathbf{k}_n).$$

Here the first – kinetic – term on the r.h.s. is the sum of parton energies, the second one is responsible for the splitting of one among $(n-1)$ partons into two, and the last term – for the fusion of two partons in the $(n + 1)$ state into one.

Let us examine the simplest, two-parton state:

$$\big[E(\mathbf{p}) - E_1(\mathbf{k}_1) - E_2(\mathbf{k}_2)\big]\psi_2(\mathbf{k}_1, \mathbf{k}_2) = h \cdot \psi_1(\mathbf{k}_1 + \mathbf{k}_2) + h' \cdot \psi_3\,. \quad (16.6)$$

Calculating the splitting term h,

$$= g\,\frac{d^3\mathbf{k}_1}{2k_{01}(2\pi)^3}\frac{d^3\mathbf{k}_2}{2k_{02}(2\pi)^3}\frac{d^3\mathbf{p}_1}{2p_0(2\pi)^3} \cdot (2\pi)^3\delta(\mathbf{p} - \mathbf{k}_1 - \mathbf{k}_2)$$

$$\Longrightarrow \frac{g}{2p} \cdot d\Gamma(\mathbf{k}_1)\,d\Gamma(\mathbf{k}_2),$$

and expanding the difference of energies,

$$E(\mathbf{p}) - E_1(\mathbf{k}_1) - E_2(\mathbf{k}_2) \simeq \frac{1}{2p}\left[m^2 - (m^2 + \mathbf{k}_{1\perp}^2)\frac{p}{k_{1\|}} - (m^2 + \mathbf{k}_{2\perp}^2)\frac{p}{k_{2\|}}\right],$$

we observe that the common factor $1/2p$ cancels, leaving us with

$$\left[m^2 - (m^2 + \mathbf{k}_{1\perp}^2)\frac{p}{k_{1\|}} - (m^2 + \mathbf{k}_{2\perp}^2)\frac{p}{k_{2\|}}\right]\psi_2 = g\,\psi_1 + \cdots.$$

We conclude that the Hamiltonian depends not on the parton longitudinal momenta themselves, but on their *ratio* to the initial momentum, $k_{i\|}/p$. In other words, the wave function is a function of the *rapidity difference*

$$\xi - \eta_i \simeq \ln\frac{p}{k_{i\|}}; \qquad \eta = \tfrac{1}{2}\ln\frac{k_0 + k_\|}{k_0 - k_\|}, \qquad \xi = \tfrac{1}{2}\ln\frac{E + p}{E - p}.$$

This is the manifestation of the (partial) relativistic invariance: as long as the partons are *fast enough* for the expansion of the energy roots E_i in (16.6) to be applicable, the parton wave function remains invariant under the *boosts* along the direction of the large initial momentum $\mathbf{p}$.

What next? Since the incident particle is a cloud of virtual particles – the partons, we can investigate the parton wave function by studying the integrals, representing the parton number density, two-parton correlations, in other words – the density matrix characterizing the multi-parton state:

$$\sum_n \int \psi_n(\{k_i\})\psi_n^*(\{k_i\}) \prod_{j \neq \ell} d\Gamma(k_j) = n(k_\ell),$$

$$\sum_n \int \psi_n(\{k_i\})\psi_n^*(\{k_i\}) \prod_{j \neq \ell,m} d\Gamma(k_j) = n(k_\ell, k_m), \quad \text{etc.}$$

$$(16.7)$$

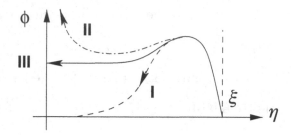

Fig. 16.2   Possible regimes of the parton-density behaviour at small momenta.

16.2.3 Parton density inside a hadron

Let us fix the momentum $\mathbf{k}$ of one of the partons and integrate over all the others, to obtain an inclusive *parton density*:

$$\phi(k_\parallel, \mathbf{k}_\perp) = \sum_n \frac{1}{(n-1)!} \int \prod_{i=1}^{n-1} \frac{d^3\mathbf{k}_i}{2k_{0i}(2\pi)^3} \cdot |\psi_n(\mathbf{k}, \mathbf{k}_1, \ldots, \mathbf{k}_{n-1})|^2.$$

One can imagine different possibilities.

(1) Multi-parton fluctuations 'bounce' from the vacuum and collapse back into the original particle, not producing enough 'wee' partons, line I in Fig. 16.2. Such a regime is typical for weakly interacting systems when the coupling is small and the perturbation theory works.

(2) A fast growth of the wee-parton density (line II) is a signal of an instability. This shows that the interaction does not stabilize the system; the density of particles in unit volume increases indefinitely, signalling an intrinsic instability. In particular, in the $\lambda\varphi^3$ theory the situation is most likely like this (no vacuum state).

(3) The vacuum plays the rôle of the boundary, but with account of production and re-absorption of partons, a certain constant density of slow partons emerges (line III).

The key hypothesis of the parton model is that, one way or another, as a result of a balance between parton emission and recombination processes, the mean parton density does occur in nature. In essence, this is the condition for a particle to exist 'independently of the vacuum', in the sense that the fast part of the parton 'comb' does not depend on the reference frame (is invariant under $\eta \to \eta + \text{const}$), and hence, does not know about the vacuum:

$$dn = \phi(\xi - \eta)\, d\eta \rightarrow \phi_\infty\, d\eta, \quad \xi - \eta \gg 1.$$

As for the transverse momentum dependence of the parton density, not much can be said about it from the first principles. At the same time, it is important to bear in mind the hypothesis which is crucial for the parton picture, namely that the transverse momentum integrals converge at some finite scale $\langle \mathbf{k}_\perp^2 \rangle \sim m^2$.

A plausible picture for the double differential distribution

$$dn(\eta, \mathbf{k}_\perp^2) = \phi(\xi - \eta, \mathbf{k}_\perp^2)\frac{d\eta\, d^2\mathbf{k}_\perp}{2(2\pi)^3}$$

can be drawn based on two observations. Firstly, if successive parton emissions are independent of each other, then, as we have discussed in Lectures 5 and 9, the random walk pattern emerges, and in the impact parameter space we have

$$\phi(\xi - \eta, \boldsymbol{\rho}^2) \propto \exp\left\{ -\frac{\rho^2}{\gamma(\xi - \eta)} \right\}. \tag{16.8a}$$

Secondly, a fluctuation with a typical lifetime $1/\mu$ may have a *total number* (multiplicity) of slow partons of order unity. Therefore, the distribution (16.8a) has to be properly normalized:

$$\phi(\Delta\eta, \boldsymbol{\rho}^2) = \frac{C}{\gamma\,\Delta\eta}\exp\left\{ -\frac{\rho^2}{\gamma\,\Delta\eta} \right\}, \quad \int d^2\boldsymbol{\rho}\,\phi(\Delta\eta, \boldsymbol{\rho}^2) = C. \tag{16.8b}$$

No surprise, this is nothing but the vacuum pole amplitude, with the pomeron slope $\alpha' = \frac{1}{4}\gamma$.

16.2.4 Partons and reggeons

Importantly, an analysis in terms of the partons essentially coincides with the analysis of high-energy scattering amplitudes that we carried out in these lectures.

Indeed, let us take the simplest ladder-type diagram, fix the momentum k of one of the partons and integrate over all the others.

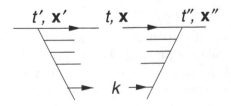

An integration of the product of the amplitude and the conjugate amplitude over the position of the parton **x** yields (in the time-ordered region, $t' < t < t''$) the Green function describing the propagation from **x**' to **x**'':

$$\int G(t - t', \mathbf{x} - \mathbf{x}') \overset{\leftrightarrow}{\partial_t} G^*(t'' - t, \mathbf{x}'' - \mathbf{x}) \, d^3\mathbf{x} \; = \; G(t'' - t, \mathbf{x}'' - \mathbf{x}').$$

Hence, integrating over all particles, we arrive at the diagram

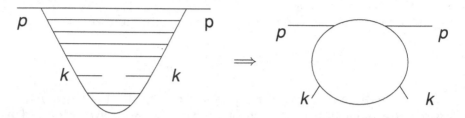

which is, essentially, the scattering amplitude. Thus, studying the parton density, we in fact investigate the scattering amplitude. If we substitute the Regge expression for the forward scattering amplitude, $s^\alpha \sim (p/k)^\alpha$, the pomeron pole, $\alpha(0) = 1$, gives us a homogeneous parton density distribution,

$$dn(k) \; \sim \; \frac{dk}{k} \; = \; \frac{dx}{x} = d\eta,$$

– the famous *Feynman plateau* in rapidity.

Our theory of the asymptotic behaviour of hadron scattering amplitudes provides an insight into the structure of the hadron wave function, and such a duality has a general physical significance.

The deuteron is a long-living object; the proton and the neutron inside the deuteron collide from time to time, but most of the time they are well separated. Does this picture apply to arbitrarily large energies?

Yes: this would mean that the deuteron always behaves as an object that consists of two independently interacting particles.

No: the parton clouds of the two nucleons eventually overlap at high-enough energies, so that by looking at the results of the collision with the target, it is impossible to tell if we had a deuteron for a projectile.

As we already know, this dilemma is related to the choice between the strong- or weak-coupling regimes of the reggeon field theory.

The knowledge of the properties of the strong interaction between slow partons is lacking. In spite of this, from the point of view of the parton picture, the asymptotic equality of the cross sections of *all* hadron processes looks rather natural.

In the laboratory reference frame a slow parton from the incident hadron interacts with the target; in the centre-of-mass frame it is slow partons from the wave functions of the colliding hadrons that interact with each another, in which case the hadron–hadron cross section is given by the product of the universal slow-parton interaction cross section and two parton densities.

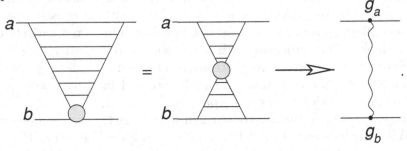

In either picture the density of slow partons at the tip of a long parton comb is universal – independent of the parent hadron (factorization). Moreover, according to the logic of the parton picture, a hadron is almost never in a sterile state: it is *always* represented by a 'comb'.

In the reggeon language this means that the probability of emitting a pomeron is one. Therefore, the pomeron–hadron residues, and thus the total hadron–hadron interaction cross sections,

$$\sigma_{ab} \propto g_a\, g_b \propto \left\langle \sum_{s,s'} \sigma_{ss'}(x, x') \right\rangle,$$

turn out to be independent of the types of hadrons.[‡]

16.2.5 Hadrons 'inside' a parton

What is new in what the parton picture tells us about the strong interaction?

When a soft parton interacts with the target, the coherence of the system breaks down. The partons of the 'comb' get released and, emitting their own 'ladders', separate from one another producing some number of hadrons in the final state.

The most interesting question is how the transition from partons to hadrons occurs. One can formulate some sort of an *uncertainty principle.*

[‡] For a detailed discussion of hadron–hadron and lepton–hadron interactions in the framework of the parton picture see Gribov's lecture entitled 'Space–time description of the hadron interactions' in Gribov (2003).

We say that a hadron consists, on average, of $\langle n \rangle \sim \ln p$ partons. It is clear that the opposite should also hold; the wave function of a single parton 'contains' $\langle n \rangle$ hadrons:

$$\Delta n_{\text{part}} \times \Delta n_{\text{hadr}} \sim \ln p.$$

How to visualize the conversion of a single parton into hadrons?

I have told you that the interaction of slow partons determines strong processes, while to explore the nature of the point-like constituents one has to employ electromagnetic and weak probes. This is not entirely true.

If we look at *rare* processes with large momentum transfers, significant simplification occur in a purely strong interaction as well. In particular, the parton $\rightarrow$ hadrons transition can be studied in a rare process of a large-angle scattering of *energetic* partons.

Such scattering produces a constituent with large transverse momentum $p_\perp$. With the increase of $p_\perp$ it becomes less and less likely that the other partons from the wave function of the incident hadron will follow suit and recombine with the struck parton. Therefore, an isolated parton will fragment on its own, producing a shower of hadrons.

Once we measure, say, a single energetic pion at 90° in the cms of the hadron collision, with a unit probability it has to be accompanied by a bunch of hadrons with logarithmic multiplicity.

Moreover, by the conservation of the transverse momentum, in the direction opposite to that of the triggered particle in the transverse plane there must be also $\ln p_\perp$ particles that originate from the recoiling parton.

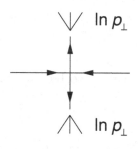

Hadrons with large transverse momenta are rarely produced in hadron collisions: the inclusive distribution falls fast with $p_\perp$. However, as soon as we have triggered one such particle, there is no additional suppression for having other large-$p_\perp$ hadrons in the final state.

Observation of hadron jets, and especially of a recoiling jet in hadron collisions with the production of large $p_\perp$ particles verifies the parton picture.

16.2.6 $e^+ e^-$ annihilation into hadrons

The annihilation of e^+ and e^- into hadrons is the cleanest process from the point of view of the parton picture. In the first order in α_{em} an electron and a positron may either scatter or *annihilate*, producing any charged particle and its antiparticle, e.g. $\mu^+\mu^-$.

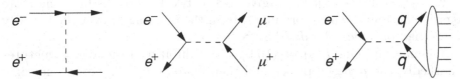

The corresponding cross section one calculates in a standard way in quantum electrodynamics. What character should have the process of e^+e^- annihilation into *hadrons*?

In quantum field theory the photon interacts with a point-like charge. If the energy is large, an intermediate-state photon has a huge virtual mass that can produce either a pair of energetic leptons or, equally well, a pair of electrically charged *partons* (quark and antiquark) flying in the opposite directions in the centre-of-mass of the collision. The strong interaction switches on, and two bare partons convert into hadrons. If transverse momenta in this transition process are, once again, limited, we will see two jets of hadrons emerging from the annihilation point.

This process is especially interesting in the sense that its cross section is easy to *calculate* since it has a purely electromagnetic nature. The cross section falls with the energy as $\sigma \propto \alpha_{\text{em}}^2/s$, but its *ratio* to the standard QED $\mu^+\mu^-$ production cross section tends to a constant given by a simple expression

$$R(s) \equiv \frac{\sigma_{e^+e^-\to\text{hadrons}}(s)}{\sigma_{e^+e^-\to\mu^+\mu^-}(s)} \simeq \frac{\sum_i e_i^2}{e^2}. \qquad (16.9)$$

The sum runs over the species of partons that can be produced at a given energy, i.e. with masses $m_i < \frac{1}{2}\sqrt{s}$.

Let us remark that at *very high energies* the production of hadrons in e^+e^- collisions is dominated by another process in which hadrons originate from the 'collision' of virtual photons belonging to electromagnetic coats of incident leptons:

$$\frac{d\sigma}{dM^2} \propto \frac{\alpha_{\text{em}}^4}{M^4} \frac{dt_1}{t_1} \frac{dt_2}{t_2}. \qquad (16.10)$$

The corresponding cross section is much smaller, of the order of α_{em}^4. However, it does fall with energy; since photon spin is one, the energy behaviour is similar to the case of the pomeron exchanges, $s_1 M^2 s_2 \sim s$. Moreover, the total hadron production cross section actually *grows* with energy. Integrations over momentum transfers t_1 and t_2 in a broad interval $t_{\min} \sim M^4/s \ll |t_i| \ll s$ give rise to two logarithms $\ln s$; one more logarithmic enhancement originates from the integration over the rapidity of the hadron block: $M^4 \, d\sigma/dM^2 \propto \ln^3 s$.

At small energies $R(s)$ of (16.9) was exhibiting resonance structures (markedly, the ρ meson peak), and then froze at $R \simeq 2$. This value came as a gift to the hypothesis of three *coloured quarks*:

$$\frac{1}{e^2}\left(e_u^2 + e_d^2 + e_s^2\right) \times 3 = \left(\left(\tfrac{2}{3}\right)^2 + \left(-\tfrac{1}{3}\right)^2 + \left(-\tfrac{1}{3}\right)^2\right) \times 3 = 2.$$

Now it seems that above $\sqrt{s} \simeq 3\,\text{GeV}$ a new threshold opens up and a new heavier quark ('charm') enters the game. Independently of the theory, in the near future we will be witnessing an avalanche of new particles. A new spectroscopy starts following that of the 1960s when the model based on three light quarks managed to classify hadrons.

16.3 Deep inelastic scattering

Now that we have gained certain knowledge about strong interactions, let us discuss some aspects of the lepton–hadron scattering that we did not touch on in Lecture 4.

Recall the essence of the deep inelastic scattering (DIS) phenomenon: a virtual photon seems to interacts with a point-like particle ('parton') inside a target hadron, which parton has spin $\tfrac{1}{2}$ and a limited transverse momentum.

16.3.1 Photon interaction with nucleons and nuclei

To penetrate deep into the proton interior, the momentum transfer $|q^2|$ has to be large. The energy transferred by the photon from the lepton to the proton should also be large enough, $W^2 = |q^2|(\omega - 1) \gg m^2$, in order to have a multi-state hadron system produced.

Imagine hadrons were point-like. Then the direct interaction process of Fig. 16.3(a) dies out at high energies as $\sigma \propto 1/s$.

It is easy to see, however, that there are certain unusual processes due to which the photon–hadron interaction cross section, although small, is *constant* in the high-energy limit.

The photon may fluctuate into a hadron pair, as shown in Fig. 16.3(b). The lifetime of such a state may be very large at high energies, $t \sim q_0/\mu^2$. This does not happen often because the electromagnetic coupling is numerically small; we may find the photon in a 'hadron state' only in one of

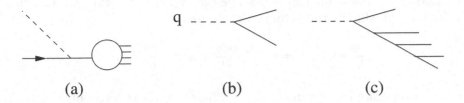

Fig. 16.3   Direct photon–proton interaction (a); energetic photon fluctuating into hadrons (b, c).

137 occasions, so to speak. However, once a fluctuation like that occurred, with respect to the *hadron multiplication* there is no perturbation theory, so that the photon will develop a full-scale parton comb, as a hadron does.

Thus, with the probability $\mathcal{O}(\alpha_{\mathrm{em}})$ the photon would approach the target as an ensemble of hadrons. Among them there are slow ones which will interact with the target providing $\sigma_{\mathrm{tot}}^{\gamma N}(s) \simeq$ const. The only difference with the hadron–hadron scattering case is in the size of the corresponding photon–pomeron coupling: $g_\gamma \propto \alpha_{\mathrm{em}}$.

There exists a rather unexpected experimental check of the validity of this picture. What would you expect for the cross section of photon interaction with a *nucleus*? A nucleus with atomic number A has a radius $R \sim A^{1/3}$. Total hadron–nucleus cross sections behave as $\sigma_{\mathrm{tot}}^{hA} \sim \pi R^2$; strongly interacting particles get stuck, with a large probability, and hence the cross section is proportional to the *surface* of the target.

On the other hand, since the electromagnetic interaction is weak, there is no large absorption and therefore the photon may freely pass through the nucleus. Therefore one would expect

$$\sigma_{\mathrm{tot}}^{\gamma A} = \sigma_{\mathrm{tot}}^{\gamma N} \cdot A \sim \sigma_{\mathrm{tot}}^{\gamma N} \cdot R^3.$$

But if our picture of a fluctuating photon is correct, at sufficiently high energy the photon must interact with the nucleus with a *hadronic* cross section $\sigma \propto R^2$!

The condition reads $q_0/\mu^2 \gg R$, that is, the lifetime of the hadron ladder fluctuation exceeding the time it takes to traverse the nucleus (Ioffe, 1969).

If we are sitting in this kinematical domain, we do not explore, with the help of a photon probe, the interior of the hadron target but simply observe a hadron–hadron interaction. We would see nothing like the Rutherford scattering in this case.

If our goal is instead to study the internal structure of the nucleon, we must cut off hadron-like configurations inside the photon. How to do that? By restricting the lifetime of the photon in such a way that it would have no time to develop a shower while passing through the nucleon, that is, by making the photon sufficiently virtual:

$$\frac{q_0}{|q^2|} \lesssim \frac{1}{\mu} \implies \frac{2mq_0}{|q^2|} = \omega \lesssim \frac{2m}{\mu} \sim 10.$$

So, the dimensionless parameter $\omega = -2pq/q^2$ that we have introduced in Lecture 4 must be of the order of unity (but not too large) in order to see the photon interacting with point-like partons inside the nucleon, to see the Bjorken scaling, etc.

16.3.2 DIS in the parton picture and quarks as partons

We are going to apply the parton model to the DIS process. Since we have a picture of the parton content of a fast hadron, let us choose a reference frame $q_0 = 0$ in which a fast hadron collides with a static electromagnetic field along some direction $\mathbf{z}$:

$$q^2 = -q_z^2, \; 2pq = -2p_z q_z, \quad \omega = -\frac{2pq}{q^2} = -\frac{2p_z}{q_z}.$$

The photon will be absorbed by one of the partons with momentum k. By which? It has to be a fast parton since otherwise an overlap of its wave function with the photon field of the size $\Delta z \sim 1/|q_z|$ would be small. The time the hadron passes through the field does not exceed $1/\mu$. For a long-living fast parton with the 'lifetime' $k/\mu^2 \gg 1/\mu$ this is an instantaneous interaction, and such interaction occurs with *conservation of the energy*. This condition selects a parton with a definite momentum:

$$k' = k + q, \; k_0' = k_0 \implies |k_z| \simeq |k_z'| \implies k_z = -k_z' = -\tfrac{1}{2}q_z. \quad (16.11)$$

After absorbing the photon, the struck parton flies in the opposite direction, $k_z' = -k_z$. What will happen with it afterwards is unimportant for the DIS cross section. It is determined simply by the product of the Born cross section for the photon absorption by a parton and the density of partons with the given rapidity η:

$$\frac{d\sigma}{dq^2\,d\omega} = \frac{d\sigma_B}{dq^2} \cdot \int \phi(\eta, \mathbf{k}_\perp^2) \frac{d^2\mathbf{k}_\perp}{(2\pi)^2}, \quad (16.12a)$$

$$\xi - \eta = \ln\frac{p_z}{k_z} = \ln\frac{2p_z}{|q_z|} = \ln\omega. \quad (16.12b)$$

Recalling that the parton density in the parton model depends only on the rapidity distance to the parent proton (16.12b), the DIS cross section takes the form

$$\frac{d\sigma}{dq^2\,d\omega} = \frac{4\pi\alpha_{\mathrm{em}}^2}{q^4}\left\{ \phi_0(\omega)\cdot e_{\mathrm{part}}^2\left[1 - \frac{pq}{pp_e}\right] \right.$$

$$\left. + \phi_{\frac{1}{2}}(\omega)\cdot e_{\mathrm{part}}^2\left[1 - \frac{pq}{pp_e} + \frac{1}{2}\left(\frac{pq}{pp_e}\right)^2\right]\right\}, \qquad (16.13)$$

where we have introduced the densities of partons with spin-$\frac{1}{2}$ and spin-zero, accompanied by the corresponding electron scattering cross sections.

Apart from simple QED factors, this expression contains two unknown functions ϕ of one variable ω. This is the *Bjorken scaling*. Experiments, as we have already discussed in Lecture 4, show no presence of spin-zero charges, $\phi_0 \simeq 0$.

Thus, from the parton model we have derived the Bjorken scaling using two hypotheses: $\phi = \phi(\xi - \eta)$, that is the existence of the stationary parton density, and, certainly, the hypothesis of limited transverse momenta of partons inside the hadron wave function.

The expression (16.13) contains, obviously, the parton charge e_{part} (in units of the electron charge). If partons are *quarks*, these charges are fractional numbers. Is there a way to measure them experimentally?

The parton density ϕ is not entirely arbitrary; it obeys certain normalization conditions. If we integrate $\phi(\eta)$ over rapidity we obtain the multiplicity of partons of a given species i:

$$\int_0^\xi d\eta\,\phi_i(\eta) = N_i.$$

More informative is another sum rule, namely

$$\sum_i \int_0^\xi d\eta\,\phi_i(\eta)\cdot e^\eta = \frac{p}{m}, \qquad (16.14)$$

which expresses the fact that the total longitudinal momentum (energy) of all the partons equals the momentum of the hadron they belong to.

The DIS cross section (16.13) contains the product $e_i^2\phi_i$, and one cannot extract the charges without knowing the normalization of ϕ. Fortunately, there is an additional source of information.

One may study *weak* processes like deep inelastic neutrino scattering, $\nu p \to \mu^- + X$. The weak interaction is insensitive to electric charges but feels the presence of spin-$\frac{1}{2}$ partons in the proton. The mass of the intermediate boson $W^\pm$ is apparently too large to make itself felt at presently available energies, so that the behaviour of partons in weak DIS processes can be described by means of the standard four-fermion Fermi interaction.

By combining the information coming from electromagnetic and weak DIS, one attempts to extract electric charges of partons. Modulo some uncertainties in the neutrino–quark scattering amplitudes, the results are consistent with

$$\sum_{i\in\text{proton}} e_i^2 \simeq 1 \qquad \left(= 2\cdot\left[\tfrac{2}{3}\right]^2 + \left[\tfrac{1}{3}\right]^2\right),$$

$$\sum_{i\in\text{neutron}} e_i^2 \simeq \frac{2}{3} \qquad \left(= 2\cdot\left[\tfrac{1}{3}\right]^2 + \left[\tfrac{2}{3}\right]^2\right).$$

Moreover, integrating and adding up the densities of quarks–partons of different species, one can estimate the total *energy-momentum* carried by charged partons inside the nucleon. The result is telling: about 50% of the nucleon momentum in (16.14) belongs to some electrically neutral fields (gluons?).

16.3.3 Structure of the final state

Let us address a very important question: what happens with the parton system after one parton is kicked off.

We turned around one of the partons of what was a coherent system. The relative invariant energy between the struck parton and its neighbours becomes large, and therefore it is unlikely to interact with the rest of the system. This means that we have prepared an isolated parton – a bare field-theoretical object. Flying in the opposite direction, it will 'decay' into $\ln(q_z/2)$ hadrons.

What about the rest of the parton comb that keeps moving in the initial direction of the nucleon?

Coherence of the *bottom part* of the parton fluctuation in Fig. 16.4(a) is not disturbed by the scattering. First slow partons with $k_i \sim \mu$ and then, successively, faster partons will be absorbed; they will revert to assembling the initial coherent proton. However, at the level of $k_i \sim k_z = \frac{1}{2}|q_z|$ the parton ensemble becomes aware that its coherence had been broken; the *upper part* of the fluctuation will get released, turning into hadrons.

The resulting rapidity distribution of final-state hadrons is sketched in Fig. 16.4(b). It has a characteristic *hole* in rapidity.

This would be the structure of the final state if quarks were *not confined*. The answer may be different if there is some specific dynamics (beyond our naive field-theoretical approach based on the locality of the interaction in rapidity) that would force the struck quark to interact with the rest of the parton ensemble in order to prevent the two separating hadron

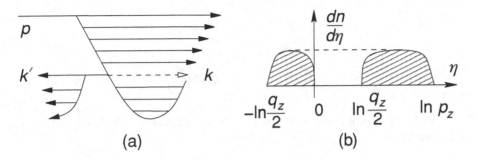

Fig. 16.4 (a) Coherent collapse of an undisturbed part of the parton wave function in DIS; (b) rapidity distribution of final-hadrons with a 'hole'.

systems from having *fractional electric charges*. Such an interaction could fill in the gap between the negative rapidity ('current fragmentation') and positive rapidity hadrons ('target fragmentation'), leading to a uniform hadron plateau – the dashed line in Fig. 16.4(b).

The existence of the hole in the hadron distribution is a key test for quarks in the rôle of partons. Experimentally, there is no sign of a hole in the proton fragmentation.

Thus we face a strange situation: on the one hand, the picture of quasi-free quark-partons works in DIS; on the other hand, the mechanism that forces the flying-away quark to 'communicate' with the rest of the proton is unclear.

16.4 The problem of quarks

There were times when quarks were thought to have large proper masses in order to explain their non-observation as free particles. Now, from the DIS experiments we know that the masses of (u, d) quarks must be smaller than 1 GeV. Thus, there has to be a special reason for quarks not appearing in the physical spectrum – the *confinement*.

Whatever the reason for quarks to be confined inside hadrons, looking at what happens in the e^+e^- annihilation it is clear that to ensure the quark confinement in such a process is not easy.

The quark and the antiquark fly too fast, and separate too far, to be able to interact with one another. Nevertheless, we expect two 'jets' of hadrons in the final state. Therefore, the production of multiple $q\bar{q}$ pairs which then recombine to form mesons, Fig. 16.5, cannot be explained by the re-interaction between the quarks created in the annihilation point. It has to be the result of the reaction of the *vacuum* on the production of two relativistic charges.

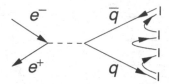

Fig. 16.5 Recombination of quarks into final-state hadrons in e^+e^- annihilation.

In quantum electrodynamics such a process is accompanied by radiation of soft photons (bremsstrahlung). Maybe in strong interactions an analogous phenomenon also plays a rôle.[‡]

16.4.1 One-dimensional electrodynamics

We have an example of a field theory with confinement – the Schwinger model. It is one-dimensional electrodynamics with massless fermions. It does not answer the question why quarks are confined in the real world, but it demonstrates nevertheless how the reaction of the vacuum results in the formation of 'hadrons' (Schwinger, 1962).

In one spatial dimension two charges cannot be separated because they interact as two infinite planes in our world, with their electromagnetic energy increasing linearly with the distance:

$$\mathcal{E} = \frac{\mathbf{E}^2 V}{8\pi}, \quad |\mathbf{E}| = \text{const.}$$

Classically, a pair of planes with opposite charges will oscillate. In the quantum case the system cannot have an arbitrary energy; it has to be quantized. So, a boson spectrum with a definite mass must appear. If we produce two planes with a large energy (as in e^+e^- annihilation), we expect that many oppositely charged pairs of planes will be produced giving rise to many bosons in the final state.

Let us study this theory in a more formal way. We have massless fermions and a photon, and the standard electromagnetic interaction between them. Since our space is a line, a massless fermion moves with the speed of light either in the positive or in the negative direction along it.

As we shall see shortly, an amplitude for the photon transfer into the fermion pair is different from zero only when the fermions move in the

---

[‡] Indeed, it is radiation of soft gluons that fills in the 'Gribov hole' (Gribov *et al.*, 1987). QCD bremsstrahlung plays a key role in the formation of final hadron states (Amati and Veneziano, 1979; Marchesini and Webber, 1984).

same direction (this is the consequence of the helicity conservation in the electromagnetic interaction).

An invariant mass of a pair of massless particles moving in the same direction is zero,

$$(k_1 + k_2)^2 = k_1^2 + k_2^2 + 2(k_{10}k_{20} - k_{1z}k_{2z}) = 0 + 0 + 0, \quad k_{1z}/k_{10} = k_{2z}/k_{20}.$$

Therefore, we have two massless states – the photon and the $q\bar{q}$ pair – that mix, so there must be a splitting of the degenerate levels. (In fact, $q\bar{q}$ is not one state but many, with different quark energies, k_{10}/k_{20}.)

As a result, the discrete state – the photon – must acquire a finite mass. Let us see how this occurs.

$$\Pi_{\mu\nu}(k) = -e^2 \int \frac{d^2p}{(2\pi)^2 i} \, \text{Tr} \left[\gamma_\mu \frac{1}{\hat{p}} \gamma_\nu \frac{1}{\hat{p} - \hat{k}} \right] = (g_{\mu\nu}k^2 - k_\mu k_\nu)\Pi(k^2).$$

Let us calculate the imaginary part of the specific component Π_{00} of the polarization tensor:

$$\text{Im}\,\Pi_{00} = -e^2 \int \frac{d^2p}{2} \, \text{Tr} \left[\gamma_0 \hat{p} \gamma_0 (\hat{p} - \hat{k}) \right] \delta_+(p^2)\delta_+((k - p)^2). \quad (16.15a)$$

The trace equals

$$-\frac{1}{2} \, \text{Tr} \left[\gamma_0 \hat{p} \gamma_0 (\hat{p} - \hat{k}) \right] = 2p_0(k_0 - p_0) - p(k - p)]$$
$$= p_0(k_0 - p_0) + p_z(k_z - p_z). \quad (16.15b)$$

Since $p_0 = |p_z|$ and $k_0 - p_0 = |k_z - p_z|$, the trace vanishes when the fermions fly in opposite directions. When the momenta are parallel, (16.15) yields

$$\text{Im}\,\Pi_{00}(k^2) = e^2 \int_0^{k_0} \frac{dp_0}{2p_0} 2p_0(k_0 - p_0)\delta(k^2) = e^2 k_0^2 \delta(k^2) \equiv k_z^2 \cdot \text{Im}\,\Pi(k^2).$$

We get

$$\text{Im}\,\Pi(k^2) = e^2 \delta(k^2) \quad \Longrightarrow \quad \Pi(k^2) = \frac{e^2/\pi}{k^2 - i\epsilon}.$$

The singularity at $k^2 = 0$ of the polarization operator results in

$$D_{\mu\nu} = \left(g_{\mu\nu} - \frac{k_\mu k_\nu}{k^2} \right) d_t(k^2), \quad d_t(k^2) = \frac{1}{k^2(1 - \Pi(k^2))} = \frac{1}{k^2 - m^2}.$$

Thus, the gauge invariance is spontaneously broken and the photon acquires a finite mass $m^2 = e^2/\pi$. (In the one-dimensional QED the charge has the dimension of mass.) Massive photon is that very bosonic state that represents a pair of planes.

Given a finite photon mass, electric fields fall exponentially with distance, $E \sim \exp\{-mr\}$. This would contradict the Maxwell equation $\mathrm{div}\,\mathbf{E} = 4\pi\rho$, unless the charge density ρ were in fact zero.

Indeed, by explicitly solving the theory one can show that the introduction of external currents provokes an appearance of the *polarization current*, which results in a local compensation of electric charges. An 'e^+e^- annihilation' process in one-dimensional electrodynamics causes the production of multiple fermion pairs, resulting in the final-state structure guessed at in Fig. 16.5.

16.4.2 A field theory for strong interactions?

The quark model has demonstrated that there is a simplicity in the spectrum of hadrons – mesons and baryons. Maybe there is a certain simplicity not only in the mass spectrum but in a deeper sense. The apparent complexity of hadron interactions does not exclude the simplicity at short distances.

In spite of the fact that the quantum mechanics of electrons in the Coulomb field is perfectly known, we would not dare to attempt to quantitatively describe the structure of a final state in, say, the collision of two atoms of mercury. We just get a mess. But at high energies, and in specific observables, the simplicity of the internal structure is manifest.

Should we approach the hadron dynamics from a short-distance side?

Although, as was mentioned in Lecture 4, the experimentally observed Bjorken scaling is not described by any field theory, we cannot imagine anything but a field theory based on simple point-like objects. Today progress (if it may be called so) goes in the direction of returning to quantum field theory with, possibly, a *small coupling constant*.

We saw in Lecture 2 that the effective pion–nucleon interaction coupling constant is large, $g_{\pi N}^2/4\pi \simeq 14$. But it may be that the coupling is large for *composite* objects – hadrons – while the quarks (which sit inside and by some reason cannot be freed) interact with a smaller constant. Otherwise, why would a parton absorb a virtual photon in the DIS process like a quasi-free particle?

Could it be possible to find a quantum field theory in which the interaction between the objects with small wavelengths is weak and the interaction with large wavelengths is strong? If such a theory will be found, we will see a leap of interest; the complexity of hadrons will cease to be the object of attention (as does the complexity of the Hg atom). Under focus will be, rather, processes in which a relatively simple internal structure of hadrons reveals itself.

16.5 Zero charge in QED and elsewhere

The best developed quantum field theory is quantum electrodynamics. What do we know about QED? On the one hand, it contains divergences, on the other hand, it is *renormalizable*. What does this, essentially, mean? Let us imagine that either the perturbation theory, or the theory itself, is wrong at short distances. Hence, integrating over momenta, we can do this correctly only up to a certain large momentum scale $k < \Lambda$.

What was the essence of progress in the 1950s? The uncertainty which appears due to the existence of such a cutoff, is contained by two quantities: by the observable charge and the mass of the electron which, therefore, remain uncalculable. It is obvious, however, that one cannot simply stop at that. Increasing the momenta of the external particles, sooner or later Λ will make itself felt.

There are two ways to formulate the problem.

(1) Let the interaction have a simple (point-like) form at $k \sim \Lambda$ (and at $k > \Lambda$ no interaction whatsoever). At momentum scales $k \sim \Lambda$ there are no corrections at all, and with the decrease of momenta, the corrections become relevant.

(2) The second approach is, though less transparent, essentially equivalent. Assume a given interaction at $k \sim m$; I want to see what happens when k increases. This is a semi-phenomenological approach; obviously, in a quantum field theory large distances are determined by small ones and not the other way round.

The basic quantities of the theory that contain ultraviolet divergences are the vertex part and the photon and electron Green functions. The integrals that determine all the other quantities are converging.[§]

16.5.1 First approach

In the first approach one introduces the bare charge e_0 at the ultraviolet momentum scale Λ.

Each subsequent correction to the diagram adds the factor e_0^2, one photon and two fermion Green functions and the momentum integration d^4k. However, it is not reasonable just to list all the corrections order by order in perturbation theory.

As we have seen in the QED course (Gribov and Nyiri, 2001), the set of Feynman diagrams can be rearranged into the series of the *skeleton*

[§] For details of the renormalization programme in quantum electrodynamics see Gribov lectures (Gribov and Nyiri, 2001) (ed.).

graphs that are built of the exact vertices (Γ) and exact photon (D) and fermion Green functions (G). The topology of the skeleton diagrams is such that they do not contain sub-diagrams that could be attributed to the vertex or propagator corrections.

Skeleton diagrams have a remarkable feature: each of them contains a single power of the logarithm of the ultraviolet cutoff, $\ln \Lambda$. The reason for this is simple: the integration momenta are mixed up, each momentum enters a large number of lines, and hence, the multiple integrals converge up to the last step. This is just the property of renormalizability of quantum electrodynamics.

For large momenta k we write $G \simeq -g(k^2)/\hat{k}$ and $D \simeq d(k^2)/k^2$, and the magnitude of the correction is determined in fact by the combination

$$e_0^2 \Gamma^2 G^2 D \, d^4k \;\to\; e_0^2 \Gamma^2(k^2) \, g^2(k^2) d(k^2) \, d\ln k^2.$$

The quantity

$$e^2(k^2) = e_0^2 \Gamma^2(k^2) g^2(k^2) \, d(k^2) \tag{16.16}$$

is the *invariant charge* which characterizes the interaction strength at the momentum scale k. This structure is common for all renormalizable theories.

How can the perturbative series be summed up? Initially we assumed the bare charge to be small, $e_0^2 \ll 1$. Let us decrease the external momenta p not too much, so that $e_0^2 \ln \Lambda^2/p^2 \lesssim 1$. In this situation all powers of the parameter $e_0^2 \ln \Lambda^2/p^2$ have to be resummed. My exact propagators and the vertex function have the following structure:

$$\Gamma(p^2) = \gamma_1 \left(e_0^2 \ln \frac{\Lambda^2}{p^2} \right) + e_0^2 \gamma_2 \left(e_0^2 \ln \frac{\Lambda^2}{p^2} \right) + e_0^4 \gamma_3 \left(e_0^2 \ln \frac{\Lambda^2}{p^2} \right) + \cdots .$$
$$\tag{16.17}$$

Neglecting the corrections of the type $e_0^4 \ln(\Lambda^2/p^2) \sim e_0^2 \ll 1$ we get the so-called leading logarithmic approximation (LLA) in which the problem was first solved by Landau *et al.* (1956).

The higher functions γ_n, with $n \geq 2$ are rather complicated; they are given by skeleton diagrams with the exact LLA vertices ($\Gamma = \gamma_1$) being propagators. Imagine that $\gamma_i(x)$ in (16.17) are of the same order not only for $x \lesssim 1$ but for any x. In this case everything seems to be ideal. If, however, γ_1 turns out to be singular or zero (e.g. at large x values), the higher-order terms have to be investigated seriously.

Actually, due to the Ward identity,

$$\frac{\partial G^{-1}(p)}{\partial p_\mu} \;=\; \Gamma_\mu(p,p,0),$$

the ultraviolet logarithms in Γ and g cancel, $\Gamma = g^{-1}$, and one is left with the photon renormalization function only in (16.16):

$$e^2(k^2) = e_0^2 \, d(k^2).$$

Corrections to the propagator and to the vertex depend on the properties of the charged particle. Therefore, if not for the Ward identity, the charge would not be such a universal quantity.

We have calculated the invariant charge in Chew and Low (1959):

$$e^2(k^2) = \frac{e_0^2}{1 + \frac{e_0^2}{12\pi^2} \ln \frac{\Lambda^2}{k^2}} \, . \tag{16.18}$$

What does this expression mean? We have a simple picture: the invariant charge decreases monotonously with the increase of the wavelength (the situation is just the opposite of what we would like to have in strong interactions).

The expression (16.18) is valid for large virtual photon momenta. At small $|k^2| \lesssim m^2$, under the logarithm the mass m appears, and for the physical charge that determines, e.g. the Coulomb scattering we get

$$e_c^2 = e^2(k^2 = 0) = \frac{e_0^2}{1 + \frac{e_0^2}{12\pi^2} \ln \frac{\Lambda^2}{m^2}} \, . \tag{16.19}$$

Suppose that I have introduced in the theory a finite charge at small distances corresponding to the large momentum scale Λ. Then at larger distances the effective charge becomes smaller. Moreover, for a point-like particle there is a *complete screening*; if we decrease the size of the region over which the bare charge e_0 is smeared, the on-mass-shell charge vanishes: $e_c^2 \to 0$ with $r_0 \sim 1/\Lambda \to 0$. The polarization of the vacuum screens the point-like charge totally, so that the observable electric charge has to be zero; $e_c^2/4\pi = 0$ instead of $1/137$.

How can this be explained?

We used the hypothesis that the bare charge e_0^2 is small. The LLA cannot be blamed for such an unphysical result. Our approximation of neglecting the higher-order corrections in (16.17) becomes even *better*, since the true measure of the interaction strength – the effective charge $e^2(k^2)$ – is *decreasing* with the decrease of the virtuality.

There remains one unsolved problem: what, if e_0^2 was large initially? More than 20 years has passed since the discovery of the running QED charge and of the zero-charge problem, but nobody has been able to explain why the large charge would be harder to screen than the small one.

May be there is a real ultraviolet cutoff Λ in nature? From the very beginning, it was assumed that Λ could be related to the *gravitational radius*. In such a scenario one would need to have $\nu = 13$ elementary fermions (presuming that only fermions are polarizing the QED vacuum).

In any case, pure quantum electrodynamics is a contradictory theory.

16.5.2 Second approach

The property of renormalizability means that all the divergences (dependence on the cutoff Λ) have to enter the observable charge e_c:

$$e_c^2 \equiv e_0^2 \cdot Z_3(e_0^2; \Lambda^2), \quad d(e_0^2; k^2, \Lambda^2) = Z_3(e_0^2; \Lambda^2) \cdot d_c(e_c^2; k^2).$$

Does our solution satisfy the renormalizability property? It certainly does:

$$
\begin{aligned}
d(k^2) &= \frac{1}{1 + \frac{e_0^2}{12\pi^2} \ln \frac{\Lambda^2}{k^2}} = \frac{1}{\left[1 + \frac{e_0^2}{12\pi^2} \ln \frac{\Lambda^2}{m^2} \right] - \frac{e_0^2}{12\pi^2} \ln \frac{k^2}{m^2}} \\
&= \frac{1}{1 + \frac{e_0^2}{12\pi^2} \ln \frac{\Lambda^2}{m^2}} \times \frac{1}{1 - \frac{e_c^2}{12\pi^2} \ln \frac{k^2}{m^2}} \equiv Z_3 \times d_c(k^2),
\end{aligned}
\tag{16.20}
$$

where e_c^2 is given by (16.19). The photon renormalization function,

$$Z_3 = \frac{1}{1 + \frac{e_0^2}{12\pi^2} \ln \frac{\Lambda^2}{m^2}},$$

embeds the whole dependence on Λ and on the bare charge e_0, while the renormalized photon Green function $d_c(k^2)$ contains only the renormalized charge $e_c^2 = Z_3 e_0^2$, as it has to be, owing to renormalizability.

Of course, this result for the Green function could have been obtained in the renormalized perturbation theory, not introducing Λ at all. We would notice that the fermion loop behaves at $k^2 \gg m^2$ as

$$(g^{\mu\nu} k^2 - k^\mu k^\nu) \cdot e_c^2 \ln \frac{k^2}{m^2},$$

i.e. perturbation theory breaks down when $e_c^2 \ln(k^2/m^2) \sim 1$. Then, repeating the logics of the logarithmic approximation,

$$e_c^2 \ll 1; \qquad e_c^2 \ln \frac{k^2}{m^2} \sim 1,$$

we would arrive at a geometrical progression, and the running coupling $e^2(k^2)$ would be expressed directly in terms of renormalized quantities:

$$e^2(k^2) = e_c^2 \Gamma_c^2 g_c^2 d_c = e_c^2 d_c = \frac{e_c^2}{1 - \frac{e_c^2}{12\pi^2} \ln \frac{k^2}{m^2}}, \tag{16.21}$$

where we have used the Ward identity.

In such a chain of thoughts we keep e_c^2 to be fixed. If so, we get a stupid result: $e^2(k^2)$ develops a singularity ('Landau pole') with the growth of k^2. The reason for that is that we have made a contradictory assumption, namely, that the observable charge e_c^2 can be taken as a finite quantity.

The effective charge $e^2(k^2)$ is the true expansion parameter. Near the pole, where $e_c^2 \ln \frac{k^2}{m^2} \simeq 1$, it becomes large, and the higher corrections become important and may help to solve the problem. The question is not cleared yet.

There is a simple relation between the two methods: $e_0^2 = e^2(\Lambda^2)$.

This situation has also influenced other theories. For a long time it seemed that the screening, and therefore the zero-charge problem, was unavoidable in any quantum field theory, and this was the reason owing to which the interest in field theories was lost.

Let us suppose, for a minute, that the sign in the denominator of (16.21) would be the opposite:

$$e^2(k^2) = \frac{e_c^2}{1 - \frac{e_c^2}{12\pi^2} \ln \frac{k^2}{m^2}} \quad \Longrightarrow \quad \frac{g_c^2}{1 + c\,g_c^2 \ln \frac{k^2}{m^2}}, \quad c > 0. \qquad (16.22)$$

How crazy is such an expression? It looks impossible at first sight. Indeed, how could it be that, when we move away from the charge, we see it not *screened* by the medium but, on the contrary, *growing* as if it pulled on other charges of the same sign?

Indeed, in electrodynamics this does not happen. However, for the *gravitational* interaction where all masses attract each other, this scenario can be easily imagined.

Such a system would be, obviously, unstable. Something must happen at *large* distances; the expression (16.22) contains an unphysical *ghost* pole at small momenta k^2. On the other hand, an instability may be a welcome feature which might help us to understand why the quarks cannot be separated at large distances but are confined inside hadrons.

16.6 Looking for a better QFT

According to deep inelastic scattering experiments, for strong interactions we need a theory with the interaction that *weakens* at small distances (e.g. as (16.22) does). We also want the theory to be renormalizable. This requirement does not follow from any physics. Nevertheless, to have a non-renormalizable theory is for us, unfortunately, the same as not having a theory at all.

In order to see whether a given theory is renormalizable, it suffices to look at the *dimension* of the coupling constant: $[g] = [m^\alpha]$. At large virtual momenta p, a higher-order correction then has the structure $(g^2/p^2)^\alpha$. If $\alpha > 0$, the correction falls in the ultraviolet region, and the theory is super-convergent (as is the $\lambda\varphi^3$ theory). If, on the contrary, $\alpha < 0$, the loop integral diverges in the ultraviolet, and the theory is non-renormalizable (as the four-fermion Fermi interaction). When g is dimensionless, corrections are logarithmic, $g^2 \ln p^2$, and the theory is renormalizable.

16.6.1 Fermion and scalar fields

There was a time when the field theory with the interaction $g\bar{\psi}\gamma_5\psi\varphi$ was considered as being related to reality, with nucleons and pions as elementary fields. We have seen already that the coupling constant g is then very large, $g^2/4\pi \simeq 14$. This is, however, on the mass shell, $g(m_\pi, m_N)$. What is the form of the effective interaction? Unfortunately, the same as in electrodynamics:

$$g^2(k^2) = \frac{g_0^2}{1 + c\, g_0^2 \ln\frac{\Lambda^2}{k^2}}, \quad c > 0.$$

At small distances the interaction is even stronger. We face the same zero-charge problem as in electrodynamics. In reality one has to add to this interaction the quartic point-like interaction between pions, $\lambda\varphi^4$. Even if we do not introduce it from the beginning, it is induced by a logarithmically divergent diagram with a four-scalar field attached to the fermion loop. Thus, the new vertex is necessary to achieve renormalizability of all amplitudes, including that of the four-scalar particle interaction.

16.6.2 Vector fields: how to construct a renormalizable theory

Let us consider now a vector particle (as in electrodynamics). Will such a theory be renormalizable? Generally speaking, vector theories are *not renormalizable*.

The Green function for a vector particle with mass m is

$$D_{\mu\nu}(k) = \frac{1}{m^2 - k^2}\left[\frac{k_\mu k_\nu}{m^2} - g_{\mu\nu}\right]. \tag{16.23a}$$

Where this structure comes from? The propagator of a vector particle,

$$-\sum_{\lambda=1}^{3} \frac{e_\mu^\lambda(k)e_\nu^\lambda(k)}{m^2 - k^2}, \tag{16.23b}$$

contains the sum over three polarizations (the overall minus sign guarantees that the residue is positive, since $(e_\mu)^2 < 0$). Three polarization vectors out of four, $e_\mu^\lambda k^\mu = 0$, $\lambda = 1, 2, 3$, correspond to a spin one particle; in the rest frame there are three spin projections, $e_\mu^\lambda = (0, \mathbf{e}^\lambda)$. The fourth vector, $e_\mu^0 = (1, \mathbf{0})$, describes a particle with spin *zero* – a scalar.

In order to eliminate the term proportional to $k_\mu k_\nu / m^2$ in (16.23a), we could try to extend the sum in (16.23b) to all four polarizations, by allowing the propagator to describe degenerate states of spin-zero and spin-one objects. In this case, however, the scalar would be a ghost (would enter with negative probability).

Thus, if we want to preserve unitarity, we must keep the $k_\mu k_\nu$ term in (16.23a), and this leads to a catastrophe: superfluous momenta multiply the vertices, and integrals diverge at large loop momenta.

Why then is electrodynamics renormalizable? There is no necessity to take special care about the 'scalar' polarization, since, owing to *current conservation*, it is not produced, and the $k_\mu k_\nu$ term can be simply dropped.

Is the fact that the photon is massless important for the renormalizability of QED? Not at all. A theory containing a massive photon, and conserved current, barely differs from electrodynamics. Unfortunately, it is also no different from QED in what concerns us, namely, the problem of zero charge; the mass in the photon propagator is unimportant in the ultraviolet momentum region, $k \to \infty$.

A few years ago it seemed that there are no other renormalizable quantum field theories. This situation looks strange: we have a theory of fermions interacting with the electromagnetic field. What about the other particles? Can they interact with the electromagnetic field? We know that the answer is yes for scalar particles, with only one subtlety: there have to be two vertices,

$$\tag{16.24}$$

However, the scattering of scalar particles contains logarithmic divergences,

and the local four-scalar interaction has to be introduced. Thus, scalar particles cannot interact with the electromagnetic field *only*.

A scalar particle cannot be considered as truly point-like with respect to the electromagnetic radiation; we are forced to introduce an additional interaction ($\lambda\phi^4$) which makes the particle somewhat 'smeared'.

One wonders, whether only fermions can be point-like? This would be in agreement with the fact that particles that are 'sterile' with respect to strong interactions (point-like with respect to the weak interaction) are only fermions (e, μ, ν).

What if we take a charged vector particle ?

$$\xrightarrow{\hspace{1.5cm}} C_\mu \qquad\qquad \Gamma \diagup k\mu$$
$$\dashrightarrow A_\mu \qquad \underset{q_1\rho}{\xrightarrow{\hspace{0.7cm}}\circ\xrightarrow{\hspace{0.7cm}}}\underset{q_2\sigma}{} \tag{16.25}$$

For renormalizability we have to require

$$q_1^\rho \Gamma_{\mu\rho\sigma}(q_1, q_2) = q_2^\sigma \Gamma_{\mu\rho\sigma}(q_1, q_2) = 0;$$

in addition, conservation of the electromagnetic current requires $k^\mu \Gamma_{\mu\rho\sigma} = 0$. It can be easily seen that no vertex can satisfy the current conservation relations simultaneously with respect to all three momenta. Does this mean that charged vector particles do not exist at all?

We can, of course, imagine a vector particle as a bound state of a fermion and an antifermion. In this case the electromagnetic interaction will not reduce to that with a point-like particle with spin $\sigma = 1$, and the renormalizability will not be broken. However, we do not know how to write down such an interaction phenomenologically; moreover, we are not able to calculate bound states of relativistic objects.

In terms of dispersion relations, how would the problem manifest itself?

We have the term $q_\mu q_\nu / m^2$ in the charged-particle propagator, and, if the vertices do not give zero, the amplitude increases. It turns out that the condition $q\Gamma = 0$ ($k\Gamma = 0$) can be satisfied if the other two vector particles

in the vertex are *real*, that is, on-mass-shell and with physical polarizations. (We shall see this later explicitly.) If so, the term $q_\mu q_\nu / m^2$ in the $C\bar{C} \to \gamma\gamma$ amplitude can be dropped as long as the external particles are real.

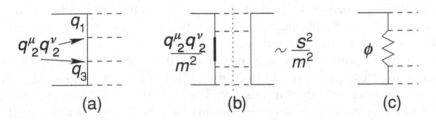

Fig. 16.6 (a) Longitudinal term in the multi-photon annihilation amplitude. (b) Energy growth of the imaginary part of the scattering amplitude. (c) Adding a scalar-particle exchange.

Let us consider more complicated processes. In the process of $C\bar{C}$ annihilation into *three* photons, everything is all right too: $q^\mu q^\nu$ pieces of the propagators of the internal virtual lines q_1 and q_2 multiply the external vertices, and their contributions vanish on the mass shell.

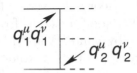

As to the production of four photons (Fig. 16.6) the situation changes. In the amplitude Fig. 16.6(a), the term $q_2^\mu q_2^\nu$ coming from the propagator surrounded by two *virtual* lines q_1 and q_3, does not disappear. As a result, the imaginary part of the amplitude of $C\bar{C}$ scattering via photons displayed in Fig. 16.6(b) grows with energy, and the restoration of the amplitude requires a subtraction term – an arbitrary constant. This means non-renormalizability of the theory, in the language of dispersion relations. Since the convolution of the momentum vector q_2^μ with the vertex $\Gamma(q_2, q_3)$ must be zero when the particle 3 is on the mass shell,

$$q_2^\mu \Gamma_{\mu\rho\sigma}(q_2, q_3) \propto (m^2 - q_3^2). \qquad (16.26)$$

Hence, the propagator of the virtual particle 3 actually cancels in the diagram, giving rise to a reduced diagram with two photons effectively emerging from one point.

As I have already mentioned, the sign of the residue coming from the badly behaving polarization is of 'ghost' type. Therefore, in principle, this unwanted contribution can be cancelled by adding to the theory a normal charged scalar particle φ as shown in Fig. 16.6(c). A simple tuning allows one to make the sum of the diagrams (a) and (c) to have good asymptotics (i.e. to cancel the divergences).

In the electrodynamics of *scalar* particles (as well as in the QFT with interacting fermions and (pseudo)scalars discussed above), in order to have a renormalizable theory, we have to introduce an *additional inter-action*. In the electrodynamics of *vector* particles the introduction of an *additional scalar field* is required.

It turns out that the class of renormalizable theories can be enlarged considerably by generalizing electrodynamics in a more serious way than by just making the photon massive. What is the idea? How could one approach the problem of renormalizability in a more general way, avoiding the necessity of inventing the counter-term diagrams?

In electrodynamics a 'scalar' component of the vector field cannot be produced even *virtually* due to the condition $k_\mu A_\mu = 0$, which selects three components of the photon field out of the four-vector A_μ. In fact a real photon has not three, but two polarizations; the 'longitudinal' compo-nent, $A_\mu(k) \propto k_\mu$, is not produced either (current conservation). Turning to a theory of a vector field with a finite mass, three components appear. Where do they come from? We inserted them into the massive Lagrangian just by hand. Would it not be possible to write the third component sep-arately, so that it would delicately join in, adding to the initially massless two-component photon?

Let us ask ourselves whether a scalar field with zero mass can exist. One may push in an arbitrary number of such particles with $\mathbf{k} = 0$ in the vacuum. The first thing that is likely to emerge is a constant field of ϕ particles – a Bose condensate.

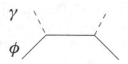

As soon as I introduce the interaction with photons, a photon propagating in a constant scalar field will experience Thomson scattering and continuously accumulate a scattering phase, in other words – it will acquire a finite mass.

Diagrammatically, a photon mixes up with the scalar field; after diagonaliza-tion the propagating states possess a non-zero mass. Two components of the photon and the (gradient of the) scalar field combine into a three-component massive vector field.

On the basis of this simple example one can try to construct a theory of massive charged vector particles.

16.6.3 Conservation of current and cancelling diagrams

- When we try to construct renormalizable field theories, for particles with spins $\sigma \geq 2$ ultraviolet divergences appear which we do not know how to deal with.

- In the case of vector particles, $\sigma = 1$, QED has taught us that the conservation of current can help to eliminate divergences. In particular, it is straightforward to construct a theory of a neutral massive vector field.

- Turning to electrodynamics of charged spin one particles, we see that there is no vertex that would provide current conservation with respect to all three vector lines simultaneously.

Maybe, there exists a deeper symmetry which provides a stronger current conservation than it can be seen just from the structure of the vertex?

Arriving at the vertices with virtual lines, $k_\mu k_\nu$ gives not zero but a quantity proportional to the departure from the mass shell, cf. (16.26). This cancels the propagators and leads to a topologically simpler diagram:

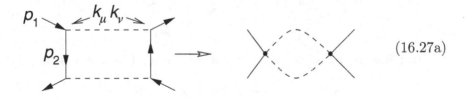

Is it not possible to invent a theory which would have such a diagram a priori, to help to cancel the 'longitudinal' term?

In electrodynamics this was just the case. Let us take, for example, a box diagram with $e^+ e^-$ annihilating into two photons and consider what happens with the 'longitudinal' part of the photon Green function:

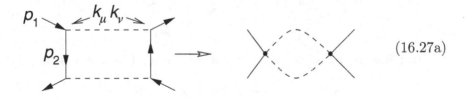

$$(16.27a)$$

The photon momentum k_μ, multiplying the vertex, produces

$$k_\mu \gamma^\mu = \hat{p}_1 - \hat{p}_2 = -(m - \hat{p}_1) + (m - \hat{p}_2);$$

the first term gives zero when it acts on a spinor describing an on-mass-shell electron, $(m - \hat{p}_1)u(p_1) = 0$ (Dirac equation), while the second term cancels the virtual fermion propagator, resulting in a reduced graph,

(16.27a). However, in QED we have also the second Feynman diagram,

$$(16.27b)$$

Two diagrams cancel in the sum.

In the case of *scalar* charges, the cancellation between reduced graphs (16.27) is incomplete (because of the momentum dependence of the $\gamma\varphi\varphi$ vertex). In this situation the diagrams with the *four-vertex*, $\gamma^2\varphi^2$, see (16.24), participate to make the $k_\mu k_\nu$ piece go away.

I wanted to demonstrate in terms of diagrams, what is required from the new theories.

16.7 Yang–Mills theory

The current conservation in quantum electrodynamics had one more very important interpretation: QED was constructed in such a way that a photon mass could not appear there due to gauge invariance.

We want to construct a theory where, again, summing the full set of diagrams, we get cancellation of dangerous $k_\mu k_\nu$ terms. This would ensure current conservation in spite of the impossibility to have conservation directly in the three-particle vertex.

What is the connection to the photon mass?

Multiplying the polarization operator by the photon momentum,

$$k_\mu \gamma^\mu = \hat{p}_1 - \hat{p}_2 = (m - \hat{p}_2) - (m - \hat{p}_1),$$

we obtain, diagrammatically,

The transversality of the photon polarization operator and, therefore, non-appearance of photon mass, is a particular case of the cancellation of divergences.

16.7.1 Electrodynamics of massless vector particles

Guided by the idea that the absence of masses is already a hint to serious cancellations, our first hope is to try to build up a theory of *massless* charged particles.

$$\mathcal{L}_{\text{int}} \sim \bar{C} C A \qquad\qquad\qquad (16.28)$$

What is a charged particle? The charged field C can be represented as a linear combination of two neutral fields, C_1 and C_2,

$$C = \frac{C_1 + C_2}{\sqrt{2}}, \quad \bar{C} = \frac{C_1 - C_2}{\sqrt{2}},$$

with positive- and negative-charge parity, correspondingly. Since a photon has negative charge parity, inserting this in the vertex (16.28), there remain the transitions $C_1 + \gamma \to C_2$ and $C_2 + \gamma \to C_1$.

If the charged particles are massless as the photon is, it is natural to consider three neutral fields C_1, C_2, C_3 ($C_3 \equiv A$) on equal footing, with the interesting interaction

$$\qquad\qquad\qquad\qquad (16.29)$$

which is just another representation for the electromagnetic vertex (16.25).

How to write an interaction so that the current is conserved? Let us construct the tensor

$$\Gamma^{\mu_1\mu_2\mu_3}(k_1, k_2, k_3) = g^{\mu_1\mu_2} p^{\mu_3}_{(3)} + g^{\mu_1\mu_3} p^{\mu_2}_{(2)} + g^{\mu_2\mu_3} p^{\mu_1}_{(1)},$$

where p is a linear combination of the momenta k_i, e.g.

$$p^{\mu}_{(1)} = a_{11} k^{\mu}_1 + a_{12} k^{\mu}_2 + a_{13} k^{\mu}_3.$$

(It is dangerous to use higher powers of momenta, since this would increase the divergence in the loop integrals.) Multiplying by the momentum, say, k_3, we get

$$k_{3,\mu_3} \Gamma^{\mu_1\mu_2\mu_3} = g^{\mu_1\mu_2}(k_3 \, p_{(3)}) + k^{\mu_1}_3 p^{\mu_2}_{(2)} + k^{\mu_2}_3 p^{\mu_1}_{(1)}. \qquad (16.30)$$

We need this expression to vanish upon multiplication by the *physical polarization* vectors $e^{\lambda_i}_{\mu_i}(k_i)$, $i = 1, 2$, which satisfy $(e^{\lambda_i}_{\mu_i} k^{\mu_i}_i) = 0$, when particles 1, 2 are on the mass shell: $k_1^2 = k_2^2 = 0$,

$$e_{\mu_1}^{\lambda_1}(k_1) e_{\mu_2}^{\lambda_2}(k_2) \cdot k_{3,\mu_3} \Gamma^{\mu_1 \mu_2 \mu_3}(k_1, k_2, k_3) = 0.$$

Since we may choose the polarization vectors $e(k_1)$, $e(k_2)$ orthogonal to the plane formed by the four-momenta $\{k_1, k_2, k_3\}$, we must have $(k_3 p_{(3)}) = 0$ in (16.30). Representing $p_{(3)}$ as a linear combination

$$p_{(3)} = a(k_1 - k_2) + b\, k_3 \qquad (-k_3 = k_1 + k_2),$$

$$(k_3 p_{(3)}) = -a[k_1^2 - k_2^2] + b\, k_3^2 = 0,$$

we conclude that $b = 0$ (recall that $k_1^2 = k_2^2 = 0$). Hence,

$$p_{(3)}^{\mu} = a(k_1^{\mu} - k_2^{\mu}),$$

and the resulting form of the vertex is (up to an overall constant)

$$\Gamma^{\mu_1 \mu_2 \mu_3} = g^{\mu_1 \mu_2}(k_1 - k_2)^{\mu_3} + g^{\mu_1 \mu_3}(k_3 - k_1)^{\mu_2} + g^{\mu_2 \mu_3}(k_2 - k_3)^{\mu_1}.$$
$$(16.31)$$

This is the sum of three electromagnetic vertices for a scalar particle.

Let us look again at the product (16.30):

$$k_{3,\mu_3} \Gamma^{\mu_1 \mu_2 \mu_3} = g^{\mu_1 \mu_2}(-k_1^2 + k_2^2) + k_3^{\mu_1}(k_3 - k_1)^{\mu_2} + k_3^{\mu_2}(k_2 - k_3)^{\mu_1}.$$

Substituting $k_3 = -(k_1 + k_2)$ we obtain

$$\begin{aligned} k_{3,\mu_3} \Gamma^{\mu_1 \mu_2 \mu_3} &= g^{\mu_1 \mu_2}(-k_1^2 + k_2^2) + k_1^{\mu_1} k_1^{\mu_2} - k_2^{\mu_2} k_2^{\mu_1} \\ &= -(G^{-1})^{\mu_1 \mu_2}(k_1) + (G^{-1})^{\mu_1 \mu_2}(k_2), \end{aligned}$$

where G^{-1} is the inverse transverse propagator of a vector particle,

$$(G^{-1})^{\mu_1 \mu_2}(k) = g^{\mu_1 \mu_2} k^2 - k^{\mu_1} k^{\mu_2}.$$

Thus, multiplying the vertex by the momentum of one of the particles ('photon line'), we have a difference of inverse propagators of two others ('charged lines'). We get the difference of the 'reduced' diagrams,

similar to that in quantum electrodynamics.

Is this everything we need? As often happens in bosonic theories, one type of the vertex, (16.29), is not enough.

Recall the second-order QED interaction amplitude:

For spinor particles, $\sigma = \frac{1}{2}$, the sum of two Born amplitudes satisfied the current conservation condition, $k^{\mu} M_{\mu\nu} = 0$. In the case of scalar charges, however, it was necessary to add a four-particle point interaction, because the triple vertex was *linear* in momenta.

Exactly the same situation is there in our vector theory: for vector currents to be conserved, one has to introduce a dimensionless four-particle vertex:

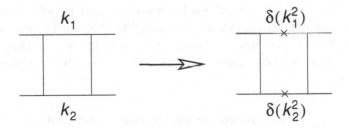

$$(16.32)$$

16.7.2 Feynman diagrams and unitarity: Faddeev–Popov ghost

Having done so, the theory is formulated. As soon as we have Born terms, perturbation theory can be constructed. How do we do this? A method which always works is to use the unitarity conditions. We may take the on-mass-shell Born amplitude on the r.h.s. of (16.32), square it, and reconstruct the higher-order scattering amplitude by its imaginary part, employing the dispersion relation (with one subtraction). In principle, there is no problem in carrying out this procedure; still, this is a rather cumbersome way. Usually, we draw a Feynman diagram instead, which coincides with the dispersion expression. However, in the present case *this is not true*.

To see what goes wrong, let us draw a diagram and take the imaginary part by putting the cut particle lines on the mass shell, replacing the cut propagators by the delta functions:

Which polarization states will emerge in the intermediate state? As a rule, we expect two transverse polarizations to appear from the left and the right, see Fig. 16.7(a), while the contributions of the *longitudinal* polarizations have to be zero, owing to current conservation. However, our four-particle amplitude (as well as the three-particle vertex Γ above) satisfies the current conservation with respect to each external line only

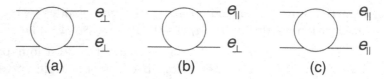

Fig. 16.7   Various combinations of intermediate state polarizations.

if *all* the other lines are on-mass-shell and with physical polarizations:

$$k_1^{\mu_1} M_{\mu_1\mu_2\mu_3\mu_4} \cdot e^{\mu_2}(k_2) e^{\mu_3}(k_3) e^{\mu_4}(k_4) = 0; \quad (e(k_i)k_i) = 0, \quad k_i^2 = 0, \quad i = 2, 3, 4.$$

Therefore, a situation like Fig. 16.7(b) (three transverse polarizations out of four) will not appear in the imaginary part. The intermediate state with *two* longitudinal polarizations shown in Fig. 16.7(c) will, however, be present. Writing unitarity conditions, we summed, of course, only over physical states, $e_\perp$. Thus, the imaginary part of the Feynman diagram differs from that in unitarity conditions. In fact we face here the first case when the Feynman diagrams are not correct literally.

What can we do? A technical problem appears: is it possible, nevertheless, to represent the correct dispersion result in terms of diagrams, which would provide us with the means to carry out calculations? The answer, at the one-loop level, was given by Feynman. He suggested to introduce a fictitious massless scalar particle (essentially, a single state $e_\parallel$ can be considered as a particle with $\sigma = 0$), and *subtract* the corresponding loop,

A general prescription was given by Faddeev and Popov, namely: one should introduce in the theory an additional field ϕ, with a normal vertex, and handle it as a *fermion*, i.e. count every ϕ loop with a minus sign.

This example demonstrates that the Feynman technique ceases to be transparent.

16.7.3 Gauge invariance and Lagrangian

Now we are going to discuss a different approach that leads to the theory of interacting vector fields – the Yang–Mills theory.

So far we have considered only vector mesons. It is easy to introduce also fermions. We should not forget, however, that up to now our construction was symmetrical with respect to the indices $1 \to 2 \to 3$ of the vector fields. The symmetry of the theory with respect to the rotation in the 'space' of the three field components can be preserved if we introduce a doublet of fermions, similar to a nucleon with two isospin states, $N = (p, n)$.

But first let us elucidate the relation of the Yang–Mills theory with electrodynamics. In the free fermion Lagrangian,

$$\mathcal{L}_{\psi 0} = \bar{\psi}(x) \left(i \gamma_\mu \frac{\partial}{\partial x_\mu} - m \right) \psi(x),$$

one can substitute

$$\psi \rightarrow \psi' = e^{i\alpha} \psi, \quad \bar{\psi} \rightarrow \bar{\psi}' = e^{-i\alpha} \bar{\psi}, \tag{16.33a}$$

with $\alpha = \text{const}$, without affecting the equation of motion. This is, essentially, the expression of the fact that a fermion and an antifermion can not transform one into the other.

A prominent feature of quantum electrodynamics is that in this theory one is allowed to carry out the transformation (16.33a) of the fermion field, with the phase depending on the space–time point, $\alpha = \alpha(x)$. (In QED one can tell a fermion from an antifermion *locally*.) To keep the action invariant under such transformation,

$$\mathcal{L}_{\psi 0} \rightarrow \mathcal{L}'_{\psi 0} = \bar{\psi}' \left(i \gamma_\mu \frac{\partial}{\partial x_\mu} - m - \gamma_\mu \frac{\partial \alpha(x)}{\partial x_\mu} \right) \psi',$$

one has to add to the Lagrangian the interaction with a *vector field*,

$$\mathcal{L}_{\psi 0} \implies \mathcal{L}_\psi = \bar{\psi} \left(i \gamma_\mu \frac{\partial}{\partial x_\mu} - m + \gamma_\mu A_\mu \right) \psi,$$

and simultaneously transform this field as

$$A_\mu(x) \rightarrow A'_\mu(x) = A_\mu(x) + \frac{\partial \alpha(x)}{\partial x_\mu}. \tag{16.33b}$$

The field A_μ changes, turns into some other field A'_μ. Hence, if A is the dynamical variable itself, its proper action must be invariant with respect to the gradient transformation (16.33b). Such an invariant photon Lagrangian, as you know, is given by the square of the antisymmetric electromagnetic stress tensor:

$$F_{\mu\nu} = \frac{\partial A_\nu}{\partial x_\mu} - \frac{\partial A_\mu}{\partial x_\nu}, \qquad \mathcal{L} = \mathcal{L}_\psi + \frac{1}{4e^2} F_{\mu\nu} F^{\mu\nu}.$$

From the point of view of such an approach the Yang–Mills theory is just an exact repetition of the same logic.

Let us introduce two fermions in the theory:

$$\mathcal{L}_\psi = \mathcal{L}_{\psi_1} + \mathcal{L}_{\psi_2} = \bar{\psi} \left(i \gamma_\mu \frac{\partial}{\partial x_\mu} - m \right) \psi \tag{16.34}$$

where ψ is a column,

$$\psi(x) = \begin{pmatrix} \psi_1(x) \\ \psi_2(x) \end{pmatrix},$$

like a nucleon with two isospin components, $N = (p, n)$.

If there is degeneracy in the system (even with account of the interaction), I can consider these two levels as one field. With the help of a unitary matrix, I can redefine (globally) who I call the 'proton' and who the 'neuteron'. Let us try to construct a theory in which such a 'redefinition' can be carried out locally.

Consider a transformation,

$$\psi'(x) = S(x)\psi(x), \quad \bar{\psi}'(x) = \bar{\psi}'(x)S^{-1}(x); \qquad S^\dagger(x) = S^{-1}(x),$$

with $S(x)$ a unitary matrix.

Let us take the Lagrangian

$$\mathcal{L}_\psi = \bar{\psi}(x)\left(i\gamma_\mu\frac{\partial}{\partial x_\mu} - m + i\hat{A}(x)\right)\psi(x), \quad \hat{A} = \gamma_\mu\mathbf{A}_\mu,$$

with $\mathbf{A}_\mu$ an *anti-hermitean* 2×2 matrix of vector fields (an analogue of photon), and rotate the spinor fields in it:

$$\mathcal{L}_\psi \to \bar{\psi}(x)\left(i\gamma_\mu\frac{\partial}{\partial x_\mu} - m + i\gamma_\mu S^{-1}(x)\frac{\partial S(x)}{\partial x_\mu} + iS^{-1}\hat{A}S\right)\psi(x). \quad (16.35)$$

The Lagrangian stays invariant if the $\mathbf{A}$ field also transforms as follows:

$$\mathbf{A}_\mu \to \mathbf{A}'_\mu = S\mathbf{A}_\mu S^{-1} - \frac{\partial S}{\partial x_\mu}S^{-1}; \qquad (16.36\text{a})$$

$$\mathbf{A}_\mu = S^{-1}\mathbf{A}'_\mu S + S^{-1}\frac{\partial S}{\partial x_\mu}. \qquad (16.36\text{b})$$

In order to have an invariant theory, the invariance of the action of the fields $\hat{A}$ is required with respect to the transformation (16.36). One observes that the field strength tensor

$$\mathbf{G}_{\mu\nu} = \frac{\partial \mathbf{A}_\nu}{\partial x_\mu} - \frac{\partial \mathbf{A}_\mu}{\partial x_\nu} + [\mathbf{A}_\mu\,\mathbf{A}_\nu] = \mathbf{F}_{\mu\nu} + [\mathbf{A}_\mu\,\mathbf{A}_\nu]$$

transforms *homogeneously*, $G'_{\mu\nu} = SG_{\mu\nu}S^{-1}$, and therefore the vector field Lagrangian

$$\mathcal{L}_A = \frac{1}{4g^2}\,\text{Tr}\left(\mathbf{G}_{\mu\nu}\mathbf{G}^{\mu\nu}\right) \qquad (16.37)$$

is invariant under the gauge transformation (16.36).

How is the field **A** built up? The 2×2 matrix has four components:

$$\mathbf{A}_\mu = I \cdot A_\mu^0 + \sum_{a=1}^{3} \tau^a A_\mu^a,$$

with τ^a the Pauli matrices. The field A^0 here corresponds to the usual electromagnetic interaction. It is generated by the (Abelian) transformation subgroup $U(1)$ contained in the group of unitary 2×2 matrices: $U(2) = U(1) \times SU(2)$. Let us exclude the photon and restrict ourselves to fields $\mathrm{Tr}\,\mathbf{A} = 0$, i.e. to A_μ^a with $a = 1, 2, 3$. The full Lagrangian of the theory will have the structure

$$\mathcal{L} = \mathcal{L}_\psi + \mathcal{L}_A = \mathcal{L}_{0\psi} + \bar{\psi}\gamma_\mu \sum_a i A_\mu^a \tau_a \, \psi + \mathcal{L}(F_{\mu\nu})$$

$$+ \frac{1}{2g^2} \mathrm{Tr}(F_{\mu\nu}[A_\mu A_\nu]) + \frac{1}{4g^2} \mathrm{Tr}([A_\mu A_\nu][A_\mu A_\nu]). \qquad (16.38)$$

The last two terms generate those two vertices that we have drawn above in (16.32). In terms of rescaled fields $\tilde{A} = A/g$, the three-boson and fermion–vector boson vertices are proportional to the coupling constant g, and the four-boson vertex proportional to g^2.

We considered here $SU(2)$ gauge symmetry. The discussed scheme can be applied, however, to any unitary group $SU(N)$. In particular, for $N = 3$ we introduce quarks of three *colours* and obtain $N^2 - 1 = 8$ *gluons*, – fundamental fields of the *quantum chromodynamics*.

16.7.4 Essential gauge invariance and cyclic variables

How can one work with such gauge theories? One of the ways to construct a quantum theory is to use the functional integral:

$$\int e^{i \int \mathcal{L}[A,\psi]dx} \, d\psi \, d\bar{\psi} \, dA_\mu. \qquad (16.39)$$

Let us say a few words about the appearing problems. Making an effort to have a gauge invariant theory, we arrived at an undefined system, since we introduced more variables than it has in reality.

We operate with fields $\mathbf{A}_\mu$ which can be substituted by other fields, $S^{-1}\mathbf{A}_\mu' S + S^{-1}\frac{\partial S}{\partial x_\mu}$, not changing the action. Hence, the Lagrangian does not depend on S, and this shows that in the functional integral (16.39) there is integration over a large number of superfluous unphysical variables. The relation (16.36a) allows us to choose new physical fields in different ways, e.g. so that $A_0' \equiv 0$ (by solving the equation $\mathbf{A}_0 = S^{-1}\frac{\partial S}{\partial x_0}$).

Let us set $A_0^a = 0$ and keep only the integration over three-vector potentials A_i^a. Then, $G_{00}^a = 0$, $G_{0i}^a = \partial A_i^a / \partial x_0 \equiv \dot{A}_i^a$, and what remains is a quite reasonable Lagrangian:

$$\mathcal{L}(A_i, \psi) = \mathcal{L}_\psi - \frac{1}{2g^2} \left[\mathrm{Tr}\big(\dot{\mathbf{A}}_i \dot{\mathbf{A}}_i\big) - \tfrac{1}{2} \mathrm{Tr}\big(\mathbf{G}_{ik} \mathbf{G}_{ik}\big) \right]; \quad (16.40\text{a})$$

$$\mathbf{G}_{ik} = \frac{\partial \mathbf{A}_k}{\partial x_i} - \frac{\partial \mathbf{A}_i}{\partial x_k} + \big[\mathbf{A}_i \mathbf{A}_k\big]. \qquad (16.40\text{b})$$

(The minus sign in front of the boson Lagrangian is due to my choice of $\mathbf{A}_i$ as anti-hermitean matrices.) Since $\mathbf{G}_{ik}$ do not contain the time derivative, the last term in (16.40a) plays the rôle of the potential energy of the self-interacting fields A_i^a; the term $(\dot{A}_i^a)^2$ is their kinetic energy.

Formally, this is not a relativistically invariant description, and usually one chooses an invariant gauge fixing condition $k_\mu A_\mu^a = 0$.

We saw already in electrodynamics that there are two types of gauge invariance.

- Firstly, the current conservation makes it possible to impose the condition $k_\mu A_\mu = 0$ in order to select three components of four.

- Secondly, a deeper consequence of the gauge invariance lies in the fact that the photon is massless, and therefore a real photon has only *two* field components.

The condition $(\mathbf{k} \cdot \mathbf{A}^\lambda) \equiv k_i A_i^\lambda = 0$ $(\lambda = 1, 2)$ applies only to *real photons*. In diagrams with *virtual particles* the longitudinal component of the electromagnetic potential, $(\mathbf{k} \cdot \mathbf{A}^\parallel) \neq 0$, is acting and represents the Coulomb field. This *essential* invariance allows us to write

$$A_i(x) = B_i(x) + \frac{\partial \varphi(x)}{\partial x_i}; \quad \mathrm{div}\,\mathbf{B} \equiv \frac{\partial}{\partial x_i} B_i = 0.$$

The field φ here is absolutely essential inside the diagrams but does not correspond to a free particle (does not 'fly away'); real photons are described solely by the two-component field B_i.

Our aim is to show that the theory with the Lagrangian (16.40) describes indeed massless vector particles. The equation $\mathbf{A}_0 = S^{-1} \frac{\partial S}{\partial x_0}$ that we have used to eliminate the scalar component of the potential, fixed S up to unitary matrices not depending on time. Arbitrary rotations depending on x_i are still at our disposal. This allows us to look for an additional invariance in our Lagrangian where no A_0 is present and everything is determined. Let us write A_i in the form

$$\mathbf{A}_i = v^{-1}\mathbf{B}_i\, v + v^{-1}\frac{\partial v}{\partial x_i} \qquad (16.41)$$

and fix B_i^a by the condition

$$\frac{\partial}{\partial x_i}B_i^a = 0 \qquad (16.42)$$

in the same way as in electrodynamics. It is important to stress that this is *not* a gauge transformation, just an attempt to separate physical degrees of freedom. Note that since (16.41) has the *form* of a gauge transformation, $\mathrm{Tr}(\mathbf{G}_{ik}^2)$ does not depend on v. From the point of view of classical mechanics, this makes v a *cyclic variable*, i.e. the one that has the kinetic but not the potential energy (n the same way as the Coulomb field in electrodynamics). A cyclic variable q in mechanics,

$$\mathcal{L} = \dot{q}^2 + \sum_k \dot{q}_k^2 - U(q_k),$$

changes linearly with time:

$$\dot{q} = \text{const}, \quad q = ct + b.$$

This is actually the real difficulty of quantizing a gauge theory: there are variables which do not oscillate but increase with time. It goes without saying that if I measure components of the *field strength*, $\dot{A}_i^a$, G_{ik}^a, all will be fine. Nevertheless, a problem remains: perturbation theory cannot be applied to such a Lagrangian possessing a cyclic variable.

Let us omit fermions and start with a free Lagrangian

$$\mathcal{L}_{A0} \propto \frac{1}{2}\sum_i \dot{A}_i^2 - \frac{1}{4}\sum_{i,k}\left(\frac{\partial A_i}{\partial x_k} - \frac{\partial A_k}{\partial x_i}\right)^2. \qquad (16.43)$$

In the momentum representation,

$$A_i(x) = \frac{1}{\sqrt{V}}\sum_{\mathbf{k}} a_i(x_0,\mathbf{k})\, e^{i\mathbf{k}\cdot\mathbf{x}}, \quad a_i(x_0,-\mathbf{k}) = a_i^*(x_0,\mathbf{k}),$$

we get

$$\mathcal{L} \propto \tfrac{1}{2}\left(\dot{a}_i\dot{a}_i^* - |[\mathbf{k}\,\mathbf{a}]|^2\right) = \tfrac{1}{2}\left(\dot{a}_i\dot{a}_i^* - (\delta_{ij}\mathbf{k}^2 - k_ik_j)a_ia_j^*\right), \qquad (16.44)$$

showing that the longitudinal component of the field, $\mathbf{a} \propto \mathbf{k}$, does not enter the potential energy. On the definite energy states, $\mathbf{a}(x_0,\mathbf{k}) = e^{-ik_0x_0}\mathbf{C}(\mathbf{k})$, the Lagrangian (16.44) turns into

$$\mathcal{L} \propto \tfrac{1}{2}\left[\delta_{ij}\,k_0^2 - (\delta_{ij}\mathbf{k}^2 - k_ik_j)\right]C_iC_j^*.$$

The expression in the square brackets is the inverse propagator:

$$D_{ij}^{-1}(k) = \delta_{ij}\, k^2 + k_i k_j, \tag{16.45a}$$

$$D_{ij}(k) = \frac{1}{k^2}\left(\delta_{ij} - \frac{k_i k_j}{k_0^2}\right). \tag{16.45b}$$

You can verify that (16.45b) is indeed an inverse tensor to (16.45a):

$$\sum_{\alpha=1}^{3} D_{i\alpha}^{-1}(k)\, D_{\alpha j}(k) = \delta_{ij}.$$

The free Green function acquires a pole in energy k_0, and all the integrals will diverge at $k_0 \to 0$, which divergence corresponds to the linear growth with time of the longitudinal component of the vector field. The Green function (16.45b) can be split into the sum of the transverse and longitudinal contributions:

$$D_{ij}(k) = \frac{1}{k^2}\left(\delta_{ij} - \frac{k_i k_j}{\mathbf{k}^2}\right) + \frac{k_i k_j}{k_0^2\,\mathbf{k}^2}. \tag{16.46}$$

The transverse part of the propagator, $k_i \cdot D_{ij}^{\perp}(k) = 0$, looks reasonable; the longitudinal part contains an infrared divergence. Evidently, the latter has to be defined additionally using some sort of $i\epsilon$ (or 'principal value') prescription of how to deal with the singularity at $k_0 = 0$.

We conclude that a formulation without superfluous degrees of freedom faces some difficulties owing to the increase of the longitudinal component of the field with time.

What can we expect from the cyclic variable φ in quantum mechanics?

In classical mechanics it can be fixed arbitrarily. Since, however, the equation of motion is $\ddot{\varphi} = \dot{\pi} = 0$, the momentum is conserved, and the wave function with a definite momentum $e^{i\pi\varphi}$ is a stationary state. The ground state corresponds to $\pi = 0$; it is the S-state in the φ variable. In other words, the ground state does not depend on the cyclic coordinate. In terms of operators,

$$\frac{\delta \mathcal{L}}{\delta \hat{v}}|\Omega\rangle = 0,$$

where Ω is any state of physical fields.

How does this S-state appear in terms of the action? We integrate over all possible field trajectories in the functional integral. The usual saying goes as follows: whichever field configuration is introduced at $t = -\infty$, in an infinite time the system arrives at the ground state. The functional integral up to $A(t)$,

$$\int_{-\infty}^{A(t)} e^{i \int \mathcal{L}[A(x)] \, dx} \, dA,$$

is just the wave function of the vacuum. If, however, a certain field A does not enter anything but the terms with *derivatives* in the Lagrangian, the integral does not depend on the value of A; the integrand contains only the *differences* over which the integral is taken. Hence, the system occurs automatically in a S-state in cyclic variables which enter only the kinetic energy.

16.7.5 Vacuum in the Yang–Mills theory

What does this tell us about the vacuum in our gauge theory? Let us calculate the time derivative of the field in the form (16.41):

$$\dot{\mathbf{A}}_i = v^{-1}\dot{\mathbf{B}}_i v + v^{-1}\mathbf{B}_i \dot{v} + (\dot{v^{-1}})\mathbf{B}_i v + \frac{\partial}{\partial t}\left(v^{-1}\frac{\partial v}{\partial x_i}\right).$$

Making use of unitarity, $v \cdot v^{-1} = 1 \Rightarrow v(\dot{v^{-1}}) + \dot{v}v^{-1} = 0$, we obtain

$$\dot{\mathbf{A}}_i = v^{-1}\big[\dot{\mathbf{B}}_i + \nabla_i(B)f\big]v, \tag{16.47}$$

where $f \equiv \dot{v}v^{-1}$ and

$$\nabla_i(B)f \equiv \frac{\partial f}{\partial x_i} + [\mathbf{B}_i f]. \tag{16.48}$$

Roughly speaking, f is the time derivative of the 'phase' of the unitary variable v. In electrodynamics where v is a number, $v = \exp i\alpha(x)$, this is literally true. The outstanding factors v, v^{-1} cancel under the trace in (16.40a), and for the kinetic energy we derive

$$T = -\tfrac{1}{2}\operatorname{Tr}(\mathbf{E}_i)^2; \qquad \mathbf{E}_i \equiv \dot{\mathbf{B}}_i + \nabla_i(B)f. \tag{16.49}$$

Here $E_i^a = \delta\mathcal{L}/\delta\dot{B}_i^a$ are analogous to 'electric fields'. Since f enters the kinetic energy (16.49) only, the momentum corresponding to the cyclic coordinate reads

$$\boldsymbol{\pi} \equiv \frac{\delta T}{\delta f} = \nabla_i(B)\big(\dot{\mathbf{B}}_i + \nabla_i(B)f\big) = \nabla_i(B)\mathbf{E}_i. \tag{16.50}$$

Here we have used the definition of the 'long' (covariant) derivative (16.48) and the transversality condition for the B field (16.42). As we have discussed above, the momentum $\boldsymbol{\pi}$ is conserved and should be set to zero in the vacuum state.

We want to extract the *transversal part* of the 'electric fields' and treat them as canonical momenta π_i conjugated to dynamical coordinates $\mathbf{B}_i$:

$$\mathbf{E}_i = \pi_i + \frac{\partial \phi}{\partial x_i}, \quad \frac{\partial}{\partial x_i}\pi_i = 0. \tag{16.51}$$

To exclude the longitudinal part, we combine (16.51) and (16.49):

$$\left(\frac{\partial}{\partial x_i}\right)^2 \phi = \frac{\partial}{\partial x_i}\mathbf{E_i} = \Box(B)f, \tag{16.52a}$$

where the operator $\Box$ is defined as

$$\Box(B) \equiv \frac{\partial}{\partial x_i}\nabla_i(B) = \nabla_i(B)\frac{\partial}{\partial x_i}. \tag{16.52b}$$

Now we are ready to set the cyclic momentum (16.50) to zero:

$$0 = \nabla_i(B)\mathbf{E}_i = \nabla_i(B)\left(\pi_i + \frac{\partial \phi}{\partial x_i}\right) = [\mathbf{B}_i\,\pi_i] + \Box(B)\varphi,$$

which gives

$$\phi = \frac{1}{\partial_i^2}\Box(B)f = -\frac{1}{\Box(B)}\rho, \quad \rho = [\mathbf{B}_i\,\pi_i]. \tag{16.53}$$

This is an analogue of the usual Coulomb equation $\phi = -(\partial_i^2)^{-1}\rho$, with ρ the charge density. When the momentum π is different from zero, after excluding the cyclic variable, an additional centrifugal energy appears. This is nothing but the Coulomb interaction energy. While in electrodynamics this is the energy of *external charges*, $\rho = \rho_{\text{ext}}$, in our case the fields A are charged themselves, and we have the Coulomb energy of the self-interacting Yang–Mills fields, produced by the 'charge density'

$$\rho = [\mathbf{B}_i\pi_i]. \tag{16.54}$$

We want to find the Hamiltonian of the system,

$$H = \int d^3x\,\mathcal{H}(x), \quad \mathcal{H}(x) = -\frac{1}{2g^2}\operatorname{Tr}\left(\mathbf{E}_i^2(x) + \tfrac{1}{2}\mathbf{G}_{ik}^2(x)\right).$$

For that we square the electric field (16.51):

$$\int d^3x\,\mathbf{E}_i^2(x) = \int d^4x\left(\pi_i^2(x) - \phi(x)\partial^2\phi(x)\right).$$

Finally, substituting (16.53) renders the Hamiltonian density:

$$\mathcal{H} = -\frac{1}{2g^2}\operatorname{Tr}\left(\pi_i^2 + \rho\,\frac{1}{\boldsymbol{\Delta}(B)}\,\rho + \tfrac{1}{2}\mathbf{G}_{ik}^2\right); \tag{16.55}$$

$$\boldsymbol{\Delta}(B) \equiv -\Box(B)\,\partial^{-2}\,\Box(B). \tag{16.56}$$

This calculation completes the usual verification of the fact that the initial Lagrangian describes interacting massless particles. This is not the end of the story, however.

Usually, the vacuum state is characterized by small oscillations of transverse fields. In our case it also contains *randomly oscillating* (*S*-state!) longitudinal Coulomb fields. We see that the effective interaction turns out to be rather unusual, non-local, because we tried to get rid of these random longitudinal fields.

If we neglect B fields in (16.56), we get the usual Coulomb interaction between charges:

$$\mathbf{\Delta}(0) = -\partial^2, \quad \frac{1}{\mathbf{\Delta}(0)} \;\Rightarrow\; \frac{1}{|\mathbf{x} - \mathbf{x}'|}.$$

The dependence on B shows that the *instantaneous* Coulomb interaction is modified by the presence of vacuum fluctuations of transverse (physical) fields. Instead of being purely dynamical, our problem of the structure of the vacuum state becomes an almost *statistical* one. It is this specificity that reverses the behaviour of the effective coupling of the theory (asymptotic freedom), see (16.22).

16.8 Asymptotic freedom

Now we are going to find out how the effective charge behaves in the Yang–Mills theory. We will follow the path suggested by Khriplovich (1969), who has calculated the invariant charge in Coulomb gauge even before the discovery of asymptotic freedom in non-Abelian theories (Gross and Wilczek, 1973; Politzer, 1973). He looked at the first $\mathcal{O}(g^2)$ correction to the vacuum energy in the presence of two infinitely heavy charges.

We add an external source $\rho_h = g\delta^3(\mathbf{x} - \mathbf{x}_0)$ to the charge density (16.54), substitute $\rho + \rho_h$ into the second term of the Hamiltonian density (16.55) and evaluate the correlator between two static sources placed at $\mathbf{x}_0 = \mathbf{x}_1$ and $\mathbf{x}_2$.

The vacuum energy of the system contains two contributions quadratic in ρ_h. The first is just 'classical' Coulomb energy of the sources given by the correlator of the external charge densities, $\rho_h \cdot \rho_h$ averaged over transverse fields $B_\perp$ in the vacuum:

$$V^c_{\text{Coul}} = \rho_h \left\langle 0 \left| \frac{1}{\mathbf{\Delta}(B)} \right| 0 \right\rangle \rho_h. \tag{16.57}$$

The second is quantum correction due to the mixing term $\rho_h \cdot \rho$ in the Hamiltonian. It leads to transition of the Coulomb field of the external charge into a pair of transverse fields (gluons with physical polarizations), and back again:

$$V_{\text{Coul}}^q = \sum_n \frac{|V_{0n}|^2}{E_0 - E_n}, \quad V_{0n} = \rho_h \left\langle 0 \left| \frac{1}{\boldsymbol{\Delta}(B)} \right| \rho \right| n \right\rangle,$$

where the sum runs over the energies of the intermediate two-gluon state. Here we can replace $\boldsymbol{\Delta}(B) \simeq \boldsymbol{\Delta}(0)$, and this contribution reduces to the Feynman diagram with two gluons in the intermediate state:

A standard calculation yields in the momentum space

$$V_{\text{Coul}}^q = \frac{g^2}{\mathbf{k}^2} \left(-\frac{C_2}{3} \cdot \frac{g^2}{16\pi^2} \ln \frac{\Lambda_{\text{UV}}^2}{\mathbf{k}^2} \right), \tag{16.58a}$$

with C_2 the number appearing from the square of the matrix commutator (the Casimir operator $C_2 = N$ for the $SU(N)$ group). If we add fermion fields (n_f families of quarks) into the game, additional quark loops appear, and the coefficient in (16.58a) gets modified as follows

$$-\frac{N}{3} \implies -\left(\frac{N}{3} + \frac{2}{3} n_f \right). \tag{16.58b}$$

This quantum correction is due to a virtual decay into physical states and corresponds to *screening* all right, having the same sign as in QED and elsewhere.

Now we return to the 'classical' piece (16.57). To find the $\mathcal{O}(g^2)$ correction due to vacuum fields we need to expand the operator $\boldsymbol{\Delta}^{-1}$ to the second order in B. First we calculate approximately the inverse of the operator $\square$,

$$\square^{-1}(B) = \partial^{-2} - \partial^{-2} [\mathbf{B}_i \, \partial_i] \partial^{-2} + \partial^{-2} [\mathbf{B}_i \, \partial_i] \partial^{-2} [\mathbf{B}_j \, \partial_j] \partial^{-2} + \cdots,$$

and substitute into (16.57):

$$\boldsymbol{\Delta}^{-1}(B) = -\square^{-1}(B) \partial^2 \square^{-1}(B)$$

$$\simeq -\partial^{-2} + 2 \, \partial^{-2} [\mathbf{B}_i \, \partial_i] \partial^{-2} - 3 \, \partial^{-2} [\mathbf{B}_i \, \partial_i] \partial^{-2} [\mathbf{B}_j \, \partial_j] \partial^{-2}.$$

The term linear in B disappears upon averaging, and we are left with the equal-time vacuum average of two transverse fields:

$$\langle 0 | B_i B_j | 0 \rangle = \frac{1}{2|\mathbf{k}'|} \left(\delta_{ij} - \frac{k_i' k_j'}{\mathbf{k}'^2} \right).$$

In the momentum space, the Coulomb energy takes the form

$$V_{\text{Coul}}^c = \frac{g^2}{\mathbf{k}^2} \left\{ 1 + 3 \cdot \int \frac{d^3\mathbf{k}'}{(2\pi)^3} \frac{N\,g^2}{2\,|\mathbf{k}'|\,(\mathbf{k}'-\mathbf{k})^2} \left(1 - \frac{(\mathbf{k}\cdot\mathbf{k}')^2}{\mathbf{k}^2\,\mathbf{k}'^2} \right) \right\}.$$

The angular integration produces

$$\int \frac{d\varphi\,d\cos\theta}{(2\pi)^3} (1 - \cos^2\theta) = \frac{1}{(2\pi)^2} \left(2 - \frac{2}{3} \right),$$

and from the large momentum region we get a logarithmically divergent correction

$$V_{\text{Coul}}^c = \frac{g^2}{\mathbf{k}^2} \left(1 + 4C_2 \cdot \frac{g^2}{4\pi^2} \ln\frac{\Lambda_{\text{UV}}^2}{\mathbf{k}^2} \right). \tag{16.59}$$

Combining with the quantum contribution (16.58), we finally obtain the first correction to the vacuum energy,

$$\frac{g^2}{\mathbf{k}^2} \implies \frac{g^2}{\mathbf{k}^2} \left\{ 1 + \left(\left[4 - \frac{1}{3} \right] C_2 - \frac{2}{3}n_f \right) \cdot \frac{g^2}{4\pi^2} \ln\frac{\Lambda_{\text{UV}}^2}{\mathbf{k}^2} \right\},$$

which corresponds to the invariant coupling of Yang–Mills fields,

$$g^2(k^2) = \frac{g^2}{1 - \beta_0 \frac{g^2}{4\pi^2} \ln\frac{\Lambda_{\text{UV}}^2}{k^2}} = \frac{4\pi^2}{\beta_0 \ln\frac{k^2}{\Lambda^2}}, \qquad \beta_0 = \frac{11}{3}N - \frac{2}{3}n_f, \tag{16.60}$$

which *decreases* with the increase of k^2.

We see that the *anti-screening* (asymptotic freedom) is entirely due to vacuum fluctuations of the gluon fields affecting the Coulomb interaction. This effect is also likely to have other serious consequences related to the infrared instability of the quantum theory of Yang–Mills fields (Gribov, 1978).

Postscript

"The theory of strong interactions,
now that is quite something"

*– Gribov's opening words to his first lecture
on hadron interactions at high energies in 1972*

In the early 1970s V. N. Gribov gave two series of lectures to students of theoretical physics at Leningrad (today St. Petersburg) State University.

The first course was devoted to Quantum Electrodynamics (QED), and Gribov completed it without mentioning the word *Lagrangian*. Such a 'bizarre' approach to QED had a hidden purpose. It aimed higher and represented, in fact, a constructive introduction to Quantum Field Theory (QFT) in general, based on the language of Feynman diagrams – 'the laboratory of theoretical physics', in Gribov's words.

The QED course of 1971 was followed by a lecture series on strong interaction physics; Gribov has later formulated his motivation: 'I wanted to tell everything I ever learnt about hadron interactions'.

The two courses had quite a different fate. The QED lectures appeared, in Russian, in the proceedings of the Winter School of the Leningrad Nuclear Physics Institute (LNPI) in 1974. Its English version, prepared with Gribov's participation, was published by the Cambridge University Press in 2001 as the first volume of the *Gribov Lectures on Theoretical Physics* series.

The second course – Strong Interactions – existed only in the form of handwritten notes taken and preserved by his students. Even in the 1990s when Gribov has been working on Quantum Chromodynamics (QCD) for already more than a decade he thought it was important to write a book based on these old notes. He wanted to do this, and, when we spent a few months with him at the University of Lund in 1991, he proposed that we should work together on the manuscript. Gribov has edited a few of the first lectures prepared by Sergey Troyan and one of us (Y.D.). Unfortunately, the lecture course as a whole started taking shape only many years after his death. The book would have certainly been different if he could have participated in its completion.

Why to return to the 'old theory' of strong interactions of the pre-QCD, even pre-quark, epoch? There are several reasons for that.

On the one hand, the Lagrangian theory of strong interactions was spectacularly successful in describing small-distance quark–gluon dynamics (*hard* processes) and became a working tool, in particular, in the search for new physics beyond the Standard Model. At the same time, our understanding of even the most general characteristics of *soft* hadron processes – like total hadron interaction cross sections in the first place – remains where the 'old theory' left it about 30 years ago.

The fact that the old theory had very limited means and, being devoid of microscopic dynamics, had to rely on the most general properties of the relativistic S-matrix theory, turns to its advantage nowadays.

The second reason is deeper, and therefore less obvious. The 'old approach' to strong interactions took off in the early 1960s when it was realized that the general properties of the relativistic S-matrix theory – crossing symmetry, unitarity, causality – put severe restrictions on the possible high-energy behaviour of hadron interactions (elastic, inelastic, total cross sections).

The effective theory describing the high-energy asymptotics of σ_{tot}, as well as fluctuations in multi-particle production, – the Gribov Reggeon Field Theory (RFT) of interacting pomerons – was constructed. It was a turning point when it has been found to be *intrinsically unstable* in the specific, but the only practically relevant, case of (nearly) constant total cross sections. Looking into possible solutions of the corresponding *infrared unstable dynamics* has helped to develop the theory of second order phase transitions in condensed matter physics (the 'scaling' solution). However, the original pomeron problem remained unsolved. Pomeron instability could not be resolved without an input from outside the S-matrix theory, namely, without understanding the structure of the hadronic vacuum.

For the answer V.N. Gribov turned to QCD. The discovery of *Gribov copies* in 1976, which has changed our view upon non-abelian QFTs, was for him, in fact, a by-product of the search for the pomeron-puzzle solution. He came to the firm belief that the pomeron instability got to be intimately related to the infrared instability of the QCD, i.e. to the physics of quark confinement.

There exists a deep relation between the 'old' and 'new' strong interaction theories, which can not be appreciated without studying both.

The new theory – QCD – has adopted many a notion of its predecessor: the Froissart regime; reggeization of quarks and gluons, and multi-regge kinematics; [QCD, BFKL, 'hard'] pomeron, and the reggeon field theory; impact parameter diffusion; parton screening and saturation; Gribov–Glauber multiple scattering theory, and the Abramovsky–Gribov–Kancheli (AGK) cutting rules. Many important issues of the old theory, its techniques, problems and achievements are, however, not exposed in existing textbooks. This book fills the gap.

Topically, this book overlaps with Gribov's *Theory of Complex Angular Momenta* (TCAM) (Gribov, 2003). There is, however, an essential difference.

The TCAM volume is based on lectures given by V.N. Gribov in 1969 to the audience of *professional theorists*, whereas the present course targeted *university students*. It does not presume, therefore, any additional knowledge beyond quantum mechanics (including the non-relativistic quantum scattering theory), relativistic kinematics and the QFT basics covered by the preceding QED lectures (Gribov and Nyiri, 2001). That is why many general topics not present in the TCAM volume are extensively discussed here: analytic properties of Feynman amplitudes, resonances, electromagnetic interaction of hadrons.

Moreover, this course covers a number of important subjects developed and understood during the 3–4 years that elapsed between the two lecture series. Among them, to name a few, are the s-channel nature of the Regge exchange (impact parameter diffusion) and the structure of multi-hadron production (Mueller–Kancheli analysis of inclusive spectra and the pattern of multiplicity fluctuations, AGK cutting rules, screening, etc.), deep inelastic lepton–hadron scattering and the quark–parton picture, as well as the basics of QCD in the original Gribov approach.

On the other hand, in the TCAM volume some technically involved themes are elaborated in greater detail. So, the two books complement each other in many ways, aiming at a comprehensive exposition of the theory of hadron interactions that preceded Quantum Chromodynamics.

In preparing the book, we used lecture notes taken by Yu. Dokshitzer, V. Petrov, S. Troyan and V. Vechernin in 1972–75.

We thank L. Frankfurt, A. Frenkel and M. Strikman for help and valuable advice. Y.D. is deeply grateful to his colleagues from the Theoretical Laboratory of High Energy Physics, University Paris VI-VII, for constant moral support and patience in the long course of preparing this book.

Yuri Dokshitzer
Julia Nyiri

References

Amati, D. and Veneziano, G. Preconfinement as a property of perturbative QCD. *Phys. Lett.* 1979, **B83**, 87.

Chew, G. F. and Low, F. E. Unstable particles as targets in scattering experiments. *Phys. Rev.* 1959, **113**, 1640.

Froissart, M. Asymptotic behavior and subtractions in the Mandelstam representation. *Phys. Rev.* 1961, **123**, 1053.

Gribov, V. N. Theory of the heavy pomeron. *Nucl. Phys.* 1976, **B106**, 189.

Gribov, V. N. Instability of non-Abelian gauge theories and impossibility of choice of Coulomb gauge. In: *Proceedings of the 12th Winter LNPI School on Nuclear and Elementary Particle Physics*, 1977, p. 147 (in Russian). English translation in: Gribov, V. N. *Gauge Theories and Quark Confinement*. PHASIS, Moscow, 2002, p. 257.

Gribov, V. N. 1978 Quantization of non-Abelian gauge theories. *Nucl. Phys.* 1978, **B139**, 1.

Gribov, V. N. *The Theory of Complex Angular Momenta*. Cambridge: Cambridge University Press, 2003.

Gribov, V. N. and Nyiri, J. *Quantum Electrodynamics*. Cambridge: Cambridge University Press, 2001.

Gribov, L. V. *et al.*, The structure of final hadronic state in deep inelastic scattering. *JETP Lett.* 1987, **45**, 515.

Gross, D. I. and Wilczek, F. Ultraviolet behavior of non-Abelian gauge theories. *Phys. Rev. Lett.* 1973, **30**, 1343.

Ioffe, B. L. Space–time picture of photon and neutrino scattering and electroproduction cross section asymptotics. *Phys. Lett.* 1969, **B30**, 123.

Kancheli, O. V. Inelastic differential cross sections at high energies and duality. *Pisma Zh. Eksp. Teor. Fiz.* 1970, **11**, 397 in Russian.

Khriplovich, I. B. *Yad. Fiz.* 1969, **10**, 409.

Landau, L. D., Abrikosov, A. and Khalatnikov, L. On the quantum theory of fields. *Nuovo Cim. Suppl.* 1956, **3**, 80.

Mandelstam, S. Determination of the pion–nucleon scattering amplitude from dispersion relations and unitarity. General theory. *Phys. Rev.* 1958 **112**, 1344.

473

Marchesini, G. and Webber, B. R. Simulation of QCD jets including soft gluon interference. *Nucl. Phys.* 1984, **B238**, 1.

Mueller, A. H. Multiplicity distributions in Regge pole dominated inclusive reactions. *Phys. Rev.* 1971, **D4**, 150.

Politzer, H. D. Reliable perturbative results for strong interactions? *Phys. Rev. Lett.* 1973 **30**, 1346.

Pomeranchuk, I. Ya. Equality of the nucleon and antinucleon total interaction cross section at high energies. *Sov. Phys. JETP* 1958, **7**, 499.

Regge, T. Introduction to complex orbital momenta. *Nuovo Cim.* 1959, **14**, 951.

Schwinger, J. S. Gauge invariance and mass. II. *Phys. Rev.* 1962, **128**, 2425.

Shifman, M. and Vainshtein, A. Comments on diquarks, strong binding and a large hidden QCD scale. *Phys. Rev.* 2005, **D71**, 074010.

Veneziano, G. Construction of a crossing-symmetric, Regge behaved amplitude for linearly rising trajectories. *Nuovo Cim.* 1968, **A57**, 190.

Index

Printed in the United States
by Baker & Taylor Publisher Services